DATE DUE

Demco, Inc. 38-293

VOLUME FIVE HUNDRED AND SEVEN

Methods in ENZYMOLOGY

Gene Transfer Vectors for Clinical Application

METHODS IN ENZYMOLOGY

Editors-in-Chief

JOHN N. ABELSON AND MELVIN I. SIMON

Division of Biology
California Institute of Technology
Pasadena, California

Founding Editors

SIDNEY P. COLOWICK AND NATHAN O. KAPLAN

VOLUME FIVE HUNDRED AND SEVEN

Methods in ENZYMOLOGY

Gene Transfer Vectors for Clinical Application

EDITED BY

THEODORE FRIEDMANN
Department of Pediatrics, School of Medicine
University of California San Diego
Center for Neural Circuits and Behavior
La Jolla, CA, USA

AMSTERDAM • BOSTON • HEIDELBERG • LONDON
NEW YORK • OXFORD • PARIS • SAN DIEGO
SAN FRANCISCO • SINGAPORE • SYDNEY • TOKYO
Academic Press is an imprint of Elsevier

Academic Press is an imprint of Elsevier
525 B Street, Suite 1900, San Diego, CA 92101-4495, USA
225 Wyman Street, Waltham, MA 02451, USA
32 Jamestown Road, London NW1 7BY, UK

First edition 2012

Copyright © 2012, Elsevier Inc. All Rights Reserved.

No part of this publication may be reproduced, stored in a retrieval system or transmitted in any form or by any means electronic, mechanical, photocopying, recording or otherwise without the prior written permission of the publisher

Permissions may be sought directly from Elsevier's Science & Technology Rights Department in Oxford, UK: phone (+44) (0) 1865 843830; fax (+44) (0) 1865 853333; email: permissions@elsevier.com. Alternatively you can submit your request online by visiting the Elsevier web site at http://elsevier.com/locate/permissions, and selecting *Obtaining permission to use Elsevier material*

Notice
No responsibility is assumed by the publisher for any injury and/or damage to persons or property as a matter of products liability, negligence or otherwise, or from any use or operation of any methods, products, instructions or ideas contained in the material herein. Because of rapid advances in the medical sciences, in particular, independent verification of diagnoses and drug dosages should be made

For information on all Academic Press publications
visit our website at elsevierdirect.com

ISBN: 978-0-12-386509-0
ISSN: 0076-6879

Printed and bound in United States of America
12 13 14   10 9 8 7 6 5 4 3 2 1

**Working together to grow
libraries in developing countries**

www.elsevier.com | www.bookaid.org | www.sabre.org

ELSEVIER   BOOK AID International   Sabre Foundation

Contents

Contributors   xi
Preface   xvii
Volume in Series   xxi

1. General Principles of Retrovirus Vector Design   1
Tammy Chang and Jiing-Kuan Yee

1. Introduction   1
2. Construction of Retroviral Vectors Carrying the Gene of Interest   4
3. Transient Production of Retroviral Vectors   6
4. Titering of Retroviral Vectors   9
5. RCR Detection   11
Acknowledgments   12
References   12

2. Strategies for Retrovirus-Based Correction of Severe, Combined Immunodeficiency (SCID)   15
Alain Fischer, Salima Hacein-Bey-Abina, and Marina Cavazzana-Calvo

1. Early Attempts   17
2. Technological Progress   18
3. Gene Therapy of SCID-X1   18
4. Gene Therapy for ADA Deficiency   22
5. Conclusions   23
References   24

3. Retrovirus and Lentivirus Vector Design and Methods of Cell Conditioning   29
Samantha Cooray, Steven J. Howe, and Adrian J. Thrasher

1. Introduction   30
2. Retrovirus and Lentivirus Vector Design   34
3. Retroviral and Lentiviral Vector Production   40
4. Transduction Procedures   45
References   53

v

4. **Analysis of the Clonal Repertoire of Gene-Corrected Cells in Gene Therapy**   59

   Anna Paruzynski, Hanno Glimm, Manfred Schmidt, and Christof von Kalle

   1. Introduction   60
   2. Analysis of the Clonal Repertoire by Performing nrLAM-PCR   61
   3. Preparation of nrLAM-PCR Samples for High-Throughput Sequencing   73
   4. Estimation of the Clonal Contribution by RT-PCR   78
   References   85

5. **Developing Novel Lentiviral Vectors into Clinical Products**   89

   Anna Leath and Kenneth Cornetta

   1. Introduction   90
   2. Developing a Development Plan   90
   3. Development of a GMP Production Method   93
   4. Certification Testing of Novel Vector Products   100
   Acknowledgments   104
   References   105

6. **Lentivirus Vectors in β-Thalassemia**   109

   Emmanuel Payen, Charlotte Colomb, Olivier Negre, Yves Beuzard, Kathleen Hehir, and Philippe Leboulch

   1. Introduction to the β-Thalassemias   110
   2. Principles of Lentiviral Gene Therapy for β-Thalassemias   111
   3. Clinical Vector Design (LentiGlobin™)   112
   4. cGMP Lentiviral Vector Production   113
   5. cGMP Transduction of $CD34^+$ Cells   113
   6. GTP and GMO Release Testing   114
   7. Clinical Protocol   114
   8. Outcome for the First Thalassemia Patient Treated by Gene Therapy (51 Months Post-transplantation)   121
   References   122

7. **Gene Therapy for Chronic Granulomatous Disease**   125

   Elizabeth M. Kang and Harry L. Malech

   1. Introduction   126
   2. Vector Design   130
   3. Vector Production   132
   4. Preclinical Testing of Vector Titer and Transgene Function (*In Vitro* Culture, Mouse Knockout (KO) CGD Models, and Mouse Models of Human CD34+ HSC Xenograft)   134
   5. Clinical Scale Production   137

6. Target Cell	139
7. Clinical Transduction	140
8. Patient Care and Monitoring	142
9. Future Endeavors	145
10. Conclusion	149
References	150

8. Alternative Splicing Caused by Lentiviral Integration in the Human Genome — 155

Arianna Moiani and Fulvio Mavilio

1. Introduction	156
2. Analysis of Aberrantly Spliced Transcripts Generated by the Integration of SIN Lentiviral Vectors in Human Genes	158
Acknowledgments	168
References	168

9. Genotoxicity Assay for Gene Therapy Vectors in Tumor Prone $Cdkn2a^{-/-}$ Mice — 171

Eugenio Montini and Daniela Cesana

1. Introduction	172
2. *In Vivo* Genotoxicity Assays Based on Transduction and Transplantation of Tumor-Prone HSPCs	173
References	183

10. Lentiviral Hematopoietic Cell Gene Therapy for X-Linked Adrenoleukodystrophy — 187

Nathalie Cartier, Salima Hacein-Bey-Abina, Cynthia C. Bartholomae, Pierre Bougnères, Manfred Schmidt, Christof Von Kalle, Alain Fischer, Marina Cavazzana-Calvo, and Patrick Aubourg

1. X-linked Adrenoleukodystrophy	188
2. Hematopoietic Stem Cell Transplantation in X-ALD	189
3. Mechanism of Allogeneic HCT Efficacy in X-ALD	189
4. Lentiviral HSC Gene Therapy in X-ALD	189
5. Design and Production of the Lentiviral Vector	190
6. Transduction of CD34+ Cells from ALD Patients	191
7. Release Testing	192
8. Clinical Protocol	192
9. Patient Biological Follow-up	193
10. Neurological Outcome of the Two Treated Patients	195
References	197

11. Retroviral Replicating Vectors in Cancer — 199

Christopher R. Logg, Joan M. Robbins, Douglas J. Jolly, Harry E. Gruber, and Noriyuki Kasahara

1. Introduction — 200
2. Virus Production by Transient Transfection — 202
3. Vector Copy Number Assay for Titer Determination and Biodistribution Studies — 205
4. Development of Novel RRV Using Molecular Evolution — 211
5. MTS Assay of RRV-Mediated Cell Killing *In Vitro* — 215
6. *In Vivo* Glioma Model for Testing RRV-Mediated Gene Therapy — 218

Acknowledgments — 225
References — 225

12. Adeno-Associated Virus Vectorology, Manufacturing, and Clinical Applications — 229

Joshua C. Grieger and R. Jude Samulski

1. Adeno-Associated Virus Biology — 230
2. AAV Vectorology — 231
3. rAAV Manufacturing Methods — 238
4. Recent Clinical Trials Utilizing rAAV Vectors — 240
5. Conclusion — 245

Acknowledgments — 246
References — 246

13. Gene Delivery to the Retina: From Mouse to Man — 255

Jean Bennett, Daniel C. Chung, and Albert Maguire

1. Introduction — 256
2. Nucleic Acid Delivery to the Outer Retina in the Mouse — 257
3. Nucleic Acid Delivery to the Outer Retina in the Dog — 261
4. Nucleic Acid Delivery to the Outer Retina in the Non-Human Primate — 265
5. Nucleic Acid Delivery to the Outer Retina in the Human — 267

Acknowledgments — 270
References — 270

14. Generation of Hairpin-Based RNAi Vectors for Biological and Therapeutic Application — 275

Ryan L. Boudreau and Beverly L. Davidson

1. Introduction — 276
2. Selecting Candidate siRNA Sequences — 278
3. Cloning shRNA Expression Cassettes — 280
4. Cloning Artificial miRNA Expression Vectors — 285

5.	Screening RNAi Vectors for Silencing Efficacy *In Vitro*	291
6.	Integration into Viral Vectors	293
7.	Summary	294
	References	294

15. Recombinant Adeno-Associated Viral Vector Reference Standards   297

Philippe Moullier and Richard O. Snyder

1.	Introduction	298
2.	Utility of Reference Standards	298
3.	Volunteer Working Groups	300
4.	The AAV2 Reference Standard Material	301
5.	The AAV8 Reference Standard Material	302
6.	Methods Used to Characterize the AAV RSMs	303
7.	Conclusions	308
	Acknowledgments	308
	References	309

16. NIH Oversight of Human Gene Transfer Research Involving Retroviral, Lentiviral, and Adeno-associated Virus Vectors and the Role of the NIH Recombinant DNA Advisory Committee   313

Marina O'Reilly, Allan Shipp, Eugene Rosenthal, Robert Jambou, Tom Shih, Maureen Montgomery, Linda Gargiulo, Amy Patterson, and Jacqueline Corrigan-Curay

1.	Introduction	314
2.	A Brief History of the *NIH Guidelines* and Recombinant DNA Research Oversight	315
3.	Role of the NIH and RAC in Review and Oversight of Human Gene Transfer Involving Retroviral, Lentiviral, and Adeno-associated Virus Vectors	322
4.	Conclusion	334
	References	334

17. Regulatory Structures for Gene Therapy Medicinal Products in the European Union   337

Bettina Klug, Patrick Celis, Melanie Carr, and Jens Reinhardt

1.	Regulatory Requirements for ATMPs—Marketing Authorization	339
2.	Annex I—Part IV to Directive 2001/83/EC	340
3.	Definition of Gene Therapy Medicinal Products	341
4.	Committee for Advanced Therapies	342

5. CAT–Stakeholder Interaction	345
6. Incentives—Small- and Medium-Sized Enterprise Office (SME Office) at EMA	346
7. National Support Structures	348
8. The role of Patients' Organizations in the Process of Development of Gene Therapy Medicinal Products	349
9. Rare Diseases and Orphan Medicinal Products	350
10. The Role of European Research Networks	351
Acknowledgments	352
References	352
Author Index	*355*
Subject Index	*377*

Contributors

Patrick Aubourg
INSERM UMR745, University Paris-Descarte; and Department of Pediatric Endocrinology and Neurology, Hôpital Bicêtre, Kremlin-Bicêtre, Paris, France

Cynthia C. Bartholomae
National Center for Tumor Diseases and German Cancer Research Center, Heidelberg, Germany

Jean Bennett
F.M. Kirby Center for Molecular Ophthalmology, Scheie Eye Institute, University of Pennsylvania; and The Center for Cellular and Molecular Therapeutics, The Children's Hospital of Philadelphia, Philadelphia, Pennsylvania, USA

Yves Beuzard
CEA, Institute of Emerging Diseases and Innovative Therapies (iMETI); Inserm U962 and University Paris 11, CEA-iMETI, Fontenay aux Roses, France and Bluebird bio, Cambridge, MA, CEA-iMETI, Fontenay aux Roses, France

Ryan L. Boudreau
Department of Internal Medicine, University of Iowa, Iowa City, Iowa, USA

Pierre Bougnères
Department of Pediatric Endocrinology and Neurology, Hôpital Bicêtre, Kremlin-Bicêtre, Paris, France

Melanie Carr
European Medicines Agency, Canary Wharf, London, United Kingdom

Nathalie Cartier
INSERM UMR745, University Paris-Descartes and Department of Pediatric Endocrinology and Neurology, Hôpital Bicêtre, Kremlin-Bicêtre, Paris, France

Marina Cavazzana-Calvo
Descartes University of Paris; INSERM Unit 768 and Department of Immunology, Necker Children's Hospital; Clinical Investigation Center in Biotherapy, Groupe Hospitalier Universitaire Ouest; and Department of Biotherapy, Necker Children's Hospital, Paris, France

Patrick Celis
European Medicines Agency, Canary Wharf, London, United Kingdom

Daniela Cesana
San Raffaele-Telethon Institute for Gene Therapy, Milan, Italy

Tammy Chang
Department of Virology, Beckman Research Institute, City of Hope National Medical Center, Duarte, California, USA

Daniel C. Chung
F.M. Kirby Center for Molecular Ophthalmology, Scheie Eye Institute, University of Pennsylvania; and The Center for Cellular and Molecular Therapeutics, The Children's Hospital of Philadelphia, Philadelphia, Pennsylvania, USA

Charlotte Colomb
CEA, Institute of Emerging Diseases and Innovative Therapies (iMETI), and Inserm U962 and University Paris 11, CEA-iMETI, Fontenay aux Roses, France

Samantha Cooray
Centre for Immunodeficiency, Molecular Immunology Unit, Institute of Child Health, University College London, and Department of Clinical Immunology, Great Ormond Street Hospital NHS Trust, London, United Kingdom

Kenneth Cornetta
Department of Medical and Molecular Genetics; Department of Medicine; and Department of Microbiology and Immunology, Indiana University School of Medicine, Indianapolis, Indiana, USA

Jacqueline Corrigan-Curay
Office of Biotechnology Activities, National Institutes of Health, Bethesda, Maryland, USA

Beverly L. Davidson
Department of Internal Medicine; Department of Neurology; and Department of Molecular Physiology & Biophysics, University of Iowa, Iowa City, Iowa, USA

Alain Fischer
Descartes University of Paris; INSERM Unit 768, Department of Immunology; and Pediatric Hematology Unit, Necker Children's Hospital, Paris, France

Linda Gargiulo
Office of Biotechnology Activities, National Institutes of Health, Bethesda, Maryland, USA

Hanno Glimm
Department of Translational Oncology, National Center for Tumor Diseases (NCT) and German Cancer Research Center (DKFZ), Im Neuenheimer Feld 581 and 460, Heidelberg, Germany

Joshua C. Grieger
Gene Therapy Center, University of North Carolina, Chapel Hill, North Carolina, USA

Harry E. Gruber
Tocagen Inc., San Diego, California, USA

Salima Hacein-Bey-Abina
Descartes University of Paris; Department of Biotherapy, Necker Children's Hospital; and Clinical Investigation Center in Biotherapy, Groupe Hospitalier Universitaire Ouest, Paris, France

Kathleen Hehir
Bluebird bio, Cambridge, MA, USA and CEA-iMETI, Fontenay aux Roses, France

Steven J. Howe
Centre for Immunodeficiency, Molecular Immunology Unit, Institute of Child Health, University College London, London, United Kingdom

Robert Jambou
Office of Biotechnology Activities, National Institutes of Health, Bethesda, Maryland, USA

Douglas J. Jolly
Tocagen Inc., San Diego, California, USA

Elizabeth M. Kang
Laboratory of Host Defenses, National Institute of Allergy and Infectious Diseases, National Institutes of Health, Bethesda, Maryland, USA

Noriyuki Kasahara
Department of Medicine; and Department of Molecular and Medical Pharmacology, University of California, Los Angeles, California, USA

Bettina Klug
Paul-Ehrlich-Institut, Langen, Germany

Anna Leath
Department of Medical and Molecular Genetics, Indiana University School of Medicine, Indianapolis, Indiana, USA

Philippe Leboulch
CEA, Institute of Emerging Diseases and Innovative Therapies (iMETI); Inserm U962 and University Paris 11, CEA-iMETI, Fontenay aux Roses, France, and Harvard Medical School and Genetics Division, Department of Medicine, Brigham & Women's Hospital, Boston, Massachusetts, USA

Christopher R. Logg
Department of Medicine, University of California, Los Angeles, California, USA

Albert Maguire
F.M. Kirby Center for Molecular Ophthalmology, Scheie Eye Institute, University of Pennsylvania; and The Center for Cellular and Molecular Therapeutics, The Children's Hospital of Philadelphia, Philadelphia, Pennsylvania, USA

Harry L. Malech
Laboratory of Host Defenses, National Institute of Allergy and Infectious Diseases, National Institutes of Health, Bethesda, Maryland, USA

Fulvio Mavilio
Division of Genetics and Cell Biology, Istituto Scientifico H. San Raffaele, Milan; and Center for Regenerative Medicine, University of Modena and Reggio Emilia, Modena, Italy

Arianna Moiani
Division of Genetics and Cell Biology, Istituto Scientifico H. San Raffaele, Milan, Italy

Maureen Montgomery
Office of Biotechnology Activities, National Institutes of Health, Bethesda, Maryland, USA

Eugenio Montini
San Raffaele-Telethon Institute for Gene Therapy, Milan, Italy

Philippe Moullier
INSERM UMR 649, CHU Hôtel Dieu, Nantes, Genethon, Evry, France; and Department of Molecular Genetics and Microbiology, College of Medicine, University of Florida, Gainesville, Florida, USA

Olivier Negre
CEA, Institute of Emerging Diseases and Innovative Therapies (iMETI); Inserm U962 and University Paris 11, CEA-iMETI, Fontenay aux Roses, France and Bluebird bio, Cambridge, MA, USA

Marina O'Reilly
Office of Biotechnology Activities, National Institutes of Health, Bethesda, Maryland, USA

Anna Paruzynski
Department of Translational Oncology, National Center for Tumor Diseases (NCT) and German Cancer Research Center (DKFZ), Im Neuenheimer Feld 581 and 460, Heidelberg, Germany

Amy Patterson
Office of Biotechnology Activities, National Institutes of Health, Bethesda, Maryland, USA

Emmanuel Payen
CEA, Institute of Emerging Diseases and Innovative Therapies (iMETI), and Inserm U962 and University Paris 11, CEA-iMETI, Fontenay aux Roses, France

Jens Reinhardt
Paul-Ehrlich-Institut, Langen, Germany

Joan M. Robbins
Tocagen Inc., San Diego, California, USA

Eugene Rosenthal
Office of Biotechnology Activities, National Institutes of Health, Bethesda, Maryland, USA

R. Jude Samulski
Gene Therapy Center; and Department of Pharmacology, University of North Carolina, Chapel Hill, North Carolina, USA

Manfred Schmidt
Department of Translational Oncology, National Center for Tumor Diseases (NCT) and German Cancer Research Center (DKFZ), Heidelberg, Germany

Tom Shih
Office of Biotechnology Activities, National Institutes of Health, Bethesda, Maryland, USA

Allan Shipp
Office of Biotechnology Activities, National Institutes of Health, Bethesda, Maryland, USA

Richard O. Snyder
INSERM UMR 649, CHU Hôtel Dieu, Nantes, France; Department of Molecular Genetics and Microbiology, College of Medicine; and Center of Excellence for Regenerative Health Biotechnology, University of Florida, Gainesville, Florida, USA

Adrian J. Thrasher
Centre for Immunodeficiency, Molecular Immunology Unit, Institute of Child Health, University College London, London, United Kingdom

Christof Von Kalle
Department of Translational Oncology, National Center for Tumor Diseases (NCT) and German Cancer Research Center (DKFZ), Im Neuenheimer Feld 581 and 460, Heidelberg, Germany

Jiing-Kuan Yee
Department of Virology, Beckman Research Institute, City of Hope National Medical Center, Duarte, California, USA

Preface

 ## The Successful Clinical use of Viral Vectors for Human Gene Therapy

The field of human gene therapy has had a long and contorted evolution but has now attained a stage of adolescence and even very early maturity. Coincident with the publication of this volume is the 40th anniversary of the publication of an early description of the potential and even the logical necessity of gene therapy for human disease; that is, the development of treatments aimed directly at the underlying genetic defects responsible for disease rather than at their biochemical and functional consequences (Friedmann and Roblin, 1972). From the outset, the development of techniques of gene therapy has centered largely on the use of disabled recombinant viruses to achieve efficient transfer of therapeutic genetic information into cells to correct genetic defects and reconstitute genetic deficiencies. That was largely because geneticists and molecular biologists realized that nature had long ago invented tool to carry out the central task of gene therapy, that is, the highly efficient introduction of foreign genetic information into mammalian cells in ways that permit stable and heritable expression of new genetic functions in the cells. Those agents are called viruses and their role in gene transfer in mammalian cells was established by the studies of Dulbecco and colleagues showing that the tumorigenic papova viruses produce neoplastic changes in cells by integrating part of their genome in stable and functional form into the genome of infected cells (Westphal and Dulbecco, 1968). Further, coincident with the emergence of the concepts of gene therapy and the discovery of viral gene transfer mechanisms came the tools for creating the kinds of recombinant DNA molecules that made possible the production of viral derivatives that retained the mechanisms that native viruses have for highly efficient cell entry of wild-type viruses but that had been disabled of their pathogenic functions by deletion of the viral replicative and cytotoxic functions (Shimotohno and Temin, 1981; Tabin *et al.*, 1982; Wei *et al.*, 1981). By providing viral replicating functions *in trans* and by ligating potentially therapeutic genes into the recombinant molecules, methods quickly became available to produce large amounts of fully infectious but disarmed viral particles that served as vectors for shuffling foreign genes into cells.

It was widely realized that eventual successful functional genetic correction would require a degree of efficient gene transfer much higher than could be achieved by nonviral techniques. Nevertheless, many nonviral gene transfer "transfection" methods continued to be developed and such chemical methods have indeed also demonstrated variable degrees of efficacy in correcting genetic defects in model studies (Friedmann and Rossi, 2006). Naked DNA, lipid-formulated complexes (liposomes), and other chemical and synthetic nucleic acid complexes, and more recently, even protein-based transduction methods, have all been shown to permit functional and even reasonably efficient gene transfer and even correction of genetic defects in human and other mammalian cells. However, to the present time, it is only the viral vector systems that have finally matured to the point of robust and clinically useful gene transfer and convincing disease therapy. Principal among these viral systems are those that utilize recombinant forms of the gamma retroviruses, lentiviruses, and adeno-associated viruses.

Retroviruses. From their inception, retrovirus vectors have been the most widely used viral vector system for transducing mammalian cells. They had the advantage of ease of construction, very high efficiency, and a capacity for carrying and delivering functional forms of most full-length cDNAs into mammalian cells. However, they were hampered by two major inadequacies, that is, their inability to transduce most quiescent and nonreplicating cells such as neurons and their genotoxic potential resulting from the relative promiscuous mechanisms of integration into the host cell genome leading to insertional mutations in the genome. The development of the lentivirus vector system has made possible the highly efficient transduction of quiescent cells and has thereby obviated the target cell deficiency of other classes of retrovirus vectors (Friedmann and Rossi, 2006).

The drawback represented by the insertional mutagenic potential of retrovirus vectors became harsh reality in the clinically successful studies using these vectors for the treatment of the human immunodeficiency disease X-SCID. Despite the remarkable therapeutic success of these studies, at least five children treated with this generation of oncoretrovirus vectors developed leukemia as a result of insertional oncogene activation (Paruzynski *et al.*, 2011). Although recent advances described in this volume have made major improvements in our understanding of the insertional properties of retrovirus vectors, much still needs to be learned about targeting these and other integrating vectors to appropriate "safe harbor" sites in the genome where integration would not cause proliferative aberrations and tumorigenesis. Valuable steps are being taken in this direction, as exemplified by the use of hybrid zinc finger-targeted endonucleases to specify the locations of double-stranded genomic breaks and subsequent repair (Porteus and Baltimore, 2003; Wilson, 2003).

Adeno-associated viruses. During the search for vectors that would avoid or correct the inadequacies of retroviruses, the parvovirus class of viruses represented by the adeno-associated viruses (AAV) became a very attractive system for the development of transducing vectors for human and other mammalian cells because of their extreme genetic simplicity and because AAV was not known to be associated with human disease and generally were thought to be nonintegrating. AAV vectors were therefore expected to be safe in the context of human clinical use. On the other hand, AAV were hampered by their very small size and therefore by their low carrying capacity for therapeutic cDNAs. Nevertheless, recent clinical studies have borne out the advantages provided by AAV vectors in selected clinical applications and have given unequivocal evidence of therapeutic efficacy (Grieger and Samulski, 2011).

It has also become clear that genetic reconstitution and gene therapy, even when clinically effective, are likely to be beset by difficult associated adverse consequences, as in the case of the leukemias in the SCID studies. Gene therapy will, like most other forms of therapy, be complicated by the unwanted but intrinsic side effects of intervening in highly interactive genetic networks and processes, both normal and pathological. Currently available tools for gene therapy are largely based on augmentation of a defective genome by addition of a complementing but largely unregulated genetic function rather than by true genetic correction of the underlying defect. That mechanism would be expected to lead to persistent aberrations of complex interactive genetic networks in the modified cell. Only true reversion of a genetic defect to a normal genotype by sequence correction would probably obviate such effects. There is reason to be optimistic that site-specific genetic correction will be feasible in the not-too-distant future.

As has been made abundantly clear by recent clinical experience with human gene transfer studies, there is no ideal vector for all applications in human gene therapy. The choice of vector systems in human gene therapy obviously depends crucially on many factors, including the replicative and functional properties and location of the target cell, the nature of the genetic and physiological deficiencies to be corrected, and the need for fine developmental and temporal control of transgene expression, at least in some cases. And so the search continues for other approaches to viral and nonviral gene transfer, other classes of vectors, other and improved methods of gene delivery. Nevertheless, despite the attractiveness and promise of many of these alternative gene delivery systems, there is a medical imperative to use available tools, even if imperfect, for the relief of human disease. We restrict the discussion in this volume to the vectors that have most clearly and most definitively produced therapeutic benefit for patients, that is, the retrovirus, lentivirus, and AAV vectors. They are not perfect, but they have proven to be therapeutically effective in a growing number of clinical settings.

THEODORE FRIEDMANN

REFERENCES

Friedmann, T., and Roblin, R. (1972). Gene therapy for human genetic disease? *Science* **175,** 949–955.
Friedmann, T., and Rossi, J. (2006). A Laboratory Manual of Gene Transfer Vectors. Cold Spring Harbor Press.
Grieger, J. C., and Samulski, R. J. (2011). Adeno-associated virus vectorology, manufacturing and clinical applications. *Methods Enzymol.* **507**(this volume).
Paruzynski, A., Glimm, H., Schmidt, M., and von Kalle, C. (2011). Analysis of the clonal repertoire of gene corrected cells in gene therapy. *Methods Enzymol.* **507**(this volume).
Porteus, M. H., and Baltimore, D. (2003). Chimeric nucleases stimulate gene targeting in human cells. *Science* **300,** 763.
Shimotohno, K., and Temin, H. M. (1981). Formation of infectious progeny virus after insertion of herpes simplex thymidine kinase gene into DNA of an avian retrovirus. *Cell* **26,** 67–77.
Tabin, C. J., Hoffmann, J. W., Goff, S. P., and Weinberg, R. A. (1982). Adaptation of a retrovirus as a eucaryotic vector transmitting the herpes simplex virus thymidine kinase gene. *Mol. Cell. Biol.* **2,** 426–436.
Wei, C. M., Gibson, M., Spear, P. G., and Scolnick, E. M. (1981). Construction and isolation of a transmissible retrovirus containing the src gene of Harvey murine sarcoma virus and the thymidine kinase gene of herpes simplex virus type 1. *J. Virol.* **39,** 935–944.
Westphal, H., and Dulbecco, R. (1968). Viral DNA in polyoma- and SV40-transformed cell lines. *Proc. Natl. Acad. Sci. USA* **59,** 1158–1165.
Wilson, J. H. (2003). Pointing fingers at the limiting step in gene targeting. *Nat. Biotechnol.* **21,** 759–760.

METHODS IN ENZYMOLOGY

VOLUME I. Preparation and Assay of Enzymes
Edited by SIDNEY P. COLOWICK AND NATHAN O. KAPLAN

VOLUME II. Preparation and Assay of Enzymes
Edited by SIDNEY P. COLOWICK AND NATHAN O. KAPLAN

VOLUME III. Preparation and Assay of Substrates
Edited by SIDNEY P. COLOWICK AND NATHAN O. KAPLAN

VOLUME IV. Special Techniques for the Enzymologist
Edited by SIDNEY P. COLOWICK AND NATHAN O. KAPLAN

VOLUME V. Preparation and Assay of Enzymes
Edited by SIDNEY P. COLOWICK AND NATHAN O. KAPLAN

VOLUME VI. Preparation and Assay of Enzymes *(Continued)*
Preparation and Assay of Substrates
Special Techniques
Edited by SIDNEY P. COLOWICK AND NATHAN O. KAPLAN

VOLUME VII. Cumulative Subject Index
Edited by SIDNEY P. COLOWICK AND NATHAN O. KAPLAN

VOLUME VIII. Complex Carbohydrates
Edited by ELIZABETH F. NEUFELD AND VICTOR GINSBURG

VOLUME IX. Carbohydrate Metabolism
Edited by WILLIS A. WOOD

VOLUME X. Oxidation and Phosphorylation
Edited by RONALD W. ESTABROOK AND MAYNARD E. PULLMAN

VOLUME XI. Enzyme Structure
Edited by C. H. W. HIRS

VOLUME XII. Nucleic Acids (Parts A and B)
Edited by LAWRENCE GROSSMAN AND KIVIE MOLDAVE

VOLUME XIII. Citric Acid Cycle
Edited by J. M. LOWENSTEIN

VOLUME XIV. Lipids
Edited by J. M. LOWENSTEIN

VOLUME XV. Steroids and Terpenoids
Edited by RAYMOND B. CLAYTON

VOLUME XVI. Fast Reactions
Edited by KENNETH KUSTIN

VOLUME XVII. Metabolism of Amino Acids and Amines (Parts A and B)
Edited by HERBERT TABOR AND CELIA WHITE TABOR

VOLUME XVIII. Vitamins and Coenzymes (Parts A, B, and C)
Edited by DONALD B. MCCORMICK AND LEMUEL D. WRIGHT

VOLUME XIX. Proteolytic Enzymes
Edited by GERTRUDE E. PERLMANN AND LASZLO LORAND

VOLUME XX. Nucleic Acids and Protein Synthesis (Part C)
Edited by KIVIE MOLDAVE AND LAWRENCE GROSSMAN

VOLUME XXI. Nucleic Acids (Part D)
Edited by LAWRENCE GROSSMAN AND KIVIE MOLDAVE

VOLUME XXII. Enzyme Purification and Related Techniques
Edited by WILLIAM B. JAKOBY

VOLUME XXIII. Photosynthesis (Part A)
Edited by ANTHONY SAN PIETRO

VOLUME XXIV. Photosynthesis and Nitrogen Fixation (Part B)
Edited by ANTHONY SAN PIETRO

VOLUME XXV. Enzyme Structure (Part B)
Edited by C. H. W. HIRS AND SERGE N. TIMASHEFF

VOLUME XXVI. Enzyme Structure (Part C)
Edited by C. H. W. HIRS AND SERGE N. TIMASHEFF

VOLUME XXVII. Enzyme Structure (Part D)
Edited by C. H. W. HIRS AND SERGE N. TIMASHEFF

VOLUME XXVIII. Complex Carbohydrates (Part B)
Edited by VICTOR GINSBURG

VOLUME XXIX. Nucleic Acids and Protein Synthesis (Part E)
Edited by LAWRENCE GROSSMAN AND KIVIE MOLDAVE

VOLUME XXX. Nucleic Acids and Protein Synthesis (Part F)
Edited by KIVIE MOLDAVE AND LAWRENCE GROSSMAN

VOLUME XXXI. Biomembranes (Part A)
Edited by SIDNEY FLEISCHER AND LESTER PACKER

VOLUME XXXII. Biomembranes (Part B)
Edited by SIDNEY FLEISCHER AND LESTER PACKER

VOLUME XXXIII. Cumulative Subject Index Volumes I–XXX
Edited by MARTHA G. DENNIS AND EDWARD A. DENNIS

VOLUME XXXIV. Affinity Techniques (Enzyme Purification: Part B)
Edited by WILLIAM B. JAKOBY AND MEIR WILCHEK

VOLUME XXXV. Lipids (Part B)
Edited by JOHN M. LOWENSTEIN

VOLUME XXXVI. Hormone Action (Part A: Steroid Hormones)
Edited by BERT W. O'MALLEY AND JOEL G. HARDMAN

VOLUME XXXVII. Hormone Action (Part B: Peptide Hormones)
Edited by BERT W. O'MALLEY AND JOEL G. HARDMAN

VOLUME XXXVIII. Hormone Action (Part C: Cyclic Nucleotides)
Edited by JOEL G. HARDMAN AND BERT W. O'MALLEY

VOLUME XXXIX. Hormone Action (Part D: Isolated Cells, Tissues, and Organ Systems)
Edited by JOEL G. HARDMAN AND BERT W. O'MALLEY

VOLUME XL. Hormone Action (Part E: Nuclear Structure and Function)
Edited by BERT W. O'MALLEY AND JOEL G. HARDMAN

VOLUME XLI. Carbohydrate Metabolism (Part B)
Edited by W. A. WOOD

VOLUME XLII. Carbohydrate Metabolism (Part C)
Edited by W. A. WOOD

VOLUME XLIII. Antibiotics
Edited by JOHN H. HASH

VOLUME XLIV. Immobilized Enzymes
Edited by KLAUS MOSBACH

VOLUME XLV. Proteolytic Enzymes (Part B)
Edited by LASZLO LORAND

VOLUME XLVI. Affinity Labeling
Edited by WILLIAM B. JAKOBY AND MEIR WILCHEK

VOLUME XLVII. Enzyme Structure (Part E)
Edited by C. H. W. HIRS AND SERGE N. TIMASHEFF

VOLUME XLVIII. Enzyme Structure (Part F)
Edited by C. H. W. HIRS AND SERGE N. TIMASHEFF

VOLUME XLIX. Enzyme Structure (Part G)
Edited by C. H. W. HIRS AND SERGE N. TIMASHEFF

VOLUME L. Complex Carbohydrates (Part C)
Edited by VICTOR GINSBURG

VOLUME LI. Purine and Pyrimidine Nucleotide Metabolism
Edited by PATRICIA A. HOFFEE AND MARY ELLEN JONES

VOLUME LII. Biomembranes (Part C: Biological Oxidations)
Edited by SIDNEY FLEISCHER AND LESTER PACKER

VOLUME LIII. Biomembranes (Part D: Biological Oxidations)
Edited by SIDNEY FLEISCHER AND LESTER PACKER

VOLUME LIV. Biomembranes (Part E: Biological Oxidations)
Edited by SIDNEY FLEISCHER AND LESTER PACKER

VOLUME LV. Biomembranes (Part F: Bioenergetics)
Edited by SIDNEY FLEISCHER AND LESTER PACKER

VOLUME LVI. Biomembranes (Part G: Bioenergetics)
Edited by SIDNEY FLEISCHER AND LESTER PACKER

VOLUME LVII. Bioluminescence and Chemiluminescence
Edited by MARLENE A. DELUCA

VOLUME LVIII. Cell Culture
Edited by WILLIAM B. JAKOBY AND IRA PASTAN

VOLUME LIX. Nucleic Acids and Protein Synthesis (Part G)
Edited by KIVIE MOLDAVE AND LAWRENCE GROSSMAN

VOLUME LX. Nucleic Acids and Protein Synthesis (Part H)
Edited by KIVIE MOLDAVE AND LAWRENCE GROSSMAN

VOLUME 61. Enzyme Structure (Part H)
Edited by C. H. W. HIRS AND SERGE N. TIMASHEFF

VOLUME 62. Vitamins and Coenzymes (Part D)
Edited by DONALD B. MCCORMICK AND LEMUEL D. WRIGHT

VOLUME 63. Enzyme Kinetics and Mechanism (Part A: Initial Rate and Inhibitor Methods)
Edited by DANIEL L. PURICH

VOLUME 64. Enzyme Kinetics and Mechanism
(Part B: Isotopic Probes and Complex Enzyme Systems)
Edited by DANIEL L. PURICH

VOLUME 65. Nucleic Acids (Part I)
Edited by LAWRENCE GROSSMAN AND KIVIE MOLDAVE

VOLUME 66. Vitamins and Coenzymes (Part E)
Edited by DONALD B. MCCORMICK AND LEMUEL D. WRIGHT

VOLUME 67. Vitamins and Coenzymes (Part F)
Edited by DONALD B. MCCORMICK AND LEMUEL D. WRIGHT

VOLUME 68. Recombinant DNA
Edited by RAY WU

VOLUME 69. Photosynthesis and Nitrogen Fixation (Part C)
Edited by ANTHONY SAN PIETRO

VOLUME 70. Immunochemical Techniques (Part A)
Edited by HELEN VAN VUNAKIS AND JOHN J. LANGONE

VOLUME 71. Lipids (Part C)
Edited by JOHN M. LOWENSTEIN

VOLUME 72. Lipids (Part D)
Edited by JOHN M. LOWENSTEIN

VOLUME 73. Immunochemical Techniques (Part B)
Edited by JOHN J. LANGONE AND HELEN VAN VUNAKIS

VOLUME 74. Immunochemical Techniques (Part C)
Edited by JOHN J. LANGONE AND HELEN VAN VUNAKIS

VOLUME 75. Cumulative Subject Index Volumes XXXI, XXXII, XXXIV–LX
Edited by EDWARD A. DENNIS AND MARTHA G. DENNIS

VOLUME 76. Hemoglobins
Edited by ERALDO ANTONINI, LUIGI ROSSI-BERNARDI, AND EMILIA CHIANCONE

VOLUME 77. Detoxication and Drug Metabolism
Edited by WILLIAM B. JAKOBY

VOLUME 78. Interferons (Part A)
Edited by SIDNEY PESTKA

VOLUME 79. Interferons (Part B)
Edited by SIDNEY PESTKA

VOLUME 80. Proteolytic Enzymes (Part C)
Edited by LASZLO LORAND

VOLUME 81. Biomembranes (Part H: Visual Pigments and Purple Membranes, I)
Edited by LESTER PACKER

VOLUME 82. Structural and Contractile Proteins (Part A: Extracellular Matrix)
Edited by LEON W. CUNNINGHAM AND DIXIE W. FREDERIKSEN

VOLUME 83. Complex Carbohydrates (Part D)
Edited by VICTOR GINSBURG

VOLUME 84. Immunochemical Techniques (Part D: Selected Immunoassays)
Edited by JOHN J. LANGONE AND HELEN VAN VUNAKIS

VOLUME 85. Structural and Contractile Proteins (Part B: The Contractile Apparatus and the Cytoskeleton)
Edited by DIXIE W. FREDERIKSEN AND LEON W. CUNNINGHAM

VOLUME 86. Prostaglandins and Arachidonate Metabolites
Edited by WILLIAM E. M. LANDS AND WILLIAM L. SMITH

VOLUME 87. Enzyme Kinetics and Mechanism (Part C: Intermediates, Stereo-chemistry, and Rate Studies)
Edited by DANIEL L. PURICH

VOLUME 88. Biomembranes (Part I: Visual Pigments and Purple Membranes, II)
Edited by LESTER PACKER

VOLUME 89. Carbohydrate Metabolism (Part D)
Edited by WILLIS A. WOOD

VOLUME 90. Carbohydrate Metabolism (Part E)
Edited by WILLIS A. WOOD

VOLUME 91. Enzyme Structure (Part I)
Edited by C. H. W. HIRS AND SERGE N. TIMASHEFF

VOLUME 92. Immunochemical Techniques (Part E: Monoclonal Antibodies and General Immunoassay Methods)
Edited by JOHN J. LANGONE AND HELEN VAN VUNAKIS

VOLUME 93. Immunochemical Techniques (Part F: Conventional Antibodies, Fc Receptors, and Cytotoxicity)
Edited by JOHN J. LANGONE AND HELEN VAN VUNAKIS

VOLUME 94. Polyamines
Edited by HERBERT TABOR AND CELIA WHITE TABOR

VOLUME 95. Cumulative Subject Index Volumes 61–74, 76–80
Edited by EDWARD A. DENNIS AND MARTHA G. DENNIS

VOLUME 96. Biomembranes [Part J: Membrane Biogenesis: Assembly and Targeting (General Methods; Eukaryotes)]
Edited by SIDNEY FLEISCHER AND BECCA FLEISCHER

VOLUME 97. Biomembranes [Part K: Membrane Biogenesis: Assembly and Targeting (Prokaryotes, Mitochondria, and Chloroplasts)]
Edited by SIDNEY FLEISCHER AND BECCA FLEISCHER

VOLUME 98. Biomembranes (Part L: Membrane Biogenesis: Processing and Recycling)
Edited by SIDNEY FLEISCHER AND BECCA FLEISCHER

VOLUME 99. Hormone Action (Part F: Protein Kinases)
Edited by JACKIE D. CORBIN AND JOEL G. HARDMAN

VOLUME 100. Recombinant DNA (Part B)
Edited by RAY WU, LAWRENCE GROSSMAN, AND KIVIE MOLDAVE

VOLUME 101. Recombinant DNA (Part C)
Edited by RAY WU, LAWRENCE GROSSMAN, AND KIVIE MOLDAVE

VOLUME 102. Hormone Action (Part G: Calmodulin and Calcium-Binding Proteins)
Edited by ANTHONY R. MEANS AND BERT W. O'MALLEY

VOLUME 103. Hormone Action (Part H: Neuroendocrine Peptides)
Edited by P. MICHAEL CONN

VOLUME 104. Enzyme Purification and Related Techniques (Part C)
Edited by WILLIAM B. JAKOBY

VOLUME 105. Oxygen Radicals in Biological Systems
Edited by LESTER PACKER

VOLUME 106. Posttranslational Modifications (Part A)
Edited by FINN WOLD AND KIVIE MOLDAVE

VOLUME 107. Posttranslational Modifications (Part B)
Edited by FINN WOLD AND KIVIE MOLDAVE

VOLUME 108. Immunochemical Techniques (Part G: Separation and Characterization of Lymphoid Cells)
Edited by GIOVANNI DI SABATO, JOHN J. LANGONE, AND HELEN VAN VUNAKIS

VOLUME 109. Hormone Action (Part I: Peptide Hormones)
Edited by LUTZ BIRNBAUMER AND BERT W. O'MALLEY

VOLUME 110. Steroids and Isoprenoids (Part A)
Edited by JOHN H. LAW AND HANS C. RILLING

VOLUME 111. Steroids and Isoprenoids (Part B)
Edited by JOHN H. LAW AND HANS C. RILLING

VOLUME 112. Drug and Enzyme Targeting (Part A)
Edited by KENNETH J. WIDDER AND RALPH GREEN

VOLUME 113. Glutamate, Glutamine, Glutathione, and Related Compounds
Edited by ALTON MEISTER

VOLUME 114. Diffraction Methods for Biological Macromolecules (Part A)
Edited by HAROLD W. WYCKOFF, C. H. W. HIRS, AND SERGE N. TIMASHEFF

VOLUME 115. Diffraction Methods for Biological Macromolecules (Part B)
Edited by HAROLD W. WYCKOFF, C. H. W. HIRS, AND SERGE N. TIMASHEFF

VOLUME 116. Immunochemical Techniques
(Part H: Effectors and Mediators of Lymphoid Cell Functions)
Edited by GIOVANNI DI SABATO, JOHN J. LANGONE, AND HELEN VAN VUNAKIS

VOLUME 117. Enzyme Structure (Part J)
Edited by C. H. W. HIRS AND SERGE N. TIMASHEFF

VOLUME 118. Plant Molecular Biology
Edited by ARTHUR WEISSBACH AND HERBERT WEISSBACH

VOLUME 119. Interferons (Part C)
Edited by SIDNEY PESTKA

VOLUME 120. Cumulative Subject Index Volumes 81–94, 96–101

VOLUME 121. Immunochemical Techniques (Part I: Hybridoma Technology and Monoclonal Antibodies)
Edited by JOHN J. LANGONE AND HELEN VAN VUNAKIS

VOLUME 122. Vitamins and Coenzymes (Part G)
Edited by FRANK CHYTIL AND DONALD B. MCCORMICK

VOLUME 123. Vitamins and Coenzymes (Part H)
Edited by FRANK CHYTIL AND DONALD B. MCCORMICK

VOLUME 124. Hormone Action (Part J: Neuroendocrine Peptides)
Edited by P. MICHAEL CONN

VOLUME 125. Biomembranes (Part M: Transport in Bacteria, Mitochondria, and Chloroplasts: General Approaches and Transport Systems)
Edited by SIDNEY FLEISCHER AND BECCA FLEISCHER

VOLUME 126. Biomembranes (Part N: Transport in Bacteria, Mitochondria, and Chloroplasts: Protonmotive Force)
Edited by SIDNEY FLEISCHER AND BECCA FLEISCHER

VOLUME 127. Biomembranes (Part O: Protons and Water: Structure and Translocation)
Edited by LESTER PACKER

VOLUME 128. Plasma Lipoproteins (Part A: Preparation, Structure, and Molecular Biology)
Edited by JERE P. SEGREST AND JOHN J. ALBERS

VOLUME 129. Plasma Lipoproteins (Part B: Characterization, Cell Biology, and Metabolism)
Edited by JOHN J. ALBERS AND JERE P. SEGREST

VOLUME 130. Enzyme Structure (Part K)
Edited by C. H. W. HIRS AND SERGE N. TIMASHEFF

VOLUME 131. Enzyme Structure (Part L)
Edited by C. H. W. HIRS AND SERGE N. TIMASHEFF

VOLUME 132. Immunochemical Techniques (Part J: Phagocytosis and Cell-Mediated Cytotoxicity)
Edited by GIOVANNI DI SABATO AND JOHANNES EVERSE

VOLUME 133. Bioluminescence and Chemiluminescence (Part B)
Edited by MARLENE DELUCA AND WILLIAM D. MCELROY

VOLUME 134. Structural and Contractile Proteins (Part C: The Contractile Apparatus and the Cytoskeleton)
Edited by RICHARD B. VALLEE

VOLUME 135. Immobilized Enzymes and Cells (Part B)
Edited by KLAUS MOSBACH

VOLUME 136. Immobilized Enzymes and Cells (Part C)
Edited by KLAUS MOSBACH

VOLUME 137. Immobilized Enzymes and Cells (Part D)
Edited by KLAUS MOSBACH

VOLUME 138. Complex Carbohydrates (Part E)
Edited by VICTOR GINSBURG

VOLUME 139. Cellular Regulators (Part A: Calcium- and Calmodulin-Binding Proteins)
Edited by ANTHONY R. MEANS AND P. MICHAEL CONN

VOLUME 140. Cumulative Subject Index Volumes 102–119, 121–134

VOLUME 141. Cellular Regulators (Part B: Calcium and Lipids)
Edited by P. MICHAEL CONN AND ANTHONY R. MEANS

VOLUME 142. Metabolism of Aromatic Amino Acids and Amines
Edited by SEYMOUR KAUFMAN

VOLUME 143. Sulfur and Sulfur Amino Acids
Edited by WILLIAM B. JAKOBY AND OWEN GRIFFITH

VOLUME 144. Structural and Contractile Proteins (Part D: Extracellular Matrix)
Edited by LEON W. CUNNINGHAM

VOLUME 145. Structural and Contractile Proteins (Part E: Extracellular Matrix)
Edited by LEON W. CUNNINGHAM

VOLUME 146. Peptide Growth Factors (Part A)
Edited by DAVID BARNES AND DAVID A. SIRBASKU

VOLUME 147. Peptide Growth Factors (Part B)
Edited by DAVID BARNES AND DAVID A. SIRBASKU

VOLUME 148. Plant Cell Membranes
Edited by LESTER PACKER AND ROLAND DOUCE

VOLUME 149. Drug and Enzyme Targeting (Part B)
Edited by RALPH GREEN AND KENNETH J. WIDDER

VOLUME 150. Immunochemical Techniques (Part K: *In Vitro* Models of B and T Cell Functions and Lymphoid Cell Receptors)
Edited by GIOVANNI DI SABATO

VOLUME 151. Molecular Genetics of Mammalian Cells
Edited by MICHAEL M. GOTTESMAN

VOLUME 152. Guide to Molecular Cloning Techniques
Edited by SHELBY L. BERGER AND ALAN R. KIMMEL

VOLUME 153. Recombinant DNA (Part D)
Edited by RAY WU AND LAWRENCE GROSSMAN

VOLUME 154. Recombinant DNA (Part E)
Edited by RAY WU AND LAWRENCE GROSSMAN

VOLUME 155. Recombinant DNA (Part F)
Edited by RAY WU

VOLUME 156. Biomembranes (Part P: ATP-Driven Pumps and Related Transport: The Na, K-Pump)
Edited by SIDNEY FLEISCHER AND BECCA FLEISCHER

VOLUME 157. Biomembranes (Part Q: ATP-Driven Pumps and Related Transport: Calcium, Proton, and Potassium Pumps)
Edited by SIDNEY FLEISCHER AND BECCA FLEISCHER

VOLUME 158. Metalloproteins (Part A)
Edited by JAMES F. RIORDAN AND BERT L. VALLEE

VOLUME 159. Initiation and Termination of Cyclic Nucleotide Action
Edited by JACKIE D. CORBIN AND ROGER A. JOHNSON

VOLUME 160. Biomass (Part A: Cellulose and Hemicellulose)
Edited by WILLIS A. WOOD AND SCOTT T. KELLOGG

VOLUME 161. Biomass (Part B: Lignin, Pectin, and Chitin)
Edited by WILLIS A. WOOD AND SCOTT T. KELLOGG

VOLUME 162. Immunochemical Techniques (Part L: Chemotaxis and Inflammation)
Edited by GIOVANNI DI SABATO

VOLUME 163. Immunochemical Techniques (Part M: Chemotaxis and Inflammation)
Edited by GIOVANNI DI SABATO

VOLUME 164. Ribosomes
Edited by HARRY F. NOLLER, JR., AND KIVIE MOLDAVE

VOLUME 165. Microbial Toxins: Tools for Enzymology
Edited by SIDNEY HARSHMAN

VOLUME 166. Branched-Chain Amino Acids
Edited by ROBERT HARRIS AND JOHN R. SOKATCH

VOLUME 167. Cyanobacteria
Edited by LESTER PACKER AND ALEXANDER N. GLAZER

VOLUME 168. Hormone Action (Part K: Neuroendocrine Peptides)
Edited by P. MICHAEL CONN

VOLUME 169. Platelets: Receptors, Adhesion, Secretion (Part A)
Edited by JACEK HAWIGER

VOLUME 170. Nucleosomes
Edited by PAUL M. WASSARMAN AND ROGER D. KORNBERG

VOLUME 171. Biomembranes (Part R: Transport Theory: Cells and Model Membranes)
Edited by SIDNEY FLEISCHER AND BECCA FLEISCHER

VOLUME 172. Biomembranes (Part S: Transport: Membrane Isolation and Characterization)
Edited by SIDNEY FLEISCHER AND BECCA FLEISCHER

VOLUME 173. Biomembranes [Part T: Cellular and Subcellular Transport: Eukaryotic (Nonepithelial) Cells]
Edited by SIDNEY FLEISCHER AND BECCA FLEISCHER

VOLUME 174. Biomembranes [Part U: Cellular and Subcellular Transport: Eukaryotic (Nonepithelial) Cells]
Edited by SIDNEY FLEISCHER AND BECCA FLEISCHER

VOLUME 175. Cumulative Subject Index Volumes 135–139, 141–167

VOLUME 176. Nuclear Magnetic Resonance (Part A: Spectral Techniques and Dynamics)
Edited by NORMAN J. OPPENHEIMER AND THOMAS L. JAMES

VOLUME 177. Nuclear Magnetic Resonance (Part B: Structure and Mechanism)
Edited by NORMAN J. OPPENHEIMER AND THOMAS L. JAMES

VOLUME 178. Antibodies, Antigens, and Molecular Mimicry
Edited by JOHN J. LANGONE

VOLUME 179. Complex Carbohydrates (Part F)
Edited by VICTOR GINSBURG

VOLUME 180. RNA Processing (Part A: General Methods)
Edited by JAMES E. DAHLBERG AND JOHN N. ABELSON

VOLUME 181. RNA Processing (Part B: Specific Methods)
Edited by JAMES E. DAHLBERG AND JOHN N. ABELSON

VOLUME 182. Guide to Protein Purification
Edited by MURRAY P. DEUTSCHER

VOLUME 183. Molecular Evolution: Computer Analysis of Protein and Nucleic Acid Sequences
Edited by RUSSELL F. DOOLITTLE

VOLUME 184. Avidin-Biotin Technology
Edited by MEIR WILCHEK AND EDWARD A. BAYER

VOLUME 185. Gene Expression Technology
Edited by DAVID V. GOEDDEL

VOLUME 186. Oxygen Radicals in Biological Systems (Part B: Oxygen Radicals and Antioxidants)
Edited by LESTER PACKER AND ALEXANDER N. GLAZER

VOLUME 187. Arachidonate Related Lipid Mediators
Edited by ROBERT C. MURPHY AND FRANK A. FITZPATRICK

VOLUME 188. Hydrocarbons and Methylotrophy
Edited by MARY E. LIDSTROM

VOLUME 189. Retinoids (Part A: Molecular and Metabolic Aspects)
Edited by LESTER PACKER

VOLUME 190. Retinoids (Part B: Cell Differentiation and Clinical Applications)
Edited by LESTER PACKER

VOLUME 191. Biomembranes (Part V: Cellular and Subcellular Transport: Epithelial Cells)
Edited by SIDNEY FLEISCHER AND BECCA FLEISCHER

VOLUME 192. Biomembranes (Part W: Cellular and Subcellular Transport: Epithelial Cells)
Edited by SIDNEY FLEISCHER AND BECCA FLEISCHER

VOLUME 193. Mass Spectrometry
Edited by JAMES A. MCCLOSKEY

VOLUME 194. Guide to Yeast Genetics and Molecular Biology
Edited by CHRISTINE GUTHRIE AND GERALD R. FINK

VOLUME 195. Adenylyl Cyclase, G Proteins, and Guanylyl Cyclase
Edited by ROGER A. JOHNSON AND JACKIE D. CORBIN

VOLUME 196. Molecular Motors and the Cytoskeleton
Edited by RICHARD B. VALLEE

VOLUME 197. Phospholipases
Edited by EDWARD A. DENNIS

VOLUME 198. Peptide Growth Factors (Part C)
Edited by DAVID BARNES, J. P. MATHER, AND GORDON H. SATO

VOLUME 199. Cumulative Subject Index Volumes 168–174, 176–194

VOLUME 200. Protein Phosphorylation (Part A: Protein Kinases: Assays, Purification, Antibodies, Functional Analysis, Cloning, and Expression)
Edited by TONY HUNTER AND BARTHOLOMEW M. SEFTON

VOLUME 201. Protein Phosphorylation (Part B: Analysis of Protein Phosphorylation, Protein Kinase Inhibitors, and Protein Phosphatases)
Edited by TONY HUNTER AND BARTHOLOMEW M. SEFTON

VOLUME 202. Molecular Design and Modeling: Concepts and Applications (Part A: Proteins, Peptides, and Enzymes)
Edited by JOHN J. LANGONE

VOLUME 203. Molecular Design and Modeling: Concepts and Applications (Part B: Antibodies and Antigens, Nucleic Acids, Polysaccharides, and Drugs)
Edited by JOHN J. LANGONE

VOLUME 204. Bacterial Genetic Systems
Edited by JEFFREY H. MILLER

VOLUME 205. Metallobiochemistry (Part B: Metallothionein and Related Molecules)
Edited by JAMES F. RIORDAN AND BERT L. VALLEE

VOLUME 206. Cytochrome P450
Edited by MICHAEL R. WATERMAN AND ERIC F. JOHNSON

VOLUME 207. Ion Channels
Edited by BERNARDO RUDY AND LINDA E. IVERSON

VOLUME 208. Protein–DNA Interactions
Edited by ROBERT T. SAUER

VOLUME 209. Phospholipid Biosynthesis
Edited by EDWARD A. DENNIS AND DENNIS E. VANCE

VOLUME 210. Numerical Computer Methods
Edited by LUDWIG BRAND AND MICHAEL L. JOHNSON

VOLUME 211. DNA Structures (Part A: Synthesis and Physical Analysis of DNA)
Edited by DAVID M. J. LILLEY AND JAMES E. DAHLBERG

VOLUME 212. DNA Structures (Part B: Chemical and Electrophoretic Analysis of DNA)
Edited by DAVID M. J. LILLEY AND JAMES E. DAHLBERG

VOLUME 213. Carotenoids (Part A: Chemistry, Separation, Quantitation, and Antioxidation)
Edited by LESTER PACKER

VOLUME 214. Carotenoids (Part B: Metabolism, Genetics, and Biosynthesis)
Edited by LESTER PACKER

VOLUME 215. Platelets: Receptors, Adhesion, Secretion (Part B)
Edited by JACEK J. HAWIGER

VOLUME 216. Recombinant DNA (Part G)
Edited by RAY WU

VOLUME 217. Recombinant DNA (Part H)
Edited by RAY WU

VOLUME 218. Recombinant DNA (Part I)
Edited by RAY WU

VOLUME 219. Reconstitution of Intracellular Transport
Edited by JAMES E. ROTHMAN

VOLUME 220. Membrane Fusion Techniques (Part A)
Edited by NEJAT DÜZGÜNEŞ

VOLUME 221. Membrane Fusion Techniques (Part B)
Edited by NEJAT DÜZGÜNEŞ

VOLUME 222. Proteolytic Enzymes in Coagulation, Fibrinolysis, and Complement Activation (Part A: Mammalian Blood Coagulation Factors and Inhibitors)
Edited by LASZLO LORAND AND KENNETH G. MANN

VOLUME 223. Proteolytic Enzymes in Coagulation, Fibrinolysis, and Complement Activation (Part B: Complement Activation, Fibrinolysis, and Nonmammalian Blood Coagulation Factors)
Edited by LASZLO LORAND AND KENNETH G. MANN

VOLUME 224. Molecular Evolution: Producing the Biochemical Data
Edited by ELIZABETH ANNE ZIMMER, THOMAS J. WHITE, REBECCA L. CANN, AND ALLAN C. WILSON

VOLUME 225. Guide to Techniques in Mouse Development
Edited by PAUL M. WASSARMAN AND MELVIN L. DEPAMPHILIS

VOLUME 226. Metallobiochemistry (Part C: Spectroscopic and Physical Methods for Probing Metal Ion Environments in Metalloenzymes and Metalloproteins)
Edited by JAMES F. RIORDAN AND BERT L. VALLEE

VOLUME 227. Metallobiochemistry (Part D: Physical and Spectroscopic Methods for Probing Metal Ion Environments in Metalloproteins)
Edited by JAMES F. RIORDAN AND BERT L. VALLEE

VOLUME 228. Aqueous Two-Phase Systems
Edited by HARRY WALTER AND GÖTE JOHANSSON

VOLUME 229. Cumulative Subject Index Volumes 195–198, 200–227

VOLUME 230. Guide to Techniques in Glycobiology
Edited by WILLIAM J. LENNARZ AND GERALD W. HART

VOLUME 231. Hemoglobins (Part B: Biochemical and Analytical Methods)
Edited by JOHANNES EVERSE, KIM D. VANDEGRIFF, AND ROBERT M. WINSLOW

VOLUME 232. Hemoglobins (Part C: Biophysical Methods)
Edited by JOHANNES EVERSE, KIM D. VANDEGRIFF, AND ROBERT M. WINSLOW

VOLUME 233. Oxygen Radicals in Biological Systems (Part C)
Edited by LESTER PACKER

VOLUME 234. Oxygen Radicals in Biological Systems (Part D)
Edited by LESTER PACKER

VOLUME 235. Bacterial Pathogenesis (Part A: Identification and Regulation of Virulence Factors)
Edited by VIRGINIA L. CLARK AND PATRIK M. BAVOIL

VOLUME 236. Bacterial Pathogenesis (Part B: Integration of Pathogenic Bacteria with Host Cells)
Edited by VIRGINIA L. CLARK AND PATRIK M. BAVOIL

VOLUME 237. Heterotrimeric G Proteins
Edited by RAVI IYENGAR

VOLUME 238. Heterotrimeric G-Protein Effectors
Edited by RAVI IYENGAR

VOLUME 239. Nuclear Magnetic Resonance (Part C)
Edited by THOMAS L. JAMES AND NORMAN J. OPPENHEIMER

VOLUME 240. Numerical Computer Methods (Part B)
Edited by MICHAEL L. JOHNSON AND LUDWIG BRAND

VOLUME 241. Retroviral Proteases
Edited by LAWRENCE C. KUO AND JULES A. SHAFER

VOLUME 242. Neoglycoconjugates (Part A)
Edited by Y. C. LEE AND REIKO T. LEE

VOLUME 243. Inorganic Microbial Sulfur Metabolism
Edited by HARRY D. PECK, JR., AND JEAN LEGALL

VOLUME 244. Proteolytic Enzymes: Serine and Cysteine Peptidases
Edited by ALAN J. BARRETT

VOLUME 245. Extracellular Matrix Components
Edited by E. RUOSLAHTI AND E. ENGVALL

VOLUME 246. Biochemical Spectroscopy
Edited by KENNETH SAUER

VOLUME 247. Neoglycoconjugates (Part B: Biomedical Applications)
Edited by Y. C. LEE AND REIKO T. LEE

VOLUME 248. Proteolytic Enzymes: Aspartic and Metallo Peptidases
Edited by ALAN J. BARRETT

VOLUME 249. Enzyme Kinetics and Mechanism (Part D: Developments in Enzyme Dynamics)
Edited by DANIEL L. PURICH

VOLUME 250. Lipid Modifications of Proteins
Edited by PATRICK J. CASEY AND JANICE E. BUSS

VOLUME 251. Biothiols (Part A: Monothiols and Dithiols, Protein Thiols, and Thiyl Radicals)
Edited by LESTER PACKER

VOLUME 252. Biothiols (Part B: Glutathione and Thioredoxin; Thiols in Signal Transduction and Gene Regulation)
Edited by LESTER PACKER

VOLUME 253. Adhesion of Microbial Pathogens
Edited by RON J. DOYLE AND ITZHAK OFEK

VOLUME 254. Oncogene Techniques
Edited by PETER K. VOGT AND INDER M. VERMA

VOLUME 255. Small GTPases and Their Regulators (Part A: Ras Family)
Edited by W. E. BALCH, CHANNING J. DER, AND ALAN HALL

VOLUME 256. Small GTPases and Their Regulators (Part B: Rho Family)
Edited by W. E. BALCH, CHANNING J. DER, AND ALAN HALL

VOLUME 257. Small GTPases and Their Regulators (Part C: Proteins Involved in Transport)
Edited by W. E. BALCH, CHANNING J. DER, AND ALAN HALL

VOLUME 258. Redox-Active Amino Acids in Biology
Edited by JUDITH P. KLINMAN

VOLUME 259. Energetics of Biological Macromolecules
Edited by MICHAEL L. JOHNSON AND GARY K. ACKERS

VOLUME 260. Mitochondrial Biogenesis and Genetics (Part A)
Edited by GIUSEPPE M. ATTARDI AND ANNE CHOMYN

VOLUME 261. Nuclear Magnetic Resonance and Nucleic Acids
Edited by THOMAS L. JAMES

VOLUME 262. DNA Replication
Edited by JUDITH L. CAMPBELL

VOLUME 263. Plasma Lipoproteins (Part C: Quantitation)
Edited by WILLIAM A. BRADLEY, SANDRA H. GIANTURCO, AND JERE P. SEGREST

VOLUME 264. Mitochondrial Biogenesis and Genetics (Part B)
Edited by GIUSEPPE M. ATTARDI AND ANNE CHOMYN

VOLUME 265. Cumulative Subject Index Volumes 228, 230–262

VOLUME 266. Computer Methods for Macromolecular Sequence Analysis
Edited by RUSSELL F. DOOLITTLE

VOLUME 267. Combinatorial Chemistry
Edited by JOHN N. ABELSON

VOLUME 268. Nitric Oxide (Part A: Sources and Detection of NO; NO Synthase)
Edited by LESTER PACKER

VOLUME 269. Nitric Oxide (Part B: Physiological and Pathological Processes)
Edited by LESTER PACKER

VOLUME 270. High Resolution Separation and Analysis of Biological Macromolecules (Part A: Fundamentals)
Edited by BARRY L. KARGER AND WILLIAM S. HANCOCK

VOLUME 271. High Resolution Separation and Analysis of Biological Macromolecules (Part B: Applications)
Edited by BARRY L. KARGER AND WILLIAM S. HANCOCK

VOLUME 272. Cytochrome P450 (Part B)
Edited by ERIC F. JOHNSON AND MICHAEL R. WATERMAN

VOLUME 273. RNA Polymerase and Associated Factors (Part A)
Edited by SANKAR ADHYA

VOLUME 274. RNA Polymerase and Associated Factors (Part B)
Edited by SANKAR ADHYA

VOLUME 275. Viral Polymerases and Related Proteins
Edited by LAWRENCE C. KUO, DAVID B. OLSEN, AND STEVEN S. CARROLL

VOLUME 276. Macromolecular Crystallography (Part A)
Edited by CHARLES W. CARTER, JR., AND ROBERT M. SWEET

VOLUME 277. Macromolecular Crystallography (Part B)
Edited by CHARLES W. CARTER, JR., AND ROBERT M. SWEET

VOLUME 278. Fluorescence Spectroscopy
Edited by LUDWIG BRAND AND MICHAEL L. JOHNSON

VOLUME 279. Vitamins and Coenzymes (Part I)
Edited by DONALD B. MCCORMICK, JOHN W. SUTTIE, AND CONRAD WAGNER

VOLUME 280. Vitamins and Coenzymes (Part J)
Edited by DONALD B. MCCORMICK, JOHN W. SUTTIE, AND CONRAD WAGNER

VOLUME 281. Vitamins and Coenzymes (Part K)
Edited by DONALD B. MCCORMICK, JOHN W. SUTTIE, AND CONRAD WAGNER

VOLUME 282. Vitamins and Coenzymes (Part L)
Edited by DONALD B. MCCORMICK, JOHN W. SUTTIE, AND CONRAD WAGNER

VOLUME 283. Cell Cycle Control
Edited by WILLIAM G. DUNPHY

VOLUME 284. Lipases (Part A: Biotechnology)
Edited by BYRON RUBIN AND EDWARD A. DENNIS

VOLUME 285. Cumulative Subject Index Volumes 263, 264, 266–284, 286–289

VOLUME 286. Lipases (Part B: Enzyme Characterization and Utilization)
Edited by BYRON RUBIN AND EDWARD A. DENNIS

VOLUME 287. Chemokines
Edited by RICHARD HORUK

VOLUME 288. Chemokine Receptors
Edited by RICHARD HORUK

VOLUME 289. Solid Phase Peptide Synthesis
Edited by GREGG B. FIELDS

VOLUME 290. Molecular Chaperones
Edited by GEORGE H. LORIMER AND THOMAS BALDWIN

VOLUME 291. Caged Compounds
Edited by GERARD MARRIOTT

VOLUME 292. ABC Transporters: Biochemical, Cellular, and Molecular Aspects
Edited by SURESH V. AMBUDKAR AND MICHAEL M. GOTTESMAN

VOLUME 293. Ion Channels (Part B)
Edited by P. MICHAEL CONN

VOLUME 294. Ion Channels (Part C)
Edited by P. MICHAEL CONN

VOLUME 295. Energetics of Biological Macromolecules (Part B)
Edited by GARY K. ACKERS AND MICHAEL L. JOHNSON

VOLUME 296. Neurotransmitter Transporters
Edited by SUSAN G. AMARA

VOLUME 297. Photosynthesis: Molecular Biology of Energy Capture
Edited by LEE MCINTOSH

VOLUME 298. Molecular Motors and the Cytoskeleton (Part B)
Edited by RICHARD B. VALLEE

VOLUME 299. Oxidants and Antioxidants (Part A)
Edited by LESTER PACKER

VOLUME 300. Oxidants and Antioxidants (Part B)
Edited by LESTER PACKER

VOLUME 301. Nitric Oxide: Biological and Antioxidant Activities (Part C)
Edited by LESTER PACKER

VOLUME 302. Green Fluorescent Protein
Edited by P. MICHAEL CONN

VOLUME 303. cDNA Preparation and Display
Edited by SHERMAN M. WEISSMAN

VOLUME 304. Chromatin
Edited by PAUL M. WASSARMAN AND ALAN P. WOLFFE

VOLUME 305. Bioluminescence and Chemiluminescence (Part C)
Edited by THOMAS O. BALDWIN AND MIRIAM M. ZIEGLER

VOLUME 306. Expression of Recombinant Genes in Eukaryotic Systems
Edited by JOSEPH C. GLORIOSO AND MARTIN C. SCHMIDT

VOLUME 307. Confocal Microscopy
Edited by P. MICHAEL CONN

VOLUME 308. Enzyme Kinetics and Mechanism (Part E: Energetics of Enzyme Catalysis)
Edited by DANIEL L. PURICH AND VERN L. SCHRAMM

VOLUME 309. Amyloid, Prions, and Other Protein Aggregates
Edited by RONALD WETZEL

VOLUME 310. Biofilms
Edited by RON J. DOYLE

VOLUME 311. Sphingolipid Metabolism and Cell Signaling (Part A)
Edited by ALFRED H. MERRILL, JR., AND YUSUF A. HANNUN

VOLUME 312. Sphingolipid Metabolism and Cell Signaling (Part B)
Edited by ALFRED H. MERRILL, JR., AND YUSUF A. HANNUN

VOLUME 313. Antisense Technology
(Part A: General Methods, Methods of Delivery, and RNA Studies)
Edited by M. IAN PHILLIPS

VOLUME 314. Antisense Technology (Part B: Applications)
Edited by M. IAN PHILLIPS

VOLUME 315. Vertebrate Phototransduction and the Visual Cycle (Part A)
Edited by KRZYSZTOF PALCZEWSKI

VOLUME 316. Vertebrate Phototransduction and the Visual Cycle (Part B)
Edited by KRZYSZTOF PALCZEWSKI

VOLUME 317. RNA–Ligand Interactions (Part A: Structural Biology Methods)
Edited by DANIEL W. CELANDER AND JOHN N. ABELSON

VOLUME 318. RNA–Ligand Interactions (Part B: Molecular Biology Methods)
Edited by DANIEL W. CELANDER AND JOHN N. ABELSON

VOLUME 319. Singlet Oxygen, UV-A, and Ozone
Edited by LESTER PACKER AND HELMUT SIES

VOLUME 320. Cumulative Subject Index Volumes 290–319

VOLUME 321. Numerical Computer Methods (Part C)
Edited by MICHAEL L. JOHNSON AND LUDWIG BRAND

VOLUME 322. Apoptosis
Edited by JOHN C. REED

VOLUME 323. Energetics of Biological Macromolecules (Part C)
Edited by MICHAEL L. JOHNSON AND GARY K. ACKERS

VOLUME 324. Branched-Chain Amino Acids (Part B)
Edited by ROBERT A. HARRIS AND JOHN R. SOKATCH

VOLUME 325. Regulators and Effectors of Small GTPases
(Part D: Rho Family)
Edited by W. E. BALCH, CHANNING J. DER, AND ALAN HALL

VOLUME 326. Applications of Chimeric Genes and Hybrid Proteins
(Part A: Gene Expression and Protein Purification)
Edited by JEREMY THORNER, SCOTT D. EMR, AND JOHN N. ABELSON

VOLUME 327. Applications of Chimeric Genes and Hybrid Proteins
(Part B: Cell Biology and Physiology)
Edited by JEREMY THORNER, SCOTT D. EMR, AND JOHN N. ABELSON

VOLUME 328. Applications of Chimeric Genes and Hybrid Proteins (Part C: Protein–Protein Interactions and Genomics)
Edited by JEREMY THORNER, SCOTT D. EMR, AND JOHN N. ABELSON

VOLUME 329. Regulators and Effectors of Small GTPases (Part E: GTPases Involved in Vesicular Traffic)
Edited by W. E. BALCH, CHANNING J. DER, AND ALAN HALL

VOLUME 330. Hyperthermophilic Enzymes (Part A)
Edited by MICHAEL W. W. ADAMS AND ROBERT M. KELLY

VOLUME 331. Hyperthermophilic Enzymes (Part B)
Edited by MICHAEL W. W. ADAMS AND ROBERT M. KELLY

VOLUME 332. Regulators and Effectors of Small GTPases (Part F: Ras Family I)
Edited by W. E. BALCH, CHANNING J. DER, AND ALAN HALL

VOLUME 333. Regulators and Effectors of Small GTPases (Part G: Ras Family II)
Edited by W. E. BALCH, CHANNING J. DER, AND ALAN HALL

VOLUME 334. Hyperthermophilic Enzymes (Part C)
Edited by MICHAEL W. W. ADAMS AND ROBERT M. KELLY

VOLUME 335. Flavonoids and Other Polyphenols
Edited by LESTER PACKER

VOLUME 336. Microbial Growth in Biofilms (Part A: Developmental and Molecular Biological Aspects)
Edited by RON J. DOYLE

VOLUME 337. Microbial Growth in Biofilms (Part B: Special Environments and Physicochemical Aspects)
Edited by RON J. DOYLE

VOLUME 338. Nuclear Magnetic Resonance of Biological Macromolecules (Part A)
Edited by THOMAS L. JAMES, VOLKER DÖTSCH, AND ULI SCHMITZ

VOLUME 339. Nuclear Magnetic Resonance of Biological Macromolecules (Part B)
Edited by THOMAS L. JAMES, VOLKER DÖTSCH, AND ULI SCHMITZ

VOLUME 340. Drug–Nucleic Acid Interactions
Edited by JONATHAN B. CHAIRES AND MICHAEL J. WARING

VOLUME 341. Ribonucleases (Part A)
Edited by ALLEN W. NICHOLSON

VOLUME 342. Ribonucleases (Part B)
Edited by ALLEN W. NICHOLSON

VOLUME 343. G Protein Pathways (Part A: Receptors)
Edited by RAVI IYENGAR AND JOHN D. HILDEBRANDT

VOLUME 344. G Protein Pathways (Part B: G Proteins and Their Regulators)
Edited by RAVI IYENGAR AND JOHN D. HILDEBRANDT

VOLUME 345. G Protein Pathways (Part C: Effector Mechanisms)
Edited by RAVI IYENGAR AND JOHN D. HILDEBRANDT

VOLUME 346. Gene Therapy Methods
Edited by M. IAN PHILLIPS

VOLUME 347. Protein Sensors and Reactive Oxygen Species (Part A: Selenoproteins and Thioredoxin)
Edited by HELMUT SIES AND LESTER PACKER

VOLUME 348. Protein Sensors and Reactive Oxygen Species (Part B: Thiol Enzymes and Proteins)
Edited by HELMUT SIES AND LESTER PACKER

VOLUME 349. Superoxide Dismutase
Edited by LESTER PACKER

VOLUME 350. Guide to Yeast Genetics and Molecular and Cell Biology (Part B)
Edited by CHRISTINE GUTHRIE AND GERALD R. FINK

VOLUME 351. Guide to Yeast Genetics and Molecular and Cell Biology (Part C)
Edited by CHRISTINE GUTHRIE AND GERALD R. FINK

VOLUME 352. Redox Cell Biology and Genetics (Part A)
Edited by CHANDAN K. SEN AND LESTER PACKER

VOLUME 353. Redox Cell Biology and Genetics (Part B)
Edited by CHANDAN K. SEN AND LESTER PACKER

VOLUME 354. Enzyme Kinetics and Mechanisms (Part F: Detection and Characterization of Enzyme Reaction Intermediates)
Edited by DANIEL L. PURICH

VOLUME 355. Cumulative Subject Index Volumes 321–354

VOLUME 356. Laser Capture Microscopy and Microdissection
Edited by P. MICHAEL CONN

VOLUME 357. Cytochrome P450, Part C
Edited by ERIC F. JOHNSON AND MICHAEL R. WATERMAN

VOLUME 358. Bacterial Pathogenesis (Part C: Identification, Regulation, and Function of Virulence Factors)
Edited by VIRGINIA L. CLARK AND PATRIK M. BAVOIL

VOLUME 359. Nitric Oxide (Part D)
Edited by ENRIQUE CADENAS AND LESTER PACKER

VOLUME 360. Biophotonics (Part A)
Edited by GERARD MARRIOTT AND IAN PARKER

VOLUME 361. Biophotonics (Part B)
Edited by GERARD MARRIOTT AND IAN PARKER

VOLUME 362. Recognition of Carbohydrates in Biological Systems (Part A)
Edited by YUAN C. LEE AND REIKO T. LEE

VOLUME 363. Recognition of Carbohydrates in Biological Systems (Part B)
Edited by YUAN C. LEE AND REIKO T. LEE

VOLUME 364. Nuclear Receptors
Edited by DAVID W. RUSSELL AND DAVID J. MANGELSDORF

VOLUME 365. Differentiation of Embryonic Stem Cells
Edited by PAUL M. WASSAUMAN AND GORDON M. KELLER

VOLUME 366. Protein Phosphatases
Edited by SUSANNE KLUMPP AND JOSEF KRIEGLSTEIN

VOLUME 367. Liposomes (Part A)
Edited by NEJAT DÜZGÜNEŞ

VOLUME 368. Macromolecular Crystallography (Part C)
Edited by CHARLES W. CARTER, JR., AND ROBERT M. SWEET

VOLUME 369. Combinational Chemistry (Part B)
Edited by GUILLERMO A. MORALES AND BARRY A. BUNIN

VOLUME 370. RNA Polymerases and Associated Factors (Part C)
Edited by SANKAR L. ADHYA AND SUSAN GARGES

VOLUME 371. RNA Polymerases and Associated Factors (Part D)
Edited by SANKAR L. ADHYA AND SUSAN GARGES

VOLUME 372. Liposomes (Part B)
Edited by NEJAT DÜZGÜNEŞ

VOLUME 373. Liposomes (Part C)
Edited by NEJAT DÜZGÜNEŞ

VOLUME 374. Macromolecular Crystallography (Part D)
Edited by CHARLES W. CARTER, JR., AND ROBERT W. SWEET

VOLUME 375. Chromatin and Chromatin Remodeling Enzymes (Part A)
Edited by C. DAVID ALLIS AND CARL WU

VOLUME 376. Chromatin and Chromatin Remodeling Enzymes (Part B)
Edited by C. DAVID ALLIS AND CARL WU

VOLUME 377. Chromatin and Chromatin Remodeling Enzymes (Part C)
Edited by C. DAVID ALLIS AND CARL WU

VOLUME 378. Quinones and Quinone Enzymes (Part A)
Edited by HELMUT SIES AND LESTER PACKER

VOLUME 379. Energetics of Biological Macromolecules (Part D)
Edited by JO M. HOLT, MICHAEL L. JOHNSON, AND GARY K. ACKERS

VOLUME 380. Energetics of Biological Macromolecules (Part E)
Edited by JO M. HOLT, MICHAEL L. JOHNSON, AND GARY K. ACKERS

VOLUME 381. Oxygen Sensing
Edited by CHANDAN K. SEN AND GREGG L. SEMENZA

VOLUME 382. Quinones and Quinone Enzymes (Part B)
Edited by HELMUT SIES AND LESTER PACKER

VOLUME 383. Numerical Computer Methods (Part D)
Edited by LUDWIG BRAND AND MICHAEL L. JOHNSON

VOLUME 384. Numerical Computer Methods (Part E)
Edited by LUDWIG BRAND AND MICHAEL L. JOHNSON

VOLUME 385. Imaging in Biological Research (Part A)
Edited by P. MICHAEL CONN

VOLUME 386. Imaging in Biological Research (Part B)
Edited by P. MICHAEL CONN

VOLUME 387. Liposomes (Part D)
Edited by NEJAT DÜZGÜNEŞ

VOLUME 388. Protein Engineering
Edited by DAN E. ROBERTSON AND JOSEPH P. NOEL

VOLUME 389. Regulators of G-Protein Signaling (Part A)
Edited by DAVID P. SIDEROVSKI

VOLUME 390. Regulators of G-Protein Signaling (Part B)
Edited by DAVID P. SIDEROVSKI

VOLUME 391. Liposomes (Part E)
Edited by NEJAT DÜZGÜNEŞ

VOLUME 392. RNA Interference
Edited by ENGELKE ROSSI

VOLUME 393. Circadian Rhythms
Edited by MICHAEL W. YOUNG

VOLUME 394. Nuclear Magnetic Resonance of Biological Macromolecules (Part C)
Edited by THOMAS L. JAMES

VOLUME 395. Producing the Biochemical Data (Part B)
Edited by ELIZABETH A. ZIMMER AND ERIC H. ROALSON

VOLUME 396. Nitric Oxide (Part E)
Edited by LESTER PACKER AND ENRIQUE CADENAS

VOLUME 397. Environmental Microbiology
Edited by JARED R. LEADBETTER

VOLUME 398. Ubiquitin and Protein Degradation (Part A)
Edited by RAYMOND J. DESHAIES

VOLUME 399. Ubiquitin and Protein Degradation (Part B)
Edited by RAYMOND J. DESHAIES

VOLUME 400. Phase II Conjugation Enzymes and Transport Systems
Edited by HELMUT SIES AND LESTER PACKER

VOLUME 401. Glutathione Transferases and Gamma Glutamyl Transpeptidases
Edited by HELMUT SIES AND LESTER PACKER

VOLUME 402. Biological Mass Spectrometry
Edited by A. L. BURLINGAME

VOLUME 403. GTPases Regulating Membrane Targeting and Fusion
Edited by WILLIAM E. BALCH, CHANNING J. DER, AND ALAN HALL

VOLUME 404. GTPases Regulating Membrane Dynamics
Edited by WILLIAM E. BALCH, CHANNING J. DER, AND ALAN HALL

VOLUME 405. Mass Spectrometry: Modified Proteins and Glycoconjugates
Edited by A. L. BURLINGAME

VOLUME 406. Regulators and Effectors of Small GTPases: Rho Family
Edited by WILLIAM E. BALCH, CHANNING J. DER, AND ALAN HALL

VOLUME 407. Regulators and Effectors of Small GTPases: Ras Family
Edited by WILLIAM E. BALCH, CHANNING J. DER, AND ALAN HALL

VOLUME 408. DNA Repair (Part A)
Edited by JUDITH L. CAMPBELL AND PAUL MODRICH

VOLUME 409. DNA Repair (Part B)
Edited by JUDITH L. CAMPBELL AND PAUL MODRICH

VOLUME 410. DNA Microarrays (Part A: Array Platforms and Web-Bench Protocols)
Edited by ALAN KIMMEL AND BRIAN OLIVER

VOLUME 411. DNA Microarrays (Part B: Databases and Statistics)
Edited by ALAN KIMMEL AND BRIAN OLIVER

VOLUME 412. Amyloid, Prions, and Other Protein Aggregates (Part B)
Edited by INDU KHETERPAL AND RONALD WETZEL

VOLUME 413. Amyloid, Prions, and Other Protein Aggregates (Part C)
Edited by INDU KHETERPAL AND RONALD WETZEL

VOLUME 414. Measuring Biological Responses with Automated Microscopy
Edited by JAMES INGLESE

VOLUME 415. Glycobiology
Edited by MINORU FUKUDA

VOLUME 416. Glycomics
Edited by MINORU FUKUDA

VOLUME 417. Functional Glycomics
Edited by MINORU FUKUDA

VOLUME 418. Embryonic Stem Cells
Edited by IRINA KLIMANSKAYA AND ROBERT LANZA

VOLUME 419. Adult Stem Cells
Edited by IRINA KLIMANSKAYA AND ROBERT LANZA

VOLUME 420. Stem Cell Tools and Other Experimental Protocols
Edited by IRINA KLIMANSKAYA AND ROBERT LANZA

VOLUME 421. Advanced Bacterial Genetics: Use of Transposons and Phage for Genomic Engineering
Edited by KELLY T. HUGHES

VOLUME 422. Two-Component Signaling Systems, Part A
Edited by MELVIN I. SIMON, BRIAN R. CRANE, AND ALEXANDRINE CRANE

VOLUME 423. Two-Component Signaling Systems, Part B
Edited by MELVIN I. SIMON, BRIAN R. CRANE, AND ALEXANDRINE CRANE

VOLUME 424. RNA Editing
Edited by JONATHA M. GOTT

VOLUME 425. RNA Modification
Edited by JONATHA M. GOTT

VOLUME 426. Integrins
Edited by DAVID CHERESH

VOLUME 427. MicroRNA Methods
Edited by JOHN J. ROSSI

VOLUME 428. Osmosensing and Osmosignaling
Edited by HELMUT SIES AND DIETER HAUSSINGER

VOLUME 429. Translation Initiation: Extract Systems and Molecular Genetics
Edited by JON LORSCH

VOLUME 430. Translation Initiation: Reconstituted Systems and Biophysical Methods
Edited by JON LORSCH

VOLUME 431. Translation Initiation: Cell Biology, High-Throughput and Chemical-Based Approaches
Edited by JON LORSCH

VOLUME 432. Lipidomics and Bioactive Lipids: Mass-Spectrometry–Based Lipid Analysis
Edited by H. ALEX BROWN

VOLUME 433. Lipidomics and Bioactive Lipids: Specialized Analytical Methods and Lipids in Disease
Edited by H. ALEX BROWN

VOLUME 434. Lipidomics and Bioactive Lipids: Lipids and Cell Signaling
Edited by H. ALEX BROWN

VOLUME 435. Oxygen Biology and Hypoxia
Edited by HELMUT SIES AND BERNHARD BRÜNE

VOLUME 436. Globins and Other Nitric Oxide-Reactive Protiens (Part A)
Edited by ROBERT K. POOLE

VOLUME 437. Globins and Other Nitric Oxide-Reactive Protiens (Part B)
Edited by ROBERT K. POOLE

VOLUME 438. Small GTPases in Disease (Part A)
Edited by WILLIAM E. BALCH, CHANNING J. DER, AND ALAN HALL

VOLUME 439. Small GTPases in Disease (Part B)
Edited by WILLIAM E. BALCH, CHANNING J. DER, AND ALAN HALL

VOLUME 440. Nitric Oxide, Part F Oxidative and Nitrosative Stress in Redox Regulation of Cell Signaling
Edited by ENRIQUE CADENAS AND LESTER PACKER

VOLUME 441. Nitric Oxide, Part G Oxidative and Nitrosative Stress in Redox Regulation of Cell Signaling
Edited by ENRIQUE CADENAS AND LESTER PACKER

VOLUME 442. Programmed Cell Death, General Principles for Studying Cell Death (Part A)
Edited by ROYA KHOSRAVI-FAR, ZAHRA ZAKERI, RICHARD A. LOCKSHIN, AND MAURO PIACENTINI

VOLUME 443. Angiogenesis: *In Vitro* Systems
Edited by DAVID A. CHERESH

VOLUME 444. Angiogenesis: *In Vivo* Systems (Part A)
Edited by DAVID A. CHERESH

VOLUME 445. Angiogenesis: *In Vivo* Systems (Part B)
Edited by DAVID A. CHERESH

VOLUME 446. Programmed Cell Death, The Biology and Therapeutic Implications of Cell Death (Part B)
Edited by ROYA KHOSRAVI-FAR, ZAHRA ZAKERI, RICHARD A. LOCKSHIN, AND MAURO PIACENTINI

VOLUME 447. RNA Turnover in Bacteria, Archaea and Organelles
Edited by LYNNE E. MAQUAT AND CECILIA M. ARRAIANO

VOLUME 448. RNA Turnover in Eukaryotes: Nucleases, Pathways and Analysis of mRNA Decay
Edited by LYNNE E. MAQUAT AND MEGERDITCH KILEDJIAN

VOLUME 449. RNA Turnover in Eukaryotes: Analysis of Specialized and Quality Control RNA Decay Pathways
Edited by LYNNE E. MAQUAT AND MEGERDITCH KILEDJIAN

VOLUME 450. Fluorescence Spectroscopy
Edited by LUDWIG BRAND AND MICHAEL L. JOHNSON

VOLUME 451. Autophagy: Lower Eukaryotes and Non-Mammalian Systems (Part A)
Edited by DANIEL J. KLIONSKY

VOLUME 452. Autophagy in Mammalian Systems (Part B)
Edited by DANIEL J. KLIONSKY

VOLUME 453. Autophagy in Disease and Clinical Applications (Part C)
Edited by DANIEL J. KLIONSKY

VOLUME 454. Computer Methods (Part A)
Edited by MICHAEL L. JOHNSON AND LUDWIG BRAND

VOLUME 455. Biothermodynamics (Part A)
Edited by MICHAEL L. JOHNSON, JO M. HOLT, AND GARY K. ACKERS (RETIRED)

VOLUME 456. Mitochondrial Function, Part A: Mitochondrial Electron Transport Complexes and Reactive Oxygen Species
Edited by WILLIAM S. ALLISON AND IMMO E. SCHEFFLER

VOLUME 457. Mitochondrial Function, Part B: Mitochondrial Protein Kinases, Protein Phosphatases and Mitochondrial Diseases
Edited by WILLIAM S. ALLISON AND ANNE N. MURPHY

VOLUME 458. Complex Enzymes in Microbial Natural Product Biosynthesis, Part A: Overview Articles and Peptides
Edited by DAVID A. HOPWOOD

VOLUME 459. Complex Enzymes in Microbial Natural Product Biosynthesis, Part B: Polyketides, Aminocoumarins and Carbohydrates
Edited by DAVID A. HOPWOOD

VOLUME 460. Chemokines, Part A
Edited by TRACY M. HANDEL AND DAMON J. HAMEL

VOLUME 461. Chemokines, Part B
Edited by TRACY M. HANDEL AND DAMON J. HAMEL

VOLUME 462. Non-Natural Amino Acids
Edited by TOM W. MUIR AND JOHN N. ABELSON

VOLUME 463. Guide to Protein Purification, 2nd Edition
Edited by RICHARD R. BURGESS AND MURRAY P. DEUTSCHER

VOLUME 464. Liposomes, Part F
Edited by NEJAT DÜZGÜNEŞ

VOLUME 465. Liposomes, Part G
Edited by NEJAT DÜZGÜNEŞ

VOLUME 466. Biothermodynamics, Part B
Edited by MICHAEL L. JOHNSON, GARY K. ACKERS, AND JO M. HOLT

VOLUME 467. Computer Methods Part B
Edited by MICHAEL L. JOHNSON AND LUDWIG BRAND

VOLUME 468. Biophysical, Chemical, and Functional Probes of RNA Structure, Interactions and Folding: Part A
Edited by DANIEL HERSCHLAG

VOLUME 469. Biophysical, Chemical, and Functional Probes of RNA Structure, Interactions and Folding: Part B
Edited by DANIEL HERSCHLAG

VOLUME 470. Guide to Yeast Genetics: Functional Genomics, Proteomics, and Other Systems Analysis, 2nd Edition
Edited by GERALD FINK, JONATHAN WEISSMAN, AND CHRISTINE GUTHRIE

VOLUME 471. Two-Component Signaling Systems, Part C
Edited by MELVIN I. SIMON, BRIAN R. CRANE, AND ALEXANDRINE CRANE

VOLUME 472. Single Molecule Tools, Part A: Fluorescence Based Approaches
Edited by NILS G. WALTER

VOLUME 473. Thiol Redox Transitions in Cell Signaling, Part A Chemistry and Biochemistry of Low Molecular Weight and Protein Thiols
Edited by ENRIQUE CADENAS AND LESTER PACKER

VOLUME 474. Thiol Redox Transitions in Cell Signaling, Part B Cellular Localization and Signaling
Edited by ENRIQUE CADENAS AND LESTER PACKER

VOLUME 475. Single Molecule Tools, Part B: Super-Resolution, Particle Tracking, Multiparameter, and Force Based Methods
Edited by NILS G. WALTER

VOLUME 476. Guide to Techniques in Mouse Development, Part A Mice, Embryos, and Cells, 2nd Edition
Edited by PAUL M. WASSARMAN AND PHILIPPE M. SORIANO

VOLUME 477. Guide to Techniques in Mouse Development, Part B Mouse Molecular Genetics, 2nd Edition
Edited by PAUL M. WASSARMAN AND PHILIPPE M. SORIANO

VOLUME 478. Glycomics
Edited by MINORU FUKUDA

VOLUME 479. Functional Glycomics
Edited by MINORU FUKUDA

VOLUME 480. Glycobiology
Edited by MINORU FUKUDA

VOLUME 481. Cryo-EM, Part A: Sample Preparation and Data Collection
Edited by GRANT J. JENSEN

VOLUME 482. Cryo-EM, Part B: 3-D Reconstruction
Edited by GRANT J. JENSEN

VOLUME 483. Cryo-EM, Part C: Analyses, Interpretation, and Case Studies
Edited by GRANT J. JENSEN

VOLUME 484. Constitutive Activity in Receptors and Other Proteins, Part A
Edited by P. MICHAEL CONN

VOLUME 485. Constitutive Activity in Receptors and Other Proteins, Part B
Edited by P. MICHAEL CONN

VOLUME 486. Research on Nitrification and Related Processes, Part A
Edited by MARTIN G. KLOTZ

VOLUME 487. Computer Methods, Part C
Edited by MICHAEL L. JOHNSON AND LUDWIG BRAND

VOLUME 488. Biothermodynamics, Part C
Edited by MICHAEL L. JOHNSON, JO M. HOLT, AND GARY K. ACKERS

VOLUME 489. The Unfolded Protein Response and Cellular Stress, Part A
Edited by P. MICHAEL CONN

VOLUME 490. The Unfolded Protein Response and Cellular Stress, Part B
Edited by P. MICHAEL CONN

VOLUME 491. The Unfolded Protein Response and Cellular Stress, Part C
Edited by P. MICHAEL CONN

VOLUME 492. Biothermodynamics, Part D
Edited by MICHAEL L. JOHNSON, JO M. HOLT, AND GARY K. ACKERS

VOLUME 493. Fragment-Based Drug Design
Tools, Practical Approaches, and Examples
Edited by LAWRENCE C. KUO

VOLUME 494. Methods in Methane Metabolism, Part A
Methanogenesis
Edited by AMY C. ROSENZWEIG AND STEPHEN W. RAGSDALE

VOLUME 495. Methods in Methane Metabolism, Part B
Methanotrophy
Edited by AMY C. ROSENZWEIG AND STEPHEN W. RAGSDALE

VOLUME 496. Research on Nitrification and Related Processes, Part B
Edited by MARTIN G. KLOTZ AND LISA Y. STEIN

VOLUME 497. Synthetic Biology, Part A
Methods for Part/Device Characterization and Chassis Engineering
Edited by CHRISTOPHER VOIGT

VOLUME 498. Synthetic Biology, Part B
Computer Aided Design and DNA Assembly
Edited by CHRISTOPHER VOIGT

VOLUME 499. Biology of Serpins
Edited by JAMES C. WHISSTOCK AND PHILLIP I. BIRD

VOLUME 500. Methods in Systems Biology
Edited by DANIEL JAMESON, MALKHEY VERMA, AND HANS V. WESTERHOFF

VOLUME 501. Serpin Structure and Evolution
Edited by JAMES C. WHISSTOCK AND PHILLIP I. BIRD

VOLUME 502. Protein Engineering for Therapeutics, Volume A
Edited by K. DANE WITTRUP AND GREGORY L. VERDINE

VOLUME 503. Protein Engineering for Therapeutics, Volume B
Edited by K. DANE WITTRUP AND GREGORY L. VERDINE

VOLUME 504. Imaging and Spectroscopic Analysis of Living Cells
Optical and Spectroscopic Techniques
Edited by P. MICHAEL CONN

VOLUME 505. Imaging and Spectroscopic Analysis of Living Cells
Live Cell Imaging of Cellular Elements and Functions
Edited by P. MICHAEL CONN

VOLUME 506. Imaging and Spectroscopic Analysis of Living Cells
Imaging Live Cells in Health and Disease
Edited by P. MICHAEL CONN

VOLUME 507. Gene Transfer Vectors for Clinical Application
Edited by THEODORE FRIEDMANN

CHAPTER ONE

General Principles of Retrovirus Vector Design

Tammy Chang *and* Jiing-Kuan Yee

Contents

1. Introduction	1
2. Construction of Retroviral Vectors Carrying the Gene of Interest	4
3. Transient Production of Retroviral Vectors	6
4. Titering of Retroviral Vectors	9
5. RCR Detection	11
Acknowledgments	12
References	12

Abstract

An understanding in the life cycle of γ-retroviruses has led to significant progress in the development of murine leukemia virus (MLV)-based vectors for gene delivery and human gene therapy. An MLV-based vector consists of the *cis*-acting sequences important for viral replication and gene expression. The sequence that encodes viral proteins is replaced with the gene of interest. To generate infectious retroviral vectors, viral-encoded proteins are supplied in *trans* for virion assembly. Here, we describe a method to rapidly generate MLV vectors from transiently transfected human 293T cells. The strategies to purify and titer the vector and to detect the presence of replication competent retrovirus (RCR) in the vector harvest are also described.

1. Introduction

Extensive knowledge gained from studying the life cycle of retroviruses such as murine leukemia virus (MLV) has led to significant progress in the development of MLV-based vectors for gene delivery and human gene therapy. Retroviruses are RNA viruses containing a nucleocapsid structure in an envelope derived from host plasma membrane during virus

Department of Virology, Beckman Research Institute, City of Hope National Medical Center, Duarte, California, USA budding. Besides viral encoded nucleocapsid (Gag) and envelope (Env) proteins, each virion contains two identical single-stranded RNAs and enzymes including reverse transcriptase and integrase essential for virus replication. Viral cell entry is mediated through the interaction between viral Env protein and its cell surface receptor (Albritton et al., 1989; Kavanaugh et al., 1994; Kim et al., 1991). In the cytoplasm of the infected cell, the viral genomic RNAs serve as the template in reverse transcription to generate a copy of the linear double-stranded DNA. This reaction is catalyzed by virion-associated reverse transcriptase. The viral DNA migrates into the nucleus and integrates into the host genome. Integration is catalyzed by virion-associated integrase which recognizes the ends of the linear viral DNA, resulting in an integrated provirus with a genetic structure always collinear with its unintegrated form. Gene delivery mediated by retroviral vector therefore avoids the problem of DNA rearrangement frequently associated with other DNA transfer technologies. Once integrated, the provirus becomes an integral part of the host genome and replicates together with the host genomic DNA. The host transcription machinery recognizes the *cis*-acting elements in the viral long terminal repeat (LTR) and mediates expression of the viral genes. In the case of γ-retroviruses such as MLV, two transcripts, one unspliced and the other spliced, are generated from the $5'$ LTR promoter (Fig. 1.1). The unspliced transcript encodes the Gag and Pol proteins, whereas the spliced transcript encodes the Env protein. Immature virions are assembled in the cytoplasm by the Gag–Pol polyprotein, and the Env protein is picked up from the plasma membrane during the virion budding process. Only the unspliced transcript is encapsidated into virions as the viral RNA packaging signal is removed from the spliced transcript during RNA splicing (Fig. 1.1). Viral maturation takes place during or after virion budding when the Gag–Pol polyprotein is processed into mature capsid and nucleocapsid proteins. To generate an MLV vector, those viral sequences important for virus life cycle are retained in the construct (Miller, 1992). They include both the $5'$ and $3'$ LTR, the initiation signals for plus- and minus-strand DNA synthesis and the RNA packaging signal (Fig. 1.1). These sequences, by coincidence, are localized on both ends of the viral genome. Other viral sequences, mainly the coding region for the three structure proteins, are removed to avoid the generation of replication competent retrovirus (RCR) and replaced with the gene of interest (GOI). To generate infectious vectors, the viral-encoded proteins will have to be supplied in *trans*. To facilitate vector production, many so-called packaging cell lines stably expressing MLV Gag, Pol, and Env proteins have been established (Miller, 1990). Packaging cell lines generally are derived from well established and transformed murine, canine, or human cell lines. MLV is originally isolated from mice, and most murine cell lines contain latent endogenous retroviruses which share sequence homology with MLV. Due to this reason, there is an increased

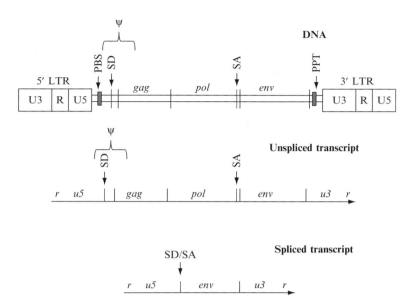

Figure 1.1 Structures of the MLV genome and its transcripts. The DNA genome with the *gag*, *pol*, and *env* genes is shown on top. Primer binding site (PBS) is the initiation site for minus-strand DNA synthesis. Polypurine tract (PPT) is the initiation site for plus-strand DNA synthesis. LTR, long terminal repeat; SD, splice donor site; SA, splice acceptor site; ψ, packaging signal. (For color version of this figure, the reader is referred to the Web version of this chapter.)

risk to generate RCR via homologous recombination when murine-based packaging cell lines are used for MLV vector production (Bodine *et al.*, 1990; Muenchau *et al.*, 1990; Otto *et al.*, 1994; Scarpa *et al.*, 1991). The risk of RCR contamination can be minimized by using nonmurine-based packaging cell lines for vector production. Introduction of the vector construct into a packaging cell line via DNA transfection or viral transduction allows the establishment of stable producer cell lines for vector production. Since MLV infection causes no obvious pathologic effect on cells, these producer cell lines continue to proliferate and secrete infectious virions into the culture media. However, the process of establishing stable producer cell lines suitable for clinical application is labor intensive and time consuming (Sheridan *et al.*, 2000). The vector titer generated from different packaging cell lines varies significantly, depending on the expression level of the retrovirus-encoded proteins in each packaging cell line, the number of the introduced vector copy, the promoter used for the expression of the vector genomic RNA, and the chromosomal location harboring the integrated vector construct. As it is difficult to optimize all these factors simultaneously, it frequently requires the screening of hundreds of individually picked clones before a stable high-titer producer line suitable for scale-up mass vector production can be identified. This laborious procedure is therefore not suitable for routine production of low quantities of retroviral vectors for laboratory research. To rapidly produce retroviral vectors, a transient transfection method is recommended (Naviaux et al., 1996). It generally takes 3–4 days for vector production, and depending on the assay used, an additional 3–4 days to up to 2 weeks for titer determination. We describe a procedure here based on a transient transfection approach to generate retroviral vectors.

2. Construction of Retroviral Vectors Carrying the Gene of Interest

Many basic retroviral vector constructs are currently available either from different laboratories or from commercial sources. Many of these constructs share similarity in their basic structure. For stable expression, the GOI usually is inserted downstream from the 5′ LTR and transcription initiates from the start of the R region in the 5′ LTR (Fig. 1.1). There are two types of retroviral vectors, N2-based and MFG-based, with a slight difference in the 5′ untranslated region (Armentano et al., 1987; Eglitis et al., 1985; Keller et al., 1985; Riviere et al., 1995). Both vectors contain the MLV 5′ splice donor site and the entire packaging signal for viral genomic RNA encapsidation. In addition, the MFG vector also contains the MLV splice acceptor site for the *env* gene (Fig. 1.2). The N2 vector lacks this splice acceptor site. A transgene inserted downstream from the splice acceptor site in the MFG vector is therefore expressed from a spliced transcript. While both vector constructs are capable of generating infectious vector at high titers, transgene expression from the MFG vector is more efficient: the absence of the complex secondary structure of the packaging signal in the spliced transcript may account for more efficient translation of the transgene in cells transduced with the MFG vector (Krall et al., 1996). To assay for the vector titer, a selectable marker such as the gene encoding neomycin phosphotransferase (neo^R) or hygromycin B phosphotransferase (hyg^R) under the control of a relatively weak promoter such as the promoter for the phosphoglycerate kinase (pgk) gene is placed downstream from the GOI (Fig. 1.2). The gene encoding green fluorescence protein (GFP) is also used frequently for titer determination. Since two promoters, the LTR promoter and the internal promoter, are present in the vector construct within a relatively short distance, they may interfere with each other (transcription interference) and reduce the expression of the GOI (Proudfoot, 1986). Using a weak house-keeping gene promoter such as the pgk promoter may reduce the effect of transcription interference and facilitate more

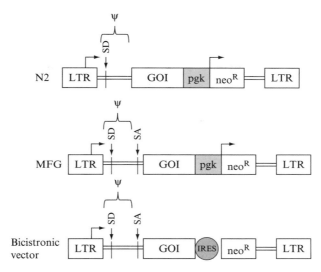

Figure 1.2 Structure of the retroviral vector. Three types of retroviral vectors are shown. The N2 vector contains the splice donor site and the full-length packaging signal. The MFG and the bicistronic vectors contain, in addition to the splice donor site and the packaging signal, the SA site and the sequence near the translation start site of the *env* gene. GOI, gene of interest; pgk, pgk promoter; neo^R, neomycin-resistance gene; IRES, internal ribosome entry site. Arrows indicate the transcription initiation sites and the direction of transcription. (For color version of this figure, the reader is referred to the Web version of this chapter.)

efficient expression of the GOI from the 5′ LTR. As drug selection only requires the antibiotics resistance gene expressed at low levels, utilization of a weak promoter to express the drug-resistance gene is usually sufficient for selection. Placing the GOI under the control of an internal promoter is not recommended as such an arrangement frequently destabilizes the expression of the GOI even under constant antibiotics selection (Xu *et al.*, 1989). To avoid the use of two promoters in a single vector, a commonly used strategy is to construct bicistronic vectors with an internal ribosome entry site (IRES) inserted between the GOI and the selectable marker (Fig. 1.2; Adam *et al.*, 1991). In this case, both genes are transcribed from the 5′ LTR, and the IRES sequence facilitates the translation of the downstream selectable marker from the transcript. Since the gene inserted immediately downstream from the IRES sequence is expressed at lower levels than that inserted downstream from the 5′ LTR (Mizuguchi *et al.*, 2000), the selectable marker is usually placed downstream from the IRES sequence in a bicistronic vector. Transient vector production generally is carried out in human kidney 293T cells. This cell line not only has extremely high-transfection efficiency but also expresses high level of adenovirus E1A

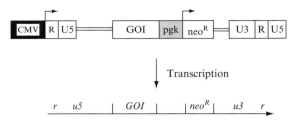

Figure 1.3 Structure of the retroviral vector containing the CMV–LTR fusion. The MLV enhancer in the 5′ LTR is replaced with the enhancer of the CMV immediate early gene. The transcript generated from such a construct is identical to that generated from a retroviral construct with the wild-type 5′ LTR.

proteins. As E1A suppresses the transactivation function of the MLV enhancer in the LTR, the intact 5′ LTR in the retroviral construct will not be able to express efficiently the vector genomic RNA in 293T cells (Naviaux et al., 1996). To address this problem, the MLV enhancer in the 5′ LTR is replaced with the enhancer in the immediate early gene of cytomegalovirus (CMV; Fig. 1.3). In contrast to the MLV enhancer, the CMV enhancer functions robustly in 293T cells. Like the wild-type LTR, transcription from the CMV–MLV hybrid LTR is initiated from the first nucleotide of the R region and thus generates exactly the same transcript as a construct containing the wild-type 5′ LTR (Fig. 1.3). This modification significantly increases the production of infectious retroviral vectors from transiently transfected 293T cells (Naviaux et al., 1996).

3. Transient Production of Retroviral Vectors

Production of retroviral vector is carried out by transient transfection into 293T cells of the required plasmids. Infectious vectors are harvested within 48–72h after transfection followed by partial purification, titer determination, and RCR detection. Partial purification takes less than 24h to complete. Depending on the assay used, the duration for titer determination can vary from 3–4 days to 2 weeks. The duration for RCR detection is longer, usually takes up to 4–5 weeks to complete the assay.

1. For vector production, culture 293T cells in Dulbecco's modified Eagle's medium (DMEM) with high glucose (4500 mg/l), supplemented with 10% fetal bovine serum (FBS), 100 units/ml penicillin, and 100 μg/ml streptomycin in a 37 °C incubator with 10% CO_2. The same culture medium is used throughout DNA transfection and vector harvest.

2. Plate 293T cells to 40% confluence in 100-mm tissue culture dishes 24h before transfection. On the day of transfection, cells should reach 80–90% confluence. Up to five plates are usually prepared for transfection to harvest sufficient vectors for the purification step.
3. On the day of transfection, carefully replace half (5 ml) of the culture medium with 5 ml of fresh medium 2 h before DNA transfection. Since 293T cells only loosely adhere to the culture dish, complete replacement of the culture medium frequently disturbs the cell monolayer, leading to cell detachment. Attempts to restore cell monolayer not only cause delay for DNA transfection but also reduce the transfection efficiency as reattached cells usually fails to form a homogenous monolayer. Replacing only half of the culture medium does not seem to cause cell detachment or induce any major change in cell morphology and does not affect the yield of vector production.
4. Prepare the transfection solution as follows:
   a. Mix 12 μg of the vector construct, 12 μg of pCMV-GP (the packaging plasmid), and 4 μg of pCMV-G (the expression plasmid for VSV-G) in a 5-ml polypropylene round-bottom tube (Peng et al., 2001). Adjust the total volume to 437 μl with TE79/10 (1 mM Tris–HCl, 0.1 mM EDTA, pH 7.9).
   b. Add 63 μl of 2 M CaCl$_2$ solution.
   c. Add 500 μl of 2× HBS solution (0.05 M Hepes, 0.28 M NaCl, 1.5 mM Na$_2$HPO$_4$, pH 7.12) drop by drop.
   d. Mix the transfection solution with pipetting.
   e. Leave the tube in a laminar flow hood for 30 min to allow the formation of DNA–calcium phosphate precipitates.
5. Mix the precipitates well with pipetting and transfer them dropwise into the culture dish while gently swirling the medium to mix thoroughly the precipitates with the culture medium.
6. Incubate the dish at 37 °C in 10% CO$_2$ for 6 h. Overnight incubation works in most cases but is not recommended as the stress 293T cells endure in the transfection solution may result in mass cell detachment. However, with extra care in manipulation, overnight transfection can lead to the production of high-titer retroviral vectors.
7. Replace the transfection solution with 6 ml fresh culture medium. Cell morphology should exhibit dramatic changes between 20 and 40 h after transfection. These changes are caused by VSV-G overexpression which induces syncytia formation and cell toxicity (Yee et al., 1994). Handle cells with extreme care from this point to avoid cell detachment from the dish. We use 6 ml instead of 10 ml of fresh medium to replace the transfection solution to reduce the collected volume for further vector processing.
8. Harvest the infectious vector by carefully collecting all the culture medium at 40 h after transfection. Add an additional 6 ml of fresh

medium into the culture dish and continue to incubate for an additional 20h for vector collection. Great care should be taken during medium change by gently adding medium from the side of the dish to avoid disturbance to the adhered cells.

9. Remove cell debris from the collected medium by spinning at $1500 \times g$ for 10 min at 4°C. Filter the supernatant through a 0.45 μm filter. Store the cleared viral supernatant at −80°C.
10. Thaw and pool the collected viral supernatant together for further purification.
11. Add 40% PEG stock solution (PEG8000 dissolved in PBS followed by autoclave) into the viral supernatant and adjust the final PEG concentration to 10%. Mix well and store the solution at 4°C for a minimum of 4h. The vector remains stable in the PEG solution at 4°C for up to a week.
12. Spin down the vector precipitate at $1500 \times g$ for 30 min at room temperature. Redissolve the vector in DMEM (without FBS) by vortex in 1/20 to 1/30 of the original vector harvest volume.
13. Once the pellet is completely dissolved, add FBS to a final concentration of 10% to stabilize the vector. The PEG treatment step not only increases the vector stability during long-term storage but also enables the vector preparation to sustain repeated freeze–thawing cycles without significant loss of its titer. The vector yield from the PEG step is high in general and is strongly recommended for any vector preparation.
14. For some studies such as animal experiments or transduction of hematopoietic progenitor cells, vector titers in the range of 10^9 infectious units/ml for efficient gene delivery are desired. Since VSV-G-pseudotyped retroviral vectors can sustain the ultracentrifugation force (Burns et al., 1993), the PEG-purified vector can be further concentrated.
15. To prepare vectors with a titer in the range between 5×10^8 and 10^9 infectious units/ml, spin down the PEG-purified vector in an SW-28 swing bucket rotor at $50,000 \times g$ for 90 min at 4°C. Presterilize the centrifuge tube overnight with UV in a laminar flow hood.
16. Decant the supernatant and place the centrifuge tube at an upside down position in the hood for a few minutes to remove the last few drops of the medium. Add DMEM at a desired volume and resuspend the vector pellet by continuing vortex at 4°C for 2h.
17. Aliquot the resuspended high-titer vector preparation and store the aliquot at −80°C.

- Several 293T-based packaging cell lines (the Phoenix system) are available from G. Nolan's laboratory (Stanford, CA; Kinsella and Nolan, 1996). When Phoenix-Ampho or Phoenix-Eco cells are used for vector production, only the vector construct needs to be transfected as these cell lines constitutively express the three retroviral encoded proteins. In the case of Phoenix-gp cells, both the vector construct and the expression plasmid for the Env protein need to be transfected since only Gag and Pol are constitutively expressed in this cell line. However, the Phoenix-gp line has the advantage that vectors with different envelope proteins can be generated, depending on the *env* gene in the expression plasmid used for vector production.

- Using 293T cells for vector production is convenient as many laboratories have already acquired this cell line. However, using the Phoenix-based cell lines for vector production has the advantage that it is not necessary to cotransfect the packaging plasmid, thereby reducing the risk of generating RCR through homologous recombination among the cotransfected plasmids.

4. Titering of Retroviral Vectors

The assay for vector titer depends on the marker in the vector. The most popular markers used include the gene encoding fluorescence proteins like GFP and antibiotics selectable markers like neo^R gene. GFP has the advantage that the titer can be determined by fluorescence-activated cell sorting (FACS) analysis within 3–4 days after transduction. Antibiotics selection usually takes 2 weeks with repeated medium changes. However, the antibiotics selection is more sensitive than FACS as even a few transduced cells in a population of 10^6 untransduced cells can be detected by this assay. Depending on the goal of a study, the two markers can be used alternately. We routinely use a human fibrosarcoma line, HT1080, for titer determination. This line adheres well to the culture dish and proliferates rapidly in culture. The cell line can be transduced by a VSV-G-pseudotyped vector at high efficiencies and is available from American Type Culture Collection (ATCC). Other fast proliferating cell lines such as 293T cells can also be used for titer determination. The disadvantage of 293T cells is its relatively poor adherence to the culture dish.

1. To determine the vector titer by GFP expression, plate out the HT1080 cells in a 6-well culture dish at a density of 2×10^5 cells/well the day before transduction. The same culture medium for 293T cells is used for HT1080 cells.
2. On the day of transduction, trypsinize the cells in a single well and determine the cell number by hemocytometer. The cell number will be used to calculate the vector titer.
3. Replace the culture medium with fresh medium supplemented with 8 µg/ml polybrene.

4. Make serial dilutions of the vector preparation to ensure accurate titer determination.
5. Add the diluted vector to HT1080 cells and incubate the transduced culture at 37 °C in 10% CO_2 for 6 h. Overnight incubation also works well with HT1080 cells. However, some cell lines, notably primary cells, are more sensitive to the toxicity of polybrene and thus the duration for transduction should be kept at a minimum. Six-hour transduction works well with most of the cell lines.
6. Replace the medium with fresh medium minus the polybrene after 6 h. Continue to culture the cells for an additional 40 h.
7. To assay for the GFP titer, harvest the transduced cells and fix them with freshly prepared 3.7% formaldehyde (in PBS) at room temperature for 20 min. Wash the cells twice with PBS and subject them to FACS analysis.
8. Calculate the vector titer as follows:

    (% of GFP^+ cells × cell count)/vector stock used for transduction (in µl) = vector titer (infectious units/µl)

    To calculate the vector titer, any transduced cell population with the percentage of GFP^+ cells exceeding 30% will be discarded for consideration as it falls outside the linear range of cell transduction.
9. To determine the vector titer by G418 selection, plate out the HT1080 cells in a 100-mm culture dish at a density of 4×10^5 cells/plate. Avoid a confluent or a near-confluent culture in the beginning of G418 selection since it takes several days for the antibiotics to completely inhibit cell proliferation and the rate of cell killing by G418 is significantly reduced in a confluent culture. Optimal cell density on the day of initial G418 administration should be adjusted to a confluence of approximately 30–40%. Other antibiotic drugs such as puromycin can kill untransduced cells at a much faster pace. Higher cell density on the day of initial drug administration can therefore be used in such cases. Higher cell density usually allows more efficient transduction, resulting in higher vector titers.
10. On the day of transduction, replace the culture medium with fresh medium supplemented with 8 µg/ml polybrene.
11. Make serial dilutions of the vector preparation and transduce HT1080 cell at 37 °C in 10% CO_2 for 6 h.
12. Replace medium with fresh medium minus polybrene and continue the incubation for an additional 40 h.
13. Replace the medium with fresh medium containing 600 µg/ml G418.
14. Change medium containing fresh G418 every 3 days. G418-resistant colonies should emerge within 2 weeks.
15. To determine the titer, wash the culture dish once with 10 ml of PBS and fix the colonies in 4 ml of methanol for 20 min at room temperature.

16. Remove methanol and stain the colonies with 5% Giemsa solution for 20 min at room temperature. Remove Giemsa stain, wash once with PBS and air dry. Count the colony and calculate the vector titer as follows:
    Colony number/vector stock used for transduction (in μl) = vector titer (colony forming units/μl)

5. RCR Detection

RCR arises from recombination between the vector construct and the packaging plasmid containing the MLV *gag*, *pol*, and *env* genes. Since the *env* gene is removed from the packaging construct and replaced with the VSV-G gene in a separate plasmid, the risk of generating RCR via homologous recombination among the three plasmids is minimized although not completely eliminated. RCR detection is a two-step process, including in the first step to amplify the small amount of RCR present in the vector preparation for 3 weeks, followed by the detection of amplified RCR in indicator cell lines (Cornetta *et al.*, 1993; Forestell *et al.*, 1996; Miller *et al.*, 1996; Printz *et al.*, 1995). Depending on the assay of detection, the whole process takes between 4 and 5 weeks to complete. Here we describe a marker rescue assay for RCR detection. It involves the use of a HT1080-based cell line, HT/hyg, with an integrated retroviral vector carrying the hyg^R gene. The vector to be tested should not contain the same selectable marker as the endogenous vector in the indicator cell line. Potential RCR in the harvested vector preparation is first amplified in HT/hyg cells for a minimum of 3 weeks. During this time, the amplified RCR can rescue the hyg^R vector and confer hygromycin resistance to naïve cells upon transduction. The entire process, including RCR amplification and marker rescue followed by the detection of the rescued hyg^R vector, takes 5 weeks to complete.

1. HT/hyg cells are cultured in the same medium as HT1080 cells. Plate HT/hyg cells in 100-mm tissue culture dishes at 20% confluence 24 h before transduction.
2. On the day of transduction, add 2% of the total vector harvest to HT/hyg cells in the presence of 8 μg/ml polybrene. Carry out the transduction at 37 °C in 10% CO_2. For positive control, the 4070A amphotropic virus is purchased from ATCC. Transduce HT/hyg cells with serially diluted 4070A virus in a similar fashion.
3. Replace the transduction medium with fresh medium minus polybrene 6 h later.
4. Passage the cells continuously for a period of 3 weeks for RCR amplification and marker rescue.
5. Harvest the rescued hyg^R vector, aliquot, and store at −80 °C.

6. To detect the rescued hyg^R vector, prepare HT1080 cells in 100-mm dishes at a density of 4×10^5 cells/dish 24 h before transduction.
7. On the day of transduction, replace the medium with fresh medium plus 8 µg/ml polybrene. Carry out the transduction with the amplified vector stock or the 4070 amphotropic virus at 37°C in 10% CO_2.
8. After 6-h transduction, replace the transduction medium with fresh medium minus polybrene. Continue to incubate the culture for an additional 40 h.
9. Replace the medium with fresh medium containing 600 µg/ml hygromycin. Continue to culture the cells with change of fresh medium containing 600 mg/ml hygromycin every 3 days.
10. After 2 weeks, fix and stain the hygromycin-resistant colonies as described above.

- The 4070A virus can be grown up to high titers in 293T or HT1080 cells. To obtain an accurate titer of the virus, the S+/L− assay is used (Bassin et al., 1982; Peebles, 1975). PG-4 cells (CRL-2032, ATCC) are infected with the serially diluted 4070A virus and plagues are counted 4–6 days later. This cell line has an endogenous murine sarcoma virus (S+) but lacks the murine leukemia virus (L−). Infection by the 4070A virus activates the endogenous murine sarcoma virus and induces cell transformation revealed by the emergence of plagues.
- To determine the sensitivity of the marker rescue assay, 4070A viruses with known titers can be spiked into the to-be-tested vector preparation followed by the amplification and marker rescue steps. This process evaluates whether the presence of the predominant recombinant retroviral vector would interfere with the detection of low-level RCR contamination in a vector preparation due to receptor competition (Printz et al., 1995).

ACKNOWLEDGMENTS

We thank Priscilla Yam for establishing the HT/hyg cell line for the marker rescue assay and Akihiro Iida for making the MFG vector construct.

REFERENCES

Adam, M. A., Ramesh, N., Miller, A. D., and Osborne, W. R. (1991). Internal initiation of translation in retroviral vectors carrying picornavirus 5′ nontranslated regions. *J. Virol.* **65,** 4985–4990.
Albritton, L. M., Tseng, L., Scadden, D., and Cunningham, J. M. (1989). A putative murine ecotropic retrovirus receptor gene encodes a multiple membrane-spanning protein and confers susceptibility to virus infection. *Cell* **57,** 659–666.

Armentano, D., Yu, S. F., Kantoff, P. W., von Ruden, T., Anderson, W. F., and Gilboa, E. (1987). Effect of internal viral sequences on the utility of retroviral vectors. *J. Virol.* **61,** 1647–1650.

Bassin, R. H., Ruscetti, S., Ali, I., Haapala, D. K., and Rein, A. (1982). Normal DBA/2 mouse cells synthesize a glycoprotein which interferes with MCF virus infection. *Virology* **123,** 139–151.

Bodine, D. M., McDonagh, K. T., Brandt, S. J., Ney, P. A., Agricola, B., Byrne, E., and Nienhuis, A. W. (1990). Development of a high-titer retrovirus producer cell line capable of gene transfer into rhesus monkey hematopoietic stem cells. *Proc. Natl. Acad. Sci. USA* **87,** 3738–3742.

Burns, J. C., Friedmann, T., Driever, W., Burrascano, M., and Yee, J. K. (1993). Vesicular stomatitis virus G glycoprotein pseudotyped retroviral vectors: Concentration to very high titer and efficient gene transfer into mammalian and nonmammalian cells. *Proc. Natl. Acad. Sci. USA* **90,** 8033–8037.

Cornetta, K., Nguyen, N., Morgan, R. A., Muenchau, D. D., Hartley, J. W., Blaese, R. M., and Anderson, W. F. (1993). Infection of human cells with murine amphotropic replication-competent retroviruses. *Hum. Gene Ther.* **4,** 579–588.

Eglitis, M. A., Kantoff, P., Gilboa, E., and Anderson, W. F. (1985). Gene expression in mice after high efficiency retroviral-mediated gene transfer. *Science (New York, NY)* **230,** 1395–1398.

Forestell, S. P., Dando, J. S., Bohnlein, E., and Rigg, R. J. (1996). Improved detection of replication-competent retrovirus. *J. Virol. Methods* **60,** 171–178.

Kavanaugh, M. P., Miller, D. G., Zhang, W., Law, W., Kozak, S. L., Kabat, D., and Miller, A. D. (1994). Cell-surface receptors for gibbon ape leukemia virus and amphotropic murine retrovirus are inducible sodium-dependent phosphate symporters. *Proc. Natl. Acad. Sci. USA* **91,** 7071–7075.

Keller, G., Paige, C., Gilboa, E., and Wagner, E. F. (1985). Expression of a foreign gene in myeloid and lymphoid cells derived from multipotent haematopoietic precursors. *Nature* **318,** 149–154.

Kim, J. W., Closs, E. I., Albritton, L. M., and Cunningham, J. M. (1991). Transport of cationic amino acids by the mouse ecotropic retrovirus receptor. *Nature* **352,** 725–728.

Kinsella, T. M., and Nolan, G. P. (1996). Episomal vectors rapidly and stably produce high-titer recombinant retrovirus. *Hum. Gene Ther.* **7,** 1405–1413.

Krall, W. J., Skelton, D. C., Yu, X. J., Riviere, I., Lehn, P., Mulligan, R. C., and Kohn, D. B. (1996). Increased levels of spliced RNA account for augmented expression from the MFG retroviral vector in hematopoietic cells. *Gene Ther.* **3,** 37–48.

Miller, A. D. (1990). Retrovirus packaging cells. *Hum. Gene Ther.* **1,** 5–14.

Miller, A. D. (1992). Retroviral vectors. *Curr. Top. Microbiol. Immunol.* **158,** 1–24.

Miller, A. D., Bonham, L., Alfano, J., Kiem, H. P., Reynolds, T., and Wolgamot, G. (1996). A novel murine retrovirus identified during testing for helper virus in human gene transfer trials. *J. Virol.* **70,** 1804–1809.

Mizuguchi, H., Xu, Z., Ishii-Watabe, A., Uchida, E., and Hayakawa, T. (2000). IRES-dependent second gene expression is significantly lower than cap-dependent first gene expression in a bicistronic vector. *Mol. Ther.* **1,** 376–382.

Muenchau, D. D., Freeman, S. M., Cornetta, K., Zwiebel, J. A., and Anderson, W. F. (1990). Analysis of retroviral packaging lines for generation of replication-competent virus. *Virology* **176,** 262–265.

Naviaux, R. K., Costanzi, E., Haas, M., and Verma, I. M. (1996). The pCL vector system: Rapid production of helper-free, high-titer, recombinant retroviruses. *J. Virol.* **70,** 5701–5705.

Otto, E., Jones-Trower, A., Vanin, E. F., Stambaugh, K., Mueller, S. N., Anderson, W. F., and McGarrity, G. J. (1994). Characterization of a replication-competent retrovirus resulting from recombination of packaging and vector sequences. *Hum. Gene Ther.* **5,** 567–575.

Peebles, P. T. (1975). An *in vitro* focus-induction assay for xenotropic murine leukemia virus, feline leukemia virus C, and the feline–primate viruses RD-114/CCC/M-7. *Virology* **67**, 288–291.

Peng, H., Chen, S. T., Wergedal, J. E., Polo, J. M., Yee, J. K., Lau, K. H., and Baylink, D. J. (2001). Development of an MFG-based retroviral vector system for secretion of high levels of functionally active human BMP4. *Mol. Ther.* **4**, 95–104.

Printz, M., Reynolds, J., Mento, S. J., Jolly, D., Kowal, K., and Sajjadi, N. (1995). Recombinant retroviral vector interferes with the detection of amphotropic replication competent retrovirus in standard culture assays. *Gene Ther.* **2**, 143–150.

Proudfoot, N. J. (1986). Transcriptional interference and termination between duplicated alpha-globin gene constructs suggests a novel mechanism for gene regulation. *Nature* **322**, 562–565.

Riviere, I., Brose, K., and Mulligan, R. C. (1995). Effects of retroviral vector design on expression of human adenosine deaminase in murine bone marrow transplant recipients engrafted with genetically modified cells. *Proc. Natl. Acad. Sci. USA* **92**, 6733–6737.

Scarpa, M., Cournoyer, D., Muzny, D. M., Moore, K. A., Belmont, J. W., and Caskey, C. T. (1991). Characterization of recombinant helper retroviruses from Moloney-based vectors in ecotropic and amphotropic packaging cell lines. *Virology* **180**, 849–852.

Sheridan, P. L., Bodner, M., Lynn, A., Phuong, T. K., Depolo, N. J., Delavega, D. J., Jr., O'Dea, J., Nguyen, K., McCormack, J. E., Driver, D. A., *et al.* (2000). Generation of retroviral packaging and producer cell lines for large-scale vector production and clinical application: Improved safety and high titer. *Mol. Ther.* **2**, 262–275.

Xu, L., Yee, J. K., Wolff, J. A., and Friedmann, T. (1989). Factors affecting long-term stability of Moloney murine leukemia virus-based vectors. *Virology* **171**, 331–341.

Yee, J. K., Miyanohara, A., LaPorte, P., Bouic, K., Burns, J. C., and Friedmann, T. (1994). A general method for the generation of high-titer, pantropic retroviral vectors: Highly efficient infection of primary hepatocytes. *Proc. Natl. Acad. Sci. USA* **91**, 9564–9568.

CHAPTER TWO

STRATEGIES FOR RETROVIRUS-BASED CORRECTION OF SEVERE, COMBINED IMMUNODEFICIENCY (SCID)

Alain Fischer,[*,†,‡] Salima Hacein-Bey-Abina,[*,†,§] *and* Marina Cavazzana-Calvo[*,†,§]

Contents

1. Early Attempts	17
2. Technological Progress	18
3. Gene Therapy of SCID-X1	18
4. Gene Therapy for ADA Deficiency	22
5. Conclusions	23
References	24

Abstract

Severe combined immunodeficiencies (SCIDs) appear as optimal disease targets to challenge potential efficacy of gene therapy. *Ex vivo*, retrovirally mediated gene transfer into hematopoietic progenitor cells has been shown to provide sustained correction of two forms of SCID, that is, SCID-X1 and adenosine deaminase deficiencies. In the former case, however, genotoxicity was observed in a minority of patients as a consequence of retroviral integration into proto-oncogenes loci and transactivation. Design of vectors in which the enhancer element of retroviral LTR has been deleted and an internal promoter added (self-inactivated vectors) could provide both safe and efficient gene transfer as being presently tested.

Proof of concept of the efficacy of gene therapy has been provided over the past 10 years by monitoring a total of 51 reported patients with one of two forms of severe, combined immunodeficiency (SCID): X-linked severe combined immunodeficiency (SCID-X1) and adenosine deaminase (ADA) deficiency (Aiuti *et al.*, 2002, 2009; Cavazzana-Calvo *et al.*, 2000; Gaspar

[*] Descartes University of Paris, Paris, France
[†] INSERM Unit 768, Department of Immunology, Necker Children's Hospital, Paris, France
[‡] Pediatric Hematology Unit, Necker Children's Hospital, Paris, France
[§] Department of Biotherapy, Necker Children's Hospital, Paris, France

Table 2.1 SCID diseases and gene therapy

Mechanisms and diseases	Affected lineages	Experimental gene therapy	Clinical trial Efficacy	Safety
Defective survival				
AK2 deficiency	T, NK, myeloid	−	−	−
Abnormal purine metabolism				
ADA deficiency	T, B, NK	+	+	+
PNP deficiency	T	+	−	−
Impairment of γc-dependent signal				
SCID-X1 (γc deficiency)	T, NK	+	+	−[a]
JAK3 deficiency	T, NK	+	−	−
IL7Rα deficiency	T	−	−	−
Defective V(D)J recombination of TCR and BCR				
RAG-1 deficiency	T, B	+[b]	−	−
RAG-2 deficiency	T, B	+	−	−
DNAPKCs deficiency	T, B	−	−	−
Artemis deficiency	T, B	+	[c]	−
DNA ligase 4 deficiency	T, B	−	−	−
Cernunnos deficiency	T, B	−	−	−
Defective (pre) TCR signaling				
CD45 deficiency	T	−	−	−
CD3, δ, ε, ζ deficiency	T	Some	−	−
Defective thymic egress				
Coronin IA deficiency	T	−	−	−

AK2, adenylate kinase deficiency; PNP, purine nucleoside phosphorylase; JAK3, Janus-associated kinase 3; IL7Ra, IL7 receptor α chain; RAG1/2, recombination activated genes 1/2; DNAPKc, catalytic subunit of DNA-dependent protein kinase; SAE, serious adverse events.
[a] Serious adverse events in 5 out of 20 patients, new vectors generated—see text.
[b] Efficacy at the expense of oncogenicity.
[c] Clinical trial in preparation.

et al., 2011a,b; Hacein-Bey-Abina et al., 2010). All but two patients are alive, whereas 39 have benefited from significant correction of their immunodeficiency by gene therapy (despite the occurrence of a severe adverse event (SAE) in five patients). This chapter will discuss why and how these results were achieved and where they are driving us for the future.

SCIDs constitute a group of rare, inherited, Mendelian disorders characterized by defective T cell development; hence adaptive immunity is abrogated in affected patients (Notarangelo et al., 2009). Seventeen distinct forms of SCID have now been identified. The estimated overall incidence of SCID is approximately 1 in 50,000 live births. The underlying genetic defects and molecular mechanisms have now been well characterized for most of SCIDs (Notarangelo et al., 2009; Table 2.1). Another aspect that is important for the perspective of gene therapy is the fact that a few T cell progenitors have a tremendous ability to proliferate during the early steps of intrathymic differentiation (thus generating a high number of T cells). Further, mature T lymphocytes are (like stem cells) essentially self-renewing cells—they can persist throughout life. This knowledge was generated (at least in part) by studying the partial correction of SCID phenotypes in patients bearing somatic, reverse mutations that either led to recovery of a wild-type sequence in the corresponding SCID-affected gene or to mutations with attenuated consequences (Bousso et al., 2000; Hirschhorn et al., 1996; Speckmann et al., 2008). These events are not rare and indicated that somatic events provided T cell precursors with a huge advantage—an effect not observed in other cell types (Davis and Candotti, 2010).

The natural outcome of an SCID is early death caused by recurrent, opportunistic infections. From the late 1960s onward, it was shown that hematopoietic stem cell transplantation (HSCT) (and, as a carryover, the transfusion of mature T cells) led to sustained correction of SCIDs (Antoine et al., 2003; Buckley et al., 1999; Gatti et al., 1968; Gennery et al., 2010). Nevertheless, when SCID patients receive a transplant from a non-HLA-identical donor, survival is hampered by the consequences of delayed immune reconstitution and the possible advent of graft versus host disease. Indeed, in a recent European survey, the 5-year survival rate after HSCT was found to be 72% (Gennery et al., 2010). Further, some of the longest surviving HSCT-treated patients appear to have limited T cell immunity and develop clinical complications (Neven et al., 2009).

In fact, ADA deficiency has also been treated since the early 1980s with enzyme replacement therapy (ERT). Although this treatment has been shown to be life-saving for most (but not all) patients, long-term immunity may not be satisfactory.

It was in this context that gene therapy was considered for SCID, on the basis of (i) a fairly good understanding of the disease mechanism and (ii) the less than fully satisfactory results achieved by conventional treatments for this lethal condition.

1. Early Attempts

The development of retroviral vectors in the 1980s prompted clinical researchers to consider treating ADA deficiency patients (initially on ERT) by targeting circulating T cells (Gaspar et al., 2009). These retroviral vectors were engineered from murine oncoretroviruses (γ retroviruses) that are able to integrate into the genome of dividing cells. Packaging cell lines were derived and appropriate envelopes were used to target human cells (Verma and Weitzman, 2005). In these vectors, the therapeutic gene (cDNA) was placed under the transcriptional control of the viral enhancer/promoter element contained in the long terminal repeats (LTRs).

The initial results demonstrated the procedure's feasibility and the persistence of transduced cells over several years *in vivo*. However, this was not sufficient to correct the immunodeficiency (Blaese *et al.*, 1995). Neither did the attempted transduction of hematopoietic progenitor cells succeed (Bordignon *et al.*, 1995; Hoogerbrugge *et al.*, 1996; Kohn *et al.*, 1998).

With the benefit of hindsight, these failures can be attributed to a combination of several factors: low retroviral titers, suboptimal conditions for *ex vivo* cell infection, and ERT's probable relief of the potential selective advantage conferred on transduced cells. The fact that myeloablation was not performed was probably also a contributory factor. This resulted in insufficient numbers of transduced progenitors *in vivo* for correcting the T cell immunodeficiency. It is noteworthy that no adverse events were observed in these patients—even in those who received hundreds of millions of transduced T cells (Blaese *et al.*, 1995).

2. Technological Progress

In the meantime, significant progress in gene transfer technology had been made. This included higher titers in virus production, the finding that adhesion of cells to a fibronectin fragment increased the retroviral infection rate and the use of a cytokine "cocktail" (including stem cell growth factor, thrombopoietin, and interleukin-3) to induce *ex vivo* cell division and increase cell fitness (Shaw and Kohn, 2011). *Ex vivo* gene transfer protocols thus consisted in (i) selection of $CD34^+$ bone marrow progenitor cells, (ii) the culture of these cells in plastic bags in the presence of cytokines, and (iii) cycles of infection with retroviral supernatants in the presence of a fibronectin fragment for 3 days. These protocols increased CD34 transduction to values of 30–40%, with a mean integration vector copy number of one.

3. Gene Therapy of SCID-X1

The first SCID disease to be treated with gene therapy was SCID-X1 (soon to be followed by ADA deficiency). The choice of SCID-X1 was based on two observations: (i) it is the most frequent SCID (accounting for ~40% of cases) and (ii) its disease mechanism and observations made in revertants (Bousso et al., 2000; Speckmann et al., 2008) made it a good model for testing the efficacy of gene therapy. SCID-X1 is caused by deficiency in the γc chain shared by six cytokine receptors (including the interleukin-7 receptor; Leonard, 1996). The latter is absolutely required for survival and expansion of T cell precursors prior to the TCRβ expression stage.

Between 1999 and 2006, two trials (performed in Paris and London, respectively) enrolled 20 patients with typical SCID-X1 but no HLA-genoidentical donor. A low number of transduced CD34 cells led to inefficient T cell development in one patient (Ginn et al., 2005), whereas the presence of splenomegaly in another led to treatment failure (because of cell trapping in the spleen; Hacein-Bey-Abina et al., 2010). In the other 18 patients, the procedure led to a burst of T cells and then sustained T cell reconstitution, after a follow-up period ranging from 4.5 to 12.5 years (median: 9.2 years), all the patients are alive and well and show the clear-cut benefits of gene therapy. Their quality of life is excellent, half of them are off other forms of therapy and the others require immunoglobulin replacement only (Gaspar et al., 2011a; Hacein-Bey-Abina et al., 2010). Even though each trial used a slightly different vector (with either an amphotropic envelope (Hacein-Bey-Abina et al., 2010) or the gibbon ape leukemia virus envelope (Gaspar et al., 2011a)), this parameter did not seem to make a difference. The presence of a cell dose threshold was clearly demonstrated, since all patients who received more than 3×10^6 transduced γc+ CD34(+) cells/kg of body weight had optimal T cell reconstitution. Over the years, it was shown that the patients' T cell populations were both polyclonal and functional. The persistence of naïve, T cell receptor excision circles–positive (TREC+) T cells over the years strongly suggests that there is persistent production of T cells from a pool of transduced progenitors. To date (12.5 years on), the peripheral T cell counts have been remarkably stable. A quantitative analysis of vector integration sites (based on ligated, mediated PCR technology and, in recent years, high-throughput DNA sequencing) has enabled researchers to track the clonal signature of the T cell precursors that effectively gave rise to T cells (Bushman, 2007; Deichmann et al., 2007; Hacein-Bey-Abina et al., 2010). Although the use of blood specimens and the presence of a sensitivity threshold can limit the scope of analysis, one remarkable result is that there are about 1000 clones on average—indicating that only a limited number of progenitor cells generated a diversified repertoire and a normal-sized blood T cell pool. This result is important because it demonstrates that the concept on which gene therapy was based, that is, transduced progenitors does indeed appear to have a selective advantage (Fischer et al., 2010). In contrast to T cell status, the lack of natural killer (NK) cells (another characteristic of SCID-X1) is not well corrected. Following an early boost in the NK count, most patients have only few NK cells in their blood (Cavazzana-Calvo et al., 2000;

Gaspar et al., 2011b; Hacein-Bey-Abina et al., 2010). Although this situation does appear to be harmful, it suggests that the NK cell population does not have such a great selective advantage—perhaps because larger number of NK cells are produced and/or the latter have a shorter life span. When other hematopoietic cell lineages were analyzed, a few transduced B and myeloid cells were detected soon after the gene therapy but then disappeared (Bushman, 2007; Deichmann et al., 2007; Gaspar et al., 2011b; Hacein-Bey-Abina et al., 2010).

These observations show that in the absence of myeloablation, transduced hematopoietic stem cells (HSCs) have limited persistence. In this setting and as above mentioned, the persistence of naïve T cells may suggest that a committed T cell precursor may persist. It may thus be argued that some degree of myeloablation (as performed for ADA deficiency; see below) might improve correction of the immune deficiency. This possibility must be weighed against the associated risk in infected patients.

Based on the efficacy of gene therapy in patients with a typical SCID (SCID-X1), a similar approach was attempted in older patients who either had a hypomorphic γc mutation or had partially failed HSCT. The results were rather disappointing, with treatment failure or a minimal improvement in the T cell immune deficiency—despite a satisfactory bone marrow cell transduction rate (Puck et al., 1990; Thrasher et al., 2005). These disappointing results probably highlight the fact that the patients no longer had a functional thymus to sustain thymopoiesis.

Five of the 20 SCID-X1 patients developed an SAE (T cell leukemia) 2–5 years after following gene therapy (Hacein-Bey-Abina et al., 2003; Howe et al., 2008). All five patients had functional T cell immunity. Chemotherapy successfully eliminated the abnormal cells and cured the disease in four patients, whereas the fifth patient died. Strikingly, polyclonal, nonmalignant, transduced (γc+) T cells reappeared in the four patients now in long-term remission from leukemia. In particular, TREC(+) naïve T cells again became detectable, reinforcing the above-mentioned hypothesis of persistent, transduced T cell precursors in these patients.

The occurrence of these SAEs was obviously a considerable setback in the context of this otherwise successful gene therapy. It naturally led to a halt in trials while major efforts were made to understand the underlying mechanism. All five cases featured clonal expansion of T cells with either an immature or mature phenotype. The key finding was the identification of a retroviral integration site (or two, in one case) located within a protooncogene: *LMO-2* in 4 cases and *CCND2* and *BMI-1* in the other. In all instances, the integration sites were in the first or second intron or in the promoter region. More importantly, active transcription of these oncogenes was detected (Hacein-Bey-Abina et al., 2003; Howe et al., 2008). It was thus clearly determined that oncogenesis was initiated by the vector-induced transactivation of oncogenes. In addition, secondary events were detected in these clonal cell populations, including Notch-activating mutations, Sil–Tal fusion, and loss of heterozygosity in key genomic regions.

These findings were explained by the fact that retroviruses do integrate within active gene loci (in the promoter region or coding regions). Recently, epigenetic signatures associated with retrovirus integration have been identified (Biasco et al., 2011; Dave et al., 2009; Santoni et al., 2010). Competent LTRs were part of these vectors. Thus, transactivation of these genes was mediated by the potent viral enhancer contained within the LTR. This, however, was not the whole story. Indeed, use of the same vector constructs to transduce mature T cells never led to leukemia—indicating a cell-type sensitivity that was probably due to the selective accessibility of some proto-oncogenes in progenitor cells. Also, this type of SAE has not occurred in any of the 20 patients successfully treated with gene therapy for SCID–ADA (see below)—despite the fact that similar vector backbones were used and a similar pattern of retroviral integration was detected (including sites inside the *LMO-2* locus) (Aiuti et al., 2007). Thus, one or more disease-related factors must be involved. Should SCID-X1 be viewed as an "at risk" disease or is ADA deficiency a "no/low risk" disease? One argument in favor of the first hypothesis is that $\gamma c(-)$ mice are more prone to develop leukemia in a gene transfer setting (Scobie et al., 2009; Shou et al., 2006). Quantitative and qualitative characteristics of differentiation-blocked $\gamma c(-)$ lymphoid progenitors may account for this sensitivity, although no clear-cut mechanism has yet been provided. It has been suggested that overexpression of γc (a proliferative receptor) by transduced cells could play a role but this was not observed in patient cells (including the absence of γc downstream signaling pathway (JAK3–STAT5) activation in leukemic cells). Conversely, the ADA–SCID might be considered to be "protective" on the following grounds. The ADA-deficient environment is toxic for cells and might partially limit the ability of transduced progenitors to develop and thus reduce the risk of leukemogenesis. It is noteworthy that T cell reconstitution in treated ADA–SCID patients after gene therapy is slower than in SCID-X1 patients (Aiuti et al., 2002, 2009; Cavazzana-Calvo et al., 2000; Gaspar et al., 2011a,b; Hacein-Bey-Abina et al., 2010). The fact that a similar leukemic event was recently observed in a patient with Wiskott–Aldrich syndrome (also based on *LMO-2* transactivation by an LTR-competent retroviral vector; Boztug et al., 2010) also suggests that there is rather a "specific, protective effect" of the ADA-deficient environment.

In any case, the obvious consequence of these findings was to modify the vectors. The LTR enhancer was deleted from the so-called "self-inactivating" (SIN) vectors. Instead, an internal promoter is used to trigger transgene activation. Several of these constructs have now been designed to induce γc expression as either retroviral (Thornhill et al., 2008) or lentiviral vectors (Huston et al., 2011; Zhou et al., 2010). The former type has entered the clinic in a multicenter trial. However, given the latency (4–5 years) of the advent of leukemia in the five SCID-X1 patients, more time will be needed before researchers can fully gauge the vector's *in vivo* safety and confirm the *in vitro* data. Other options could rely on the so-called "safe harbor" integration of vectors. This strategy is based on the using of endonucleases to target integration at sites known to be safe (Papapetrou *et al.*, 2011). Direct gene repair can be achieved by combining the same methodology with a homologous template for gene mutation replacement (Lombardo *et al.*, 2007). However, these technologies have not yet proven to be effective in stem cells.

4. Gene Therapy for ADA Deficiency

ADA deficiency is an autosomal recessive SCID with an estimated frequency of 1 in 250,000 live births (Gaspar *et al.*, 2009). It leads to an accumulation of adenosine and, more importantly, deoxyadenosine. The latter is transformed inside cells into deoxy-ADP and deoxy-ATP. Excess intracellular deoxy-ATP induces the premature death of lymphocyte precursors and result in virtual alymphocytosis when null mutations of the ADA gene occur. ADA deficiency has been treated with much the same methodology used in SCID-X1. Three trials have been performed to date (in Italy, the United Kingdom, and the United States). A significant difference is the addition of a mild, myeloablative chemotherapy regimen consisting of 4 mg/kg busulfan (in most patients) or melphalan (140 mg/m^2), in order to achieve a higher rate of engraftment by transduced stem cells (Aiuti *et al.*, 2002; Gaspar *et al.*, 2011b). The rationale was to increase the overall number of ADA-expressing cells and thus reduce the accumulation of deoxyadenosine metabolites. A major difference between these three trials and the previous ADA–SCID trials consisted in the exclusion of ERT, in order to leverage the growth advantage of transduced cells and maximize the chances of success.

Reports have been published on the treatment of 30 patients over the past 10.5 years. In 20 cases, correction of the immunodeficiency was good enough to allow withdrawal of ERT. The ERT was (re)initiated in the other patients. All 20 patients remain free of infections. Sustained populations of T cells have been observed (over 500/m^2) (time range: 2–10.5 years; median: 4.5 years) (Aiuti *et al.*, 2009; Ferrua *et al.*, 2010; Gaspar *et al.*, 2011b) and include naïve T cells. The T cells have a diverse repertoire and can be activated to proliferate and exert various functions upon antigen stimulation. It is noteworthy that the time course of T cell reconstitution was significantly slower than in SCID-X1 patients and, overall, T cell blood counts are lower. This is probably due to persistent ADA deficiency in nonhematopoietic cell lineages (including thymic epithelial cells, in particular). The fact that significant proportions of NK and B lymphocytes and myeloid cells were found to express ADA provides evidence of the persistence of transduced HSCs. This was also demonstrated by the detection of a common integration site signature for the different lymphoid and myeloid cell lineages and the bone marrow CD34 cells. As mentioned above, no adverse events have occurred in these 20 patients—despite the detection of RV integration sites located within proto-oncogenes like *LMO-2* (Aiuti *et al.*, 2009; Gaspar *et al.*, 2011b).

Nevertheless, most future trials are likely to be based on SIN vectors. The use of lentiviral vectors appears to be logical, since they enable the transduction of a higher number of HSCs. In turn, this may further boost T cell reconstitution.

5. Conclusions

Over the past 12 years, the five clinical trials of retrovirus-mediated gene transfer into hematopoietic progenitor cells for the treatment of SCID-X1 and ADA deficiency have undoubtedly provided proof of concept for the efficacy of gene therapy. The results have validated the hypothesis whereby the transduced cells' selective advantage enables sustained T cell immunity in treated patients. The advent of oncogenicity in the SCID-X1 trial was not anticipated. Further developments in vector design (as discussed above) are hopefully one way of overcoming this risk.

In this context, one can expect to see more extensive use of gene therapy to treat inherited disorders of hematopoiesis. Other SCID conditions are on the front line but gene therapy will not turn out to be equally feasible for all diseases. Gene therapy in RAG-1 deficiency in a murine model has provided relatively disappointing results, for reasons that are not yet very clear (Lagresle-Peyrou *et al.*, 2006).

Nevertheless, a clinical trial of the SIN-lentivector-based treatment of Artemis deficiency (see Table 2.1) is under preparation. Treatment of other primary immunodeficiencies has already started, with relatively promising results for Wiskott–Aldrich syndrome (Boztug *et al.*, 2010) but not chronic granulomatous disease (Grez *et al.*, 2011) (because transduced neutrophils were only observed concomitantly with the toxic transactivation of an oncogene). However, LTR-competent retroviruses were used in this setting. There is thus hope that the combination of SIN vector use and myeloablation could be safer and produce efficient transduction of a few percent of neutrophils, which would be of clinical benefit. This is not an unrealistic assumption, in view of the results reported for this type of protocol in the adrenoleukodystrophy trial (Cartier *et al.*, 2009).

Further technological advances in gene targeting (by using engineered endonucleases—see above—and/or stem cell manipulation) are likely to significantly transform this field in the years to come. In particular, the identification of human HSCs enables the latter to be directly manipulated (Notta *et al.*, 2011). Their potential for expansion (Boitano *et al.*, 2010) or their production from other cell types (Szabo *et al.*, 2010) should create exciting new avenues for therapy.

REFERENCES

Aiuti, A., Slavin, S., Aker, M., Ficara, F., Deola, S., Mortellaro, A., Morecki, S., Andolfi, G., Tabucchi, A., Carlucci, F., *et al.* (2002). Correction of ADA-SCID by stem cell gene therapy combined with nonmyeloablative conditioning. *Science* **296**(5577), 2410–2413.

Aiuti, A., Cassani, B., Andolfi, G., Mirolo, M., Biasco, L., Recchia, A., Urbinati, F., Valacca, C., Scaramuzza, S., Aker, M., *et al.* (2007). Multilineage hematopoietic reconstitution without clonal selection in ADA-SCID patients treated with stem cell gene therapy. *J. Clin. Invest.* **117**(8), 2233–2240.

Aiuti, A., Cattaneo, F., Galimberti, S., Benninghoff, U., Cassani, B., Callegaro, L., Scaramuzza, S., Andolfi, G., Mirolo, M., Brigida, I., *et al.* (2009). Gene therapy for immunodeficiency due to adenosine deaminase deficiency. *N. Engl. J. Med.* **360**(5), 447–458.

Antoine, C., Muller, S., Cant, A., Cavazzana-Calvo, M., Veys, P., Vossen, J., Fasth, A., Heilmann, C., Wulffraat, N., Seger, R., *et al.* (2003). Long-term survival and transplantation of haemopoietic stem cells for immunodeficiencies: Report of the European experience 1968–99. *Lancet* **361**(9357), 553–560.

Biasco, L., Ambrosi, A., Pellin, D., Bartholomae, C., Brigida, I., Roncarolo, M. G., Di Serio, C., von Kalle, C., Schmidt, M., and Aiuti, A. (2011). Integration profile of retroviral vector in gene therapy treated patients is cell-specific according to gene expression and chromatin conformation of target cell. *EMBO Mol. Med.* **3**(2), 89–101.

Blaese, R. M., Culver, K. W., Miller, A. D., Carter, C. S., Fleisher, T., Clerici, M., Shearer, G., Chang, L., Chiang, Y., Tolstoshev, P., *et al.* (1995). T lymphocyte-directed gene therapy for ADA-SCID: Initial trial results after 4 years. *Science* **270**(5235), 475–480.

Boitano, A. E., Wang, J., Romeo, R., Bouchez, L. C., Parker, A. E., Sutton, S. E., Walker, J. R., Flaveny, C. A., Perdew, G. H., Denison, M. S., *et al.* (2010). Aryl hydrocarbon receptor antagonists promote the expansion of human hematopoietic stem cells. *Science* **329**(5997), 1345–1348.

Bordignon, C., Notarangelo, L. D., Nobili, N., Ferrari, G., Casorati, G., Panina, P., Mazzolari, E., Maggioni, D., Rossi, C., Servida, P., *et al.* (1995). Gene therapy in peripheral blood lymphocytes and bone marrow for ADA-immunodeficient patients. *Science* **270**(5235), 470–475.

Bousso, P., Wahn, V., Douagi, I., Horneff, G., Pannetier, C., Le Deist, F., Zepp, F., Niehues, T., Kourilsky, P., Fischer, A., *et al.* (2000). Diversity, functionality, and stability of the T cell repertoire derived in vivo from a single human T cell precursor (In process citation). *Proc. Natl. Acad. Sci. USA* **97**(1), 274–278.

Boztug, K., Schmidt, M., Schwarzer, A., Banerjee, P. P., Diez, I. A., Dewey, R. A., Bohm, M., Nowrouzi, A., Ball, C. R., Glimm, H., *et al.* (2010). Stem-cell gene therapy for the Wiskott–Aldrich syndrome. *N. Engl. J. Med.* **363**(20), 1918–1927.

Buckley, R. H., Schiff, S. E., Schiff, R. I., Markert, L., Williams, L. W., Roberts, J. L., Myers, L. A., and Ward, F. E. (1999). Hematopoietic stem-cell transplantation for the treatment of severe combined immunodeficiency. *N. Engl. J. Med.* **340**(7), 508–516.

Bushman, F. D. (2007). Retroviral integration and human gene therapy. *J. Clin. Invest.* **117**(8), 2083–2086.

Cartier, N., Hacein-Bey-Abina, S., Bartholomae, C. C., Veres, G., Schmidt, M., Kutschera, I., Vidaud, M., Abel, U., Dalcortivo, L., Caccavelli, L., et al. (2009). Hematopoietic stem cell gene therapy with a lentiviral vector in X-linked adrenoleukodystrophy. *Science* **326**(5954), 818–823.

Cavazzana-Calvo, M., Hacein-Bey, S., De Saint Basile, G., Gross, F., Yvon, E., Nusbaum, P., Selz, F., Hue, C., Certain, S., Casanova, J. L., et al. (2000). Gene therapy of human severe combined immunodeficiency (SCID)-X1 disease. * Equal contribution. *Science* **288**, 669–672.

Dave, U. P., Akagi, K., Tripathi, R., Cleveland, S. M., Thompson, M. A., Yi, M., Stephens, R., Downing, J. R., Jenkins, N. A., and Copeland, N. G. (2009). Murine leukemias with retroviral insertions at Lmo2 are predictive of the leukemias induced in SCID-X1 patients following retroviral gene therapy. *PLoS Genet.* **5**(5), e1000491.

Davis, B. R., and Candotti, F. (2010). Genetics. Mosaicism—Switch or spectrum? *Science* **330**(6000), 46–47.

Deichmann, A., Hacein-Bey-Abina, S., Schmidt, M., Garrigue, A., Brugman, M. H., Hu, J., Glimm, H., Gyapay, G., Prum, B., Fraser, C. C., et al. (2007). Vector integration is nonrandom and clustered and influences the fate of lymphopoiesis in SCID-X1 gene therapy. *J. Clin. Invest.* **117**(8), 2225–2232.

Ferrua, F., Brigida, I., and Aiuti, A. (2010). Update on gene therapy for adenosine deaminase-deficient severe combined immunodeficiency. *Curr. Opin. Allergy Clin. Immunol.* **10**(6), 551–556.

Fischer, A., Hacein-Bey-Abina, S., and Cavazzana-Calvo, M. (2010). 20 years of gene therapy for SCID. *Nat. Immunol.* **11**(6), 457–460.

Gaspar, H. B., Aiuti, A., Porta, F., Candotti, F., Hershfield, M. S., and Notarangelo, L. D. (2009). How I treat ADA deficiency. *Blood* **114**(17), 3524–3532.

Gaspar, H. B., Cooray, S., Gilmour, K. C., Parsley, K. L., Adams, S., Howe, S. J., Al Ghonaium, A., Bayford, J., Brown, L., Davies, E. G., et al. (2011a). Long-term persistence of a polyclonal T cell repertoire after gene therapy for x-linked severe combined immunodeficiency. *Sci. Transl. Med.* **3**(97), 97ra79.

Gaspar, H. B., Cooray, S., Gilmour, K. C., Parsley, K. L., Zhang, F., Adams, S., Bjorkegren, E., Bayford, J., Brown, L., Davies, E. G., et al. (2011b). Hematopoietic stem cell gene therapy for adenosine deaminase-deficient severe combined immunodeficiency leads to long-term immunological recovery and metabolic correction. *Sci. Transl. Med.* **3**(97), 97ra80.

Gatti, R. A., Meuwissen, H. J., Allen, H. D., Hong, R., and Good, R. A. (1968). Immunological reconstitution of sex-linked lymphopenic immunological deficiency. *Lancet* **2**(7583), 1366–1369.

Gennery, A. R., Slatter, M. A., Grandin, L., Taupin, P., Cant, A. J., Veys, P., Amrolia, P. J., Gaspar, H. B., Davies, E. G., Friedrich, W., et al. (2010). Transplantation of hematopoietic stem cells and long-term survival for primary immunodeficiencies in Europe: Entering a new century, do we do better? *J. Allergy Clin. Immunol.* **126**(3), 602–610 e1–11.

Ginn, S. L., Curtin, J. A., Kramer, B., Smyth, C. M., Wong, M., Kakakios, A., McCowage, G. B., Watson, D., Alexander, S. I., Latham, M., et al. (2005). Treatment of an infant with X-linked severe combined immunodeficiency (SCID-X1) by gene therapy in Australia. *Med. J. Aust.* **182**(9), 458–463.

Grez, M., Reichenbach, J., Schwable, J., Seger, R., Dinauer, M. C., and Thrasher, A. J. (2011). Gene therapy of chronic granulomatous disease: The engraftment dilemma. *Mol. Ther.* **19**(1), 28–35.

Hacein-Bey-Abina, S., Hauer, J., Lim, A., Picard, C., Wang, G. P., Berry, C. C., Martinache, C., Rieux-Laucat, F., Latour, S., Belohradsky, B. H., et al. (2010). Efficacy of gene therapy for X-linked severe combined immunodeficiency. *N. Engl. J. Med.* **363**(4), 355–364.

Hacein-Bey-Abina, S.*, Von Kalle, C.*, Schmidt, M.*, McCormack, M. P., Wulffraat, N., Leboulch, P., Lim, A., Osborne, C. S., Pawliuk, R., Morillon, E., et al. (2003). LMO2-

associated clonal T cell proliferation in two patients after gene therapy for SCID-X1. *Science* **302**(5644), 415–419 (* Equal contribution).

Hirschhorn, R., Yang, D. R., Puck, J. M., Huie, M. L., Jiang, C. K., and Kurlandsky, L. E. (1996). Spontaneous in vivo reversion to normal of an inherited mutation in a patient with adenosine deaminase deficiency (see comments). *Nat. Genet.* **13**(3), 290–295.

Hoogerbrugge, P. M., van Beusechem, V. W., Fischer, A., Debree, M., le Deist, F., Perignon, J. L., Morgan, G., Gaspar, B., Fairbanks, L. D., Skeoch, C. H., et al. (1996). Bone marrow gene transfer in three patients with adenosine deaminase deficiency. *Gene Ther.* **3**(2), 179–183.

Howe, S. J., Mansour, M. R., Schwarzwaelder, K., Bartholomae, C., Hubank, M., Kempski, H., Brugman, M. H., Pike-Overzet, K., Chatters, S. J., de Ridder, D., et al. (2008). Insertional mutagenesis combined with acquired somatic mutations causes leukemogenesis following gene therapy of SCID-X1 patients. *J. Clin. Invest.* **118**(9), 3143–3150.

Huston, M. W., van Til, N. P., Visser, T. P., Arshad, S., Brugman, M. H., Cattoglio, C., Nowrouzi, A., Li, Y., Schambach, A., Schmidt, M., et al. (2011). Correction of murine SCID-X1 by lentiviral gene therapy using a codon-optimized IL2RG gene and minimal pretransplant conditioning. *Mol. Ther.* **19**(10), 1867–1877.

Kohn, D. B., Hershfield, M. S., Carbonaro, D., Shigeoka, A., Brooks, J., Smogorzewska, E. M., Barsky, L. W., Chan, R., Burotto, F., Annett, G., et al. (1998). T lymphocytes with a normal ADA gene accumulate after transplantation of transduced autologous umbilical cord blood $CD34^+$ cells in ADA-deficient SCID neonates. *Nat. Med.* **4**(7), 775–780.

Lagresle-Peyrou, C., Yates, F., Malassis-Seris, M., Hue, C., Morillon, E., Garrigue, A., Liu, A., Hajdari, P., Stockholm, D., Danos, O., et al. (2006). Long-term immune reconstitution in RAG-1-deficient mice treated by retroviral gene therapy: A balance between efficiency and toxicity. *Blood* **107**(1), 63–72.

Leonard, W. J. (1996). The molecular basis of X-linked severe combined immunodeficiency: Defective cytokine receptor signaling. *Annu. Rev. Med.* **47**, 229–239.

Lombardo, A., Genovese, P., Beausejour, C. M., Colleoni, S., Lee, Y. L., Kim, K. A., Ando, D., Urnov, F. D., Galli, C., Gregory, P. D., et al. (2007). Gene editing in human stem cells using zinc finger nucleases and integrase-defective lentiviral vector delivery. *Nat. Biotechnol.* **25**(11), 1298–1306.

Neven, B., Leroy, S., Decaluwe, H., Le Deist, F., Picard, C., Moshous, D., Mahlaoui, N., Debre, M., Casanova, J. L., Dal Cortivo, L., et al. (2009). Long-term outcome after hematopoietic stem cell transplantation of a single-center cohort of 90 patients with severe combined immunodeficiency. *Blood* **113**(17), 4114–4124.

Notarangelo, L. D., Fischer, A., Geha, R. S., Casanova, J. L., Chapel, H., Conley, M. E., Cunningham-Rundles, C., Etzioni, A., Hammartrom, L., Nonoyama, S., et al. (2009). Primary immunodeficiencies: 2009 update. *J. Allergy Clin. Immunol.* **124**(6), 1161–1178.

Notta, F., Doulatov, S., Laurenti, E., Poeppl, A., Jurisica, I., and Dick, J. E. (2011). Isolation of single human hematopoietic stem cells capable of long-term multilineage engraftment. *Science* **333**(6039), 218–221.

Papapetrou, E. P., Lee, G., Malani, N., Setty, M., Riviere, I., Tirunagari, L. M., Kadota, K., Roth, S. L., Giardina, P., Viale, A., et al. (2011). Genomic safe harbors permit high beta-globin transgene expression in thalassemia induced pluripotent stem cells. *Nat. Biotechnol.* **29**(1), 73–78.

Puck, J. M., Siminovitch, K. A., Poncz, M., Greenberg, C. R., Rottem, M., and Conley, M. E. (1990). Atypical presentation of Wiskott–Aldrich syndrome: Diagnosis in two unrelated males based on studies of maternal T cell X chromosome inactivation. *Blood* **75**(12), 2369–2374.

Santoni, F. A., Hartley, O., and Luban, J. (2010). Deciphering the code for retroviral integration target site selection. *PLoS Comput. Biol.* **6**(11), e1001008.

Scobie, L., Hector, R. D., Grant, L., Bell, M., Nielsen, A. A., Meikle, S., Philbey, A., Thrasher, A. J., Cameron, E. R., Blyth, K., *et al.* (2009). A novel model of SCID-X1 reconstitution reveals predisposition to retrovirus-induced lymphoma but no evidence of gammaC gene oncogenicity. *Mol. Ther.* **17**(6), 1031–1038.

Shaw, K. L., and Kohn, D. B. (2011). A tale of two SCIDs. *Sci. Transl. Med.* **3**(97), 97ps36.

Shou, Y., Ma, Z., Lu, T., and Sorrentino, B. P. (2006). Unique risk factors for insertional mutagenesis in a mouse model of XSCID gene therapy. *Proc. Natl. Acad. Sci. USA* **103**(31), 11730–11735.

Speckmann, C., Pannicke, U., Wiech, E., Schwarz, K., Fisch, P., Friedrich, W., Niehues, T., Gilmour, K., Buiting, K., Schlesier, M., *et al.* (2008). Clinical and immunologic consequences of a somatic reversion in a patient with X-linked severe combined immunodeficiency. *Blood* **112**(10), 4090–4097.

Szabo, E., Rampalli, S., Risueno, R. M., Schnerch, A., Mitchell, R., Fiebig-Comyn, A., Levadoux-Martin, M., and Bhatia, M. (2010). Direct conversion of human fibroblasts to multilineage blood progenitors. *Nature* **468**(7323), 521–526.

Thornhill, S. I., Schambach, A., Howe, S. J., Ulaganathan, M., Grassman, E., Williams, D., Schiedlmeier, B., Sebire, N. J., Gaspar, H. B., Kinnon, C., *et al.* (2008). Self-inactivating gammaretroviral vectors for gene therapy of X-linked severe combined immunodeficiency. *Mol. Ther.* **16**(3), 590–598.

Thrasher, A. J., Hacein-Bey-Abina, S., Gaspar, H. B., Blanche, S., Davies, E. G., Parsley, K., Gilmour, K., King, D., Howe, S., Sinclair, J., *et al.* (2005). Failure of SCID-X1 gene therapy in older patients. *Blood* **105**(11), 4255–4257.

Verma, I. M., and Weitzman, M. D. (2005). Gene therapy: Twenty-first century medicine. *Annu. Rev. Biochem.* **74,** 711–738.

Zhou, S., Mody, D., DeRavin, S. S., Hauer, J., Lu, T., Ma, Z., Hacein-Bey Abina, S., Gray, J. T., Greene, M. R., Cavazzana-Calvo, M., *et al.* (2010). A self-inactivating lentiviral vector for SCID-X1 gene therapy that does not activate LMO2 expression in human T cells. *Blood* **116**(6), 900–908.

CHAPTER THREE

Retrovirus and Lentivirus Vector Design and Methods of Cell Conditioning

Samantha Cooray,*,† Steven J. Howe,* *and* Adrian J. Thrasher*

Contents

1. Introduction	30
2. Retrovirus and Lentivirus Vector Design	34
2.1. The promoter	36
2.2. Variations in vector design	36
2.3. Selectable markers	37
2.4. Marker genes	37
2.5. The virus vector envelope protein	38
2.6. Additional regulatory sequences	38
2.7. Safety considerations	39
2.8. Gene silencing	39
3. Retroviral and Lentiviral Vector Production	40
3.1. Production of retrovirus vectors using Phoenix producer cell lines	41
3.2. Production of VSV-G pseudotyped lentiviral vectors by transient transfection	42
3.3. Harvesting virus vector	43
3.4. Concentration of virus vector particles by ultracentrifugation	43
3.5. Determination of infectious vector titer	44
4. Transduction Procedures	45
4.1. Transduction of human HSCs	46
4.2. Determination of transgene copy number by real-time qPCR analysis	48
4.3. Determination of protein expression	51
References	53

* Centre for Immunodeficiency, Molecular Immunology Unit, Institute of Child Health, University College London, London, United Kingdom
† Department of Clinical Immunology, Great Ormond Street Hospital NHS Trust, London, United Kingdom

Abstract

Retroviruses are useful tools for the efficient delivery of genes to mammalian cells, owing to their ability to stably integrate into the host cell genome. Over the past few decades, retroviral vectors have been used in gene therapy clinical trials for the treatment of a number of inherited diseases and cancers. The earliest retrovirus vectors were based on simple oncogenic gammaretroviruses such as Moloney murine leukemia virus (MMLV) which, when pseudotyped with envelope proteins from other viruses such as the gibbon ape leukemia virus envelope protein (GALV) or vesicular stomatitis virus G protein (VSV-G), can efficiently introduce genes to a wide range of host cells. However, gammaretroviral vectors have the disadvantage that they are unable to efficiently transduce nondividing or slowly dividing cells. As a result, specific protocols have been developed to activate cells through the use of growth factors and cytokines. In the case of hematopoietic stem cells, activation has to be carefully controlled so that pluripotency is maintained. For many applications, gammaretroviral vectors are being superseded by lentiviral vectors based on human immunodeficiency virus type-1 (HIV-1) which has additional accessory proteins that enable integration in the absence of cell division. In addition, retroviral and lentiviral vector design has evolved to address a number of safety concerns. These include separate expression of the viral genes *in trans* to prevent recombination events leading to the generation of replication-competent viruses. Further, the development of self-inactivating (SIN) vectors reduces the potential for transactivation of neighboring genes and allows the incorporation of regulatory elements that may target gene expression more physiologically to particular cell types.

1. INTRODUCTION

Retroviruses are RNA viruses which possess the unique ability to reverse transcribe their RNA genomes into a double-stranded DNA (proviral) intermediate, which can stably integrate into the genome of the host cell (Dezzutti *et al.*, 2007). This feature has been exploited in the development of vectors for therapeutic applications such as gene therapy of inherited diseases, where stable long-term integration of the gene is required. In addition, retrovirus vectors can infect a wide variety of cell types and have a relatively low toxicity profile when compared to the other virus vectors. As a result, they have been extensively represented in clinical trials. Gammaretroviruses such as Moloney murine leukemia virus (MMLV) and lentiviruses such as human immunodeficiency virus type-1 (HIV-1) have been used extensively in the engineering of derivative viral vectors.

The proviral DNA of simple gammaretroviruses like MMLV is composed of *gag*, *pro*, *pol*, and *env* genes flanked by two identical noncoding long-terminal repeat sequences (LTRs) (Fig. 3.1A). The *gag* gene encodes

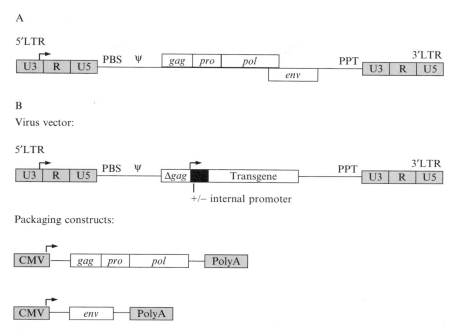

Figure 3.1 Schematic representation of the proviral genomic structure of gammaretroviruses and derivative vectors. (A) Proviral wild-type DNA of a typical gammaretrovirus: The *gag*, *pro*, *pol*, and *env* coding regions are indicated together with the U3, R, and U5 regions of the 3′- and 5′-long terminal repeats (LTRs). The primer binding site (PBS), packaging signal (Ψ), and polypurine tract (PPT) regulatory regions in *cis* are also shown. (B) General structure of gammaretroviral-derived vectors showing the PBS, Ψ, and mutated and truncated gag sequence (Δgag) upstream, and the PPT downstream of the transgene. Arrows indicate the U3 and internal promoter regions. Packaging constructs containing *gag*, *pro*, and *pol* sequences and the *env* sequence with polyA tracts under the control of the cytomegalovirus (CMV) promoter are shown below.

the virus structural proteins: matrix (MA), capsid (CA), and nucleocapsid (NC). The *pol* gene encodes the reverse transcriptase (RT) which carries out the reverse transcription of viral RNA to DNA, as well as the integrase (IN) which catalyzes the integration of proviral DNA into the host's genome. The *gag*, *pro*, and *pol* coding sequences are expressed as a Gag–Pro–Pol fusion protein which is self-cleaved by the viral protease Pro. The protease further processes the Gag and Pol polyproteins to MA, CA, and NC, and RT and IN, respectively. The *env* gene encodes the surface (SU) glycoprotein and transmembrane (TM) protein of the viral envelope (Vogt, 1997). The LTRs comprise three regions U3, R, and U5 (Fig. 3.1A). The U3 and U5 regions in both LTRs in the proviral DNA are derived from the 3′- and 5′-LTR of the viral RNA genome, respectively. The R region forms a repeat at both ends and has the sequence homology necessary for strand transfer during the reverse transcription process (Palu et al., 2000). In the proviral DNA, the 5'-U3 region contains the promoter and enhancer sequences required for viral gene transcription in *cis* and the 3'-U5 region serves as a site for transcriptional termination. The 3'-R region contains the polyadenylation (pA) signal (Palu et al., 2000).

Other elements present in *cis* are also necessary for virus replication. Immediately, downstream of the 5'-LTR is a tRNA primer binding site (PBS), which is required for the initiation of minus strand DNA synthesis during reverse transcription (Fig. 3.1A). Immediately, upstream of the 3'-LTR is a polypurine tract (PPT) which functions as the site for the initiation of plus strand DNA synthesis. Adjacent to the PBS is a virus packaging signal or psi (Ψ) which extends into the 5'-*gag* sequence and is required for the packaging of the RNA genome in newly formed virions (Miller, 1997).

In terms of vector design, virus replication and production of replication-competent viruses are obviously undesirable in a therapeutic setting, after integration of the desired gene has occurred. Thus, virus vectors are designed so that once integration has taken place, further rounds of replication cannot occur and they are "replication defective." To distinguish this process from productive virus infection by replication-competent retroviruses, it is usually referred to as *transduction*. The gene of choice that has been transferred to the host cell genome is referred to as the *transgene*. These terms will be used herein to describe gene transfer using virus vectors.

The generation of replication-defective virus vectors is achieved by replacing retroviral coding regions with the transgene(s) of choice, under the control of the *cis*-acting transcriptional regulatory elements (Fig. 3.1B). As the Ψ sequence extends into the 5' of *gag*, the *gag* ATG start codon is mutated so that Gag polyprotein synthesis does not occur in the vector and is denoted Δgag (Miller and Rosman, 1989). The viral *gag*, *pro*, *pol*, and *env* genes required for replication and virus particle assembly are provided in *trans* under the control of high-level expression promoters from other viruses such as cytomegalovirus (CMV) or respiratory syncytial virus (RSV) (Fig. 3.1B). This *trans*-complementation is accomplished either by using a packaging cell line that stably expresses the viral genes or by using a transient expression of the genes from plasmid constructs (Fig. 3.1B). Virus vector particles created in this way can be used to transduce cells without further rounds of virus replication occurring. In most vector systems currently used, the *gag–pro–pol* and *env* together with pA signals are expressed from different expression constructs. This is a safety feature which minimizes the likelihood of homologous recombination occurring between the vector construct and the viral genes in *trans* leading to the generation of replication-competent viruses.

A disadvantage of virus vectors generated from gammaretroviruses is that transduction is dependent on cell division because the viral preintegration complex (PIC) cannot penetrate the nuclear membrane until it is disassembled during mitosis (Lewis and Emerman, 1994; Roe et al., 1993). Thus, the use of these vectors to transduce slowly dividing or quiescent cells such as hematopoietic stem cells (HSCs) has required stimulation with cytokines and growth factors (see Section 4). Vectors based on lentiviruses like HIV-1 have the advantage that they are able to transduce both dividing and nondividing cells, although in HSCs progression to at least G1 is still required (Case et al., 1999; Sutton et al., 1999). This ability is due to the fact that the PIC of lentiviruses is able to enter the nucleus without disrupting the nuclear membrane. Nuclear translocation is facilitated by the MA and IN proteins together with an accessory protein Vpr. The MA and IN proteins contain nuclear localization signals and remain associated with the PIC. The MA protein has also been shown to bind directly to importin-α, facilitating entry of the PIC through nuclear pores (Bukrinsky et al., 1993; Gallay et al., 1995, 1997; Heinzinger et al., 1994). Vpr binds directly to the nuclear pore complex and causes transient herniations in the nuclear membranes (de Noronha et al., 2001).

In addition to Vpr, lentiviruses encode a number of other accessory proteins Vif, Vpu, Tat, Rev, and Nef (Fig. 3.2A). Only Tat and Rev are strictly required for virus replication. Tat activates the promoter of the HIV LTR so that viral RNA is produced more efficiently, and Rev interacts with a region of RNA known as the Rev response element (RRE) to promote the transport of viral RNA from the nucleus into the cytoplasm (Brenner and Malech, 2003; Luciw, 1996). The larger genomic size of HIV-1 and other lentiviruses means that large or multiple transgenes can be incorporated into vectors developed from these viruses.

Lentiviral vectors based on HIV-1 are more complex than gammaretroviral vectors. This is not only due to the additional requirement of Tat, Rev, and RRE but also because the *pol* coding region in HIV-1 contains a central PPT (cPPT) sequence which facilitates nuclear import of the viral PIC (Follenzi et al., 2000; Zennou et al., 2000). In lentiviral vectors, the *cis*-acting elements are similar to that of retroviruses, but the additional cPPT and RRE regions are also present upstream of the transgene (Fig. 3.2B). The presence of these additional elements increases the risk of homologous recombination events, thus the *gag–pro–pol*, *rev*, *tat*, and *env* genes are all provided in *trans* on separate expression constructs to reduce the risk of such events. The *gag–pro–pol* is expressed together with its RRE. The functions of MA, IN, and *Vpr* are redundant, thus the presence of the *gag–pro–pol* sequence encoding MA and IN is sufficient for nuclear import of the viral PIC (Trono, 2000). Third-generation lentivirus vectors currently in use have a self-inactivating (SIN) design which negates the use of Tat (Fig. 3.3C) (see Section 2.7). The generation of helper cell lines for lentivirus vectors has proved difficult because some of the lentivirus proteins are toxic to cells leading to low vector titers. Lentivirus vectors are usually produced by transient transfection methods, and pseudotyping lentivirus vectors with the surface glycoprotein from vesicular stomatitis virus G

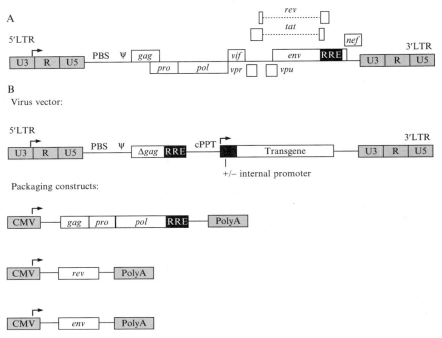

Figure 3.2 Schematic representation of the proviral genomic structure of lentiviruses and derivative vectors. (A) Proviral wild-type DNA of a typical lentivirus: The *gag*, *pro*, *pol*, *env*, and associated accessory genes *vif*, *vpr*, *vpu*, *rev*, *tat*, and *nef* coding regions are indicated together with the U3, R, and U5 regions of the 3′- and 5′-long terminal repeats (LTRs). The primer binding site (PBS), packaging signal (Ψ), central polypurine tract (cPPT), and rev response element (RRE) regulatory regions in *cis* are also shown. (B) General structure of lentiviral-derived vectors showing the mutated gag sequence (Δgag), cPPT, and RRE upstream of the transgene. Arrows indicate the U3 and internal promoter regions. Packaging constructs containing *gag*, *pro*, and *pol* sequences together with the RRE, the *rev* sequence, and the *env* sequence with polyA tracts under the control of the cytomegalovirus (CMV) promoter are shown below.

protein (VSV-G) have helped to increase stability and titer, as well as broaden the tropism (see Section 2.5).

2. Retrovirus and Lentivirus Vector Design

A number of factors must be considered for virus vector design and these are outlined in this section. The most important consideration is the choice of transgene. Transgenes selected for therapeutic applications

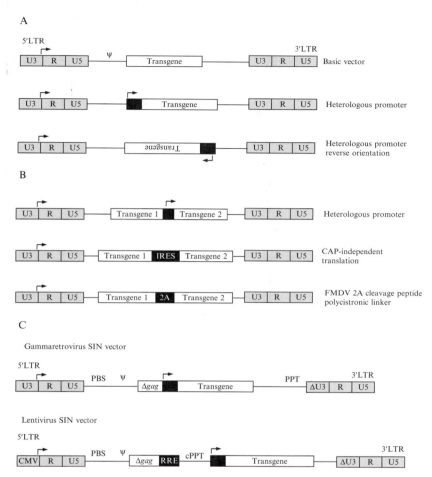

Figure 3.3 Different strategies for gammaretroviral and lentiviral vector design. Vector design options for expression of (A) a single transgene; (B) two transgenes. Regulatory elements have been omitted for simplicity. (C) Structures of gammaretroviral and lentiviral self-inactivating (SIN) vectors showing the modified U3 region (ΔU3) in the 3′-LTR. Alternative promoter such as the CMV promoter can be used in place of the 5′-U3 region to drive expression of genomic RNA.

normally have very defined functions, and mutations in such genes are associated with loss or gain of very specific phenotypic characteristics. Genes involved in cell cycle regulation, cellular proliferation, and apoptosis pose particular problems for therapeutic applications in humans as they are often involved in multiple signaling pathways. Nonphysiological expression may lead to deleterious effects such as tissue damage or cancer, so-called phenotoxicity. Other factors to consider are whether the gene is ubiquitously expressed or only expressed in specific cells and tissues, and the level of protein expression in different tissues.

2.1. The promoter

The promoter of choice will ideally drive expression of the transgene to a level similar to that of the endogenous protein, and in clinical applications must enable phenotypic correction of the disease or disorder. Early retroviral vectors utilized the endogenous retroviral 5′-U3 regulatory regions which provided good levels of gene expression but were found unsurprisingly to cause aberrant *trans*-activation of neighboring proto-oncogenes, and to be susceptible to epigenetic silencing (see Sections 2.7 and 2.8). If the protein encoded by the gene of interest is known to be ubiquitously expressed, a number of different internal heterologous promoters can be used to drive transgene expression in a wide variety of cell types, including the promoters of the spleen focus-forming virus LTR, elongation factor 1 alpha (EF1α), and phosphoglycerate kinase (PGK) (Weber and Cannon, 2007). However, if the protein is not ubiquitously expressed but found in specific cell types, for example, cells of the hematopoietic system, then it may be preferable to use a promoter that is cell type specific, such as that described for the Wiskott–Aldrich syndrome gene (Charrier et al., 2007). Specificity and activity can also be achieved through the construction of artificial promoters. This has been demonstrated by the chimeric c-Fes and Cathepsin G promoter used to achieve high-level expression of the *CYBB* gene (encoding gp91^{phox}) gene in myeloid cells for the treatment of chronic granulomatous disease (CGD) (Santilli et al., 2011). New methods employing artificial microRNAs to regulate endogenous miRNA levels are also being developed to improve cell-specific targeting for gene therapy studies (Brown and Naldini, 2009).

2.2. Variations in vector design

The viral vectors shown in Figs. 3.1B and 3.2B illustrate the simplest type of vector design. In the case of vectors containing a single transgene, expression can be driven either from the 5′-U3 LTR promoter or an internal heterologous promoter as mentioned above (Fig. 3.3A). Heterologous internal promoters can be used to drive expression of the transgene in the reverse as well as the forward orientation (Fig. 3.3A). This reverse orientation allows the intron structure of a transgene to be maintained, as introns are spliced out in the forward orientation. It also permits the use of a particular pA site as the viral vector polyA tail is added during packaging in the forward orientation. When simultaneous expression of two to three genes is required, a number of different strategies can be adopted. Two genes can be expressed from a single transcript by using a heterologous promoter placed between the two genes to drive expression of the downstream gene (Fig. 3.3B). A further possibility is to use an internal ribosomal entry site (IRES) between the two genes. The IRES sequence was first identified in picornaviruses and allows cap-independent translation of the downstream gene (Adam et al., 1991). However, the use of the IRES can result in a low expression level of the downstream gene (Flasshove et al., 2000; Mizuguchi et al., 2000). An alternative to using an IRES sequence is to engineer the vector so that the two genes are separated by a sequence which encodes a 2A cleavage peptide encoded by foot-and-mouth disease virus (Fig. 3.3B). The proteins are expressed as a single polyprotein which is cleaved at the 2A cleavage peptide through a posttranslational nonenzymatic cleavage mediated by ribosome stuttering, hypothesized to be caused by impairment of normal peptide bond formation (de et al., 2003; Donnelly et al., 2001). Although the use of the 2A signal peptide as a polycistronic linker allows for equivalent expression of both transgenes, the posttranslational cleavage event leaves the upstream protein with 19 residual amino acids from the 2A sequence on its carboxy terminus, and the downstream protein with two to five amino acids on its amino terminus. This may have consequences for protein localization and function and should be tested for each transgene.

2.3. Selectable markers

Virus vectors and packaging constructs often contain antibiotic resistance genes such as ampicillin or kanamycin. These antibiotic resistance genes permit the transformation of electrically or chemically competent *Escherichia coli* for the propagation of constructs. Selectable markers for other types of antibiotic resistance such as G418 are used for selection of mammalian cells transfected with packaging constructs to generate stably expressing packaging cell lines. Details of molecular cloning techniques, plasmid preparation, and generation of helper cell lines will not be discussed further in this chapter.

2.4. Marker genes

When constructing virus vectors, it is often advisable to initially generate a vector containing a *marker gene* whose expression can be easily monitored before testing the transgene of choice. This enables troubleshooting of any problems associated with the vector backbone, promoter, *cis*-elements, or packaging constructs prior to testing the transgene itself. The enhanced green fluorescent protein (eGFP) is one of the most commonly used marker proteins as its expression can be easily monitored by flow cytometry or confocal microscopy (Bierhuizen et al., 1997). Truncated versions of cell surface proteins such as CD34 and low-affinity nerve growth factor receptor can also be used as cell surface markers. These proteins retain their extracellular and TM regions but lack intracellular domains and therefore do not affect cell function. When live imaging is required, the luciferase reporter gene is a useful marker for tracking gene expression *in vivo* in real time.

2.5. The virus vector envelope protein

Retroviruses have quite a narrow host cell tropism due to the fact that the surface glycoprotein encoded by the *env* gene only binds to receptors expressed on a limited number of cell types and from a small number of species. Lentiviruses have an even narrower host cell tropism, exemplified by HIV-1 which predominantly infects human $CD4^+$ T helper cells, dendritic cells, and macrophages. To improve host cell tropism, retroviral and lentiviral vectors can be pseudotyped with surface glycoproteins from other viruses. Early retroviral vectors were pseudotyped with glycoproteins from the amphotropic 4070 and 10A1 strains of MMLV which permitted transduction of a wide range of mammalian cell types. The glycoprotein from the cat endogenous virus (RD114) and gibbon ape leukemia virus (GALV) has been used for efficient transduction in human HSCs (Brenner *et al.*, 2003; Kiem *et al.*, 1997; Relander *et al.*, 2005; Sandrin *et al.*, 2002). Many virus vectors, particularly lentivirus vectors, are now pseudotyped with the envelope glycoprotein from VSV-G which not only has a very broad host range but also imparts stability to the viral envelope, permitting ultracentrifugation and concentration of virus particles for the generation and storage of high-titer virus stocks (10^7–10^{10} infectious virus particles/ml) (Gallardo *et al.*, 1997). Expression of the VSV-G envelope, however, is toxic to cells. Therefore, generation of VSV-G virus vectors is best achieved by transient transfection of the VSV-G on a separate packaging expression construct or by using a regulatable packaging cell line such as the tetracycline-inducible system developed by Yang and colleagues (Schambach *et al.*, 2000; Yang *et al.*, 1995).

2.6. Additional regulatory sequences

Enhancer regions can be inserted into virus vectors to increase transgene expression. The woodchuck hepatitis virus posttranscriptional regulatory element (WPRE), for example, is a hepadnavirus sequence that is widely used as a *cis*-acting regulatory module (Schambach *et al.*, 2000; Zufferey *et al.*, 1999). When the WPRE is placed in the 3′-untranslated region of virus vectors, the WPRE enhances the expression of the transgene by increasing both nuclear and cytoplasmic RNA levels (Schambach *et al.*, 2000; Zufferey *et al.*, 1999).

2.7. Safety considerations

Gammaretroviral vectors have a propensity to integrate near regulatory regions (including transcription start sites) of active genes (De et al., 2005; Wu et al., 2003). The activity of promoter/enhancer elements within the vector LTRs or aberrant splicing events can lead to the activation of genes flanking the integration site (Hacein-Bey-Abina et al., 2008; Howe et al., 2008). This is commonly referred to as *insertional mutagenesis*. If the genes near the integration sites are proto-oncogenes, this may lead to increased proliferation of clonal populations of cells. This has been observed in several patients with SCID-X1 and Wiskott–Aldrich syndrome treated with gammaretroviral vectors who developed T cell leukemias, and in the patients with CGD who developed dominant myeloid clones (Hacein-Bey-Abina et al., 2008; Howe et al., 2008; Ott et al., 2006). Analysis of clonal populations of lymphoid cells from four of the SCID-X1 patients and one WAS patient revealed that the mutagenic retrovirus integration site was located near the promoter of the LMO2 gene, a proto-oncogene associated with acute T cell leukemias (Hacein-Bey-Abina et al., 2008; Howe et al., 2008).

These unwanted effects have been addressed through the development of SIN viral vectors. These vectors incorporate a deletion in the U3 region of the 3′-viral LTR (designated ΔU3) resulting in inactivation of the viral promoter/enhancer elements, including the CAAT and TATA boxes, following reverse transcription (Fig. 3.3C) (Thornhill et al., 2008). The U3 region of the 5′-LTR can be replaced with a strong promoter from another virus such as the CMV or RSV promoter employed in pCCL vectors or pRRL vectors, respectively (Fig. 3.3C). This drives high-level expression of the genomic RNA and permits Tat-independent transcription in lentivirus vectors without affecting viral titers (Miyoshi et al., 1998) (Fig. 3.3C). In addition to the SIN design, the use of internal promoters with weak to moderate activity to drive transgene expression is also less likely to lead to activation of neighboring genes. Thus, these vectors are currently considered to have an improved safety profile, although the true genotoxicity can only be properly evaluated in clinical trials with these vectors. Future safety measures may include the incorporation of chromatin remodeling elements such as insulators that shield neighboring genes from influence of the vector itself (Naldini, 2011). In clinical protocols, a maximum transgene copy number of one per cell is desirable if sufficient expression can be achieved, as higher numbers increase the risk of insertional mutagenesis (see Section 4).

2.8. Gene silencing

A number of studies have shown that the promoter regions of gammaretroviral and lentiviral vectors may be subject to epigenetic silencing which has been associated with DNA methylation of CpG sequences (Bestor,

2000; Ellis, 2005; Klug *et al.*, 2000; Pannell *et al.*, 2000). Such silencing events gradually reduce the expression of transgenes under the control of both viral LTR and internal promoters (Zhang *et al.*, 2010). Vector designs have been modified to address this problem, for example, by flanking transgenes with DNA insulators such as the chicken β-globin locus control region HS4 element (cHS4). However, such elements have only conferred partial protection against silencing and can lead to reduced vector titers (Gaszner and Felsenfeld, 2006; Jakobsson *et al.*, 2004; Urbinati *et al.*, 2009). The novel enhancer-less ubiquitous chromatin opening element has recently been demonstrated to be resistant to DNA methylation and silencing *in vitro* and *in vivo* and may offer a viable alternative for gene therapy applications (Zhang *et al.*, 2007, 2010).

3. Retroviral and Lentiviral Vector Production

A number of packaging cell lines are commercially available for the generation of gammaretrovirus vectors and the choice of system depends on the envelope protein required (Miller, 1997). Phoenix packaging cells lines, based on HEK 293T cells, express the MLV *gag–pro–pol* and *env* proteins and offer the advantage that these cells have high protein expression levels of these viral coding sequences, allowing for higher virus vector titers to be achieved. Two variations are available that allow pseudotyping with either amphotropic (Phoenix-Ampho) or ectotropic (Phoenix-Eco) envelope proteins. There is also a Phoenix-gp cell line which expresses only *gag–pro–pol* enabling pseudotyping of the retroviral vector with other envelop proteins such as GALV or VSV-G (Nolan laboratory www.stanford.edu/group/nolan). Packaging cell lines based on human HT1080 fibroblast cells are also available for the generation of recombinant retroviral vectors with amphotropic MLV envelope (FLYA13) or the cat endogenous virus RD114 envelope (FLYRD18 line) (Cosset *et al.*, 1995). Only one stable lentivirus packaging system (STAR) exists to date. STAR packaging cells were generated by transducing 293T cells with a gammaretroviral vector encoding a codon-optimized *gag–pol* gene followed by introduction of the genes for the HIV-1 regulatory proteins Tat and Rev, an envelope protein, and an HIV-1 lentivirus vector (Ikeda *et al.*, 2003). However, the disadvantage of this system is that generation of lentivirus vector in this way takes longer than transient transfection methods.

Transient transfection using the calcium phosphate precipitation method (Graham *et al.*, 1977) or other commercially available transfection reagents is used to introduce virus vector constructs to packaging cell lines for the generation of vector particles, followed by a selection method to produce permanent packaging cell lines. Transient transfection of the vector construct together with packaging constructs for viral genes can also be used to generate vector particles. This is the method of choice for lentivirus vectors due to the toxicity of lentiviral proteins and the difficulty in establishing packaging cell lines. Large-scale production of virus vectors for clinical use can be achieved by the use of cell factories, and purification using ion-exchange and size exclusion chromatography (Merten et al., 2011).

3.1. Production of retrovirus vectors using Phoenix producer cell lines

The method below describes the use of the Phoenix packaging cell lines for the generation of retrovirus vectors.

3.1.1. Required materials

- *Equipment*: Bench top centrifuge and plate adaptors; ultracentrifuge (SW41 rotor or similar) and ultracentrifuge tubes; safety cabinet, humidified 37 °C CO_2 incubator. 10-, 20-, 200-, and 1000-µl pipettes; pipetman; flow cytometry tubes.
- *Cell lines*: Phoenix cell lines are commercially available from the National Gene Vector Laboratory Biorepository (NGVB). HT1080 fibrosarcoma cells are available for the American Tissue Culture Collection (ATCC).
- *Retrovirus transfer vector and helper envelope glycoprotein construct*: DNA of a high quality (endotoxin-free) and concentration is recommended.
- *Growth medium*: The growth medium consists of DMEM supplemented with 10% FCS, $2 mM$ L-glutamine, 100 IU/ml penicillin, and 100 µg/ml streptomycin.
- *Transfection reagent*: $2M$ $CaCl_2$ solution.
- *Transfection medium*: Growth medium supplemented with $25 \mu M$ chloroquine.
- *Other reagents*: 1× PBS, 2× Hepes-buffered saline (HBS) solution (150 mM NaCl, 100 mM Hepes, 1.5 mM Na_2HPO_4); sterile water; 10 mg/ml polybrene stock.
- *Disposables*: 10-, 20-, 200-, and 1000–µl filter tips, 5- to 25-ml pipettes, 10-cm and 12-well tissue culture plates, 15- and 50-ml Falcon tubes, 0.45-μM filters.

3.1.2. Transient transfection procedure for gammaretroviral vector production

The day prior to transient transfection: Seed 5×10^6 Phoenix cells (or equivalent) in a 10-cm tissue culture plate in 10 ml fresh prewarmed growth medium and place in a humidified incubator at 37 °C, 5% CO_2. The cells should be 70–80% confluent on the day of transient transfection.

The day of transient transfection: Two hours prior to the transfection procedure, replace the medium with 10 ml of fresh growth medium. In a 15-ml Falcon, dilute 10 μg viral vector DNA in sterile water to give a final volume of 450 μl for each dish to be transfected. Add 50 μl $2M$ $CaCl_2$ to the diluted DNA and mix by tapping gently. If using Phoenix-gp cells and pseudotyping with an alternative envelop protein, use 10 μg viral vector DNA and 10 μg of the envelope construct and double the amount of water and $CaCl_2$. Add 500 μl 2× HBS solution to a 15-ml Falcon tube and add the DNA/$CaCl_2$ solution drop-wise, tapping the tube with each drop. Incubate the transfection mixture for 20 min at room temperature and vortex briefly. While the transfection complex is incubating, remove the media from the plates and replace with transfection medium. Chloroquine inhibits lysosomal degradation of the DNA to be transfected. Add the transfection mixture drop-wise to the cells, gently swirling the media in the plates at the same time to evenly distribute the DNA/$CaCl_2$ precipitate and place the cells back in the incubator at 37 °C with 5% CO_2.

The day after transient transfection: Replace the cell culture medium with 6 ml fresh growth medium. This reduced volume will help to increase the concentration of virus particles produced.

3.2. Production of VSV-G pseudotyped lentiviral vectors by transient transfection

Below, we describe a transient transfection procedure for the small-scale production of a VSV-G pseudotyped lentivirus vector.

3.2.1. Required materials

- *Equipment*: Equipment as listed in Section 3.1.1.
- *Cell line*: HEK 293T cells are available for the ATCC.
- *Lentivirus HIV-1 transfer vector and helper plasmid constructs encoding Gag–Pol, rev, and VSV-G envelope gene*: DNA of a high quality (endotoxin-free) and concentration is recommended. Plasmids can be obtained from www.addgene.com or the NGVB.
- Growth medium, transfection medium, and other reagents as listed in Section 3.1.1.
- *Transfection reagents*: Opti-Mem media (Invitrogen); 10 mM polyethylenimine, (PEI; MW—25,000) (Sigma Aldrich).
- *Disposables*: 10-, 20-, 200-, and 1000-μl filter tips, 5- to 25-ml pipettes, 175-cm^2 flasks tissue culture plates, 15- and 50-ml Falcon tubes, 0.45-μM filters.

3.2.2. Transient transfection procedure

The day prior to transient transfection: Seed 1.5×10^7 293T cells in a 175-cm² tissue culture flasks in 40 ml fresh prewarmed growth medium and place in a humidified incubator at 37°C, 5% CO_2. The cells should be 70–80% confluent on the day of transient transfection.

The day of transient transfection: Two hours prior to the transfection procedure, replace the medium with 15 ml of fresh growth medium. In a 15-ml Falcon tube, dilute 40 μg viral transfer vector DNA, 30 μg gag–pol construct, 5 μg of *rev* construct, and 10 μg of VSV-G construct into Opti-Mem to a final volume of 5 ml for each flask. The relative amounts of DNA may need to be optimized as transfection efficiency can vary depending on the backbone and size of construct used. Dilute 1 μl PEI in 5 ml Opti-Mem per flask and mix together with the DNA solution (do not filter after this stage as the DNA–PEI complexes will not pass through the filter). Incubate the transfection mixture for 20 min at room temperature. Add the 10 ml transfection mixture to each flask and place the cells back in the incubator at 37°C with 5% CO_2.

3.3. Harvesting virus vector

After 24–48 h of cell culture, collect and pool cell culture supernatants containing virus particles from the plates into 50-ml Falcon tubes. This harvesting time is a rough guide based on the experience of our laboratory. However, the time chosen for harvesting depends on when the peak of viral particle production occurs and can vary with vector and cell type. Therefore, the optimal time may have to be empirically determined by harvesting supernatants 24–96 h after transfection. VSV-G pseudotyped vectors can be harvested 24–96 h after transfection; after this time, cells will begin to die due to the associated toxicity of the protein. After harvesting, clarify the supernatants by centrifugation at $400 \times g$ in a bench top centrifuge for 10 min at room temperature to remove any dead cells and cell debris and filter the supernatant through 0.45 μM filters. Large-scale virus vector production may require overnight treatment with a DNase enzyme solution to remove any free vector DNA from the virus particles.

3.4. Concentration of virus vector particles by ultracentrifugation

Depending on the titer required, unconcentrated supernatant can be used directly, but virus particles can also be concentrated by centrifugation at $18,500 \times g$ at 4°C for 14–16 h. Although this is a relatively slow centrifugation speed, there may be some loss of virus particles depending on the envelope glycoproteins chosen for pseudotyping and the associated fragility.

If the retrovirus vector or pseudotyped equivalent is known to be fragile, avoid ultracentrifugation. VSV-G pseudotyped vectors can be concentrated using speeds of up to $50,000 \times g$ (retroviruses) and $100,000 \times g$ (lentiviruses) at 4°C for 1.5–2h. Resuspend the pellet in ice-cold growth medium to 0.5% of the original volume, aliquot, and store at −80°C. Lentiviral vectors can be stored long-term at this temperature, but freeze–thaw cycles have a detrimental effect on virus titers.

3.5. Determination of infectious vector titer

The day prior to vector titration: Seed 1×10^5 HT1080 cells in a 12-well tissue culture plate in 500 µl fresh prewarmed growth medium and place in a humidified incubator at 37°C, 5% CO_2 overnight to adhere. The cells used for titration must be permissible to virus entry with the natural or pseudotyped envelope glycoprotein used.

The day of vector titration: Add 1:1000–1:1,000,000 dilutions of vector directly to the cell monolayer in a final volume of 500 µl growth medium. Leave at least one well free of vector for a negative control. If using a retrovirus or a lentivirus that is likely to have a low titer (e.g., not pseudotyped with VSV-G), supplement the media with 4 µg/ml polybrene to assist glycoprotein–cell receptor binding (see Section 4). If using a gammaretrovirus vector, spin the plates at $950 \times g$ in a bench top centrifuge for 1h at room temperature and then incubate in a humidified incubator at 37°C, 5% CO_2 for 6h. If working with a VSV-G pseudotyped lentivirus vector, this step is usually not required. Top up the wells with 1.5 ml growth media and place back in the incubator for 2 days. Wash the cells in PBS and remove cells by incubation in 0.5 ml trypsin–EDTA. Inactivate the trypsin by adding 0.5 ml growth medium and transfer the cells to flow cytometry tubes. Centrifuge cells for 5 min at $950 \times g$, remove supernatants, resuspend in PBS with 5% FCS, and stain with an antibody directed against the transgene of interest. A protocol for staining cell surface proteins with fluorescent-conjugated antibodies and analysis by flow cytometry is given in Section 4.3. If the transgene encodes an intracellular protein, the cells will need to be fixed and permeabilized prior to staining and may require primary, secondary, and even tertiary antibodies for proper detection. Detailed flow cytometry protocols will not be discussed further here. If using a fluorescent marker such as eGFP, positive cells can be viewed by flow cytometry directly without the need for staining with fluorescent-conjugated antibodies. In order to determine virus titer, it is best to select a sample where approximately 30% of the cells are positive for transgene expression. Samples with higher levels of expression may have more than one copy of the transgene per cell, leading to an underestimation of virus titer.

The virus titer (transducing units/ml) is determined by multiplying the number of transgene-expressing positive cells detected by the dilution factor as follows:

Virus titer (transducing units/ml)
$$= \frac{(\% \text{positive cells}/100) \times \text{number of cells transduced} \times \text{dilution factor}}{\text{Volume (ml)}}$$

Virus titers can also be determined by real-time quantitative PCR (qPCR) as infectious genomes/ml (see Section 4.2) or by ELISA to indirectly measure virus concentration through the amount of viral MA protein (often denoted p24).

4. Transduction Procedures

Transduction of mammalian cell lines is easily achieved in the presence of normal growth media supplemented with growth factors normally present in FCS (as described in Section 3.5). In a related manner, primary human or mouse cells can be transduced in growth media supplemented with growth factors, lipids, amino acids, and cytokines/chemokines in which they are normally grown. However, human HSCs, which have been the target cells transduced *ex vivo* in the vast majority of clinical trials undertaken, have more specific culture requirements. It is desirable to maintain the pluripotent properties of HSCs for the duration of the transduction procedure. This is particularly important if transgene expression in different lymphoid and myeloid lineages is required for proper phenotypic correction of the genetic defect. At the same time, though, it is necessary to induce mitosis or cycling to at least G1 to allow efficient transduction with retroviral vectors and lentivirus vectors, respectively. However, extensive cell cycle activation can lead to stem cell differentiation and loss of pluripotency. Thus, a fine balance between cellular activation and maintenance of stem cell characteristics is required. This has been achieved through the development of specific stem cell culture media containing the minimum required nutrients and growth factors to maintain cell viability. Cells are activated to divide prior to transduction through the use of a cocktail of cytokines, which have been selected based on their ability to improve transduction and also balance activation so that if differentiation occurs, it will be to a similar level in all immune cell progenitors (Zielske and Braun, 2004).

The type of transduction protocol used depends largely on the stock virus titer. This is because virus entry is facilitated by the likelihood of cell-to-virus interactions and the proximity of virus particles to cells rather the overall number of virus particles present. Therefore, a small volume of high-titer vector is preferable to large volume of low-titer vector. The transduction efficiency can be increased by using agents which help to enhance virus–cell interactions such as polycations and fibronectins. Polycations such as polybrene (hexadimethrine bromide) and protamine sulfate are small positively charged molecules that neutralize the charge repulsion between negatively charged salicylic acid molecules on the cell surface and negatively charged glycoproteins on the virus envelope, thus facilitating virus entry. Polybrene and protamine sulfate can increase transduction efficiency by 2- to 10-fold depending on the target cell. Fibronectins are multifunctional cell adhesion glycoproteins, which when coated on the surfaces of flasks help to colocalize cells and virus particles through protein–protein interactions. Commercial chimeric fibronectins such as RetroNectin (Takeda) have been developed which have been modified further to improved interactions. However, such fibronectins do not increase the transduction efficiency of VSV-G pseudotyped virus vectors. In addition to these factors, repeated rounds of transduction may be required to increase the efficiency of the procedure particularly when lower-titer vectors and larger cell numbers are used. Large numbers of HSCs are needed to achieve successful engraftment at high proportions (Perez-Simon et al., 1998; Schwella et al., 1995). Apart from virus vector concentration, cell viability can also affect the efficiency of the transduction and should be monitored throughout the transduction procedure. Poor viability is often associated with a low efficiency of transduction. The ideal transduction efficiency would be to have one copy of the transgene in each cell (i.e., 100%). It is not desirable to have unnecessarily high transgene copies per cell as this may increase the risk insertional mutagenesis. Handling of human HSCs and clinical transduction protocols is carried out in isolator cabinets in designated clean rooms to ensure the cells remain free from contamination with microorganisms.

4.1. Transduction of human HSCs

HSCs are isolated from bone marrow, cord blood, or mobilized peripheral blood through the use of antibodies specific to the $CD34^+$ surface marker and column purification using magnetic beads (Miltenyi Biotech). The cells are placed in culture for 24 h in the presence of cytokines. This serves as a prestimulation step to induce the cells to cycle so that the virus vector can enter the nucleus and efficiently transduce the cells. A standard protocol for transduction of HSCs is described below. A prestimulation step is not usually necessary for small-scale transductions.

4.1.1. Required materials
- *Equipment*: Bench top centrifuge, centrifuge adaptors for flasks and plates, safety cabinet, humidified 37 °C CO_2 incubator; 10-, 20-, 200-, and 1000-µl pipettes; pipetman.

- *Donor CD34-positive HSCs*: Cells should be purified from peripheral blood, cord blood, or bone marrow using magnetic beads conjugated with anti-CD34 antibody (Miltenyi Biotech).
- *Culture medium*: Phenol-red free X-Vivo-10 or X-Vivo-20 media (Lonza) supplemented with 2% human serum albumin (HSA), 300 µg/ml SCF, 300 µg/ml Flt-3, 100 µg/ml TPO, 4 µg/ml IL-3 (Peprotech).
- *Virus vector*: High-titer retrovirus (10^6–10^8 transducing particles/ml) or lentivirus (10^7–10^{10} transducing particles/ml).
- *Other reagents*: 1 mg/ml RetroNectin (Takara); PBS supplemented with 2% HSA; 10 mg/ml Polybrene; 4 mg/ml trypan blue.
- *Disposables*: Gilson filter tips, 5- to 25-ml pipettes, 25- and 75-cm^2 flasks, and 24-well tissue culture plates.

4.1.2. Transduction procedure

4.1.2.1. The day before the transduction procedure If using a low-titer vector ($<10^7$ transducing units/ml), coat the surface of the flasks with 5 µg/cm^2 RetroNectin in an appropriate volume of PBS. A 25-cm^2 flask requires 125 µl RetroNectin in 2 ml of PBS and 75-cm^2 flasks will need 375 µl in 6 ml PBS. Place the flasks on a rocker for 2 h at room temperature and store in the fridge overnight. If you expect to have less than 3×10^6 cells, then smaller culture vessels such as 6-well plates will need to be coated.

4.1.2.2. Prestimulation of cells Centrifuge the CD34$^+$ stem cells at $400 \times g$ for 5–10 min at room temperature and resuspend them in 5–10 ml culture media. If working with frozen cells, it is advisable to resuspend them in 5–10 ml media without cytokines and allow them to recover overnight at 37 °C, 5% CO_2. Take a 20 µl sample of cells and determine the number of live cells and their viability using the trypan blue method (Shapiro, 1988). Fresh cells should have a viability of >90%, whereas the viability of frozen cells can vary between 50% and 90%. DNA dyes such as 7AAD can also be used to assess cell viability. Remove excess RetroNectin and rinse with PBS. Seed cells at a density of 0.5–0.7 $\times 10^6$ cells/ml in 25-cm^2 (7-ml) or 75-cm^2 flasks (20-ml) and place flasks in the incubator overnight at 37 °C, 5% CO_2.

4.1.2.3. First round of transduction If using a lentivirus vector, the first round of transduction can be carried out after 12–24 h of prestimulation. This length of time should be sufficient to ensure that the majority of CD34$^+$ HSCs are in G1. If using a retrovirus vector, it is preferable to prestimulate the cells for 48 h to ensure the majority of cells are actively undergoing mitosis.

Prewarm the cell culture medium and thaw out stock vials of virus vector. Remove the cells from the flasks, place in 15- or 50-ml Falcon tubes. If RetroNectin has been used, rinse the surface of the flask twice with PBS and add to the cells in the Falcon tubes. Pellet the cells by centrifugation at $400 \times g$ for 5–10 min at room temperature and resuspend in 5–10 ml media. Take a 20 μl small sample and perform a cell count using trypan blue. Seed cells for an untransduced control in fresh media at a density of $0.5-1.0 \times 10^6$ cells/ml. Resuspend the cells and dilute them into the appropriate volumes of media with vector to achieve a virus-to-cell ratio of 50–100 infectious virus particles per cell (known as multiplicity of infection, MOI), and a cell density of $0.5-0.7 \times 10^6$ cells/ml. Seed the cells back into the same RetroNectin-coated culture vessels and place in the incubator overnight at 37 °C, 5% CO_2.

If using a low-titer virus, add the volume of vector to the flasks to give the amount of virus particles required for infection and spin at $400 \times g$ for 30 min at 4 °C to facilitate virus entry. Remove excess virus vector and seed the cells in fresh media at a cell density of $0.5-0.7 \times 10^6$ cells/ml in the presence of polybrene 6 μg/ml.

4.1.2.4. Second round of transduction
After 24 h, repeat the transduction procedure as described above for the first round of transduction. Examine the untransduced cells under the microscope, and if the cell density has increased, split them 1:2 by dilution in fresh media and place into fresh culture vessels.

4.1.2.5. Third round of transduction
When working with gammaretrovirus vectors, a third round of transduction may be necessary to achieve a copy number of one copy of transgene per cell, particularly with low-titer vectors. Cells can be harvested 6–24 h after this final transduction.

4.1.2.6. Harvesting of transduced cells
After a further 24–72 h, remove the transduced cells and untransduced control from the flasks and pellet by centrifugation at $400 \times g$ for 5–10 min at room temperature. Wash and pellet the cells twice by resuspending them in 10–20 ml PBS and centrifuging at $400 \times g$ for 5–10 min at room temperature. Determine cell count and viability using trypan blue, the copies of transgene in each cell (see Section 4.2), and the protein expression levels of the encoded protein. To determine transgene copy numbers, it is advisable to culture the cells for 2–3 days after the final round of transduction as this eliminates the presence of free vector which may be present immediately after transduction.

4.2. Determination of transgene copy number by real-time qPCR analysis

Following the transduction procedure, the efficiency of the transduction can be assessed by real-time qPCR amplification of the transgene or integrated vector sequences such as the packaging signal (Ψ). The transduced cells should be cultured for 24–72 h following transduction to ensure that any free virus vector in the culture media is degraded and does not interfere with detection of integrated sequences. Amplification of vector sequences is useful, as the same qPCR can be used for virus vectors developed from the same backbone but carrying different transgenes. This method can be also used to calculate the titer of virus vectors, which are referred to as infectious genomes (ig)/ml of cell supernatant, and the titer can be up to 10-fold higher than that achieved by analyzing the level of transduction by flow cytometry (Section 4.3).

In this section, we describe the dual amplification of the HIV Ψ region present in the pCCLSINcPPTWPRE-mut6 vector backbone and an albumin internal control gene (Charrier et al., 2007; Dull et al., 1998; Zanta-Boussif et al., 2009). This vector is an advanced generation SIN and Tat-independent HIV-LV system which produces high titers and has been used successfully to transduce various types of cells (Dull et al., 1998). Standard curves can be generated by two methods: dilutions of plasmid(s) containing known copy numbers of the transgene of interest and known copies of a house-keeping gene such as albumin, or cell lines containing a known copy number of the viral backbone. The plasmid method will be discussed below.

4.2.1. Required materials

- *Equipment*: ABI PRISM 7300 sequence detector (Applied Biosystems) or similar, 10-, 20-, 200-, and 1000-μl pipettes.
- Primers and TaqMan probes, $10\,\mu M$ each of
  HIV Ψ Forward Primer: 5'-CAG GAC TCG GCT TGC TGA AG-3'
  HIV Ψ Reverse Primer: 5'-TCC CCC GCT TAA TAC TGA CG-3'
  HIV Ψ Probe: 5'-FAM-CGC ACG GCA AGA GGC GAG G-TAMRA-3'
  Albumin Forward Primer: 5'-GCT GTC ATC TCT TGT GGG CTG T-3'
  Albumin Reverse Primer: 5'-ACT CAT GGG AGC TGC TGG TTC-3'
  Albumin Probe: 5'-VIC-CCT GTC ATG CCC ACA CAA ATC TCT CC-TAMRA-3'.
- Dual expression plasmid DNA standards: Plasmid encoding the Ψ sequence and the albumin house-keeping gene. A serial dilution of the plasmid is used to generate a standard curve for qPCR.

The number of plasmid molecules/μl can be calculated using the equation below:

Number of copies of plasmid/μl

$$= \frac{\text{Plasmid concentration}(\text{ng}/\mu\text{l}) \times 10^{-9} \times \text{molecular weight}(\text{mol})}{\text{Number of base pairs in the plasmid} \times 650\,\text{g}/\text{mol}}$$

As 1 Mole = Avogadro's constant (6.022×10^{23})

Number of copies of plasmid/µl

$$= \frac{\text{Plasmid concentration}(\text{ng}/\mu l) \times 10^{-9} \times (6.022 \times 10^{23})}{\text{bp} \times 650\,\text{g/mol}}$$

- Make an initial dilution of your stock down to 10^7 and then serial 10-fold dilutions from 10^7 down to 10^1 copies per 10 µl in a DNAase-free Tris buffer, pH 8, or Tris–EDTA (TE).
- *Genomic DNA extracted from transduced cells*: DNA of a high quality (endotoxin, DNAse, RNAase free), eluted in a Tris buffer pH 8 without EDTA.
- *Other reagents*: 1× TaqMan Universal PCR master mix, no AmpErase UNG (Applied Biosystems).
- *Disposables*: 10-, 20-, 200-, and 1000-µl filter tips, 96-well optical qPCR plates (Applied Biosystems).

4.2.1.1. Setting up the qPCR reaction Set up the qPCR reaction mix for each sample as follows:

Reagents	Initial concentration (µM)	Final concentration (µM)	Volume (µl)/well
Master Mix	–	–	12.5
HIV Forward Primer	10	0.1	0.25
HIV Reverse Primer	10	0.1	0.25
Albumin Forward Primer	10	0.1	0.25
Albumin Reverse Primer	10	0.1	0.25
HIV Probe FAM	10	0.1	0.25
Albumin Probe VIC	10	0.1	0.25
Sterile H_2O	–	–	1
Total volume			15

Add 15 µl of the reaction mix to each of the required wells for standards and samples in an optical 96-well qPCR plate. Add 10 µl of each standard dilution (e.g., 10^6–10^1) and DNA sample at a concentration of 10 ng/µl in triplicate to wells containing the reaction mix. To other wells, add water as a nontemplate control and positive and negative controls where appropriate. Seal the plate with plate sealer film and centrifuge briefly in a bench top centrifuge to ensure all the reaction components are at the bottom of each well and any bubbles are removed. Place the plate in the ABI machine and perform 1 cycle 95 °C for 2 min followed by 40 cycles of amplification consisting of 95 °C for 15 s and 60 °C for 1 min. Ensure that both FAM-TAMRA and VIC-TAMRA detections are selected for each standard and sample well.

4.2.1.2. Calculating the copy number Once the reaction is complete, set the baseline to auto for the HIV Ψ and albumin and set the threshold bar in the linear part of the logarithmic curves. Ensure that the standard curves have a slope value of between -3.1 and -3.6 and the R^2 value is at least 0.98. To obtain the sample mean copy number per cell, divide the mean HIV Ψ signal by the mean albumin signal and multiply by 2 (as there are two albumin genes in each cell). Copy numbers should range between 0.5 and 1.5 copies/cell after two or three rounds of transduction with lentivirus or retrovirus vectors, respectively (Fig. 3.4A). Copy numbers can also be expressed as % transgene. A mean copy number per cell of 1 is equivalent to 100% transgene expression.

4.3. Determination of protein expression

The efficiency of transduction can also be determined by analysis of transgene expression by flow cytometry. Fluorochrome-conjugated primary antibodies specific to the expressed protein can be easily detected using flow cytometry machines with specific lasers to excite the fluorochrome and detectors to pick up the emission spectra. This allows you to visualize the protein on whole cell populations. The method below describes a standard staining protocol for the detection of a cell surface protein. In addition to the determination of protein expression, if a functional assay exists, this can be used to compare the function of the transduced gene and encoded protein compared to a normal untransduced control.

4.3.1. Required materials

- *Equipment*: Flow cytometer machine (BD Biosciences or Roche), bench top centrifuge; 10-, 20-, 200-, and 1000-µl pipettes.
- *Antibodies*: Fluorochrome-conjugated antibody and isotype control.

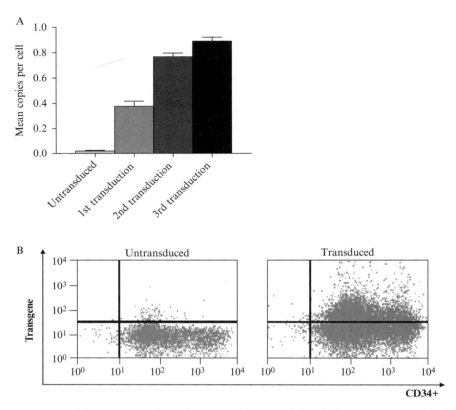

Figure 3.4 Measurement of transduction efficiency. (A) Graph showing an example of the relative increases in the mean copy number per cell that can be obtained after one, two, and three successive rounds of transduction with a gammaretroviral vector. (B) Analysis of transgene expression by FACS in untransduced and transduced $CD34^+$ hematopoietic stem cells. (See Color Insert.)

- *Other reagents*: Cell wash buffer (PBS with 5% FCS); fixation buffer (4% paraformaldehyde in PBS).
- *Disposables*: 2-, 10-, 20-, 200-, and 1000 µl filter tips, FACS tubes.

4.3.1.1. Cell surface staining and cell preparation Add 5–10 µl of fluorochrome-conjugated antibody or isotype control to two FACS tubes. Add at least 100,000 untransduced cells to the tubes containing the antibody or the isotype control and do the same for the transduced cells. Mix the cells and the antibodies by gently tapping the tubes. If analyzing live cells or for detection of transgene expressed protein where the levels of expression are known to be low, incubate the tubes on ice for 40 min in the dark. Otherwise, incubate the flow cytometry tubes in the dark at room temperature for 5 min. Wash the cells by using wash buffer and centrifuge at $400 \times g$ for 5 min at room temperature. Remove the wash buffer using a pipette or pour off; being careful not to disturb the cell pellet, repeat twice more. If analyzing live cells, resuspend the cell pellet in 250 µl wash buffer and analyze immediately, otherwise resuspend the cells in 250 µl fixative and analyze within 24 h.

4.3.1.2. Flow cytometry analysis Analyze your stained cells by looking at a dot plot of the forward (size) on the *x* axis versus side scatter (shape) on the *y* axis and then the detection of the fluorochrome stained cells using a histogram. Adjust the voltages so that your cell population can easily be seen. Run your isotype control and your untransduced control followed by your transduced sample. To calculate the transduction efficiency, overlay your transduced histogram over your isotype or untransduced control histograms and calculate the percentage expression above the background using a histogram gate. Alternatively, look at the side scatter or other fluorescent marker signal on one axis and the transduced fluorescent signal on the other axis; set a gate to detect positive signal being anything above the background, for example, above 10^1 or 10^2. Use appropriate gates to calculate the % positive cells above the background (Fig. 3.4B).

REFERENCES

Adam, M. A., Ramesh, N., Miller, A. D., and Osborne, W. R. (1991). Internal initiation of translation in retroviral vectors carrying picornavirus 5′ nontranslated regions. *J. Virol.* **65**, 4985–4990.

Bestor, T. H. (2000). Gene silencing as a threat to the success of gene therapy. *J. Clin. Invest.* **105**, 409–411.

Bierhuizen, M. F., Westerman, Y., Visser, T. P., Dimjati, W., Wognum, A. W., and Wagemaker, G. (1997). Enhanced green fluorescent protein as selectable marker of retroviral-mediated gene transfer in immature hematopoietic bone marrow cells. *Blood* **90**, 3304–3315.

Brenner, S., and Malech, H. L. (2003). Current developments in the design of oncoretrovirus and lentivirus vector systems for hematopoietic cell gene therapy. *Biochim. Biophys. Acta* **1640**, 1–24.

Brenner, S., Whiting-Theobald, N. L., Linton, G. F., Holmes, K. L., Anderson-Cohen, M., Kelly, P. F., Vanin, E. F., Pilon, A. M., Bodine, D. M., Horwitz, M. E., and Malech, H. L. (2003). Concentrated RD114-pseudotyped MFGS-gp91phox vector achieves high levels of functional correction of the chronic granulomatous disease oxidase defect in NOD/SCID/beta-microglobulin−/− repopulating mobilized human peripheral blood CD34$^+$ cells. *Blood* **102**, 2789–2797.

Brown, B. D., and Naldini, L. (2009). Exploiting and antagonizing microRNA regulation for therapeutic and experimental applications. *Nat. Rev. Genet.* **10**, 578–585.

Bukrinsky, M. I., Haggerty, S., Dempsey, M. P., Sharova, N., Adzhubel, A., Spitz, L., Lewis, P., Goldfarb, D., Emerman, M., and Stevenson, M. (1993). A nuclear localization signal within HIV-1 matrix protein that governs infection of non-dividing cells. *Nature* **365,** 666–669.

Case, S. S., Price, M. A., Jordan, C. T., Yu, X. J., Wang, L., Bauer, G., Haas, D. L., Xu, D., Stripecke, R., Naldini, L., Kohn, D. B., and Crooks, G. M. (1999). Stable transduction of quiescent CD34(+)CD38(−) human hematopoietic cells by HIV-1-based lentiviral vectors. *Proc. Natl. Acad. Sci. USA* **96,** 2988–2993.

Charrier, S., Dupre, L., Scaramuzza, S., Jeanson-Leh, L., Blundell, M. P., Danos, O., Cattaneo, F., Aiuti, A., Eckenberg, R., Thrasher, A. J., Roncarolo, M. G., and Galy, A. (2007). Lentiviral vectors targeting WASp expression to hematopoietic cells, efficiently transduce and correct cells from WAS patients. *Gene Ther.* **14,** 415–428.

Cosset, F. L., Takeuchi, Y., Battini, J. L., Weiss, R. A., and Collins, M. K. (1995). High-titer packaging cells producing recombinant retroviruses resistant to human serum. *J. Virol.* **69,** 7430–7436.

de Noronha, C. M., Sherman, M. P., Lin, H. W., Cavrois, M. V., Moir, R. D., Goldman, R. D., and Greene, W. C. (2001). Dynamic disruptions in nuclear envelope architecture and integrity induced by HIV-1 Vpr. *Science* **294,** 1105–1108.

de, F. P., Hughes, L. E., Ryan, M. D., and Brown, J. D. (2003). Co-translational, intraribosomal cleavage of polypeptides by the foot-and-mouth disease virus 2A peptide. *J. Biol. Chem.* **278,** 11441–11448.

De, P. M., Montini, E., Santoni de Sio, F. R., Benedicenti, F., Gentile, A., Medico, E., and Naldini, L. (2005). Promoter trapping reveals significant differences in integration site selection between MLV and HIV vectors in primary hematopoietic cells. *Blood* **105,** 2307–2315.

Dezzutti, C. S., Heneine, W., Boneva, R. S., and Folks, T. M. (2007). Retroviruses and associated human diseases. In "Topley & Wilson's Microbiology and Microbial Infections: Virology," (B. W. Mahy, ed.)10th edn. , pp. 1284–1303. Hodder Arnold, London.

Donnelly, M. L., Luke, G., Mehrotra, A., Li, X., Hughes, L. E., Gani, D., and Ryan, M. D. (2001). Analysis of the aphthovirus 2A/2B polyprotein 'cleavage' mechanism indicates not a proteolytic reaction, but a novel translational effect: A putative ribosomal 'skip'. *J. Gen. Virol.* **82,** 1013–1025.

Dull, T., Zufferey, R., Kelly, M., Mandel, R. J., Nguyen, M., Trono, D., and Naldini, L. (1998). A third-generation lentivirus vector with a conditional packaging system. *J. Virol.* **72,** 8463–8471.

Ellis, J. (2005). Silencing and variegation of gammaretrovirus and lentivirus vectors. *Hum. Gene Ther.* **16,** 1241–1246.

Flasshove, M., Bardenheuer, W., Schneider, A., Hirsch, G., Bach, P., Bury, C., Moritz, T., Seeber, S., and Opalka, B. (2000). Type and position of promoter elements in retroviral vectors have substantial effects on the expression level of an enhanced green fluorescent protein reporter gene. *J. Cancer Res. Clin. Oncol.* **126,** 391–399.

Follenzi, A., Ailles, L. E., Bakovic, S., Geuna, M., and Naldini, L. (2000). Gene transfer by lentiviral vectors is limited by nuclear translocation and rescued by HIV-1 pol sequences. *Nat. Genet.* **25,** 217–222.

Gallardo, H. F., Tan, C., Ory, D., and Sadelain, M. (1997). Recombinant retroviruses pseudotyped with the vesicular stomatitis virus G glycoprotein mediate both stable gene transfer and pseudotransduction in human peripheral blood lymphocytes. *Blood* **90,** 952–957.

Gallay, P., Swingler, S., Song, J., Bushman, F., and Trono, D. (1995). HIV nuclear import is governed by the phosphotyrosine-mediated binding of matrix to the core domain of integrase. *Cell* **83,** 569–576.

Gallay, P., Hope, T., Chin, D., and Trono, D. (1997). HIV-1 infection of nondividing cells through the recognition of integrase by the importin/karyopherin pathway. *Proc. Natl. Acad. Sci. USA* **94,** 9825–9830.

Gaszner, M., and Felsenfeld, G. (2006). Insulators: Exploiting transcriptional and epigenetic mechanisms. *Nat. Rev. Genet.* **7,** 703–713.

Graham, F. L., Smiley, J., Russell, W. C., and Nairn, R. (1977). Characteristics of a human cell line transformed by DNA from human adenovirus type 5. *J. Gen. Virol.* **36,** 59–74.

Hacein-Bey-Abina, S., Garrigue, A., Wang, G. P., Soulier, J., Lim, A., Morillon, E., Clappier, E., Caccavelli, L., Delabesse, E., Beldjord, K., Asnafi, V., MacIntyre, E., et al. (2008). Insertional oncogenesis in 4 patients after retrovirus-mediated gene therapy of SCID-X1. *J. Clin. Invest.* **118,** 3132–3142.

Heinzinger, N. K., Bukinsky, M. I., Haggerty, S. A., Ragland, A. M., Kewalramani, V., Lee, M. A., Gendelman, H. E., Ratner, L., Stevenson, M., and Emerman, M. (1994). The Vpr protein of human immunodeficiency virus type 1 influences nuclear localization of viral nucleic acids in nondividing host cells. *Proc. Natl. Acad. Sci. USA* **91,** 7311–7315.

Howe, S. J., Mansour, M. R., Schwarzwaelder, K., Bartholomae, C., Hubank, M., Kempski, H., Brugman, M. H., Pike-Overzet, K., Chatters, S. J., de, R. D., Gilmour, K. C., Adams, S., et al. (2008). Insertional mutagenesis combined with acquired somatic mutations causes leukemogenesis following gene therapy of SCID-X1 patients. *J. Clin. Invest.* **118,** 3143–3150.

Ikeda, Y., Takeuchi, Y., Martin, F., Cosset, F. L., Mitrophanous, K., and Collins, M. (2003). Continuous high-titer HIV-1 vector production. *Nat. Biotechnol.* **21,** 569–572.

Jakobsson, J., Rosenqvist, N., Thompson, L., Barraud, P., and Lundberg, C. (2004). Dynamics of transgene expression in a neural stem cell line transduced with lentiviral vectors incorporating the cHS4 insulator. *Exp. Cell Res.* **298,** 611–623.

Kiem, H. P., Heyward, S., Winkler, A., Potter, J., Allen, J. M., Miller, A. D., and Andrews, R. G. (1997). Gene transfer into marrow repopulating cells: Comparison between amphotropic and gibbon ape leukemia virus pseudotyped retroviral vectors in a competitive repopulation assay in baboons. *Blood* **90,** 4638–4645.

Klug, C. A., Cheshier, S., and Weissman, I. L. (2000). Inactivation of a GFP retrovirus occurs at multiple levels in long-term repopulating stem cells and their differentiated progeny. *Blood* **96,** 894–901.

Lewis, P. F., and Emerman, M. (1994). Passage through mitosis is required for oncoretroviruses but not for the human immunodeficiency virus. *J. Virol.* **68,** 510–516.

Luciw, P. A. (1996). Human immunodeficiency viruses and their replication. In "Field's Virology," (D. M. Knipe and P. M. Howely, eds.), pp. 1881–1952. Raven Press, New York.

Merten, O. W., Charrier, S., Laroudie, N., Fauchille, S., Dugue, C., Jenny, C., Audit, M., Zanta-Boussif, M. A., Chautard, H., Radrizzani, M., Vallanti, G., Naldini, L., et al. (2011). Large-scale manufacture and characterization of a lentiviral vector produced for clinical *ex vivo* gene therapy application. *Hum. Gene Ther.* **22,** 343–356.

Miller, A. D. (1997). Development and applications of retroviral vectors. In "Retroviruses," (J. M. Coffin, S. H. Hughes, and H. Varmus, eds.), pp. 437–473. 2nd edn. Cold Spring Harbor Laboratory, New York.

Miller, A. D., and Rosman, G. J. (1989). Improved retroviral vectors for gene transfer and expression. *Biotechniques* **7,** 980–986.

Miyoshi, H., Blomer, U., Takahashi, M., Gage, F. H., and Verma, I. M. (1998). Development of a self-inactivating lentivirus vector. *J. Virol.* **72,** 8150–8157.

Mizuguchi, H., Xu, Z., Ishii-Watabe, A., Uchida, E., and Hayakawa, T. (2000). IRES-dependent second gene expression is significantly lower than cap-dependent first gene expression in a bicistronic vector. *Mol. Ther.* **1,** 376–382.

Naldini, L. (2011). Ex vivo gene transfer and correction for cell-based therapies. *Nat. Rev. Genet.* **12**, 301–315.

Ott, M. G., Schmidt, M., Schwarzwaelder, K., Stein, S., Siler, U., Koehl, U., Glimm, H., Kuhlcke, K., Schilz, A., Kunkel, H., Naundorf, S., Brinkmann, A., et al. (2006). Correction of X-linked chronic granulomatous disease by gene therapy, augmented by insertional activation of MDS1-EVI1, PRDM16 or SETBP1. *Nat. Med.* **12**, 401–409.

Palu, G., Parolin, C., Takeuchi, Y., and Pizzato, M. (2000). Progress with retroviral gene vectors. *Rev. Med. Virol.* **10**, 185–202.

Pannell, D., Osborne, C. S., Yao, S., Sukonnik, T., Pasceri, P., Karaiskakis, A., Okano, M., Li, E., Lipshitz, H. D., and Ellis, J. (2000). Retrovirus vector silencing is de novo methylase independent and marked by a repressive histone code. *EMBO J.* **19**, 5884–5894.

Perez-Simon, J. A., Caballero, M. D., Corral, M., Nieto, M. J., Orfao, A., Vazquez, L., Amigo, M. L., Berges, C., Gonzalez, M., Del, C. C., and San Miguel, J. F. (1998). Minimal number of circulating $CD34^+$ cells to ensure successful leukapheresis and engraftment in autologous peripheral blood progenitor cell transplantation. *Transfusion* **38**, 385–391.

Relander, T., Johansson, M., Olsson, K., Ikeda, Y., Takeuchi, Y., Collins, M., and Richter, J. (2005). Gene transfer to repopulating human $CD34^+$ cells using amphotropic-, GALV-, or RD114-pseudotyped HIV-1-based vectors from stable producer cells. *Mol. Ther.* **11**, 452–459.

Roe, T., Reynolds, T. C., Yu, G., and Brown, P. O. (1993). Integration of murine leukemia virus DNA depends on mitosis. *EMBO J.* **12**, 2099–2108.

Sandrin, V., Boson, B., Salmon, P., Gay, W., Negre, D., Le, G. R., Trono, D., and Cosset, F. L. (2002). Lentiviral vectors pseudotyped with a modified RD114 envelope glycoprotein show increased stability in sera and augmented transduction of primary lymphocytes and $CD34^+$ cells derived from human and nonhuman primates. *Blood* **100**, 823–832.

Santilli, G., Almarza, E., Brendel, C., Choi, U., Beilin, C., Blundell, M. P., Haria, S., Parsley, K. L., Kinnon, C., Malech, H. L., Bueren, J. A., Grez, M., et al. (2011). Biochemical correction of X-CGD by a novel chimeric promoter regulating high levels of transgene expression in myeloid cells. *Mol. Ther.* **19**, 122–132.

Schambach, A., Wodrich, H., Hildinger, M., Bohne, J., Krausslich, H. G., and Baum, C. (2000). Context dependence of different modules for posttranscriptional enhancement of gene expression from retroviral vectors. *Mol. Ther.* **2**, 435–445.

Schwella, N., Siegert, W., Beyer, J., Rick, O., Zingsem, J., Eckstein, R., Serke, S., and Huhn, D. (1995). Autografting with blood progenitor cells: Predictive value of pre-apheresis blood cell counts on progenitor cell harvest and correlation of the reinfused cell dose with hematopoietic reconstitution. *Ann. Hematol.* **71**, 227–234.

Shapiro, H. M. (1988). Practical Flow Cytometry. John Wiley & Sons, New York.

Sutton, R. E., Reitsma, M. J., Uchida, N., and Brown, P. O. (1999). Transduction of human progenitor hematopoietic stem cells by human immunodeficiency virus type 1-based vectors is cell cycle dependent. *J. Virol.* **73**, 3649–3660.

Thornhill, S. I., Schambach, A., Howe, S. J., Ulaganathan, M., Grassman, E., Williams, D., Schiedlmeier, B., Sebire, N. J., Gaspar, H. B., Kinnon, C., Baum, C., and Thrasher, A. J. (2008). Self-inactivating gammaretroviral vectors for gene therapy of X-linked severe combined immunodeficiency. *Mol. Ther.* **16**, 590–598.

Trono, D. (2000). Lentiviral vectors: Turning a deadly foe into a therapeutic agent. *Gene Ther.* **7**, 20–23.

Urbinati, F., Arumugam, P., Higashimoto, T., Perumbeti, A., Mitts, K., Xia, P., and Malik, P. (2009). Mechanism of reduction in titers from lentivirus vectors carrying large inserts in the 3′LTR. *Mol. Ther.* **17**, 1527–1536.

Vogt, V. M. (1997). Retroviral virons and genomes. *In* "Retroviruses," (J. M. Coffin, S. H. Hughes, and H. Varmus, eds.), .2nd edn. Cold Spring Harbor Laboratory, New York.

Weber, E. L., and Cannon, P. M. (2007). Promoter choice for retroviral vectors: Transcriptional strength versus trans-activation potential. *Hum. Gene Ther.* **18,** 849–860.

Wu, X., Li, Y., Crise, B., and Burgess, S. M. (2003). Transcription start regions in the human genome are favored targets for MLV integration. *Science* **300,** 1749–1751.

Yang, Y., Vanin, E. F., Whitt, M. A., Fornerod, M., Zwart, R., Schneiderman, R. D., Grosveld, G., and Nienhuis, A. W. (1995). Inducible, high-level production of infectious murine leukemia retroviral vector particles pseudotyped with vesicular stomatitis virus G envelope protein. *Hum. Gene Ther.* **6,** 1203–1213.

Zanta-Boussif, M. A., Charrier, S., Brice-Ouzet, A., Martin, S., Opolon, P., Thrasher, A. J., Hope, T. J., and Galy, A. (2009). Validation of a mutated PRE sequence allowing high and sustained transgene expression while abrogating WHV-X protein synthesis: Application to the gene therapy of WAS. *Gene Ther.* **16,** 605–619.

Zennou, V., Petit, C., Guetard, D., Nerhbass, U., Montagnier, L., and Charneau, P. (2000). HIV-1 genome nuclear import is mediated by a central DNA flap. *Cell* **101,** 173–185.

Zhang, F., Thornhill, S. I., Howe, S. J., Ulaganathan, M., Schambach, A., Sinclair, J., Kinnon, C., Gaspar, H. B., Antoniou, M., and Thrasher, A. J. (2007). Lentiviral vectors containing an enhancer-less ubiquitously acting chromatin opening element (UCOE) provide highly reproducible and stable transgene expression in hematopoietic cells. *Blood* **110,** 1448–1457.

Zhang, F., Frost, A. R., Blundell, M. P., Bales, O., Antoniou, M. N., and Thrasher, A. J. (2010). A ubiquitous chromatin opening element (UCOE) confers resistance to DNA methylation-mediated silencing of lentiviral vectors. *Mol. Ther.* **18,** 1640–1649.

Zielske, S. P., and Braun, S. E. (2004). Cytokines: Value-added products in hematopoietic stem cell gene therapy. *Mol. Ther.* **10,** 211–219.

Zufferey, R., Donello, J. E., Trono, D., and Hope, T. J. (1999). Woodchuck hepatitis virus posttranscriptional regulatory element enhances expression of transgenes delivered by retroviral vectors. *J. Virol.* **73,** 2886–2892.

CHAPTER FOUR

Analysis of the Clonal Repertoire of Gene-Corrected Cells in Gene Therapy

Anna Paruzynski, Hanno Glimm, Manfred Schmidt, *and* Christof von Kalle

Contents

1. Introduction	60
2. Analysis of the Clonal Repertoire by Performing nrLAM-PCR	61
2.1. Linear PCR	62
2.2. Concentration with Microcon-50 (optional step)	64
2.3. Magnetic capture	64
2.4. Ligation of a single-stranded oligonucleotide	66
2.5. First exponential PCR	68
2.6. Magnetic capture of the first exponential PCR	69
2.7. Second exponential PCR	71
2.8. Agarose gel electrophoresis	72
3. Preparation of nrLAM-PCR Samples for High-Throughput Sequencing	73
3.1. Purification of the nrLAM-PCR products	73
3.2. Third exponential PCR	75
3.3. Agarose gel electrophoresis	76
3.4. Purification of the third exponential PCR products	77
3.5. High-throughput sequencing by using the Roche 454 platform	78
4. Estimation of the Clonal Contribution by RT-PCR	78
4.1. Standard preparation by exponential PCR	79
4.2. Agarose gel electrophoresis	80
4.3. Purification of the PCR product	81
4.4. Dilution of the standard and the samples for RT-PCR	82
4.5. Absolute quantification—RT-PCR	83
References	85

Abstract

Gene therapy-based clinical phase I/II studies using integrating retroviral vectors could successfully treat different monogenetic inherited diseases.

Department of Translational Oncology, National Center for Tumor Diseases (NCT) and German Cancer Research Center (DKFZ), Im Neuenheimer Feld 581 and 460, Heidelberg, Germany

However, with increased efficiency of this therapy, severe side effects occurred in various gene therapy trials. In all cases, integration of the vector close to or within a proto-oncogene contributed substantially to the development of the malignancies. Thus, the in-depth analysis of integration site patterns is of high importance to uncover potential clonal outgrowth and to assess the safety of gene transfer vectors and gene therapy protocols. The standard and nonrestrictive linear amplification-mediated PCR (nrLAM-PCR) in combination with high-throughput sequencing exhibits technologies that allow to comprehensively analyze the clonal repertoire of gene-corrected cells and to assess the safety of the used vector system at an early stage on the molecular level. It enables clarifying the biological consequences of the vector system on the fate of the transduced cell. Furthermore, the downstream performance of real-time PCR allows a quantitative estimation of the clonality of individual cells and their clonal progeny. Here, we present a guideline that should allow researchers to perform comprehensive integration site analysis in preclinical and clinical studies.

1. INTRODUCTION

Gene therapy enables the treatment of inherited monogenetic diseases by incorporating therapeutic genes into the cells of a patient (Scollay, 2001). This can lead to the replacement of the defective gene by a functional gene copy. Furthermore, gene therapy also allows directed alterations of the treated cell (Anderson, 1984; Brown *et al.*, 2006; Friedmann, 1992). The effective incorporation of the therapeutic gene and the stable expression is an important requirement to provide long-term efficacy of this kind of treatment. Today, there are several different nonviral and viral gene transfer strategies available that allow the insertion of therapeutic genes into several cell types, tissues, and organs by *ex vivo* and *in vivo* strategies.

Replication-deficient viral vectors are used for the viral transfer of a therapeutic gene to the host cell. Nonintegrating vectors lead to the transient expression of the transgene in dividing cell populations, whereas integrating vectors allow a stable long-term expression of the transgene by the integration into the host genome (Ehrhardt *et al.*, 2008; High, 1999; von Kalle *et al.*, 1999).

By using integrating gammaretroviral and lentiviral vectors, different primary human immunodeficiencies were cured (Aiuti *et al.*, 2002, 2007, 2009; Boztug *et al.*, 2010; Cartier-Lacave *et al.*, 2009; Cavazzana-Calvo *et al.*, 2000; Gaspar *et al.*, 2004, 2006; Hacein-Bey-Abina *et al.*, 2002; Ott *et al.*, 2006). However, the integration of the vector may lead to mutagenic events in the host genome. For instance, vector-mediated activation of proto-oncogenes triggered clonal selection of the affected cells and, together with the accumulation of additional mutations, led to the development of leukemias and myelodysplastic syndromes in patients of three clinical gene therapy trials, respectively (Hacein-Bey-Abina et al., 2008; Howe et al., 2008; Stein et al., 2010). Therefore, the comprehensive analysis of retroviral integration sites to monitor the fate of individual gene-corrected cells and to assess vector biosafety is of high importance.

Ligation-mediated PCR (LM-PCR, Mueller and Wold, 1989; Pfeifer et al., 1989), inverse PCR (Silver and Keerikatte, 1989), and linear amplification-mediated PCR (LAM-PCR, Schmidt et al., 2001, 2007, 2009) are methods that allow the detection of integration sites by amplification of the vector–genome junction. Out of those methods, LAM-PCR is the most sensitive one that enables the analysis of integration sites down to the single cell level (Schmidt et al., 2001, 2007, 2009). For all methods, restriction enzymes are used to fragment the DNA. As the competitive amplification in subsequent PCRs is much more efficient for shorter DNA fragments, the use of restriction enzymes can lead to an unbalanced description of the clonal integration repertoire from a transduced sample (Gabriel et al., 2009; Harkey et al., 2007; Wang et al., 2008). Therefore, we have recently established a nonrestrictive LAM-PCR (nrLAM-PCR) (Gabriel et al., 2009; Paruzynski et al., 2010). This method allows to circumvent the use of a restriction digest. Here, we provide a detailed protocol for the analysis of integration sites by performing nrLAM-PCR in combination with high-throughput sequencing from preclinical and clinical samples. Furthermore, we also describe the quantitative estimation of individual clonal contributions by clone specific real-time (RT) PCR.

2. Analysis of the Clonal Repertoire by Performing nrLAM-PCR

nrLAM-PCR allows the comprehensive analysis of the integration repertoire without using restriction enzymes for the fragmentation of the DNA as it is common for other methods. Thus, this analysis is not biased due to the differences in the amplification efficiency of different sized fragments (Gabriel et al., 2009; Paruzynski et al., 2010). In brief, the nrLAM-PCR starts with a linear amplification of the vector–genome junctions by using vector-specific primers. After enrichment of the linear amplicons by magnetic capture, a single-stranded oligonucleotide is directly ligated to the unknown genomic sequence part of the amplicons. This allows subsequent exponential amplification by nested PCR. The nrLAM-PCR is not as sensitive as the standard LAM-PCR due to the lower efficiency of the single-stranded ligation but reaches the sensitivity of other conventional methods (LM-PCR, inverse PCR).

In the following, we will exemplarily describe integration site analysis via nrLAM-PCR of gammaretrovirally transduced cells (MLV-based vector).

2.1. Linear PCR

2.1.1. Required materials

Reagents

Taq DNA Polymerase (Genaxxon Bioscience)
10× PCR buffer (Qiagen)
dNTPs (Genaxxon Bioscience)
Aqua ad injectabilia (B.Braun)
Human genomic DNA (negative control, Roche Diagnostics)
Primers 5′-biotinylated (MWG Biotech)

Disposables

0.5 and 1.5 ml tubes (Eppendorf)
Pipette tips (Starlab GmbH)

Equipment

Vortexer (IKA Labortechnik)
Picofuge (Stratagene)
Pipettes (Eppendorf)
PCR cycler (Biometra)

2.1.2. Procedure

The first step of the nrLAM-PCR is a linear amplification of the vector–genome junctions. The primers for this PCR are biotinylated at the 5′-end to allow further enrichment of the amplicons by using paramagnetic streptavidin beads (steptavidin–biotin binding). The sequences of the primers are shown in Table 4.1. nrLAM-PCR can be performed starting from the 5′- or 3′-LTR of the vector (see Note [1]).

1. For a 50 µl PCR, use between 100 and 1000 ng of DNA, 5 µl 10× buffer, 1 µl dNTPs (0.5 mM), 0.25 µl primer 1 and primer 2 (0.17 pmol/µl), and 0.5 µl Taq polymerase. Fill up the reaction to 50 µl with ddH$_2$O. Use untransduced human genomic DNA and an additional ddH$_2$O as negative controls for the PCR. Always use the same amount of untransduced DNA and transduced sample (e.g., if 1000 ng of the sample is used, also use 1000 ng of untransduced DNA as negative control) (see Note [2]).

---

[1] Note 1. Primers should bind stringently to the vector part; avoid any unspecific binding to the host genome. Primers should be located close to the end of the vector LTR (≤120 bp) to ensure the amplification of the vector–genome junctions. If the fragments are too long, amplification will be inefficient.
[2] Note 2. Use 100 ng DNA from a mono—oligoclonal sample (100% transduced cells) and 1000 ng from a polyclonal sample (100% transduced cells).

Table 4.1 Primers for the nonrestrictive linear amplification-mediated PCR (nrLAM-PCR)

Viral vector region	Vector	Linear PCR primers (5′–3′)	First Exp. PCR primers (5′–3′)	Second Exp. PCR primers (5′–3′)
MLV 5′-LTR		(B)ATCCTGTTTGGCCCATATTC (B)TGCTTACCACAGATATCCTG	(B)GACCTTGATCTGAACTTCTC (B)GACCCGGGAGATCTGAATTC	TTCCATGCCTTGCAAAATGGC GATCTGAATTCAGTGGCACAG
MLV 3′-LTR		(B)TCCGATTGACTGCGTCGC (B)GGTACCCGTATTCCCAATA	(B)TCTTGCTGTTTGCATCCGAA (B)GACCCGGGAGATCTGAATTC	GTGGACTCGCTGATCCTT GATCTGAATTCAGTGGCACAG

Exp., exponential; LTR, long terminal repeat; MLV, murine leukemia virus. The primers for the linear and the first exponential PCR are biotinylated at the 5′-end (**B**).

2. Incubate the reaction for 50 cycles in the thermocyler using the following program: initial denaturation: 2 min at 95°C; denaturation: 45 s at 95°C; annealing: 45 s at 58°C; elongation: 5 s at 72°C; no final elongation step.
3. Add 0.5 μl of Taq polymerase and incubate the reaction for another 50 cycles using the program described above (see Note [3]).

2.2. Concentration with Microcon-50 (optional step)

2.2.1. Required materials

Reagents

Microcon-50 Kit (Millipore)
Aqua ad injectabilia (B.Braun)

Disposables

1.5 ml tubes (Eppendorf)
Pipette tips (Starlab GmbH)

Equipment

Pipettes (Eppendorf)
Centrifuge (Eppendorf)

2.2.2. Procedure

This optional step can be performed to get rid of short PCR fragments. However, each additional step can increase the risk of contaminations.

1. Add 450 μl of ddH$_2$O to the reaction (total volume of 500 μl).
2. Transfer the reaction to a Microcon-50 column.
3. Centrifuge for 12 min at 15,000 × g, room temperature.
4. Reverse the column and place it into a fresh 1.5 ml tube.
5. Centrifuge for 3 min at 1,000 × g, room temperature (see Note [4]).
6. Adjust the volume to 50 μl with ddH$_2$O.

2.3. Magnetic capture

2.3.1. Required materials

Reagents

Phosphate-buffered saline (PBS, pH 7.4, Gibco)
Bovine serum albumin (BSA, Sigma)
Lithium chloride (Sigma)

---

[3] Note 3. Products can be stored at −20°C.
[4] Note 4. If there is no eluate after centrifugation, add 10 μl of ddH$_2$O on the column and centrifuge again.

Dynabeads M-280 Streptavidin (Invitrogen)
Aqua ad injectabilia (B.Braun)

Disposables

1.5 ml tubes (Eppendorf)
96-well plates (Greiner bio-one)
Pipette tips (Starlab GmbH)

Equipment

Pipettes (Eppendorf)
Magnetic separation units (MSUs) for 1.5 ml tubes (Invitrogen)
MSUs for 96-well plates (Invitrogen)
Horizontal shaker (IKA Labortechnik)

2.3.2. Procedure

The magnetic capture allows enriching the amplified linear products by biotin–streptavidin binding, while reagents and unspecific products are extracted.

Before starting, prepare the following solutions:

PBS/0.1% BSA: Dissolve 1g of BSA in 1l PBS.
> Solution can be stored at room temperature for up to 3 months.

3M/6M lithium chloride (LiCl) solution: Dissolve 6.36g (3M) or 12.72g (6M) of LiCl in 0.5ml 1M Tris–HCl (pH 7.5) and 0.1ml 0.5M EDTA (pH 8.0) and adjust the volume with ddH$_2$O to 50ml. Filtrate the solution using a 0.45-µm filter.
> Solutions can be stored at room temperature for several months (see Note [5]).

1. Use 20µl of magnetic bead solution (200µg magnetic beads) per reaction.
2. Expose the magnetic beads to a MSU until the solution is clear.
3. Discard the supernatant and place the beads containing tube into a normal rack.
4. Resuspend the magnetic beads in 40µl PBS/0.1% BSA.
5. Expose the magnetic bead solution to the MSU until the solution is clear.
6. Discard the supernatant and place the beads containing tube into a normal rack.

---

[5] Note 5. Alternatively to lithium chloride solution, binding solution can be used that is provided by the manufacturer.

7. Repeat steps 4–6.
8. Resuspend the magnetic beads in 20 μl 3 M lithium chloride solution.
9. Expose the magnetic bead solution to the MSU until the solution is clear.
10. Discard the supernatant and place the beads containing tube into a normal rack.
11. Resuspend the magnetic beads in 50 μl 6 M lithium chloride solution.
12. Add 50 μl of prepared magnetic beads to the linear PCR product (also 50 μl) (see Note [6]).
13. Incubate the reaction on a horizontal shaker at 300 rpm for at least 2 h, room temperature (see Note [7]).

In all the following steps, the amplified DNA fragments are bound to the beads by streptavidin–biotin binding→DNA–beads complex!

14. Transfer the reaction into a 96-well plate.
15. Expose the DNA–beads complex to the MSU until the solution is clear.
16. Discard the supernatant and place the DNA–beads complex containing tube into a normal rack.
17. Resuspend the DNA–beads complex in 100 μl of ddH$_2$O.

2.4. Ligation of a single-stranded oligonucleotide

2.4.1. Required materials

Reagents

Aqua ad injectabilia (B.Braun)
10× ligation buffer (New England Biolabs)
Single-stranded oligonucleotide (MWG Biotech)
Phosphoethyleneglycol (PEG) 8000 (Genaxxon)
Hexammine-cobalt (III) chloride (Fluka)
T4 RNA ligase 1 (New England Biolabs)

Disposables

1.5 ml tubes (Eppendorf)
Pipette tips (Starlab GmbH)

Equipment

Pipettes (Eppendorf)
MSUs for 96-well plates (Invitrogen)
Horizontal shaker (IKA Labortechnik)

---

[6] Note 6. The ratio between magnetic beads and PCR product should always be 1:1.
[7] Note 7. The Magnetic Capture reaction step can be prolonged to 48 h.

2.4.2. Procedure

This step leads to the ligation of a single-stranded oligonucleotide to the unknown part of the DNA amplicons. As both ends after the ligation consist of known sequences, a subsequent exponential amplification of the PCR products is possible. The sequence of the oligonucleotide and the modifications are shown in Table 4.2.

Before starting, prepare the following solutions:

50% PEG 8000: Dissolve 50 g of PEG 8000 in 50 ml ddH$_2$O and adjust the volume to 100 ml.
   Solution can be stored for several months at 4 °C.

Hexammine-cobalt (III) chloride (0.1 M): Dissolve 2.68 g of Hexammine-cobalt (III) chloride in 80 ml ddH$_2$O. Adjust the volume to 100 ml with ddH$_2$O. Filtrate the solution using a 0.45-μm filter.
   Solution can be stored for several months at 4 °C.

1. Set up a 10 μl reaction with 1 μl 10× ligation buffer, 1 μl single-stranded oligonucleotide (10 pmol/μl), 5 μl PEG 8000 (50%), 1 μl Hexammine-cobalt (III) chloride (10 mM), 1 μl T4 RNA ligase 1, and 1 μl ddH$_2$O.
2. Expose the DNA–beads complex (*Magnetic Capture, step 17*) to the MSU until the solution is clear.
3. Discard the supernatant and place the DNA–beads complex containing tube into a normal rack.
4. Resuspend the DNA–beads complex in 10 μl of prepared ligation reaction (*step 1*) (see Note [8]).
5. Incubate the reaction for at least 16 h and not longer than 24 h at 300 rpm on a horizontal shaker, room temperature.
6. Add 90 μl of ddH$_2$O to the reaction to adjust the volume to 100 μl.
7. Expose the DNA–beads complex to the MSU until the solution is clear.
8. Discard the supernatant and place the DNA–beads complex containing tube into a normal rack.

Table 4.2 Single-stranded oligonucleotide for ligation

Sequence (5′–3′)	Modifications
CCTAACTGCTGTGCCACTGAATT CAGATCTCCCGGGTC	Phosphate modification at the 5′-end Didesoxynucleotide modification at the 3′-end

The phosphate modification is needed for the ligation and the didesoxynucleotid modification ensures the ligation of a single linker.

---

[8] Note 8. 50% PEG 8000 is very viscous. Take care while pipetting.

9. To wash the DNA–beads complex, resuspend it in 100 µl ddH$_2$O.
10. Expose the DNA–beads complex to the MSU until the solution is clear.
11. Discard the supernatant and place the DNA–beads complex containing tube into a normal rack.
12. Resuspend the DNA–beads complex in 10 µl ddH$_2$O and transfer it to a fresh 0.5 ml tube (see Note 3).

2.5. First exponential PCR

2.5.1. Required materials

Reagents

Taq DNA Polymerase (Genaxxon Bioscience)
10× PCR buffer (Qiagen)
dNTPs (Genaxxon Bioscience)
Aqua ad injectabilia (B.Braun)
Primers 5′-biotinylated (MWG Biotech)

Disposables

0.5 and 1.5 ml tubes (Eppendorf)
Pipette tips (Starlab GmbH)

Equipment

Vortexer (IKA Labortechnik)
Picofuge (Stratagene)
Pipettes (Eppendorf)
PCR cycler (Biometra)

2.5.2. Procedure

The known sequences on both sides of the DNA fragments allow for further exponential amplification. For the first exponential PCR, the LTR-specific primer is biotinylated to enable the enrichment of the amplified products via magnetic capture. The sequences of the primers are shown in Table 4.1.

1. Use 5 µl 10× buffer, 1 µl dNTPs (10 mM), 0.5 µl of primer 1 and primer 2 (16.7 pmol/µl), 0.5 µl Taq polymerase, 2 µl of the template DNA (*ligation, step 12*) and adjust the volume to 50 µl with ddH$_2$O. Use an additional ddH$_2$O as negative control for the first exponential PCR (in total three negative controls).
2. Incubate the reaction for 10 cycles in the thermocycler using the following program: initial denaturation: 2 min at 95 °C; denaturation: 45 s at

95°C; annealing: 45s at 58°C; elongation: 5s at 72°C. Afterwards, incubate the reaction for another 25 cycles in the thermocyler using the following program: initial denaturation: 2min at 90°C; denaturation: 45s at 90°C; annealing: 45s at 58°C; elongation: 5s at 72°C; no final elongation (see Note 3).

2.6. Magnetic capture of the first exponential PCR
2.6.1. Required materials

Reagents

Phosphate-buffered saline (PBS, pH 7.4, Gibco)
Bovine serum albumin (BSA, Sigma)
Lithium chloride (Sigma)
Dynabeads M-280 Streptavidin (Invitrogen)
Aqua ad injectabilia (B.Braun)
Sodium hydroxide solution ($1\,M$ NaOH, Fluka)

Disposables

1.5 ml tubes (Eppendorf)
96-well plates (Greiner bio-one)
Pipette tips (Starlab GmbH)

Equipment

Pipettes (Eppendorf)
MSUs for 1.5 ml tubes (Invitrogen)
MSUs for 96-well plates (Invitrogen)
Horizontal shaker (IKA Labortechnik)

2.6.2. Procedure

The procedure for the magnetic capture of the first exponential PCR product is the same as for the linear PCR product (Section 2.3). Again, this step allows for enrichment of the amplified products. Instead of 50 µl, 20 µl of the first exponential PCR product is used for the magnetic capture. After magnetic capture, the nonbiotinylated strands are separated from the biotinylated strands bound to the beads by denaturation with 0.1 N NaOH. The separated single strands are used for a second exponential PCR.

Before starting, prepare the following solutions:

PBS/0.1% BSA: Dissolve 1 g of BSA in 1 l PBS.
    Solution can be stored at room temperature for up to 3 months.

3M/6M lithium chloride (LiCl) solution: Dissolve 6.36 g (3M) or 12.72 g (6M) of LiCl in 0.5 ml 1M Tris–HCl (pH 7.5) and 0.1 ml 0.5M EDTA (pH 8.0) and adjust the volume with ddH$_2$O to 50 ml. Filtrate the solution using a 0.45-μm filter.

Solutions can be stored for several months at room temperature (see Note 5).

0.1 N NaOH: Dilute 1M NaOH 1:10 with ddH$_2$O to get a 0.1N NaOH solution (see Note [9]).

1. Use 20 μl magnetic bead solution (200 μg magnetic beads) per reaction.
2. Expose the magnetic beads to a MSU until the solution is clear.
3. Discard the supernatant and place the beads containing tube into a normal rack.
4. Resuspend the magnetic beads in 40 μl PBS/0.1% BSA.
5. Expose the magnetic bead solution to the MSU until the solution is clear.
6. Discard the supernatant and place the beads containing tube into a normal rack.
7. Repeat steps 4–6.
8. Resuspend the magnetic beads in 20 μl 3M lithium chloride solution.
9. Expose the magnetic bead solution to the MSU until the solution is clear.
10. Discard the supernatant and place the beads containing tube into a normal rack.
11. Resuspend the magnetic beads in 20 μl 6M lithium chloride solution.
12. Transfer 20 μl of the first exponential PCR product to a 96-well plate.
13. Add 20 μl of prepared magnetic beads to the first exponential PCR product (see Note 6).
14. Incubate the reaction on a horizontal shaker at 300 rpm for at least 1 h, room temperature (see Note 7).
15. Adjust the volume to 100 μl by adding 60 μl of ddH$_2$O to the reaction.
16. Expose the DNA–beads complex to the MSU until the solution is clear.
17. Discard the supernatant and place the DNA–beads complex containing tube into a normal rack.
18. Resuspend the DNA–beads complex in 100 μl of ddH$_2$O.
19. Expose the DNA–beads complex to the MSU until the solution is clear.
20. Discard the supernatant and place the DNA–beads complex containing tube into a normal rack.

---

[9] Note 9. Use always freshly prepared 0.1 N NaOH for each experiment. Do not store it.

21. Repeat steps 18–19 to wash again the DNA–beads complex.
22. Resuspend the DNA–beads complex in 10 μl 0.1 N NaOH.
23. Incubate the reaction for 10 min at 300 rpm on a horizontal shaker, room temperature.
24. Expose the DNA–beads complex to the MSU until the solution is clear.
25. Transfer the supernatant (contains single-stranded nonbiotinylated strands) to a fresh 0.5 ml tube (see Note 3).
26. Discard the magnetic beads.

2.7. Second exponential PCR

2.7.1. Required materials

Reagents

Taq DNA Polymerase (Genaxxon Bioscience)
10× PCR buffer (Qiagen)
dNTPs (Genaxxon Bioscience)
Aqua ad injectabilia (B.Braun)
Primers (MWG Biotech)

Disposables

0.5 and 1.5 ml tubes (Eppendorf)
Pipette tips (Starlab GmbH)

Equipment

Vortexer (IKA Labortechnik)
Picofuge (Stratagene)
Pipettes (Eppendorf)
PCR cycler (Biometra)

2.7.2. Procedure

The single-stranded nonbiotinylated strands consist of a vector-specific part (LTR), flanking genomic DNA, and the ligated single-stranded oligonucleotide. Thus, the product from the magnetic capture can be used for a second exponential PCR. The sequences of the primers are shown in Table 4.1 (see Note [10]).

1. Use 5 μl 10× buffer, 1 μl dNTPs (10 mM), 0.5 μl of primer 1 and primer 2 (16.7 pmol/μl), 0.5 μl Taq polymerase, 2 μl of the template DNA

---

[10] Note 10. Primers for the second exponential PCR should be located at least 20 bp in distance to the LTR end to guarantee the detection of "true" integration sites.

(*Magnetic Capture, step 25*) and adjust the volume to 50 µl with ddH$_2$O. Use an additional ddH$_2$O as negative control for the second exponential PCR (in total four negative controls) (see Note [11]).
2. Incubate the reaction for 10 cycles in the thermocyler using the following program: initial denaturation: 2 min at 95°C; denaturation: 45 s at 95°C; annealing: 45 s at 58°C; elongation: 5 s at 72°C. Afterwards, incubate the reaction for another 25 cycles in the thermocyler using the following program: initial denaturation: 2 min at 90°C; denaturation: 45 s at 90°C; annealing: 45 s at 58°C; elongation: 5 s at 72°C; final elongation: 2 min at 72°C (see Note 3).

2.8. Agarose gel electrophoresis

2.8.1. Required materials

Reagents

Agarose (Sigma)
Loading buffer (Elchrom Scientific)
100 bp ladder (New England Biolabs)
Ethidium bromide (Applichem)
Tris–Borat–EDTA (TBE) Buffer 10× (Amresco)

Disposables

0.5 ml tubes (Eppendorf)
Pipette tips (Starlab GmbH)

Equipment

Pipettes (Eppendorf)
Gel electrophoresis equipment (Biometra)
Gel documentation system (Peqlab)

2.8.2. Procedure

The visualization of the nrLAM-PCR products on an agarose gel allows monitoring the successful amplification of the integration sites prior to sequencing. For every integration site, fragments of different length will be generated during nrLAM-PCR. Thus, in case of a successful amplification, a smear between 100 and 300 bp should be visible.

Before starting, prepare the following solutions:
TBE buffer 1×: Dilute the 10×solution 1:10 to get a 1× TBE solution.

[11] Note 11. Do not use more than 2 µl of template DNA.

1. Prepare a 2% agarose gel by dissolving 2g agarose in 100ml 1×TBE solution.
2. Heat the agarose–TBE mixture until no cords are visible anymore.
3. Cool down the gel for 5min at room temperature.
4. Add 0.5 μg/ml ethidium bromide (see Note [12]).
5. Insert a comb into a gel electrophoresis sledge and cast the gel.
6. After polymerization of the gel, remove the comb and transfer the sledge into an electrophoresis tank filled with 1× TBE buffer.
7. Mix 10 μl of the nrLAM-PCR product (*after the second exponential PCR*) with 2 μl loading buffer.
8. Load the samples on the prepared 2% agarose gel. For size control, load 5 μl of a 100 bp ladder into the first well of the agarose gel. Also load the negative controls to detect possible contaminations.
9. Connect the gel unit to a power supply and run the gel at 10V/cm electrode gap until the dye front runs 7cm below the upper edge of the gel.
10. Transfer the gel to a gel documentation system and control if the nrLAM-PCR was successful.

3. Preparation of nrLAM-PCR Samples for High-Throughput Sequencing

The nrLAM-PCR samples are further prepared for high-throughput sequencing to yield the precise localization of the integration sites in the host genome. In the following, we will give a guideline how to proceed optimally with the nrLAM-PCR samples to allow subsequent high-throughput sequencing. In brief, 40ng of purified nrLAM-PCR products are used to perform a third exponential PCR. This PCR allows adding the 454 specific amplification and sequencing adaptors on both sides of the nrLAM-PCR amplicons. Another round of purification follows after the third exponential PCR. By incorporating a 6–10 bp barcode into one of the sequencing adaptors, different samples can be pooled for sequencing.

3.1. Purification of the nrLAM-PCR products

3.1.1. Required materials

Reagents

QIAquick® PCR-Purification Kit (Qiagen)
Aqua ad injectabilia (B.Braun)

---

[12] Note 12. Be careful while working with ethidium bromide as it is mutagenic.

Ethanol (VWR)
Guanidine hydrochloride (Sigma)

Disposables

1.5 ml tubes (Eppendorf)
Pipette tips (Starlab GmbH)

Equipment

Centrifuge (Eppendorf)
Pipettes (Eppendorf)
Qubit 2.0 Fluorometer (Invitrogen)

3.1.2. Procedure

The purification of the nrLAM-PCR products removes the PCR reagents (buffer, primers, etc.) from the second exponential PCR. The purified products will be used for the third exponential PCR.

Before starting, prepare the following solutions:

35% guanidine hydrochloride: Dissolve 35 g of guanidine hydrochloride in 100 ml ddH$_2$O (see Note [13]).

1. Mix the PCR products with 5× buffer PB (e.g., 40 µl PCR product + 200 µl buffer PB).
2. Load the mixture on a QIAquick purification column.
3. Centrifuge for 1 min at 15,700×g, room temperature.
4. Discard the flow-through.
5. Load 750 µl 35% guanidine hydrochloride on the column.
6. Centrifuge for 1 min at 15,700×g, room temperature.
7. Discard the flow-through.
8. Repeat steps 5–7.
9. Wash the column with 750 µl PE buffer, centrifuge for 1 min at 15,700×g, room temperature.
10. Centrifuge for 1 min at 15,700×g, room temperature to dry the column.
11. Place the column into a fresh 1.5 ml tube.
12. For elution, load 30 µl of ddH$_2$O on the column.
13. Incubate for 1 min at room temperate and centrifuge for 1 min at 15,700×g, room temperature (see Note 3).
14. Measure the concentration of the purified products with the Qubit Fluorometer.

---

[13] Note 13. Use always freshly prepared 35% guanidine hydrochloride. Do not store it.

Comprehensive Integration Site Analysis

3.2. Third exponential PCR

3.2.1. Required materials

Reagents

Taq DNA Polymerase (Genaxxon Bioscience)
10× PCR buffer (Qiagen)
dNTPs (Genaxxon Bioscience)
Aqua ad injectabilia (B.Braun)
Primers (MWG Biotech)

Disposables

0.5 and 1.5 ml tubes (Eppendorf)
Pipette tips (Starlab GmbH)

Equipment

Vortexer (IKA Labortechnik)
Picofuge (Stratagene)
Pipettes (Eppendorf)
PCR cycler (Biometra)

3.2.2. Procedure

For the third exponential PCR, fusion primers are used that enable the addition of the adaptors needed for 454 pyrosequencing to the nrLAM-PCR products. After purification, the PCR products can be directly used for sequencing. The incorporation of a 6–10 bp barcode into the sequencing adaptor allows the pooling of different samples. This increases the number of samples that can be sequenced in parallel and reduces the sequencing costs enormously. The sequences of the fusion primers are shown in Table 4.3.

Table 4.3 Primers for the third exponential PCR

Viral vector	Vector region	Third Exp. PCR primers (5′–3′)
MLV	5′-LTR	GCCTCCCTCGCGCCATCAG$(N)_{6-10bp}$ CCTTGCAAAATGGCGTTACT GCCTTGCCAGCCCGCTCAGAGT GGCACAGCAGTTAGG
MLV	3′-LTR	GCCTCCCTCGCGCCATCAG$(N)_{6-10bp}$ GTCTCCTCTGAGTGATTGAC GCCTTGCCAGCCCGCTCAGAG TGGCACAGCAGTTAGG

Exp., exponential; LTR, long terminal repeat; MLV, murine leukemia virus; $(N)_{6-10bp}$, 6–10 bp barcode to allow the parallel sequencing of different samples.

1. Use 5 µl 10× buffer, 1 µl dNTPs (10 mM), 0.5 µl of primer 1 and primer 2 (10 pmol/µl), 0.5 µl Taq polymerase, 40 ng of the template DNA (*Purification, step 13*) and adjust the volume to 50 µl with ddH$_2$O. Use an additional ddH$_2$O as negative control for the third exponential PCR.
2. Incubate the reaction for 12 cycles in the thermocyler using the following program: initial denaturation: 2 min at 95 °C; denaturation: 45 s at 95 °C; annealing: 45 s at 58 °C; elongation: 1 min at 72 °C; final elongation: 5 min at 72 °C (see Note 3).

3.3. Agarose gel electrophoresis

3.3.1. Required materials

Reagents

Agarose (Sigma)
Loading buffer (Elchrom Scientific)
100 bp ladder (New England Biolabs)
Ethidium bromide (Applichem)
TBE buffer 10× (Amresco)

Disposables

0.5 ml tubes (Eppendorf)
Pipette tips (Starlab GmbH)

Equipment

Pipettes (Eppendorf)
Gel electrophoresis equipment (Biometra)
Gel documentation system (Peqlab)

3.3.2. Procedure

The visualization of the third exponential PCR products is performed in the same way as for the nrLAM-PCR products (Section 2.8)

Before starting, prepare the following solutions:

TBE buffer 1×: Dilute the 10× solution 1:10 to get a 1× TBE solution.

11. Prepare a 2% agarose gel by dissolving 2 g agarose in 100 ml 1× TBE solution.
12. Heat the agarose–TBE mixture until no cords are visible anymore.
13. Cool down the gel for 5 min at room temperature.
14. Add 0.5 µg/ml ethidium bromide (see Note 12).
15. Insert a comb into a gel electrophoresis sledge and cast the gel.

16. After polymerization of the gel, remove the comb and transfer the sledge into an electrophoresis tank filled with 1× TBE buffer.
17. Mix 10 µl of the third exponential PCR product with 2 µl loading buffer.
18. Load the samples on the prepared 2% agarose gel. For size control, load 5 µl of a 100 bp ladder into the first well of the agarose gel. Also load the negative controls to detect possible contaminations.
19. Connect the gel unit to a power supply and run the gel at 10 V/cm electrode gap until the dye front runs 7 cm below the upper edge of the gel.
20. Transfer the gel to a gel documentation system and control if the third exponential PCR was successful.

3.4. Purification of the third exponential PCR products

3.4.1. Required materials

Reagents

QIAquick® PCR-Purification Kit (Qiagen)
Aqua ad injectabilia (B.Braun)
Ethanol (VWR)
Guanidine hydrochloride (Sigma)

Disposables

1.5 ml tubes (Eppendorf)
Pipette tips (Starlab GmbH)

Equipment

Centrifuge (Eppendorf)
Pipettes (Eppendorf)
Qubit 2.0 Fluorometer (Invitrogen)

3.4.2. Procedure

The purification of the third exponential PCR products comprises the same steps as for the nrLAM-PCR products (Section 3.1). PCR reagents (buffer, primers, etc.) are removed and the purified products can be used directly for 454 pyrosequencing.

Before starting, prepare the following solutions:

35% guanidine hydrochloride: Dissolve 35 g of guanidine hydrochloride in 100 ml ddH$_2$O (see Note 13).

1. Mix the PCR products with 5× buffer PB (e.g., 40 μl PCR product + 200 μl buffer PB).
2. Load the mixture on a QIAquick purification column.
3. Centrifuge for 1 min at 15,700 ×g, room temperature.
4. Discard the flow-through.
5. Load 750 μl 35% guanidine hydrochloride on the column.
6. Centrifuge for 1 min at 15,700 ×g, room temperature.
7. Discard the flow-through.
8. Repeat steps 5–7.
9. Wash the column with 750 μl PE buffer, centrifuge for 1 min at 15,700 ×g, room temperature.
10. Centrifuge for 1 min at 15,700 ×g, room temperature to dry the column.
11. Place the column into a fresh 1.5 ml tube.
12. For elution, load 30 μl of ddH_2O on the column.
13. Incubate for 1 min at room temperate and centrifuge for 1 min at 15,700 ×g, room temperature (see Note 3).
14. Measure the concentration of the purified products with the Qubit Fluorometer.
15. Due to the barcode incorporated into the sequencing adaptor, different samples can now be pooled for subsequent sequencing.

3.5. High-throughput sequencing by using the Roche 454 platform

The combination of nrLAM-PCR with high-throughput sequencing allows a comprehensive integration site analysis of transduced preclinical and clinical samples. With the 454 platform from Roche, more than 1,000,000 reads for a single sequencing run and a sequencing length of 400 bp can be reached. The incorporation of a barcode into the 454 adaptors enables the sequencing of hundreds of samples in parallel and thus decreases time and costs enormously. The samples should be pooled according to the sequencing reads that needs to be reached for every single sample. After pooling, the samples can directly be sequenced and the retrieved sequences can be analyzed by public available bioinformatics tools like Seqmap 2.0 (Hawkins et al., 2011), QuickMap (Appelt et al., 2009), or own developed bioinformatics programs (Paruzynski et al., 2010).

4. ESTIMATION OF THE CLONAL CONTRIBUTION BY RT-PCR

The assessment of the vector safety is of high importance as severe side effects already occurred in clinical gene therapy trials (Hacein-Bey-Abina et al., 2008; Howe et al., 2008; Stein et al., 2010). One important tool is the analysis of the quantitative contribution of single integration sites or cell clones, respectively. In all clinical gene therapy trials with severe side effects, the contribution of one affected cell clone increased over time and led to clonal dominance and even malignant transformation. A possibility to analyze the contribution of single clones is the specific amplification of the integration site by Realtime (RT)-PCR. In the following, we will give a detailed RT-PCR protocol for the assessment of individual clonal contributions by absolute quantification. Two RT-PCRs need to be done. One RT-PCR is used to amplify the integration site of interest and to estimate the copies of this specific clone, and another RT-PCR is measuring the total vector copy number that is present. To calculate the exact contribution of the integration site, the integration site copies have to be adjusted with the total vector copies.

4.1. Standard preparation by exponential PCR

4.1.1. Required materials

Reagents

Taq DNA Polymerase (Genaxxon Bioscience)
10× PCR buffer (Qiagen)
dNTPs (Genaxxon Bioscience)
Aqua ad injectabilia (B.Braun)
Primers (MWG Biotech)

Disposables

0.5 and 1.5 ml tubes (Eppendorf)
Pipette tips (Starlab GmbH)

Equipment

Vortexer (IKA Labortechnik)
Picofuge (Stratagene)
Pipettes (Eppendorf)
PCR cycler (Biometra)

4.1.2. Procedure

By performing an exponential PCR covering the integration site sequence or part of the vector sequence, a specific fragment of a defined size is generated. After purification of this fragment and measurement of the concentration, it can be used as standard for the absolute quantification via RT-PCR.

1. For the exponential PCR, use a transduced sample containing the integration site that needs to be analyzed (e.g., transduced patient sample or transduced cell culture sample that was already sequenced). If the amount of the material is limited (e.g., in case of patient material), an

amplification of the DNA can be performed prior to preparation of the RT-PCR standard (e.g., by using the Genomiphi V2 DNA Amplification Kit from GE Healthcare). Plasmids containing the integration site or part of the vector can directly be used as standard.
2. Two independent exponential PCRs have to be performed. For the amplification of the integration site fragment, one primer has to be located in the LTR of the vector and the other primer in the flanking genomic DNA (standard to estimate the integration site copies). For the amplification of the vector fragment, one primer has to be located in the LTR of the vector and the other one in the backbone of the vector (standard to estimate the total vector copies) (see Note [14]).
3. Use 5 µl 10× buffer, 1 µl dNTPs (10 mM), 0.5 µl of primer 1 and primer 2 (16.7 pmol/µl), 0.5 µl Taq polymerase, 200 ng of the template DNA and adjust the volume to 50 µl with ddH_2O. Use an additional ddH_2O as negative control for each exponential PCR.
4. Incubate the reaction for 35 cycles in the thermocyler using the following program: initial denaturation: 2 min at 95°C; denaturation: 45 s at 95°C; annealing: 45 s at 58°C; elongation: 1 min at 72°C; final elongation: 5 min at 72°C (see Notes 3 and [15]).

4.2. Agarose gel electrophoresis

4.2.1. Required materials

Reagents

Agarose (Sigma)
Loading buffer (Elchrom Scientific)
100 bp ladder (New England Biolabs)
Ethidium bromide (Applichem)
TBE buffer 10× (Amresco)

Disposables

0.5 ml tubes (Eppendorf)
Pipette tips (Starlab GmbH)

Equipment

Pipettes (Eppendorf)
Gel electrophoresis equipment (Biometra)
Gel documentation system (Peqlab)

---

[14] Note 14. Fragment size should be around 300 bp.
[15] Note 15. Annealing temperature may vary.

4.2.2. Procedure

It is very important to control if the PCR was successful before further preparing the standard. The procedure comprises the same steps as in Sections 2.8 and 3.3. On the agarose gel, only one defined fragment should be visible.

Before starting, prepare the following solutions:
TBE buffer 1×: Dilute the 10× solution 1:10 to get a 1× TBE solution.

1. Prepare a 2% agarose gel by dissolving 2g agarose in 100ml 1× TBE solution.
2. Heat the agarose–TBE mixture until no cords are visible anymore.
3. Cool down the gel for 5 min at room temperature.
4. Add 0.5 µg/ml ethidium bromide (see Note 12).
5. Insert a comb into a gel electrophoresis sledge and cast the gel.
6. After polymerization of the gel, remove the comb and transfer the sledge into an electrophoresis tank filled with 1× TBE buffer.
7. Mix 10 µl of the PCR product with 2 µl loading buffer.
8. Load the samples on the prepared 2% agarose gel. For size control, load 5 µl of a 100 bp ladder into the first well of the agarose gel. Also load the negative controls to detect possible contaminations.
9. Connect the gel unit to a power supply and run the gel at 10 V/cm electrode gap until the dye front runs 7 cm below the upper edge of the gel.
10. Transfer the gel to a gel documentation system and control if the PCR was successful.

4.3. Purification of the PCR product

4.3.1. Required materials

Reagents

High Pure PCR Product Purification Kit (Roche)
Ethanol (VWR)
Aqua ad injectabilia (B.Braun)

Disposables

1.5 ml tubes (Eppendorf)
Pipette tips (Starlab GmbH)

Equipment

Centrifuge (Eppendorf)
Pipettes (Eppendorf)
Qubit 2.0 Fluorometer (Invitrogen)
Vortexer (IKA Labortechnik)

4.3.2. Procedure

The purification of the amplified fragment is an important step for the subsequent estimation of the concentration. As for previous purification steps (Sections 3.1 and 3.4), it helps to remove PCR reagents (buffer, polymerase, etc.).

1. Adjust the volume of the reaction to 100 µl with ddH$_2$O.
2. Add 500 µl binding buffer to the reaction and load it on a column.
3. Centrifuge for 1 min at 13.000 rpm, room temperature.
4. Discard the flow-through.
5. Wash the column first with 500 µl and afterwards with 200 µl wash buffer by centrifugation for 1 min at 13,000 rpm, room temperature.
6. Discard the flow-through.
7. Place the column into a fresh 1.5 ml tube.
8. Load 30 µl elution buffer (*not ddH$_2$O!*) on the column.
9. Incubate the column for 1 min at room temperature.
10. Centrifuge for 1 min at 13,000 rpm, room temperature.
11. Vortex the eluted sample and centrifuge again for 1 min at 13,000 rpm, room temperature.
12. Transfer the sample into a fresh 1.5 ml tube (see Note 3).
13. Measure the concentration by using a Qubit Fluorometer.

4.4. Dilution of the standard and the samples for RT-PCR

4.4.1. Required materials

Reagents

MS2RNA (Roche)
ddH$_2$O PCR Grade (Roche)

Disposables

1.5 ml tubes (Eppendorf)
Pipette tips (Starlab GmbH)

Equipment

Picofuge (Stratagene)
Vortexer (IKA Labortechnik)
Pipettes (Eppendorf)

4.4.2. Procedure

The purified standard and the samples for RT-PCR need to have a defined concentration. Thus, dilution steps are necessary prior to RT-PCR.

1. Calculate the exact copy number of the standard by using the following formula:
   1 copy = x bp × 660 Da/bp × 1.66 × 10^{-24} g/Da
   (x = exact length of the fragment)

   Example

   Length: 321 bp
   Concentration 37.77 ng/µl
   321 bp × 660 Da/bp × 1.66 × 10^{-24} g/Da = 3.51688 × 10^{-19} g (1 copy)
   Calculate the copies in 1 ng and use the concentration of the standard to calculate the copies in 1 µl of the standard:
   1 g / 3.51688 × 10^{-19} g = 2.84343 × 10^{18} copies in 1 g
   1 ng / 3.51688 × 10^{-10} ng = 2.84343 × 10^{9} copies in 1 ng
   1 µl = 37.77 ng
   37.77 ng × 2.84343 × 10^{9} copies/ng = 1.07396 × 10^{11} copies in 37.77 ng (= 1 µl)

2. For each standard, mix 60 µl of MS2RNA with 2340 µl of ddH$_2$O (dilution of 1:40).
   Dilute the standard to 2 × 10^{10} copies/µl with MS2RNA/ddH$_2$O
   Prepare a dilution series with the purified standard from 2 × 10^{10}–2 × 10^{-1} copies/µl. Also include a 1 × 10^{0} dilution.
   Dilute the samples to 4 ng/µl with ddH$_2$O (*not* MS2RNA/ddH$_2$O) (see Note [16]).

4.5. Absolute quantification — RT-PCR

4.5.1. Required materials

Reagents

SYBRGREEN I Mix (Roche)
ddH$_2$O PCR Grade (Roche)
Primers (MWG Biotech)

Disposables

1.5 ml tubes (Eppendorf)
96-well plates (Roche)
Sealing foils (Roche)
Pipette tips (Starlab GmbH)

---

[16] Note 16. Always prepare new dilutions for every experiment. Use the PCR Grade ddH$_2$O from Roche.

Equipment

Centrifuge (Eppendorf)
Picofuge (Stratagene)
Vortexer (IKA Labortechnik)
Pipettes (Eppendorf)
LightCycler LC480 System (Roche)
LightCycler LC480 Software (Roche)

4.5.2. Procedure

The absolute quantification by using the LightCycler LC480 system allows a quantitative estimation of the contribution of single integration sites. Therefore, two PCRs need to be performed. One PCR that estimates the integration site copy number and another PCR measuring the total vector copy number.

1. Create nested primers for the RT-PCR based on the primers for the generation of the standard (Section 4.1).
2. Set up a 15 µl reaction for every sample by using 4 µl ddH$_2$O, 0.5 µl primer 1 and primer 2 (10 pmol/µl), and 10 µl SYBR GREEN I Mix (see Note [17]).
3. Pipette 15 µl of reaction mix for every sample into a 96-well plate.
4. Use the following standard dilutions for the RT-PCR: 2×10^5–2×10^{-1} copies/µl and 1×10^0 copies/µl. Use the 4 ng/µl dilution for the samples. Use at least two ddH$_2$O controls, one in the beginning and one in the end of the 96-well plate.
5. Pipette 5 µl of the standard, sample, or ddH$_2$O into the 96-well plate (total volume of the reaction is 20 µl). Standards are analyzed in triplets–octets. Samples are analyzed in quartets. Use the following scheme as template.

	1	2	3	4	5	6	7	8	9	10	11	12
A	H$_2$O	1×10^6				1×10^5				1×10^4		
B		1×10^3						1×10^2				
C				1×10^1						5×10^0		
D		5×10^0				1×10^0						
E		Sample				Sample				Sample		
F		Sample				Sample				Sample		
G		Sample				Sample				Sample		
H		Sample				Sample		H$_2$O				

---

[17] Note 17. Take care that you generate DNA fragments of similar length by both RT-PCRs. Amplified fragments should not be longer than 150 bp. It is recommended to use the same conditions for both RT-PCRs. Use the PCR Grade ddH$_2$O from Roche.

5 μl of standard is equivalent to 1×10^6–1×10^0 and 5×10^0 copies. 5 μl of sample is equivalent to 20 ng.
6. Close the 96-well plate with the sealing foil and centrifuge for 2 min at $1,500\times g$, room temperature.
7. Transfer the 96-well plate to the LightCycler LC480 system and start the RT-PCR.
8. Incubate the reaction for 45 cycles in the LightCycler using the following amplification program: initial denaturation: 15 min at 95 °C; denaturation: 10 s at 95 °C; annealing: 5 s at 58 °C; elongation: 25 s at 72 °C. Melting curve program for 1 cycle: 5 s at 95 °C; 1 min at 65 °C and 97 °C continuous. Cooling for 1 cycle: 30 s at 40 °C.
9. Analyze the data by using the LightCycler LC480 software. Control for the melting curve and calculate the copy numbers for both PCRs. Adjust the copy number from the integration site with the total vector copy number as follows:

Example

Total vector copy number: 5000
Copy number integration site: 1350
5000 = 100%
1350 = x
x = 27% (see Note [18]).

[18] Note 18. Take care that the efficiency of the two RT-PCRs is similar and in the range between 1.8 and 2. The primer design is a very crucial point for RT-PCR.

REFERENCES

Aiuti, A., Slavin, S., Aker, M., Ficara, F., Deola, S., Mortellaro, A., Morecki, S., Andolfi, G., Tabucchi, A., Carlucci, F., et al. (2002). Correction of ADA-SCID by stem cell gene therapy combined with nonmyeloablative conditioning. *Science* **296,** 2410–2413.

Aiuti, A., Cassani, B., Andolfi, G., Mirolo, M., Biasco, L., Recchia, A., Urbinati, F., Valacca, C., Scaramuzza, S., Aker, M., et al. (2007). Multilineage hematopoietic reconstitution without clonal selection in ADA-SCID patients treated with stem cell gene therapy. *J. Clin. Invest.* **117**(8), 2233–2240.

Aiuti, A., Cattaneo, F., Galimberti, S., Benninghoff, U., Cassani, B., Callegaro, L., Scaramuzza, S., Andolfi, G., Mirolo, M., Brigida, I., et al. (2009). Gene therapy for immunodeficiency due to adenosine deaminase deficiency. *N. Engl. J. Med.* **360**(5), 447–458.

Anderson, W. F. (1984). Prospects for human gene therapy. *Science* **226**(4673), 401–409.

Appelt, J. U., Giordano, F. A., Ecker, M., Roeder, I., Grund, N., Hotz-Wagenblatt, A., Opelz, G., Zeller, W. J., Allgayer, H., Fruehauf, S., and Laufs, S. (2009). QuickMap:

A public tool for large-scale gene therapy vector insertion site mapping and analysis. *Gene Ther.* **16,** 885–893.

Boztug, K., Schmidt, M., Schwarzer, A., Banerjee, P. P., Díez, I. A., Dewey, R. A., Böhm, M., Nowrouzi, A., Ball, C. R., Glimm, H., Naundorf, S., et al. (2010). Stem-cell gene therapy for the Wiskott-Aldrich syndrome. *N. Engl. J. Med.* **363**(20), 1918–1927.

Brown, B. D., Venneri, M. A., Zingale, A., Sergi Sergi, L., and Naldini, L. (2006). Endogenous microRNA regulation suppresses transgene expression in hematopoietic lineages and enables stable gene transfer. *Nat. Med.* **12**(5), 585–591.

Cartier-Lacave, N., Hacein-Bey-Abina, S., Bartholomae, C. C., Veres, G., Schmidt, M., Kutschera, I., Vidaud, M., Dal-Cortivo, L., Caccavelli, L., Malhaoui, N., et al. (2009). Hematopoietic stem cell gene therapy with lentiviral vector in X-adrenoleukodystrophy. *Science* **326,** 818–823.

Cavazzana-Calvo, M., Hacein-Bey, S., de Saint Basile, G., Gross, F., Yvon, E., Nusbaum, P., Selz, F., Hue, C., Certain, S., Casanova, J. L., et al. (2000). Gene therapy of human severe combined immunodeficiency (SCID)-X1 disease. *Science* **288,** 669–672.

Ehrhardt, A., Haase, R., Schepers, A., Deutsch, M. J., Lipps, H. J., and Baiker, A. (2008). Episomal vectors for gene therapy. *Curr. Gene Ther.* **8,** 147–161.

Friedmann, T. (1992). A brief history of gene therapy. *Nat. Genet.* **2**(2), 93–98.

Gabriel, R., Eckenberg, R., Paruzynski, A., Bartholomae, C. C., Nowrouzi, A., Arens, A., Howe, S. J., Recchia, A., Cattoglio, C., Wang, W., et al. (2009). Comprehensive genomic access to vector integration in clinical gene therapy. *Nat. Med.* **15**(12), 1431–1436.

Gaspar, H. B., Parsley, K. L., Howe, S., King, D., Gilmour, K. C., Sinclair, J., Brouns, G., Schmidt, M., Von Kalle, C., Barington, T., et al. (2004). Gene therapy of X-linked severe combined immunodeficiency by use of a pseudotypedgammaretroviral vector. *Lancet* **364,** 2181–2187.

Gaspar, H. B., Bjorkegren, E., Parsley, K., Gilmour, K. C., King, D., Sinclair, J., Zhang, F., Giannakopoulos, A., Adams, S., Fairbanks, D., et al. (2006). Successful reconstitution of immunity in ADA-SCID by stem cell gene therapy following cessation of PEG-ADA and use of mild preconditioning. *Mol. Ther.* **14,** 505–513.

Hacein-Bey-Abina, S., Le Deist, F., Carlier, F., Bouneaud, C., Hue, C., De Villartay, J. P., Thrasher, A. J., Wulffraat, N., Sorensen, R., Dupuis-Girod, S., et al. (2002). Sustained correction of X-linked severe combined immunodeficiency by ex vivo gene therapy. *N. Engl. J. Med.* **346**(16), 1185–1193.

Hacein-Bey-Abina, S., Garrigue, A., Wang, G. P., Soulier, J., Lim, A., Morillon, E., Clappier, E., Caccavelli, L., Delabesse, E., Beldjord, K., et al. (2008). Insertionaloncogenesis in 4 patients after retrovirus-mediated gene therapy of SCID-X1. *J. Clin. Invest.* **118,** 3132–3142.

Harkey, M. A., Kaul, R., Jacobs, M. A., Kurre, P., Bovee, D., Levy, R., and Blau, C. A. (2007). Multiarm high-throughput integration site detection: Limitations of LAM-PCR technology and optimization for clonalanalysis. *Stem Cells Dev.* **16,** 381–392.

Hawkins, T. B., Dantzer, J., Peters, B., Dinauer, M., Mockaitis, K., Mooney, S., and Cornetta, K. (2011). Identifying viral integration sites using SeqMap 2.0. *Bioinformatics* **27,** 720–722.

High, K. A. (1999). Gene therapy for disorders of hemostasis. *Hematology* **1,** 438–446.

Howe, S. J., Mansour, M. R., Schwarzwaelder, K., Bartholomae, C., Hubank, M., Kempski, H., Brugman, M. H., Pike-Overzet, K., Chatters, S. J., de Ridder, D., et al. (2008). Insertional mutagenesis combined with acquired somatic mutations causes leukemogenesis following gene therapy of SCID-X1 patients. *J. Clin. Invest.* **118,** 3143–3150.

Mueller, P. R., and Wold, B. (1989). In vivo footprinting of a muscle specific enhancer by ligation mediated PCR. *Science* **246,** 780–786.

Ott, M., Schmidt, M., Schwarzwaelder, K., Stein, S., Siler, U., Koehl, U., Glimm, H., Kühlcke, K., Schilz, A., Kunkel, H., et al. (2006). Correction of X-linked chronic granulomatous disease by gene therapy, augmented by insertional activation of *MDS1-EVI1, PRDM16* or *SETBP1*. Nat. Med. **12,** 401–409.

Paruzynski, A., Arens, A., Gabriel, R., Bartholomae, C. C., Scholz, S., Wolf, S., Glimm, H., Schmidt, M., and von Kalle, C. (2010). Genome-wide high-throughput integrome analyses by nrLAM-PCR and next-generation sequencing. Nat. Protoc. **5,** 1379–1395.

Pfeifer, G. P., Steigerwald, S. D., Mueller, P. R., Wold, B., and Riggs, A. D. (1989). Genomic sequencing and methylation analysis by ligation mediated PCR. Science **246** (4931), 810–813.

Schmidt, M., Hoffmann, G., Wissler, M., Lemke, N., Mussig, A., Glimm, H., Williams, D. A., Ragg, S., Hesemann, C. U., and von Kalle, C. (2001). Detection and direct genomic sequencing of multiple rare unknown flanking DNA in highly complex samples. Hum. Gene Ther. **12**(7), 743–749.

Schmidt, M., Schwarzwaelder, K., Bartholomae, C., Zaoui, K., Ball, C., Pilz, I., Braun, S., Glimm, H., and von Kalle, C. (2007). High-resolution insertion-site analysis by linear amplification-mediated PCR (LAM-PCR). Nat. Methods **4**(12), 1051–1057.

Schmidt, M., Schwarzwaelder, K., Bartholomae, C. C., Glimm, H., and von Kalle, C. (2009). Detection of retroviral integration sites by linear amplification-mediated PCR and tracking of individual integration clones in different samples. Methods Mol. Biol. **506,** 363–372.

Scollay, R. (2001). Gene therapy: A brief overview of the past, present, and future. Ann. NY. Acad. Sci. **953,** 26–30.

Silver, J., and Keerikatte, V. (1989). Novel use of polymerase chain reaction to amplify cellular DNA adjacent to an integrated provirus. J. Virol. **63**(5), 1924–1928.

Stein, S., Ott, M. G., Schultze-Strasser, S., Jauch, A., Burwinkel, B., Kinner, A., Schmidt, M., Krämer, A., Schwäble, J., Glimm, H., et al. (2010). Genomic instability and myelodysplasia with monosomy 7 consequent to EVI1 activation after gene therapy for chronic granulomatous disease. Nat. Med. **16**(2), 198–204.

von Kalle, C., Veelken, H., and Rosenthal, F. M. (1999). Gentherapie in der Onkologie. Der Onkologe **5,** 898–909.

Wang, G. P., Garrigue, A., Ciuffi, A., Ronen, K., Leipzig, J., Berry, C., Lagresle-Peyrou, C., Benjelloun, F., Hacein-Bey-Abina, S., Fischer, A., et al. (2008). DNA bar coding and pyrosequencing to analyze adverse events in therapeutic gene transfer. Nucleic Acids Res. **36**(9), e49.

CHAPTER FIVE

Developing Novel Lentiviral Vectors into Clinical Products

Anna Leath* *and* Kenneth Cornetta*,†,‡

Contents

1. Introduction	90
2. Developing a Development Plan	90
2.1. Start at the finish line	90
2.2. Titer assessment	92
3. Development of a GMP Production Method	93
3.1. Optimizing plasmid concentrations: Small-scale transfection procedures	93
3.2. Large-scale transfection procedures	95
3.3. Purification and concentration	96
3.4. Evaluating envelope pseudotypes	100
4. Certification Testing of Novel Vector Products	100
4.1. Selecting the amplification phase cell line	101
4.2. Options for indicator phase assays	102
4.3. Selection of the positive control	103
4.4. Maximizing assay sensitivity	104
Acknowledgments	104
References	105

Abstract

Gene therapy vectors based on murine retroviruses have now been in clinical trials for over 20 years. During that time, a variety of novel vector pseudotypes were developed in an effort to improve gene transfer. Lentiviral vectors are now in clinical trials and a similar evolution of vector technology is anticipated. These modifications present challenges for those producing large-scale clinical materials. This chapter discusses approaches to process development for novel lentiviral vectors, highlight considerations, and methods to be incorporated into the development schema.

---

* Department of Medical and Molecular Genetics, Indiana University School of Medicine, Indianapolis, Indiana, USA
† Department of Medicine, Indiana University School of Medicine, Indianapolis, Indiana, USA
‡ Department of Microbiology and Immunology, Indiana University School of Medicine, Indianapolis, Indiana, USA

1. Introduction

The receptor and coreceptors for the HIV-1 envelope limit the virus's host range to a small number of human cell types (e.g., CD4+ cells and macrophages). To broaden the potential clinical applications, HIV-1 based vectors replace the naïve envelope with an alternative envelope, a process referred to as "pseudotyping." HIV-1 vectors often use the VSV-G envelope glycoprotein (env) due to its wide host range, its ability to pseudotype murine retroviruses (Emi et al., 1991), its relative stability, and its ability to be concentrated to titers of 10^8–10^9 IU/mL (Bartz and Vodicka, 1997; Kafri et al., 1999; Miyoshi et al., 1997; Reiser, 2000; Schonely et al., 2003; Slepushkin et al., 2003; Yamada et al., 2003). Due to the fusogenic nature of VSV-G env glycoprotein, this env presents certain technical challenges in generating stable packaging cell line. Therefore, the majority of investigators use the transient transfection method of vector production. In this method, plasmids expressing the transgene vector, the HIV-1 gag/pol, the VSV-G envelope, and HIV-1 Rev are introduced into the HEK293 or HEK293T cell line (Dull et al., 1998b; Gasmi et al., 1999; Kim et al., 1998; Mochizuki et al., 1998).

While VSV-G env has certain advantages, investigators are evaluating alternative envelopes. Investigators have generated "pseudotyped" vectors with the amphotropic, RD114, ecotropic, Gibbon Ape Leukemia Virus (Beyer et al., 2002; Christodoulopoulos and Cannon, 2001; Hanawa et al., 2002; Kelly et al., 2000; Kowolik and Yee, 2002; Landau et al., 1991; Lewis et al., 2001; Stitz et al., 2000), and a large number of other glycoproteins. Unfortunately, vectors that provide exciting advantages in laboratory and small animals models may be problematic when attempting to scale-up vector production. This paper describes an approach to developing a large-scale vector production method for a novel vector pseudotype.

2. Developing a Development Plan

2.1. Start at the finish line

When developing a vector with a novel pseudotype for large-scale clinical production, the initial step is to assess feasibility by defining the amount of vector required for the clinical trial. Knowing how much materials are required at the end of production drives the design of the manufacturing process.

First, one must determine the amount of vector that will be used for each patient, the number of patients to be treated, and an estimate of the vector titer at the time of harvest. To calculate the vector dose per patient (infectious unit per patient, IU/pt) for an *ex vivo* product, one must determine the number of cells that will be exposed to vector and the ratio of infectious particles per cell (multiplicity of infection, MOI). In most cases, the number of cells will be based on the patient weight (e.g., 5×10^6 CD34+ cells/kg), so the age of trial participants can significantly alter the required amount of vector. If multiple transductions will be performed this must also be considered. A formula for determining the vector dose per patient is

$$\text{IU/pt} = \text{MOI}(\text{IU/cell}) \times \text{no. cells treated/kg}$$
$$\times \text{ mean pt weight(kg)} \times \text{no. transductions}$$

Next, the number of IUs required for a study (X) is calculated. This considers dose per patient, number of patients, and also the amount required to conduct replication competent lentivirus (RCL) and other release testing. Per FDA, RCL testing requires 5% of the final product, and an additional 5% is usually held in reserve in case the test needs to be repeated. The remaining release test will be product specific but usually includes sterility, mycoplasma, the *in vitro* virus assay for adventitious agents, titer, residual DNA, vector function, and integrity. A formula for determining the total no. of particles needed is:

$$X = (\text{IU/pt} \times \text{pt}) + (X \times 0.1 \,\text{for RCL}) + \text{release testing needs}$$

Once the total number of particles required for treatment and testing is calculated, the total volume of unconcentrated vector that must be manufactured can be calculated. Key in this calculation is the anticipated titer (IU/ml) of the unprocessed vector. The challenge in determining this value is discussed in Section 2.2. The product volume is based on the number of IUs required for a study (X) and considers the loss due to processing (purification and concentration). For example, diafiltration and concentration using tangential flow filters can yield approximately 70–80% of the unconcentrated particles; adding purification with ion exchange chromatography will decrease the yield to 30–50%, and adding an additional size chromatography can further decrease yields to around 15% (Merten *et al.*, 2011). Second, how the vector will be vialed must be considered. US FDA requires sterility testing on the first, last, and every tenth vial of material so too many vials will increase the loss of material to sterility testing. Too large a vial is also wasteful. It is unlikely that a patient dose will be precisely equal to the amount available in a vial; partially used vials will be discarded and the excess material will be lost. Therefore, an aliquoting strategy that minimizes this loss is prudent. The anticipated loss should then be calculated into the vector needs. A formula for determining this is:

$$\text{Production volume(mL)} = \frac{X(\text{IU})}{\text{yield}(\%) \times \text{unconcentrated IU/mL} \times (\text{estimated vialing loss})}$$

Once the total volume of unconcentrated material is calculated, an assessment can be made to determine if the scale and cost of vector production is technically feasible.

2.2. Titer assessment

At this writing, there is no standardized assay for lentiviral vector titer, nor is there an agreed upon standard to be used as a positive control. If a research laboratory is collaborating with a GMP facility to scale-up vector production, it is critical that the laboratories standardized the measurement of vector potency (titer). It is well documented that the method of vector detection (e.g., RNA, DNA, or protein), the cell type used, and the cell density all significantly alter the estimated titer (Sanburn and Cornetta, 1999; Sastry et al., 2002). Therefore, vector should be tested in both the developers' and the production site so each may interpret the titer results of the other.

There are two approaches to assessing vector potency. Infectious titer is a functional measure of how many IUs are present in the supernatant. This requires assessment of vector content in transduced cells and remains the gold standard for documenting vector potency. Physical titer estimates virion number based on measurement of vector protein or RNA without assessing infectivity. For HIV-1, physical titer is relatively easy to assess using commercially available p24 HIV-1 viral capsid antigen kits. Other methods include viral supernatant for RNA genomes of the transgene vector or by measuring reverse transcriptase activity (Miskin et al., 2006; Pyra et al., 1994; Rohll et al., 2002; Sanburn and Cornetta, 1999; Sastry et al., 2005). Comparison of the infectious and physical titers has been proposed as a means for documenting the effectiveness of purification.

Special care must be taken when utilizing an existing titer assay to assess the potency of a novel vector pseudotype. A physical titer can be useful in assessing whether the product has a similar number of particles to controls (e.g., VSV-G pseudotyped vector). Comparing the infectious titer is often more problematic as titer assays utilize immortalized cell lines that may have different levels of infectivity depending on the pseudotype. Also, many titer assays utilize polycations to increase gene transfer, but polycations can inhibit infection of certain envelopes. Finally, titer on an immortalized cell line may not reflect the potency of vector on the primary target cell. When comparing one vector pseudotype to another, care should be taken to utilize optimized assays for each envelope to accurately gage the potency of the vector product.

3. Development of a GMP Production Method

In this section, we present a method for generating VSV-G pseudotyped lentiviral vector using a third generation packaging system. We will use this method to illustrate how to develop a large-scale production method for novel pseudotypes. It should be stated that different packaging plasmids, transgene vectors, and source of HEK293T cells will vary so some optimization will be required. The focus of this discussion is on optimization and the methods described are illustrations of starting points, rather than hard and fast parameters for vector production.

3.1. Optimizing plasmid concentrations: Small-scale transfection procedures

For this discussion, we are assuming that the novel envelope will be the variable with the transgene and packaging plasmids remaining the same. We currently use third generation HIV-1 based lentiviral vectors; production involves a combination of four plasmids, containing the transgene plasmid, gag/pol plasmid, Rev expressing plasmid, and the envelope plasmid (Dull et al., 1998b). When evaluating a novel envelope plasmid, first define the optimal envelope concentration in small-scale productions and then translate into larger scale methods. Vector pseudotyped with VSV-G envelope is generated simultaneously to serve as a positive control for the transfection reaction.

3.1.1. Required materials
Equipment: Lentiviral vectors require special handling. The level of containment will vary depending on the transgene and packaging system and is determined by the institutional biosafety committee at most academic institutions. At our laboratory, work must be performed in a biosafety cabinet, technicians wear gloves, lab coats, and eye protection, and avoid the use of needles or other sharps. Cells are maintained in a CO_2 incubator and a tissue culture centrifuge is required for cell passage.

3.1.1.1. Additional materials
- HEK293T cells can be obtained from the American Type Culture Collection or academic investigators may obtain cells from the National Gene Vector Biorepository (www.NCVBCC.org).
- Growth medium. Dulbecco's modified eagle medium (DMEM) supplemented with 10% fetal calf serum and $2\,mM$ L-glutamine (D10 medium). Antibiotics (100 units/mL penicillin and 100 μg/mL streptomycin) can be used during the development phase but are not permitted during GMP production.

- Transfection reagents. Calcium phosphate reagents are inexpensive to produce but consistent performance can be an issue. Commercial kits (e.g., Promega or SIGMA) do provide consistent performance.
- Packaging and transgene plasmids. Plasmids should be generated using an endotoxin free protocol (e.g., Qiagen Endotoxin-Free Purification kit) and prepared in a sterile environment to avoid contamination of cell culture.
- Sterile tissue culture supplies include 75 cm^2 tissue-culture treated flasks, 2 mL and 10 mL disposable pipettes, Ca and Mg free phosphate buffered saline (PBS), and Trypsin-EDTA (trypsin 0.25%, EDTA 0.05%).

3.1.2. Transfections

One day prior to transfection: Prepare a single cell suspension of HEK293T cells by trypsinizing cells, centrifuge, and resuspend in fresh media. Perform a cell count and plate cells at a density of 5×10^6 cells per 75 cm^2 tissue culture flask. Return cells to the incubator (5% CO_2, 37°C).

Day of transfection: Cells should be approximately 80–90% confluent at the time of transfection. (*Note*: Different lots of HEK293T cells have different growth characteristics, and different lots of serum may also affect the growth. If cell density is not at the desired level, the number of plated cells may need to be adjusted. Cell density is critical to optimizing vector titer.)

Perform calcium phosphate transfection according to manufacturer's instructions. The optimal concentration of the transgene plasmid will vary with the specific plasmid for our development work. We utilized the GFP-expressing lentiviral vector pcDNA-HIV-CS-CGW (kindly provided by Phil Zoltick) at a concentration of 13.2 μg/75 cm^2 flask. For each 75 cm^2 flask the packaging plasmid concentrations are: gag/pol pMDLg (6.6 μg), pRSV/Rev (3.3 μg), and pMD.G (4.5 μg) (Dull *et al.*, 1998b). For novel envelopes, a starting point concentration can be made from prior work with the envelope or if this is unknown start with the VSV-G concentration. Conduct transfections with the pseudotyped env at concentrations above and below the starting point concentration. Incubate from approximately 20 h.

Day 1: Aspirate medium from flasks and discard. Rinse with 5 mL of PBS then aspirate. Add 12 mL D10 media to each flask and return to the incubator.

Day 2: After 20–24 h, remove supernatants from flasks. Filter through a 0.45 μm syringe filter to remove cells. Freeze in aliquots based on anticipated need and store at $<-70°C$. Fresh media may be added to the flask and a second harvest performed 12–24 h later, although titer will be approximately one-third of that obtained on the initial harvest.

3.1.3. Analysis

Perform an infectious titer assay appropriate for the transgene vector (see Section 2.2). Compare titer results of novel envelope and VSV-G pseudotyped vectors. Identify the concentration providing the highest titer and then conduct repeat transfections with concentrations above and below this level; repeat as needed until the optimal titer is identified.

Further refinements in vector production should be evaluated once the plasmid concentrations are optimized. For example, the addition of sodium butyrate to the culture can improve titer of some vectors by two- to threefold. Also, collection of vector in serum free medium will result in a drop in titer. Assessment of modifications in small-scale is prudent prior to conducting large-scale vector productions.

3.2. Large-scale transfection procedures

In general, HEK293T cells generate the highest titer material when adherent to a cell surface. While there are a number of options, most investigators utilize multitier cell factories (CF) for clinical productions.

3.2.1. Required materials

Equipment: Biosafety precautions described in Section 3.1.1 apply. NIH Guidelines recommend increased containment measures if generating more than 10 L of vector. Larger CO_2 incubators may be required to accommodate CF. A sterile tubing welder (e.g., Terumo) is also used to seal tubing to maintain a closed system for media exchanges.

Additional materials

- The same media and cells discussed in Section 3.1.1 can be utilized.
- CF are available from a variety of vendors (including Corning, Nunc). Systems should be purchased that have one cap for gas exchange and another that can be connected to tubing via cell stack cap (Corning). Media and other reagents should be introduced into bags (Stedim) that can be connected through tubing to minimize open manipulations.
- Peristaltic pumps are extremely helpful when processing volumes over 2 L.
- Fluid transfer tube set (Baxa), Cryocyte manifold kit, Transfer pack container with coupler-2000 mL (Fenwal), 40/150 µm dual screen filter (Baxter), Leukocyte reduction filter for red cells (such as Sepacell filters, Fenwall), and 0.45 or 0.22 µm filters.

3.2.2. Large Scale transfection procedure

Prepare for transfection: Considerable time will be required to expand cells to a sufficient number for plating in CF. In our laboratory, we plate 5×10^7 cells per layer of a cell factor with 100 mL of D10 media approximately 24 h prior to transfection. Return cells to incubator (5% CO_2, 37 °C).

Day of transfection: Cells should reach a confluence of approximately 80% (±10%) prior to commencing transfection. If cells are below the desired confluence the transfection should be delayed and rechecked every 8 h. If consistently outside of the desired range, the cell number plated may need to be optimized.

When cells are ready, prepare for calcium phosphate transfection using the procedure above in Section 3.1.2 except increase the plasmid concentration for the increase surface area. For example, for generating the lentiviral vector pcDNA-HIV-CS-CGW, we add 555 μg per layer of cell factory (636 cm^2), 277.5 μg of gag/pol pMDLg, 137.5 μg of pRSV/Rev, and 190.8 μg of VSV-G pMD.G (Kahl et al., 2004). For novel envelopes, the concentration will be derived from data generated in Section 3.1.3. Add the DNA/transfection mixture and return CF to the incubator.

Remove transfection media: Unclamp and drain transfection media from CF 6–24 h after transfection. Attached new bag with fresh media and refeed cells. If the production is to be collected in serum-free media (e.g., X-VIVO 15, Fisher), this should be added in the place of D10. Return to the incubator.

Day 2 after transfection: Approximately 48 h post-transfection, remove supernatants from CF. Examine each CF for contamination prior to removing media. There are a number of options for clarification of the supernatant, but a step process is important to minimize clogging and titer loss (Fig. 5.1; Feldman et al., 2011; Reeves and Cornetta, 2000). Fresh media may be added to the flask and a second harvest performed 12–24 h later. If a second harvest is performed, the clarified vector from the first harvest may be stored at 4 °C and then combined for processing.

3.3. Purification and concentration

To obtain highly concentrated material, most investigators utilize ultracentrifugation; unfortunately, this approach is impractical when processing clinically relevant amounts of material. Alternative methods include column purification, ultrafiltration, selection with immunomagnetic beads, and affinity columns (Chan et al., 2005; Lu et al., 2004; Slepushkin et al., 2003; Transfiguracion et al., 2003; Yu and Schaffer, 2006). One of the advantages of the VSV-G envelope is the capacity to tolerate a number of different manipulations without significant loss in titer. This may not be the case

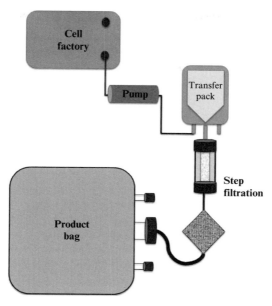

Figure 5.1 *Schematic of viral supernatant clarification.* Viral supernatant is pumped from the cell factory to a transfer pack. The supernatant is then filtered using multiple progressively smaller filters and collected in the product bag. This step filtration process assists in the prevention of filter clogging, which can lead to significant titer loss. (See Color Insert.)

when developing a large-scale production method for vectors pseudotyped with other envelopes.

Concentrating lentiviral vectors generated by transient transfection appears to also concentrate materials that can affect the viability of certain cell types. Therefore, many laboratories utilize purification by Mustang Q ion exchange chromatography (Merten *et al.*, 2011; Slepushkin *et al.*, 2003). This purification step is combined with tangential flow filters in a closed system to provide purification, diafiltration, and concentration. Below, we describe one option for purification and concentration of VSV-G pseudotyped vector.

3.3.1. Required materials

Equipment: Peristaltic pump to circulate vector. Pressure transducers to monitor pressure across tangential flow filter.

3.3.1.1. Additional materials

- Pall Mustang Q ion exchange capsules (for concentrating 10–20 L, 60 mL may be sufficient; larger volumes may require larger capsules)
- Tangential flow filters (Spectrum MicroKros® filters or GE filter systems)

- Autoclave for sterilization of the Mustang Q capsule
- Luer locks, clamps, and adaptors (Cole Palmer) and tubing (Fisher) can be used to construct a closed system for vector processing.

3.3.2. Purification by ion exchange

Preparation: Vector product material should be clarified to remove cellular debris and aggregates to prevent clogging of the Mustang Q capsule. We begin with a 40/150 µm dual screen filter followed by a leukocyte reduction filter for red cells. Both are FDA approved for blood banking needs. The manufacturer recommends final filtration through a 0.22 µm filter, although a significant loss in yield has been observed when using this filter size. For most applications, the product should be treated with Benzonase to remove residual plasmids prior to Mustang Q purification (Merten *et al.*, 2011; Sastry *et al.*, 2004).

Capsule equilibration: Close the vent and waste ports on the side of the Mustang Q capsule. Place the assembled unit in an upright position with the aid of a support stand. Attach a waste collection bag to the outlet end of the Mustang Q unit. Position the tubing on the inlet side of the Mustang Q unit in a peristaltic pump. Connect a bag with equilibration buffer (25 mM Hepes, 1 mM MgCl$_2$, 5% glycerol, 0.2 M NaCl, pH 7.5) to the inlet port. Turn on the peristaltic pump, and ensure that the equilibration buffer fills the capsule and then passes through into the waste bag. From this point on the capsule should remain filled with fluid. Pass 20 column volumes (1200 mL for a 60 mL capsule) of equilibration buffer through the capsule.

Load vector onto capsule: Connect the bag containing vector to the capsule and open all inlet and outlet clamps, then turn on the peristaltic pump. Pump the entire supernatant through the Mustang Q ensuring the pressure does not exceed 10 psi. Switch off the pump and close all inlet and outlet clamps.

Elute vector: Attach a bag containing equilibration buffer. Wash with 25 column volumes (1500 mL for a 60 mL capsule) of equilibration buffer by opening all inlet and outlet clamps, then turn on pump. Switch the pump off and close all inlet and outlet clamps when done. Attach a bag containing the Mustang Q elution buffer. The buffer should contain 25 mM Hepes, 1 mM MgCl$_2$, 5% glycerol, plus hypertonic sodium chloride. The salt concentration can vary between 0.75 and 2 M NaCl and should be evaluated for novel pseudotypes. Lower salt concentration will result in incomplete elution and higher concentrations will result in inactive vector particles. The final buffer should have a pH of 7.5. Open all inlet and outlet clamps, then switch the pump on. Elute the vector with 15 column volumes (900 mL) of Mustang Q elution buffer into the final product medium. Potency will decrease if vector is allowed to remain at a high salt concentration: elute into a volume that dilutes eluate 5–10-fold (3600–8100 mL).

3.3.3. Concentration

Vector concentration and diafiltration: Wet the tangential flow filter membrane according to manufacturer's specifications. Assemble the filter in a sterile apparatus that allows bags containing vector and diafiltration media to enter the filter in a closed system (Fig. 5.2). Attach bags to collect dialysate. Connect pressure transducers according to the manufacturer's instructions. The apparatus weight should be recorded. Connect the apparatus to a peristaltic pump. The optimal flow rate depends on the filter used (e.g., between 40 and 60 mL/min when using a 3100 cm^2 spectrum TFF unit). The inlet pressure should be maintained at less than 10 psi by adjusting the pump rate. Begin circulating vector product in the filter and concentrate the product to approximately 1 L. Begin diafiltration by connecting a bag containing diafiltration medium (PBS or serum-free medium) at a volume of approximately five times the volume of the product to be diafiltered. Diafilter by maintaining equal pressure across the filter; the product bag volume should remain constant during the diafiltration procedure. Once

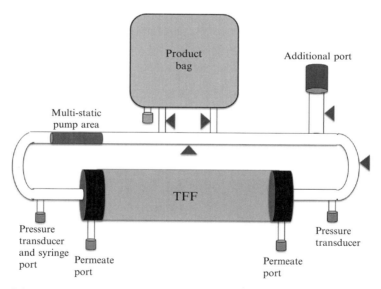

Figure 5.2 *Schematic of vector concentration and diafiltration apparatus.* The tangential flow filter (TFF) and product bag are connected using tubing with the pump and pressure transducers at the indicated locations. The product is then concentrated to approximately 1 L with waste being removed via one of the permeate ports. The diafiltration then begins by connecting the media to be diafiltered at the additional port. Once all of the diafiltration media has entered the system, this port will be clamped off and concentration will continue until the desired final volume is remaining. The green arrows indicate locations where clamps may be attached at different times during the process. (For interpretation of the references to color in this figure legend, the reader is referred to the Web version of this chapter.)

all the diafiltrate has been delivered to the system, stop the pump and clamp off the diafiltrate line. Continue concentration by increasing the inlet pressure, again maintaining it at less than 10 psi by adjusting the pump rate. If maximal concentration is desired, continue circulating vector until there is no dead volume in the vector reservoir. Remove concentrated vector with a 60 mL syringe connected to a Luer lock port. Vial final product in volumes suited to the intended use, freeze, and store at $\leq -70\,^\circ$C.

3.4. Evaluating envelope pseudotypes

While the process described above works well for VSV-G pseudotyped HIV-1 vectors, gamma retroviral vectors do not tolerate the process without significant loss of titer. We note HIV-1 vectors pseudotyped with foamy viral envelopes will tolerate the procedure, but other lentiviral vectors demonstrate significant loss of function after tangential flow filter processing (unpublished results). Therefore, moderate scale productions should be performed prior to full-scale production to ensure a novel vector will tolerate processing and concentration.

4. CERTIFICATION TESTING OF NOVEL VECTOR PRODUCTS

In addition to altering vector production, novel pseudotypes require development of new certification assays, including assays that can screen vector products for RCL. For lentiviral vectors, a paper discussing RCL assays has been published that highlights a number of important issues (Kiermer et al., 2005), but the US FDA has yet to publish specific guidance. Existing guidances are useful starting points (FDA, 1998, 2006a,b), but the FDA recognizes the challenge of publishing regulations given the rapid changes in vector technology. When developing a new assay, the amount of material that will be required to screen is also important. For example, US FDA guidelines state that 5% of a vector product and up to 10^8 end-of-production cells should be screened for RCL. For clinical productions, this requires a large number of vector particles to be screened.

Currently, most investigators utilize RCL detection methods that combine transduction of a permissive cell line followed by a molecular or immunologic detection assays (see Fig. 5.3). Ideally, one IU can be detected in a test article to ensure patients are not exposed to RCL. Unfortunately, assays such as PCR or ELISA, require 100 to over 1000 IU before virus is detected unequivocally. Further, molecular assays such as PCR have limited usefulness in direct testing of the clinical product because vector preparations are often contaminated with DNA from packaging cell lines or

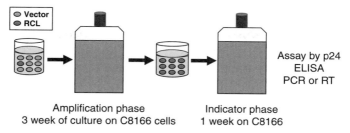

Figure 5.3 *Schematic of replication competent lentivirus detection assay.* Viral supernatant is used to transduce a permissive cell line such as C8166 cells. Replication competent lentivirus (RCL) will amplify during the 3-week culture, if present. The cell-free supernatant is then collected and used to transduce C8166 cells, which are cultured for 1 week, known as the indicator phase. Presence or absence of virus is then determined using an assay such as ELISA for p24 or detection of reverse transcriptase (RT) activity. (For color version of this figure, the reader is referred to the Web version of this chapter.)

production plasmids leading to false positive results (Chen et al., 2001b). Similarly, ELISA assay for p24 are not useful because the vector particles themselves contain this antigen. Also, the amount of DNA in a PCR reaction or the amount of vector in an ELISA assay represents a relatively small amount of DNA compared to the large amount of material that must be screened. To meet these challenges, vector is first introduced into a culture assay that is permissive for RCL growth. In the remainder of this chapter, we will illustrate the rational and basic methodology used in designing a RCL detection method for a HIV-1 vector with a novel pseudotype.

4.1. Selecting the amplification phase cell line

The amplification phase is the cell culture phase performed to increase sensitivity of RCL detection. Large amounts of material can be screened and virus expansion within the culture should increase the concentration of virus to a level that is easily detected in an indicator phase assay (i.e., PCR or ELISA). When developing a RCL assay for a new vector pseudotype, we begin by identifying potential cell lines that produce virus at high titer and are highly infectable. From a reproducibility standpoint, the cell line should be stable, easy to maintain in culture, and be able to efficiently expand RCL to high titer.

4.1.1. Selecting a highly infectable cell line

A major concern with any RCL assay is the level of sensitivity, a property that will be directly related to the infectivity of a particular cell line by the novel vector pseudotype. One way to compare infectivity among different cell lines is to conduct a viral titer assay, using the same method and vector stock on different candidate cell lines. In conducting this assessment, we generally use a RCL-free lot of GFP-expressing vector that is pseudotyped with the novel envelope. The GFP-vector is then analyzed by dilutional titer on each of the candidate cell line using standard titering methods (Sastry et al., 2002). The cell line demonstrating the highest "titer" indicates a line that will become infected with the least number of particles; a property that will translate into maximal sensitivity for RCL infection.

4.1.2. Assessing RCL amplification

A second criteria for any cell line used in a RCL detection assay is the ability of the line to amplify virus. To assess this, candidate cell lines should be infected with low levels of a RCL virus (typically the assay positive control) that is below the detection level of PCR and ELISA assays (e.g., 1–100 IU). Monitor the level of virus expansion in the culture over a 3-week period. Normally, virus concentration in the media will plateau within 1–2 weeks; cell lines that quickly expand virus and reach a high plateau are candidates for the RCL assay. When choosing between a line that is highly infectable or generated RCL at the highest titer, we generally prefer the line that is highly infectable as long at the plateau is over 10^4 IU/mL. This is because PCR and ELISA assays can easily detect RCL that is in excess of 10^4 IU/mL; once this plateau is reached, the assay is positive irrespective of how high the RCL concentration plateaus. A number of laboratories have identified the C8166 cell line as a line that is highly infectable by HIV-1 virus and VSV-G pseudotypes vector (Cornetta et al., 2011; Escarpe et al., 2003). This line may be appropriate when designing a RCL assay for a novel pseudotype but this must be tested experimentally.

4.2. Options for indicator phase assays

Once virus has been amplified, an indicator assay is required to document the presence (or absence) of a RCL. For detection, replication competent retroviruses based on gamma retroviruses, a wide variety of detection assays have been developed (Bassin et al., 1971, 1982; Cornetta et al., 1993; Phillips et al., 1973) and subsequently adapted as investigators developed novel vector pseudotypes (Chen et al., 2001a; Cornetta et al., 1993; Duffy et al., 2003; Reeves et al., 2002).

HIV-1 based RCL can be detected by utilizing commercially available ELISA kits for the p24 capsid antigen (Escarpe et al., 2003; Sastry et al., 2003). Since viral vectors have very high content of p24, one must exclude carryover of p24 from the test material leading to false positives. Some investigators have suggested that a decreasing p24 over the culture period is sufficient evidence to exclude RCL. Unfortunately, if p24 is above the level of detection a slow growing RCL cannot be excluded and therefore, we have required passage of virus from the amplification phase cells to naïve C8166 cells before concluding a true RCL is present (Fig. 5.3).

Molecular assays such as PCR are sensitive and easy (Sanburn and Cornetta, 1999; Sastry et al., 2002), but PCR primers and probe should target a region that is critical for maintaining virus infectivity. PCR primers that target the viral envelope will provide specificity to RCL detection; primers that target viral genes can detect RCL irrespective of the envelope. Combining PCR with p24 ELISA minimizes the chance of a false negative RCL assay resulting in a mutation that alters PCR primer binding.

Measurements of reverse transcriptase activity using the product enhanced reverse transcriptase (PERT) assay can also serve to identify RCL (Miskin et al., 2006; Pyra et al., 1994; Sanburn and Cornetta, 1999; Sastry et al., 2005). These assays have the advantage of detecting other retroviruses in the test material. One of the challenges is the variability of these assays, which can be effected by different cell lines and serum (Sastry et al., 2005). This variability can be problematic when attempting to maximize sensitivity and reducing false positives.

4.3. Selection of the positive control

A major challenge for developing RCL testing is selection of the positive control. First, wild-type HIV-1 is a complex virus with a number of accessory genes that are normally deleted in lentiviral vectors (Dull et al., 1998a; Zufferey et al., 1998). The FDA has generally requested the RCL positive control also be one in which accessory genes are deleted (Escarpe et al., 2003). Another major challenge can be a marked difference in the wild-type HIV-1 and the virus used to pseudotype the vector. For example, VSV-G is a vastly different, cytopathic virus that is not suitable as a positive control for RCL assays. Some investigators have utilized a recombinant HIV-1 that combined the HIV-1 LTR, gag/pol with the VSV-G envelope glycoprotein (Escarpe et al., 2003; Segall et al., 2003). Our laboratory has chosen not to utilize such a construct as it expands the infectivity of the HIV-1 virus to almost any cell type, placing unnecessary risk to laboratory workers who are conducting the RCL assay. We instead have chosen to utilize an attenuated HIV-1 virus that contains the wild-type HIV-1 envelope. This does require that the amplification cells express the CD4 and HIV-1 coreceptors.

Ideally, the assay positive control should mimic the growth kinetics of the RCL. Unfortunately, RCL remains a theoretical adverse event; therefore, the growth kinetics of any RCL is unknown. Data from retroviral vectors suggest a RCL could have a significantly different structure and growth kinetics compared to any positive control, regardless if the positive control is pseudotyped with the HIV-1 or another viral envelope (Bonham et al., 1997; Chong et al., 1998; Garrett et al., 2000; Miller et al., 1996).

Therefore, the positive control is not utilized in order to be a direct comparison to any potential RCL; rather the positive control demonstrates assay sensitivity. Specifically, the positive control should be plated at a very low inoculum (≤5IU) and should expand to detectable levels over the 3 week amplification phase to document the cells and culture conditions are suitable for virus expansion.

4.4. Maximizing assay sensitivity

After identifying cells that are highly infectable and amplify virus to high titer, care should be taken in designing the parameters of the assay to maximize sensitivity. A major concern will be the ratio of viral particles to amplification cells. Since the test material may have vector particles in great excess of RCL, insufficient number of cells will decrease sensitivity as vector particles will outcompete RCL for the limited number of receptors (receptor interference). It is also possible that the vector product itself may contain factors (including unbound envelope protein) that might inhibit RCL detection. This can be assessed by introducing limiting amount of the positive control into cultures with the test article. For example, in our RCL detection assay, we establish cultures containing 50IU of attenuated HIV-1 and the test article (Cornetta et al., 2011). Failure to detect RCL would indicate that a significant inhibitor exists in the culture and we would be required to adjust the sensitivity level of virus detection accordingly. Finally, a particular challenge for RCL detection is certification of vectors that are intended for treatment of HIV-1. In this case, the vector itself has the potential to inhibit growth of any RCL within the culture. Particular care should be taken to ensure the cell to vector ratio is high allowing cells to be infected by the RCL without concomitant infection with the anti-HIV-1 vector (Cornetta et al., 2011).

Development of RCL testing should begin early in the clinical product development as not to delay the onset of clinical trials. As new assays are developed, it is prudent to engage regulatory agencies early in the development process to maximize the chance that the new assay will be acceptable for screening clinical products.

ACKNOWLEDGMENTS

Indiana University is the home of the NHLBI Gene Therapy Resources Program (HHSN26820078204) and the NCRR National Gene Vector Biorepository (P40 RR024928), K. C. serves as principle investigator on both these efforts. A. L. is funded by NIH grant number T32 HL007910, Basic Science Studies on Gene Therapy of Blood Diseases.

REFERENCES

Bartz, S. R., and Vodicka, M. A. (1997). Production of high-titer human immunodeficiency virus type 1 pseudotyped with vesicular stomatitis virus glycoprotein. *Methods* **12,** 337–342.

Bassin, R. H., Phillips, L. A., Kramer, M. J., Haapala, D. K., Peebles, P. T., Nomura, S., and Fischinger, P. J. (1971). Transformation of Mouse 3T3 cells by murine sarcoma virus: Release of virus-like particles in the absence of replicating murine leukemia helper virus. *Proc. Natl. Acad. Sci. USA* **68,** 1520–1524.

Bassin, R. H., Ruscetti, S., Ali, I., Haapala, D., and Rein, A. (1982). Normal DBA/2 mouse cells synthesize a glycoprotein which interferes with MCF virus infection. *Virology* **123,** 139–151.

Beyer, W. R., Westphal, M., Ostertag, W., and von Laer, D. (2002). Oncoretrovirus and lentivirus vectors pseudotyped with lymphocytic choriomeningitis virus glycoprotein: Generation, concentration, and broad host range. *J. Virol.* **76,** 1488–1495.

Bonham, L., Wolgamot, G., and Miller, A. D. (1997). Molecular cloning of *Mus dunni* Endogenous Virus: An unusual retrovirus in a new murine viral interference group with a wide host range. *J. Virol.* **71,** 4663–4670.

Chan, L., Nesbeth, D., MacKey, T., Galea-Lauri, J., Gaken, J., Martin, F., Collins, M., Muftin, G., Farzaneh, F., and Darling, D. (2005). Conjugation of lentivirus to paramagnetic particles via nonviral proteins allow efficient concentration and infection of primary acute myeloid leukemia cells. *J. Virol.* **79,** 13190–13194.

Chen, J., Reeves, L., and Cornetta, K. (2001a). Safety testing for replication-competent retrovirus (RCR) associated with Gibbon Ape Leukemia Virus pseudotyped retroviral vectors. *Hum. Gene Ther.* **12,** 61–70.

Chen, J., Reeves, L., Sanburn, N., Croop, J., Williams, D. A., and Cornetta, K. (2001b). Packaging cell line DNA contamination of vector supernatants: Implication for laboratory and clinical research. *Virology* **282,** 186–197.

Chong, H., Starkey, W., and Vile, R. G. (1998). A replication-competent retrovirus arising from a split-function packaging cell line was generated by recombination events between the vector, one of the packaging constructs, and endogenous retroviral sequences. *J. Virol.* **72,** 2663–2670.

Christodoulopoulos, I., and Cannon, P. (2001). Sequences in the cytoplasmic tail of the Gibbon Ape Leukemia Virus envelope protein that prevent its incorporation into lentivirus vectors. *J. Virol.* **75,** 4129–4138.

Cornetta, K., Nguyen, N., Morgan, R. A., Muenchau, D. D., Hartley, J., and Anderson, W. F. (1993). Infection of human cells with murine amphotropic replication-competent retroviruses. *Hum. Gene Ther.* **4,** 579–588.

Cornetta, K., Yao, J., Jasti, A., Koop, S., Douglas, M., Hsu, D., Couture, L. A., Hawkins, T., and Duffy, L. (2011). Replication competent lentivirus analysis of clinical grade vector products. *Mol. Ther.* **19,** 557–566.

Duffy, L., Koop, S., Fyffe, J., and Cornetta, K. (2003). Extended S+/L- assay for detecting replication competent retroviruses (RCR) pseudotyped with the RD114 viral envelope. *Preclinica* May/June, 53–59.

Dull, T., Zufferey, R., Kelly, M., Mandel, R. J., Nguyen, M., Trono, D., and Naldini, L. (1998a). A third generation lentivirus with a conditional packaging system. *J. Virol.* **72,** 8463–8471.

Dull, T., Zufferey, R., Kelly, M., Mandel, R. J., Nguyen, M., Trono, D., and Naldini, L. (1998b). A third-generation lentivirus vector with a conditional packaging system. *J. Virol.* **72,** 8463–8471.

Emi, N., Friedmann, T., and Yee, J. K. (1991). Pseudotype formation of murine leukemia virus with the G protein of vesicular stomatitis virus. *J. Virol.* **65,** 1202–1207.

Escarpe, P., Zayek, N., Chin, P., Borellini, F., Zufferey, R., Veres, G., and Kiermer, V. (2003). Development of a sensitive assay for detection of replication-competent recombinant lentivirus in large-scale HIV-based vector preparations. *Mol. Ther.* **8,** 332–341.

FDA (1998). Guidance for Human Somatic Cell Therapy and Gene Therapy. U.S. Food and Drug Administration. http://www.fda.gov/downloads/BiologicsBloodVaccines/ GuidanceComplianceRegulatoryInformation/Guidances/CellularandGeneTherapy/ ucm081670.pdf.

FDA (2006a). Guidance for Industry Gene Therapy Clinical Trials—Observing Subjects for Delayed Adverse Events. U.S. Food and Drug Administration. http://www.fda.gov/ downloads/BiologicsBloodVaccines/GuidanceComplianceRegulatoryInformation/ Guidances/CellularandGeneTherapy/ucm078719.pdf.

FDA (2006b). Guidance for Industry—Supplemental Guidance on Testing for Replication Competent Retrovirus in Retroviral Vector Based Gene Therapy Products and During Follow-up of Patients in Clinical Trials Using Retroviral Vectors. U.S. Food and Drug Administration. http://www.fda.gov/downloads/BiologicsBloodVaccines/GuidanceCompliance RegulatoryInformation/Guidances/CellularandGeneTherapy/ucm078723.pdf.

Feldman, S. A., Goff, S. L., Xu, H., Black, M. A., Kochenderfer, J. N., Johnson, L. A., Yang, J. C., Wang, Q., Parkhurst, M. R., Cross, S. B., Morgan, R. A., Cornetta, K., et al. (2011). Rapid production of clinical-grade gammaretroviral vectors in expanded surface roller bottles using a "modified" step-filtration process for clearance of packaging cells. *Hum. Gene Ther.* **22,** 107–115.

Garrett, E., Miller, A. R.-M., Goldman, J., Apperley, J. F., and Melo, J. V. (2000). Characterization of recombinant events leading to the production of an ecotropic replication-competent retrovirus in a GP+envAM12-derived producer cell line. *Virology* **266,** 170–179.

Gasmi, M., Glynn, J., Jin, M. J., Jolly, D. J., Yee, J. K., and Chen, S. T. (1999). Requirements for efficient production and transduction of human immunodeficiency virus type 1-based vectors. *J. Virol.* **73,** 1828–1834.

Hanawa, H., Kelly, P. F., Nathwani, A. C., Persons, D. A., Vandergriff, J., Hargrove, P., Vanin, E. F., and Nienhuis, A. W. (2002). Comparison of various envelope proteins for their ability to pseudotype lentiviral vectors and transduce primitive hemaopoietic cells from human blood. *Mol. Ther.* **5,** 242–251.

Kafri, T., van Praag, H., Ouyang, L., Gage, F. H., and Verma, I. M. (1999). A packaging cell line for lentivirus vectors. *J. Virol.* **73,** 576–584.

Kahl, C. A., Marsh, J., Fyffe, J., Sanders, D. A., and Cornetta, K. (2004). Human immunodeficiency virus type 1-derived lentivirus vectors pseudotyped with envelope glycoproteins derived from Ross River virus and Semliki Forest virus. *J. Virol.* **78,** 1421–1430.

Kelly, P. F., Vandergriff, J., Nathwani, A., Nienhuis, A. W., and Vanin, E. F. (2000). Highly efficient gene transfer into cord blood nonobese diabetic/severe combined immunodeficiency repopulating cells by oncoretroviral vector particles pseudotyped with the feline endogenous retrovirus (RD114) envelope protein. *Blood* **96,** 1206–1214.

Kiermer, V., Borellini, F., Lu, X., Slepushkin, V., Binder, G., Dropulic, B., Audit, M., Engel, B., Cornetta, K., Wilson, C., Takefman, D., Zhao, Y., et al. (2005). Report from the lentivirus vector working group: Issues for developing assays and reference materials for detecting replication-competent lentivirus in production lots of lentivirus vectors. *Bioprocessing* March/April, 39–42.

Kim, V. N., Mitrophanous, K., Kingsman, S. M., and Kingsman, A. J. (1998). Minimal requirement for a lentivirus vector based on human immunodeficiency virus type 1. *J. Virol.* **72,** 811–816.

Kowolik, C. M., and Yee, J.-K. (2002). Preferential transduction of human hepatocytes with lentiviral vectors pseudotyped by Sendai virus F protein. *Mol. Ther.* **5,** 762–769.

Landau, N. R., Page, K. A., and Littman, D. R. (1991). Pseudotyping with human T-cell leukemia virus type I broadens the human immunodeficiency virus host range. *J. Virol.* **65,** 162–169.

Lewis, B. C., Chinnasamy, N., Morgan, R. A., and Varmus, H. E. (2001). Developmnet of an avian leukosis-sarcoma virus subgroup a pseudotyped lentiviral vector. *J. Virol.* **75,** 9339–9344.

Lu, X., Humeau, L., Slepushkin, V., Binder, G., Yu, Q., Slepushkina, T., Chen, Z., Merling, R., Davis, B., Chang, Y. N., and Dropulic, B. (2004). Safe two-plasmid production for the first clinical lentivirus vector that achieves >99% transduction in primary cells using a one-step protocol. *J. Gene Med.* **6,** 963–973.

Merten, O., Charrier, S., Laroudie, N., Fauchille, S., Dugue, C., Jenny, C., Audit, M., Zanta-Boussif, M., Chautard, H., Radrizzani, M., Vallanti, G., Naldini, L., *et al.* (2011). Large scale manufacture and characterisation of a lentiviral vector produced for clinical *ex vivo* gene therapy application. *Hum. Gene Ther.* **22,** 343–356.

Miller, A. D., Bonham, L., Alfano, J., Kiem, H. P., Reynolds, T., and Wolgamot, G. (1996). A novel murine retrovirus identified during testing for helper virus in human gene transfer trials. *J. Virol.* **70,** 1804–1809.

Miskin, J., Chipchase, D., Rohll, J., Beard, G., Wardell, T., Angell, D., Roehl, H., Jolly, D., Kingsman, S., and Mitrophanous, K. (2006). A replication competent lentivirus (RCL) assay for equine infectious anaemia virus (EIAV)-based lentiviral vectors. *Gene Ther.* **13,** 196–205.

Miyoshi, H., Takahashi, M., Gage, F. H., and Verma, I. M. (1997). Stable and efficient gene transfer into the retina using an HIV-based lentiviral vector. *Proc.Natl. Acad. Sci. USA* **94,** 10319–10323.

Mochizuki, H., Schwartz, J. P., Tanaka, K., Brady, R. O., and Reiser, J. (1998). High-titer human immunodeficiency virus type 1-based vector systems for gene delivery into nondividing cells. *J. Virol.* **72,** 8873–8883.

Phillips, L. A., Hollis, V. W., Jr., Bassin, R. H., and Fischinger, P. J. (1973). Characterization of RNA from noninfectious virions produced by sarcoma positive-leukemia negative transformed 3T3 cells. *Proc. Natl. Acad. Sci. USA* **70,** 3002–3006.

Pyra, H., Boni, J., and Schupbach, J. (1994). Ultrasensitive retrovirus detection by a reverse transcriptase assay based on product enhancement. *Proc. Natl. Acad. Sci. USA* **191,** 1544–1548.

Reeves, L., and Cornetta, K. (2000). Clinical retroviral vector production: Step filtration using clinically approved filters improves titers. *Gene Ther.* **7,** 1993–1998.

Reeves, L., Duffy, L., Koop, S., Fyffe, J., and Cornetta, K. (2002). Detection of ecotropic replication-competent retroviruses: Comparison of S+/L- and marker rescue assays. *Hum. Gene Ther.* **13,** 1783–1790.

Reiser, J. (2000). Production and concentration of pseudotyped HIV-1-based gene transfer vectors. *Gene Ther.* **7,** 910–913.

Rohll, J. B., Mitrophanous, K. A., Martin-Rendon, E., Ellard, F. M., Radcliffe, P. A., Mazarakis, N., and Kingsman, S. M. (2002). Design, production, safety, evaluation, and clinical applications of nonprimate lentiviral vectors. *Methods Enzymol.* **346,** 466–500.

Sanburn, N., and Cornetta, K. (1999). Rapid titer determination using quantitative real-time PCR. *Gene Ther.* **6,** 1340–1345.

Sastry, L., Johnson, T., Hobson, M. J., Smucker, B., and Cornetta, K. (2002). Titering lentiviral vectors: Comparison of DNA, RNA and marker expression methods. *Gene Ther.* **9,** 1155–1162.

Sastry, L., Xu, J., Johnson, T., Desai, K., Rissing, D., Marsh, J., and Cornetta, K. (2003). Certification assays for HIV-1-based vectors: Frequent passage of gag sequences without evidence of replication competent viruses. *Mol. Ther.* **8,** 830–839.

Sastry, L., Xu, Y., Cooper, R., Pollok, K., and Cornetta, K. (2004). Evaluation of plasmid DNA removal from lentiviral vector by benzonase treatment. *Hum. Gene Ther.* **15,** 221–226.

Sastry, L., Xu, Y., Marsh, J., and Cornetta, K. (2005). Product enhanced reverse transcriptase (PERT) assay for detection of RCL associated with HIV-1 vectors. *Hum. Gene Ther.* **16,** 1227–1236.

Schonely, K., Afable, C., SLepushkin, V., Lu, X., Andre, K., Boehmer, J., Bengtson, K., Doub, M., Cohen, R., Berlinger, D., Slepushkina, T., Chen, Z., *et al.* (2003). QC release testing of an HIV-1 based lentiviral vector lot and transduced cellular product. *Bioprocess. J.* July/August, 39–47.

Segall, H. I., Yoo, E., and Sutton, R. E. (2003). Characterization and detection of artificial replication-competent lentivirus of altered host range. *Mol. Ther.* **8,** 118–129.

Slepushkin, V., Chang, N., Cohen, R., Gan, Y., Jiang, B., Deausen, E., Berlinger, D., Binder, G., Andre, K., Humeau, L., and Dropulic, B. (2003). Large-scale purification of a lentiviral vector by size exclusion chromatography or mustang Q ion exchange capsule. *Bioprocess. J.* Sep/Oct, 89–95.

Stitz, J., Buchholz, C. J., Engelstadter, M., Uckert, W., Bleimer, U., Schmitt, I., and Cichutek, K. (2000). Lentiviral vectors pseudotyped with envelope glycoproteins derived from Gibbonr Ape Leukemia Virus and Murine Leukemia Virus 10A1. *Virology* **273,** 16–20.

Transfiguracion, J., Jaalouk, D. E., Ghani, K., Galipeau, J., and Kamen, A. (2003). Size-exclusion chromatography purification of high-titer vesicular stomatitis virus G glyco-protein-pseudotyped retrovectors for cell and gene therapy applications. *Hum. Gene Ther.* **14,** 1139–1153.

Yamada, K., McCarty, D. M., Madden, V. J., and Walsh, C. E. (2003). Lentivirus vector purification using anion exchange HPLC leads to improved gene transfer. *Biotechniques* **34,** 1074–1078.

Yu, J. H., and Schaffer, D. V. (2006). Selection of novel vesicular stomatitis virus glycoproteins from a peptide insertion library for enhanced purification of retroviral and lentiviral vectors. *J. Virol.* **80,** 3285–3292.

Zufferey, R., Dull, T., Mandel, R. J., Bukovsky, A., Quiroz, D., Naldini, L., and Trono, D. (1998). Self-inactivating lentivirus vector for safe and efficiency *in vivo* gene delivery. *J. Virol.* **72,** 9873–9880.

CHAPTER SIX

Lentivirus Vectors in β-Thalassemia

Emmanuel Payen,[*,†] Charlotte Colomb,[*,†] Olivier Negre,[*,†,‡,¶] Yves Beuzard,[*,†,‡,¶] Kathleen Hehir,[‡,¶] and Philippe Leboulch[*,†,§]

Contents

1. Introduction to the β-Thalassemias	110
2. Principles of Lentiviral Gene Therapy for β-Thalassemias	111
3. Clinical Vector Design (LentiGlobin™)	112
4. cGMP Lentiviral Vector Production	113
5. cGMP Transduction of CD34⁺ Cells	113
6. GTP and GMO Release Testing	114
7. Clinical Protocol	114
7.1. Synopsis	114
7.2. Inclusion criteria for β-thalassemia patients	115
7.3. Exclusion criteria for β-thalassemia patients	116
7.4. Patient follow-up	117
8. Outcome for the First Thalassemia Patient Treated by Gene Therapy (51 Months Post-transplantation)	121
References	122

Abstract

Patients with β-thalassemia major require lifelong transfusions and iron chelation, regardless of the type of causative mutations (e.g., β^0, β^E/β^0). The only available curative therapy is allogeneic hematopoietic transplantation, although most patients do not have an HLA-matched, geno-identical donor, and those who do still risk graft-versus-host disease. Hence, gene therapy by ex vivo transfer of a functional β-globin gene is an attractive novel therapeutic modality. In β-thalassemia, transfer of a therapeutic globin gene does not confer a selective advantage to transduced stem cells, and complex DNA regulatory sequences have to be present within the transfer vector for proper expression. This is why lentiviral vectors have proven especially suited for this application, and the first Phase I/II human clinical trial was initiated. Here, we report on the

[*] CEA, Institute of Emerging Diseases and Innovative Therapies (iMETI), Fontenay aux Roses, France
[†] Inserm U962 and University Paris 11, CEA-iMETI, Fontenay aux Roses, France
[‡] Bluebird bio, Cambridge, MA, USA
[§] Harvard Medical School and Genetics Division, Department of Medicine, Brigham & Women's Hospital, Boston, Massachusetts, USA
[¶] CEA-iMETI, Fontenay aux Roses, France

first gene therapy patient with severe β^E/β^0-thalassemia, who has become transfusion-independent, and provide methods and protocols used in the context of this clinical trial.

1. Introduction to the β-Thalassemias

The β-hemoglobinopathies, β-thalassemias, and sickle cell disease are the most prevalent monogenic diseases worldwide. A mutated β-globin gene variant is present in approximately 5% of the world's population, with uneven geographical distribution, resulting in ~3/1000 newborns per year suffering from a severe form of these inherited diseases (Modell and Darlison, 2008).

The β-thalassemias are caused by a number of mutations of the β-globin gene or locus (Giardine *et al.*, 2007) that result in either no β-globin production (β^0) or reduced levels of synthesis (β^+) that lead to imbalanced α/non-α-globin chain ratios. Excess α-chain damages the cell membrane, leading to profound anemia as a result of dyserythropoisis, upon apoptosis of erythroid precursors (Mathias *et al.*, 2000), together with hemolysis and red blood cell (RBC) destruction in the spleen (Vigi *et al.*, 1969). β-Thalassemias have a wide range of clinical severity, from the severe transfusion-dependent thalassemia major to the highly variable transfusion-independent thalassemia intermedia. In addition to residual β-globin expression when present (e.g., β^+ or β^E alleles), the major genetic modulators contributing to reduced severity of the disease are inheritance of mild α-thalassemia (Camaschella *et al.*, 1995, 1997; Winichagoon *et al.*, 1985) or the ability to produce fetal Hb after birth (Cappellini *et al.*, 1981; Winichagoon *et al.*, 1993).

A subtype of special interest is compound β^E/β^0-thalassemia, which is the most common form of severe thalassemia in many Asian countries and Western immigrants from these regions (Fucharoen and Winichagoon, 2000; Olivieri *et al.*, 2008). The β^E-globin allele comprises a point mutation that causes alternative splicing. The abnormally spliced form is noncoding, whereas the correctly spliced mRNA expresses a mutated β^E-globin of decreased stability (Fucharoen and Winichagoon, 2000; Olivieri *et al.*, 2008). When this is combined with a nonfunctional β^0 allele, β-globin synthesis is very low, and approximately half of β^E/β^0-thalassemia patients require regular transfusions for survival (Fucharoen and Winichagoon, 2000; Olivieri *et al.*, 2008).

Current therapeutic options in β-thalassemias are limited. Severe cases require RBC transfusions on a monthly basis and iron chelation for survival. Pharmacological agents capable of stimulating γ-globin gene expression are of major interest for this disease (El-Beshlawy *et al.*, 2009; Perrine, 2008). However, the increased Hb levels achieved with this class of therapeutics are limited and insufficient for a number of patients (Perrine *et al.*, 2010).

Allogeneic hematopoietic stem cell (HSC) transplantation can be curative for severely affected patients who have a healthy, HLA-matched, sibling donor (Isgro et al., 2010), although only approximately 25% of patients have such a suitable donor. Further, severe complications that include graft rejection and acute or chronic graft-versus-host disease (GVHD) remain major issues (Caocci et al., 2011; Luznik et al., 2010).

2. Principles of Lentiviral Gene Therapy for β-Thalassemias

For severely affected patients who do not have a suitable donor, hematopoietic gene therapy is a promising approach, although the risk of insertional mutagenesis (Hacein-Bey-Abina et al., 2008) requires careful consideration of the risk/benefit ratio. If proven safe and effective, gene therapy may be later extended to patients with a possible allogeneic donor, as gene transfer to autologous patient's cells is not expected to cause the complications, which include GVHD, that are associated with allogeneic transplantation.

Gene addition by vector-based transfer and chromosomal integration of a normal globin gene together with suitable *cis*-linked regulatory elements into HSCs remains the approach of choice. However, efficient transduction of HSCs and high expression of globin genes restricted to erythroid cells represent major challenges. The genetic elements required for high and erythroid-specific expression while decreasing the occurrence of position effect variegation (PEV) include the β^A-globin gene with its introns, promoter, and β-locus control region (β-LCR), the discovery of which having being essential to achieving position-independent expression in transgenic mice (Grosveld et al., 1987; Tuan et al., 1985). β-LCR elements have been subsequently reduced to smaller sizes suitable for inclusion in gene transfer vectors (Leboulch et al., 1994; Sadelain et al., 1995). In contrast to retroviral vectors derived from oncovirus, such as Moloney Murine Leukemia Virus, which have posed major issues of genetic instability and low titers, lentiviral vectors have proven capable of transferring these elaborate structures with fidelity and high titers (May et al., 2000; Pawliuk et al., 2001). Lentiviral vectors have the additional property of better transducing quiescent cells that include HSCs resting in G1 (Naldini et al., 1996a,b). In an effort to decrease PEV further while shielding cellular chromatin in the vicinity of the sites of chromosomal integration from enhancer effects brought upon by the vector, genetic elements with chromatin insulator properties (Arumugam et al., 2007; Puthenveetil et al., 2004) are included in certain vectors. Hence, several mouse models of the β-hemoglobinopathies have been corrected, long-term, by *ex vivo* transduction of HSCs with β-globin lentiviral vectors (Hanawa et al., 2004; Levasseur et al., 2003; May et al., 2000; Pawliuk et al., 2001; Puthenveetil et al., 2004), including the clinical grade vector batch as used in our current clinical study (Ronen et al., 2011).

These preclinical findings have prompted the prudent initiation of a human clinical trial in France entitled "A Phase I/II Open Label Study with Anticipated Clinical Benefit Evaluating Genetic Therapy of the β-Hemoglobinopathies (Sickle Cell Anemia and β-Thalassemia Major) by Transplantation of Autologous CD34$^+$ Stem Cells Modified Ex Vivo with a Lentiviral $β^{A-T87Q}$-globin (LentiGlobin^{TM}) Vector." Approval was granted by the French regulatory agency (Afssaps) in 2005 concurrently to that also granted to a lentiviral clinical trial for childhood adrenoleukodystrophy (Cartier et al., 2009). These two trials are the first using lentiviral vectors having received regulatory approval worldwide for the gene therapy of inherited disorders. The first patient, an 18-year-old male with severe, transfusion-dependent $β^E/β^0$-thalassemia without suitable donor (only child), was engrafted on June 6, 2007. It was decided jointly with the regulatory agency to perform a thorough analysis of the outcome of the first patient before including other patients. The second patient will thus be transplanted with transduced autologous cells in October 2011.

3. Clinical Vector Design (LentiGlobin^{TM})

The general structure of the β-globin expressing lentiviral vector, referred to as LentiGlobin^{TM}, is shown in Fig. 6.1. It is a self-inactivating (SIN) vector with two copies of the 250 bp core of the cHS4 chromatin

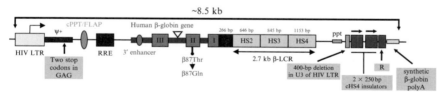

Figure 6.1 LentiGlobin^{TM} vector and experimental design. (A) Diagram of the human β-globin ($β^{A-T87Q}$) lentiviral vector (LentiGlobin^{TM}, LG). The 3′β-globin enhancer, the 372-bp IVS2 deletion, the $β^{A-T87Q}$ mutation (ACA[Thr] to CAG[Gln]) and DNase I Hypersensitive Sites (HS) 2, HS3 and HS4 of the human β-globin locus control region (LCR) are indicated. Safety modifications including the two stop codons in the ψ+ signal, the 400-bp deletion in the U3 of the right HIV LTR, the rabbit β-globin polyA signal, and the 2×250bp cHS4 chromatin insulators are indicated. HIV LTR, human immunodeficiency type-1 virus long terminal repeat; ψ+, packaging signal; cPPT/flap, central polypurine tract; RRE, Rev-responsive element; βp, human β-globin promoter; ppt, polypurine tract. (See Color Insert.)

insulator implanted in the U3 region. Because this vector has the tendency to eliminate 1 of the 250 bp elements from one or more of the long terminal repeats (LTRs) during the reverse transcription/chromosomal integration processes, the next generation of vectors will comprise more stable insulator variants or will not include any. This clinical vector encodes a mutated β-globin ($\beta^{A\text{-}T87Q}$) with anti-sickling properties that can be distinguished from normal β^A by high-performance liquid chromatography (HPLC) analysis in individuals receiving RBC transfusions and/or β^+-thalassemia patients.

4. cGMP Lentiviral Vector Production

The clinical grade LentiGlobin™ vector (the gene therapy product, GTP) is produced in a cGMP facility (e.g., Indiana University Vector Production Facility, IN, USA) by transfection of HEK293T cells expanded from a qualified master cell bank. The five LentiPak™ plasmids described in Westerman et al. (2007), that is, HPV569 (Lentiglobin™ transfer vector containing the modified globin gene; Fig. 6.1), HPV275 (gag–pol), ΨN 15 (VSV-G env), p633 (rev), and HPV601 (tat)) are manufactured and purified to qualify for use in cGMP manufacturing (Puresyn, PA, USA). The calcium phosphate-mediated transient transfection (Promega, WI, USA) in DMEM media (Invitrogen) supplemented with New Zealand sourced 5% fetal bovine serum (HyClone) is followed by harvest in serum-free media (OptiPRO; Invitrogen). Clarification, concentration, Benzonase™ treatment (Novagen®), and diafiltration (TFF/UF, Spectrum) into X Vivo 20 medium (Lonza) produced the Lentiglobin™ GTP which is subsequently cryopreserved until release testing is completed. The infectious titer is determined by transducing NIH 3 T3 mouse fibroblasts with serial dilutions of the viral supernatant in the presence of polybrene (8 μg/mL). After 7 days in culture, the total DNA is prepared and integrated copies are quantified by TaqMan assay using GAG primers and normalized by Southern blot analyses. The titer must be $\geq 1.0 \times 10^8$ TU/mL.

5. cGMP Transduction of CD34$^+$ Cells

Prestimulation of cells (>90% CD34$^+$) is conducted for 24–36 h in RetroNectin™-coated (50 μg/mL, 2 h RT; Takara) X-fold (Miltenyi) cell culture bags in ex vivo-20 media (Lonza) containing the recombinant human cytokines SCF, Flt-3L, IL-3, and TPO (Peprotech) at 300, 300, 100, and 10 ng/mL, respectively, as well as protamine sulfate at 4 μg/mL (Choay; AMM# 310 117.3 (1955/1998)). The transduction of cells (100% CD34$^+$)

is performed in undiluted GTP for 18–24h at ~2×10^6 CD34$^+$ cells/mL in the same media with cytokines. The post-transduction cells are then washed and cryopreserved until release testing is completed.

6. GTP and GMO Release Testing

The LentiGlobin^{TM} GTP is tested and released according to specifications agreed upon for potency (transducing titer of the GTP on NIH3T3 cells followed by qPCR analysis, and GTP particle concentration by p24 ELISA), identity (β^A-globin protein production using HPLC analysis of erythroid cells differentiated from GTP-transduced human CD34$^+$ cells, and restriction digest/Southern blot analysis of GTP-transduced cells), safety (sterility, endotoxin, mycoplasma, adventitious viruses, and replication competent lentivirus using amplification/coculture with readout assays as described in Sastry et al., 2003), purity (residual host and plasmid DNA, residual Benzonase^{TM}, and residual EGF), appearance, and pH. The LentiGlobin^{TM} GMO is also tested and released according to specifications agreed upon for potency and identity (transduction efficiency of the GTP-transduced patient CD34$^+$ cells using qPCR analyses and restriction digest/Southern blot analysis of GTP-transduced patient CD34$^+$ cells, the expression of β^{A-T87Q}-globin in patient erythroid progenitors using BFU-E culture followed by HPLC analysis, and the level of hemoglobin containing β^A-globin in bulk erythroid cell progeny using HPLC analysis), safety (sterility, endotoxin, mycoplasma, and replication competent lentivirus using amplification/coculture of 1% of the patients GTP-transduced CD34$^+$ cells with readout assays as described in Sastry et al., 2003 as well as RT-PCR for gag sequences), and preinfusion viability.

7. Clinical Protocol

7.1. Synopsis

Patients are selected from a population with at least 2 years of follow-up at a specialized center. Selected patients must meet the inclusion criteria and have none of the exclusion criteria (see below) and have given their informed consent. Following a screening period, patients undergo a period of hypertransfusion followed by either peripheral blood mobilization or bone marrow harvest under general anesthesia. The bone marrow is processed for CD34$^+$ cell selection. If sufficient cells are not obtained, additional mobilizations and/or a second bone marrow harvest may be performed. A portion of CD34$^+$ cells is cryopreserved for rescue therapy, and a portion transduced with the Lentiviral vector encoding the β-chain of hemoglobin (i.e., the GTP). A sample of transduced cells (i.e., the genetically modified organism or GMO) undergoes release testing, while the remainder is cryopreserved for approximately 2 months. Genetically altered cells are not infused into patients until all release testing, including extensive screening for replication competent lentivirus (RCL), is completed and meets specifications (see GTP and GMO release testing below). Patients are admitted to the bone marrow transplant unit of the hospital and undergo conditioning with Busulfex™ to induce myelosuppression at the fully myeloablative dose. If the absolute neutrophil count is $<500/\mu L$, patients are kept in isolation. The transduced cells are washed, counted, and checked for viability. The LentiGlobin™-transduced $CD34^+$ cells are infused into the patients via a central catheter at a dose of at least 3×10^6 $CD34^+$ cells/kg over 20 min while vital signs are monitored. Patients are followed daily in the transplant unit for adverse events and laboratory parameters to monitor bone marrow engraftment.

7.2. Inclusion criteria for β-thalassemia patients

- Patients must be at least 15 years of age and must be sexually mature at the time of study entry to allow for the option of sperm or ovarian, oocyte and/or embryo preservation banking. Sexual maturity in females is defined as the onset on menses. Male patients must be able to donate or undergo sperm extraction; the resulting sample for banking must have a density of at least 20×10^6/mL and $>40\%$ motility.
- Patients must have one of the genetic forms of β-thalassemia confirmed by Hb studies and genomic DNA analysis.
- Patients must have permanent transfusion dependency defined as requiring at least 100 mL/kg/year of packed RBCs and a requirement for iron chelation therapy. The diagnosis must be confirmed by Hb studies.
- Patients must be candidates for allogeneic bone marrow transplant and not have a suitable, willing HLA-identical sibling donor.
- Patients must meet Lansky performance status $\geq 70\%$ for children or Karnofsky index $\geq 80\%$ for adults.
- Patients must have survival expectancy >6 months.
- Patients must have been treated and followed up for at least the past 2 years in a specialized center where they have undergone full evaluation of their disease, including psychiatric evaluation. Detailed medical records of this care for at least the past 2 years must be available, as each patient will serve as their own control.
- Patients ≥ 18 years of age must be able to provide written informed consent and have had a psychiatric evaluation to establish their motivation, the clarity of their consent, and the absence of severe psychiatric disease. For patients <18 years of age, both parents and legal guardian

must be able to provide written informed consent and must be willing to undergo a psychiatric evaluation to establish their motivation for enrolling the minor, the clarity of their consent, and the absence of severe psychiatric disease. When possible, involvement of the child in the decision is highly recommended and should be clearly documented.
- Discontinuation of any hydroxyurea therapy at least 3 months prior to infusion of stem cells.

7.3. Exclusion criteria for β-thalassemia patients

- Having a suitable, willing HLA-identical sibling.
- HIV seropositivity.
- Positivity for Hepatitis B surface antigen.
- Positivity for Hepatitis C antibody.
- Active bacterial, viral (e.g., hepatitis B or C), or fungal infection.
- A contraindication to anesthesia for bone marrow harvesting.
- A history of prior malignancy (excluding basal cell carcinoma of the skin or *in situ* carcinoma of the cervix, curatively resected), even if considered cured.
- A white blood cell (WBC) count <3000/μL and/or platelet count <120,000/μL.
- A family member with a known or a suspected familial cancer syndrome (including but not limited to breast, colorectal, ovarian, prostate, and pancreatic cancers).
- Prior bone marrow transplantation.
- A history of psychosis, any psychiatric disorder, severe mental retardation, or current drug or alcohol abuse, which, in the opinion of the investigator, would make the patient unsuitable for participation in the study.
- A history of malaria relapses in the absence of recent infestation.
- The presence of antibodies against vesicular stomatitis virus (VSV).
- A history of complex allo-immunization which could cause difficulty administering transfusions.
- Pregnancy or breastfeeding in a postpartum female or absence of adequate contraception for fertile patients. Females of child-bearing potential must agree to use a medically acceptable method of birth control such as oral contraceptive, intrauterine device, barrier, and spermicide, or contraceptive implant/injection throughout the 26-month study period. Premenarchal females must agree to use a medically acceptable method of birth control if their child-bearing status should change during the study period. A female will be considered postmenopausal if it has been at least 12 months since her last menstrual cycle or if she has undergone surgical sterilization.
- A history of major organ damage including:

- Cerebral vascular disease with severe neurological or cognitive-mental sequelae, excluding isolated hemiplegia
- Liver disease with ALT >3× upper limit of normal or the presence of histopathological evidence of liver cirrhosis on liver biopsy.
- Heart disease with ejection fraction <25%.
- Kidney disease with a calculated creatinine clearance <30% normal value.
- Lung disease with a substantial alteration in pulmonary function tests (i.e., $pO_2 < 90$ and/or carbon dioxide diffusion coefficient <60%).
- The presence of pulmonary fibrosis or pulmonary hypertension.
- Hormonal disorders including insulin-dependent diabetes, hyper- or hypo-thyroidism, or hypo- or hyper-parathyroidism.
- Participation in another clinical study with an investigational drug within 30 days of screening.
- Presence of chromosomal abnormalities in bone marrow detected after bone marrow harvest.
- Absence of informed consent.
- Presence of Lucarelli Class III (inadequacy of iron chelation therapy, presence of hepatomegaly, and portal fibrosis on liver biopsy).
- Patients who are compliant with iron chelation therapy and who have young children and/or a desire to become a parent in the short term.
- Patients whose transfusion requirement is decreased by erythropoietin treatment.

7.4. Patient follow-up

Once engraftment occurs and patients are stable, they are discharged from hospital and followed monthly for 6 months then at least every 3 months for a total of 3 years and then every year. Evaluations must include routine hematology and chemistry and special hematologic testing (see below), bone marrow examination, collection of adverse events and concomitant medications, and evaluation of disease-specific hematologic and clinical parameters.

7.4.1. Determination of vector copy numbers in peripheral blood cells

Granulocytes ($CD15^+$), monocytes ($CD14^+$) T-Lymphocytes ($CD3^+$), B-lymphocytes ($CD19^+$), and erythroid ($CD45^-$) cells are purified by two rounds of magnetic cell sorting using whole blood purification kits (Miltenyi Biotec). Specific antibodies against CD3, CD20, CD15, CD71 (Miltenyi Biotec), GlycophorinA, CD45 (eBiosciences), and a flow cytometer are used to check for purity (it must be more than 98%). Genomic DNA is prepared using Nucleospin blood kits (Macherey-Nagel). Quantitative PCR using TaqMan probes with primers amplifying the GAG region of the vector and the endogenous human *β-actin* gene is used to

Table 6.1 Primers and probes used for qPCR

Assays	Amplicons	Name	Sequence (5′–3′) or TaqMan gene expression assay number*	Modification	Conc.
Vector copy number in human and mouse cells	Vector	GAG F	GGAGCTAGAACGATTCGCAGTTA	–	720 nM
		GAG R	GGTTGTAGCTGTCCCAGTATTTGTC	–	720 nM
		GAG P	ACAGCCTTCTGATGTCTCTAAAAGGCCAGG	5′FAM 3′TAMRA	140 nM
	hβ-actin	hbAct F	TCCGTGTGGATCGGCGGCTCCA	–	900 nM
		hbAct R	CTGCTTGCTGATCCACATCTG	–	900 nM
		hbAct P	CCTGGCCTCGCTGTCCACCTTCCA	5′FAM 3′TAMRA	250 nM
	mβ-actin		Mm00607939_S1*	5′FAM 3′NFQ-MGB	1×

FAM, 6-carboxyfluorescein ester; TAMRA, tetramethyl-6-carboxyrhodamine; NFQ, nonfluorescent quencher; MGB, minor groove binder; Conc., concentration
* Applied Biosystems.

Table 6.2 Primers used for PCR on DNA of CFC colonies

Assays	Amplicons	Size (bp)	Name	Sequence (5'–3')
Detection of the vector in human colonies	LG vector	431	LGF10	GAGAGCGTCGGTATTAAGC
			LGR11	TGGCCTGATGTACCATTTGC
	Human Epo	322	Epo24	CGCTTTGGAGGCGATTTACC
			Epo26	CATTTCCCGGACCTGGACC

quantify vector copy number (Table 6.1). Results are compared with those obtained after serial dilutions of genomic DNA from a cell line containing one copy of the integrated globin lentiviral vector per haploid genome. The percent of specific insertion sites relative to total provirus is determined by comparative CT method ($\Delta\Delta$CT).

7.4.2. Proportion of transduced hematopoietic progenitors

Peripheral blood mononuclear cells or CD34$^+$ bone marrow cells are cultured over 14 days in α-MEM medium (PAA)-based methylcellulose medium (Methocult H4230, Stemcell Technologies) supplemented with 2 mM L-glutamine, antibiotics, hEpo (3 U/mL; Assay design), IL3 (10 ng/mL; Peprotech), and hSCF (50 ng/mL; Peprotech) at 3×10^5 and 2500 mL^{-1}, respectively. The colonies are counted and collected after 14 days incubation at 37 °C and 5% CO_2. They are washed with PBS and kept frozen for subsequent analysis. DNA is obtained upon proteinase K lysis and purified using the Qiaex II gel extraction beads (Qiagen). Detection of the LG vector is carried out by PCR with primers amplifying the vector. The presence of DNA is monitored using primers amplifying the human erythropoietin gene (Table 6.2).

7.4.3. LTC-IC assays

LTC-IC assays are performed in StemSpan SFEM medium (Stemcell technologies) on irradiated MS5 monolayers at several dilutions of CD34$^+$ cells (2000–16 per well) in 96-well plates with 12–24 replicate wells per concentration. After 5 weeks with weekly change of one-half medium volume, all cells are transferred in α-MEM-based methylcellulose medium (GF H4434, Stemcell technologies) to determine the total clonogenic cell content of each LTC. The frequency of LTC-IC is determined using the L-Calc software (Stemcell technologies). The mean number of colonies produced by LTC-IC, or proliferative potential, is calculated by dividing the total number of LTC-IC by the total number of CFCs (Rizzo et al., 2002). In order to assess the percentage of vector modified LTC-IC during the readout phase of the assay, a maximum of one clonogenic colony per well is submitted to PCR-based scoring to ensure that only independent LTC-ICs are analyzed.

7.4.4. Globin chain production

Globin chains from whole blood, pooled erythroid colonies and reticulocytes are separated by reverse phase HPLC using a 4×250 mm Nucleosil 300-5 C4 column (Macherey-Nagel). Samples are eluted with a gradient mixture of solution A (water/acetonitrile/trifluoroacetic acid/heptafluorobutyric acid, 700:300:0.7:0.1) and solution B (water/acetonitrile/trifluoroacetic acid/heptafluorobutyric acid, 450:550:0.5:0.1). The absorbance is measured at 220 nm. Hemoglobins from individual erythroid colonies are separated by ion exchange HPLC on a PolyCAT A column (PolyLC Inc.). Elution is achieved with a linear gradient mixture of buffer C (Tris 40mM, KCN 3mM; pH adjusted at 6.5 with acetic acid) and buffer D (Tris 40mM, KCN 3mM, NaCl 200mM; pH adjusted at 6.5 with acetic acid) of different ionic strength. The detection wavelength is 418 nm.

7.4.5. Bone marrow karyotype and high-resolution array-CGH analysis

Total BM cells are cultured for 17, 24, and 48h. Metaphases are treated for reverse heat and giemsa (RHG) banding, and 30 mitoses are fully analyzed following the recommendations of the International System for Human Cytogenetic Nomenclature (Mitelman, 1995). Genomic copy number analysis is performed using high-density CGH Arrays technology as described (Clappier et al., 2007; Hacein-Bey-Abina et al., 2008). Five hundred nanograms of genomic DNA is labeled and cohybridized with control DNA on the 244K Human Genome CGH Microarray (Agilent Technologies). Scanned data are processed using Feature Extraction software and DNA Analytics software (Agilent Technologies). The analysis tools ADM1, ADM2, and visual inspection are used to search for copy number abnormalities.

7.4.6. Integration site analysis (LM-PCR and DNA pyrosequencing)

Aliquots of genomic DNA extracted from patient samples are digested using two or three different restriction enzymes. The digested samples are ligated to DNA linkers, then digested to cleave the internal fragments derived from the vector, and amplified by nested PCR as previously described (Hacein-Bey-Abina et al., 2008). Each second-round LTR-specific primer contains a unique 8nt barcode which indexes the amplification products (Hacein-Bey-Abina et al., 2008; Wang et al., 2008). The PCR products are gel purified, pooled, and sequenced using a 454/Roche GS FLX platform. Pyrosequencing reads are decoded, trimmed to remove LTR and linker sequences, and then mapped to the human genome to yield integration sites (ISs) using criteria as previously described (Hacein-Bey-Abina et al., 2008; Wang et al., 2007).

8. Outcome for the First Thalassemia Patient Treated by Gene Therapy (51 Months Post-transplantation)

As indicated above, the first treated patient was 18 years of age at the time of treatment on June 6, 2007. He suffered from severe β^E/β^0-thalassemia and required monthly RBC transfusions since early childhood to stay alive. He was splenectomized at age 6 years, did not have a positive response to hydroxyurea, and did not have a possible sibling donor. Bone marrow $CD34^+$ cells were harvested, prestimulated, and transduced with the LentiGlobin vector. The transduction efficiency, evaluated on myeloid progenitors 1 week after transduction, was 0.6 vector copies per cell. Transduced $CD34^+$ cells (3.9×10^6 cells/kg) were injected after conditioning. The percentage of modified peripheral blood cells in this patient increased slowly and stabilized to $\sim$10% at 30 months after transplantation. In the bone marrow 3 years post-transplant, the hematopoietic compartment with the highest proportion of modified cells was the erythroid lineage (about 30%). This result was expected as genetically modified cells are likely to have a selective advantage over unmodified cells because they are less prone to dyserythropoiesis (premature death of abnormal cells). Remarkably, 1 year after gene therapy and BMT, the patient became transfusion independent and the Hb level stabilized above 9 g/dL. The number of peripheral blood erythroblasts decreased and RBC survival improved. The MCH level is within the normal range, and the therapeutic β^{A-T87Q}-globin output on a per gene basis is between 70% and 100% of the normal value. The patient is now phlebotomized monthly to reduce iron overload, and plasma ferritin levels have decreased significantly (unpublished data). In spite of these repeated phlebotomies, the patient has not required nor received any blood transfusion for now over 3 years (39 months), this 4 years (51 months) after transplantation and gene therapy (Cavazzana-Calvo et al., 2010).

Chromosomal IS analysis of the vector in peripheral blood and in purified hematopoietic cell compartments showed the relative dominance of an IS in the high-mobility group AT-hook 2 (*HMGA2*) gene. This specific IS was detected in myeloid cells but not in B- nor in T-cells. The corresponding myeloid cell clone was first detected 4 months post-BMT, stabilized since month 15, and now represents $\sim$2–3% of the circulating nucleated cells. As translocation events in the *HMGA2* gene and overexpression of a truncated form of HMGA2 have been involved in neoplasia (Cleynen and Van de Ven, 2008), though mostly benign, careful and regular follow-up of this patient is pursued. So far, no breach of hematopoietic homeostasis has been observed.

Long-term clinical benefit (complete transfusion independence for over 3 years) has thus been achieved in the first treated patient with severe, transfusion-dependent β^E/β^0-thalassemia, and other patients are now in the process of being treated. This study has also increased our knowledge of the occurrence of endogenous gene activation upon lentiviral vector integration and the dynamics of human hematopoiesis. Larger series of patients will ultimately shed light on risk–benefit ratios for the gene therapy of the β-hemoglobinopathies and for hematopoietic gene therapy as a whole.

REFERENCES

Arumugam, P. I., Scholes, J., Perelman, N., Xia, P., Yee, J. K., and Malik, P. (2007). Improved human beta-globin expression from self-inactivating lentiviral vectors carrying the chicken hypersensitive site-4 (cHS4) insulator element. *Mol. Ther.* **15,** 1863–1871.

Camaschella, C., Mazza, U., Roetto, A., Gottardi, E., Parziale, A., Travi, M., Fattore, S., Bacchiega, D., Fiorelli, G., and Cappellini, M. D. (1995). Genetic interactions in thalassemia intermedia: Analysis of beta-mutations, alpha-genotype, gamma-promoters, and beta-LCR hypersensitive sites 2 and 4 in Italian patients. *Am. J. Hematol.* **48,** 82–87.

Camaschella, C., Kattamis, A. C., Petroni, D., Roetto, A., Sivera, P., Sbaiz, L., Cohen, A., Ohene-Frempong, K., Trifillis, P., Surrey, S., and Fortina, P. (1997). Different hematological phenotypes caused by the interaction of triplicated alpha-globin genes and heterozygous beta-thalassemia. *Am. J. Hematol.* **55,** 83–88.

Caocci, G., Efficace, F., Ciotti, F., Roncarolo, M. G., Vacca, A., Piras, E., Littera, R., Dawood Markous, R. S., Collins, G. S., Ciceri, F., Mandelli, F., Marktel, S., *et al.* (2011). Prospective assessment of health-related quality of life in pediatric beta-thalassemia patients following hematopoietic stem cell transplantation. *Biol. Blood Marrow Transplant.* **17,** 861–866.

Cappellini, M. D., Fiorelli, G., and Bernini, L. F. (1981). Interaction between homozygous beta (0) thalassaemia and the Swiss type of hereditary persistence of fetal haemoglobin. *Br. J. Haematol.* **48,** 561–572.

Cartier, N., Hacein-Bey-Abina, S., Bartholomae, C. C., Veres, G., Schmidt, M., Kutschera, I., Vidaud, M., Abel, U., Dal-Cortivo, L., Caccavelli, L., Mahlaoui, N., Kiermer, V., *et al.* (2009). Hematopoietic stem cell gene therapy with a lentiviral vector in X-linked adrenoleukodystrophy. *Science* **326,** 818–823.

Cavazzana-Calvo, M., Payen, E., Negre, O., Wang, G., Hehir, K., Fusil, F., Down, J., Denaro, M., Brady, T., Westerman, K., Cavallesco, R., Gillet-Legrand, B., *et al.* (2010). Transfusion independence and HMGA2 activation after gene therapy of human beta-thalassaemia. *Nature* **467,** 318–322.

Clappier, E., Cuccuini, W., Kalota, A., Crinquette, A., Cayuela, J. M., Dik, W. A., Langerak, A. W., Montpellier, B., Nadel, B., Walrafen, P., Delattre, O., Aurias, A., *et al.* (2007). The C-MYB locus is involved in chromosomal translocation and genomic duplications in human T-cell acute leukemia (T-ALL), the translocation defining a new T-ALL subtype in very young children. *Blood* **110,** 1251–1261.

Cleynen, I., and Van de Ven, W. J. (2008). The HMGA proteins: A myriad of functions (Review). *Int. J. Oncol.* **32,** 289–305.

El-Beshlawy, A., Hamdy, M., and El Ghamarawy, M. (2009). Fetal globin induction in beta-thalassemia. *Hemoglobin* **33,** 197–203.

Fucharoen, S., and Winichagoon, P. (2000). Clinical and hematologic aspects of hemoglobin E beta-thalassemia. *Curr. Opin. Hematol.* **7,** 106–112.

Giardine, B., van Baal, S., Kaimakis, P., Riemer, C., Miller, W., Samara, M., Kollia, P., Anagnou, N. P., Chui, D. H., Wajcman, H., Hardison, R. C., and Patrinos, G. P. (2007). HbVar database of human hemoglobin variants and thalassemia mutations: 2007 update. *Hum. Mutat.* **28,** 206.

Grosveld, F., van Assendelft, G. B., Greaves, D. R., and Kollias, G. (1987). Position-independent, high-level expression of the human beta-globin gene in transgenic mice. *Cell* **51,** 975–985.

Hacein-Bey-Abina, S., Garrigue, A., Wang, G. P., Soulier, J., Lim, A., Morillon, E., Clappier, E., Caccavelli, L., Delabesse, E., Beldjord, K., Asnafi, V., MacIntyre, E., et al. (2008). Insertional oncogenesis in 4 patients after retrovirus-mediated gene therapy of SCID-X1. *J. Clin. Invest.* **118,** 3132–3142.

Hanawa, H., Hargrove, P. W., Kepes, S., Srivastava, D. K., Nienhuis, A. W., and Persons, D. A. (2004). Extended beta-globin locus control region elements promote consistent therapeutic expression of a gamma-globin lentiviral vector in murine beta-thalassemia. *Blood* **104,** 2281–2290.

Isgro, A., Gaziev, J., Sodani, P., and Lucarelli, G. (2010). Progress in hematopoietic stem cell transplantation as allogeneic cellular gene therapy in thalassemia. *Ann. N. Y. Acad. Sci.* **1202,** 149–154.

Leboulch, P., Huang, G. M., Humphries, R. K., Oh, Y. H., Eaves, C. J., Tuan, D. Y., and London, I. M. (1994). Mutagenesis of retroviral vectors transducing human beta-globin gene and beta-globin locus control region derivatives results in stable transmission of an active transcriptional structure. *EMBO J.* **13,** 3065–3076.

Levasseur, D. N., Ryan, T. M., Pawlik, K. M., and Townes, T. M. (2003). Correction of a mouse model of sickle cell disease: Lentiviral/antisickling beta-globin gene transduction of unmobilized, purified hematopoietic stem cells. *Blood* **102,** 4312–4319. Epub 2003 Aug 4321.

Luznik, L., Jones, R. J., and Fuchs, E. J. (2010). High-dose cyclophosphamide for graft-versus-host disease prevention. *Curr. Opin. Hematol.* **17,** 493–499.

Mathias, L. A., Fisher, T. C., Zeng, L., Meiselman, H. J., Weinberg, K. I., Hiti, A. L., and Malik, P. (2000). Ineffective erythropoiesis in beta-thalassemia major is due to apoptosis at the polychromatophilic normoblast stage. *Exp. Hematol.* **28,** 1343–1353.

May, C., Rivella, S., Callegari, J., Heller, G., Gaensler, K. M., Luzzatto, L., and Sadelain, M. (2000). Therapeutic haemoglobin synthesis in beta-thalassaemic mice expressing lentivirus-encoded human beta-globin. *Nature* **406,** 82–86.

Mitelman, F. (ed.) (1995). An International System for Human Cytogenetic Nomenclature, S. Karger, Basel, Switzerland.

Modell, B., and Darlison, M. (2008). Global epidemiology of haemoglobin disorders and derived service indicators. *Bull. World Health Organ.* **86,** 480–487.

Naldini, L., Blomer, U., Gage, F. H., Trono, D., and Verma, I. M. (1996a). Efficient transfer, integration, and sustained long-term expression of the transgene in adult rat brains injected with a lentiviral vector. *Proc. Natl. Acad. Sci. USA* **93,** 11382–11388.

Naldini, L., Blomer, U., Gallay, P., Ory, D., Mulligan, R., Gage, F. H., Verma, I. M., and Trono, D. (1996b). In vivo gene delivery and stable transduction of nondividing cells by a lentiviral vector. *Science* **272,** 263–267.

Olivieri, N. F., Muraca, G. M., O'Donnell, A., Premawardhena, A., Fisher, C., and Weatherall, D. J. (2008). Studies in haemoglobin E beta-thalassaemia. *Br. J. Haematol.* **141,** 388–397.

Pawliuk, R., Westerman, K. A., Fabry, M. E., Payen, E., Tighe, R., Bouhassira, E. E., Acharya, S. A., Ellis, J., London, I. M., Eaves, C. J., Humphries, R. K., Beuzard, Y., et al. (2001). Correction of sickle cell disease in transgenic mouse models by gene therapy. *Science* **294,** 2368–2371.

Perrine, S. P. (2008). Fetal globin stimulant therapies in the beta-hemoglobinopathies: Principles and current potential. *Pediatr. Ann.* **37,** 339–346.

Perrine, S. P., Castaneda, S. A., Chui, D. H., Faller, D. V., Berenson, R. J., Siritanaratku, N., and Fucharoen, S. (2010). Fetal globin gene inducers: Novel agents and new potential. *Ann. N. Y. Acad. Sci.* **1202,** 158–164.

Puthenveetil, G., Scholes, J., Carbonell, D., Xia, P., Qureshi, N., Zeng, L., Li, S., Yu, Y., Hiti, A. L., Yee, J. K., and Malik, P. (2004). Successful correction of the human beta-thalassemia major phenotype using a lentiviral vector. *Blood* **3**, 3.

Rizzo, S., Scopes, J., Elebute, M. O., Papadaki, H. A., Gordon-Smith, E. C., and Gibson, F. M. (2002). Stem cell defect in aplastic anemia: Reduced long term culture-initiating cells (LTC-IC) in CD34+ cells isolated from aplastic anemia patient bone marrow. *Hematol. J.* **3**, 230–236.

Ronen, K., Negre, O., Roth, S., Colomb, C., Malani, N., Denaro, M., Brady, T., Fusil, F., Gillet-Legrand, B., Hehir, K., Beuzard, Y., Leboulch, P., et al. (2011). Distribution of lentiviral vector integration sites in mice following therapeutic gene transfer to treat beta-thalassemia. *Mol. Ther.* **19**, 1273–1286.

Sadelain, M., Wang, C. H., Antoniou, M., Grosveld, F., and Mulligan, R. C. (1995). Generation of a high-titer retroviral vector capable of expressing high levels of the human beta-globin gene. *Proc. Natl. Acad. Sci. USA* **92**, 6728–6732.

Sastry, L., Xu, Y., Johnson, T., Desai, K., Rissing, D., Marsh, J., and Cornetta, K. (2003). Certification assays for HIV-1-based vectors: Frequent passage of gag sequences without evidence of replication-competent viruses. *Mol. Ther.* **8**, 830–839.

Tuan, D., Solomon, W., Li, Q., and London, I. M. (1985). The "beta-like-globin" gene domain in human erythroid cells. *Proc. Natl. Acad. Sci. USA* **82**, 6384–6388.

Vigi, V., Volpato, S., Gaburro, D., Conconi, F., Bargellesi, A., and Pontremoli, S. (1969). The correlation between red-cell survival and excess of alpha-globin synthesis in beta-thalassemia. *Br. J. Haematol.* **16**, 25–30.

Wang, G. P., Ciuffi, A., Leipzig, J., Berry, C. C., and Bushman, F. D. (2007). HIV integration site selection: Analysis by massively parallel pyrosequencing reveals association with epigenetic modifications. *Genome Res.* **17**, 1186–1194.

Wang, G. P., Garrigue, A., Ciuffi, A., Ronen, K., Leipzig, J., Berry, C., Lagresle-Peyrou, C., Benjelloun, F., Hacein-Bey-Abina, S., Fischer, A., Cavazzana-Calvo, M., and Bushman, F. D. (2008). DNA bar coding and pyrosequencing to analyze adverse events in therapeutic gene transfer. *Nucleic Acids Res.* **36**, e49.

Westerman, K. A., Ao, Z., Cohen, E. A., and Leboulch, P. (2007). Design of a trans protease lentiviral packaging system that produces high titer virus. *Retrovirology* **4**, 96.

Winichagoon, P., Fucharoen, S., Weatherall, D., and Wasi, P. (1985). Concomitant inheritance of alpha-thalassemia in beta 0-thalassemia/Hb E disease. *Am. J. Hematol.* **20**, 217–222.

Winichagoon, P., Thonglairoam, V., Fucharoen, S., Wilairat, P., Fukumaki, Y., and Wasi, P. (1993). Severity differences in beta-thalassaemia/haemoglobin E syndromes: Implication of genetic factors. *Br. J. Haematol.* **83**, 633–639.

CHAPTER SEVEN

Gene Therapy for Chronic Granulomatous Disease

Elizabeth M. Kang *and* Harry L. Malech

Contents

1. Introduction	126
2. Vector Design	130
3. Vector Production	132
3.1. Producer cell line description	132
4. Preclinical Testing of Vector Titer and Transgene Function (*In Vitro* Culture, Mouse Knockout (KO) CGD Models, and Mouse Models of Human CD34+ HSC Xenograft)	134
5. Clinical Scale Production	137
6. Target Cell	139
7. Clinical Transduction	140
8. Patient Care and Monitoring	142
9. Future Endeavors	145
9.1. Lentivectors	147
9.2. Zinc finger/TALENs	148
9.3. iPSC	149
10. Conclusion	149
References	150

Abstract

Mutations in phagocyte NADPH oxidase cause CGD, resulting in recurrent infections and granulomatous inflammation. Hematopoietic stem cell (HSC) transplant can cure CGD, but most patients lack a suitable donor. We conducted a clinical trial of *ex vivo* autologous HSC gene transfer as salvage therapy for three patients with X-linked CGD (X-CGD) who had incurable infection. Patients received nonmyeloablative busulfan conditioning and then were infused with amphotropic MFGS-gp91phox murine retrovirus vector-transduced autologous HSC, resulting in early gene marking and high-level oxidase function correction of 24%, 5%, and 4% of circulating neutrophils. Subjects #1 and #3 fully resolved infection and have maintained gene marking at 5 years at 0.7% and

Laboratory of Host Defenses, National Institute of Allergy and Infectious Diseases, National Institutes of Health, Bethesda, Maryland, USA

0.03% oxidase-normal neutrophils. Subject #2 lost gene marking by 4 weeks and at 6 months succumbed to his infection. The two surviving subjects have normal blood count and bone marrow exam, with no evidence for clonal dominance of vector inserts. We conclude that gene therapy salvage treatment for severe infection unresponsive to conventional therapy can provide life-saving clinical benefit to CGD patients lacking a suitable donor. We are developing lentivectors for our next generation gene therapy of CGD. We are also exploring novel alternate approaches to gene therapy using zinc finger nuclease-mediated gene targeting of induced pluripotent stem cells derived from CGD patients.

1. INTRODUCTION

Chronic granulomatous disease (CGD) comprises a set of genetic disorders affecting the production of NADPH oxidase by phagocytic cells, including neutrophils, monocytes, eosinophils, dendritic cells, and many types of tissue macrophages (Kang and Malech, 2009; Kang et al., 2011; Malech and Hickstein, 2007; Segal et al., 2011). Normal production of superoxide by the NADPH oxidase requires the translocation of four cytoplasmic based oxidase components, that is p47phox, p40phox, p67phox, and rac2, to the cell membrane where they complex with the a transmembrane heteromeric flavocytochrome b558 consisting of peptides from p22phox and gp91phox to form a functional phagocytic oxidase (phox). Although p40phox does not seem to be required for enzyme activity in cell-free preparations of the enzyme, in the intact cell, this component promotes the oxidase activation by increasing the affinity of the p47phox for the enzyme (Matute et al., 2009). Patients with deleterious mutations in any of the five phox components may manifest some or all of the clinical features of CGD. CGD due to deficiency in rac2 has not been observed, but heterozygous carriers of dominant-negative function mutations in rac2 have been described who have defects in neutrophil migration and recurrent infections but appear to have intact oxidative burst (Williams et al., 2000). Clinical manifestations of CGD may include recurrent bacterial and/or fungal infections as well as hyperinflammation and autoimmune disorders, though in some patients the problems with inflammation/autoimmune complications may predominate.

The majority of the cases (~70%) are X chromosome-linked CGD (X-CGD) involving a mutation in or deletion of the *CYBB* gene that encodes gp91phox. Mutations are located throughout *CYBB* in different kindreds. The most common autosomal form of CGD (AR-CGD), accounting for ~25% of cases, results from the deficiency of p47phox encoded by *NCF1* on chromosome 7. Curiously, >90% of mutant alleles causing p47phox AR-CGD are the same mutation of *NCF1* consisting of a GT deletion at the exon 2 splice acceptor site resulting in a frameshift with early stop. There are two nearby pseudogenes of *NCF1* that contain this GT deletion, and it has been postulated that gene conversion errors from misalignment during meiosis are the cause of the relatively high incidence of this specific mutation (Vázquez et al., 2001). Mutations in p22phox (*CYBA*, chromosome 16), p67phox (*NCF2*, chromosome 1), and p40phox (*NCF4*, chromosome 22) account for the remaining 5–6% of cases. Only two cases of p40phox deficiency CGD have been observed, and in both cases there is significant residual oxidase activity, and inflammatory bowel disease is the predominant clinical manifestation (Matute et al., 2009). Notably, all the p47phox and 67phox and the two p40phox mutations described to date as well as the majority of the p22 mutations result in a null function—complete absence of protein; however, there are patients with the gp91phox form, and a few of the p22phox form, that have point mutations resulting in production of normal amounts of a nonfunctional protein. Despite the presence of this nonfunctional protein in some cases, there has been no evidence of a dominant-negative effect, which would otherwise be a concern when considering gene therapy-based treatments for these patients.

It has been shown that allogeneic donor hematopoietic stem cell (HSC) transplant for CGD is curative of the infectious complications as well as of the autoimmune-associated complications of CGD and is a treatment modality being considered more frequently by various centers, particularly as advances in the HLA typing of unrelated donors has improved the likelihood of finding suitable donors for patients with this group of monogenetic disorders (Horwitz et al., 2001; Kang et al., 2011; Seger, 2010; Seger et al., 2002). However, allogeneic transplant still carries risks of both graft rejection as well as graft versus host disease and still requires the use of moderate doses of immunosuppressive chemotherapy.

Female carriers of X-CGD are phenotypic mosaics with variable portion of their circulating neutrophils oxidase normal and the remainder oxidase negative. Those carriers who have at least10% or more of their neutrophils oxidase normal are for the most part phenotypically normal, though there are rare reported cases of CGD-type infections in carriers with >10% of neutrophils oxidase normal. Experience with carriers provides compelling evidence that complete correction of all cells is not needed for either significant clinical benefit or for potential cure of CGD. Because the genes encoding the various components of the oxidase that are deficient in CGD have been identified and sequenced, because it has been shown that allogeneic HSC is curative, and because it appears that achieving correction >10% of neutrophils is sufficient for significant clinical benefit; gene therapy is an attractive treatment goal for this disease and efforts to improve gene therapy techniques have been ongoing for more than 15 years, with the most significant advances made in the past 5 years.

In 1995 and then in 1997, we initiated trials of clinical gene therapy for the treatment of adults and older teenage patients with p47phox AR-CGD and then p91phox X-CGD, respectively (Malech et al., 1997, 2004). These early studies used high titer amphotropic-pseudotyped MFGS vector (derived from murine Molony retrovirus) encoding p47phox or gp91phox cDNA. These trials pioneered the use of closed gas-permeable bag culture/transduction systems and culture media free of fetal calf serum or other animal proteins to grow the patients' autologous CD34+ HSCs, that have become standard for most subsequent trials of this type (Malech, 2000). Although rates of *ex vivo* transduction in these early trials exceeded 40% (range ~40–80%) with significantly >10 million transduced HSC infused in each case, the number of oxidase normal neutrophils appearing in the peripheral blood never exceeded 0.1% in any patient and the effect was transient lasting only a few months at most. These early trials did not use any marrow conditioning chemotherapeutic agent to enhance engraftment of the autologous gene-corrected HSC. As will be discussed in more detail below, the very low marking and failure of persistence of marking in these early CGD gene therapy studies without conditioning is typical of the pattern of marking to be expected where there is no selective cell growth or survival advantage conferred by the gene therapy. However, these early studies had to be done in this fashion without chemotherapy in order to be certain that chemotherapy was a necessary component of future studies of gene therapy for CGD. No short- or long-term toxicities have been observed in the CGD patients who participated in these early studies.

Subsequent to those early CGD studies in young adults and older teenagers, there appeared reports of substantial immune reconstitution of lymphocyte production and function with long-term clinical benefit from gene therapy in 18 of 20 infants and very young children treated for X-linked severe combined immune deficiency (X-SCID; Gaspar et al., 2004; Hacein-Bey-Abina et al., 2002). It should be noted that an almost identical clinical vector (MFG) of similar titer, very similar transduction conditions with similar *ex vivo* transduction efficiencies, was used for these X-SCID trials, and no conditioning was given, yet the outcome was dramatically different than we had observed in our patients in these early CGD trials. It appeared that gene correction of X-SCID provides a very strong selective advantage to outgrowth of functional T lymphocytes as the underlying basis for the successful gene therapy outcome in this disease. It is also possible that the very young age of the patients in those X-SCID trials also contributed as well to the successful outcome, since subsequent reports of clinical trials of gene therapy in older children with X-SCID have failed to replicate the remarkable results seen in infants (Chinen et al., 2007; Thrasher et al., 2005).

Vector insertional genotoxicity was observed in the gene transfer treated X-SCID infants that resulted in development of T-cell leukemias in 5 of the 18 infants achieving significant immune correction, with one subject succumbing to the leukemia and the others successfully treated for the leukemia. Despite these reports of vector insertion-related toxicities, this was overall a major advance in achieving clinical benefit from gene therapy (Hacein-Bey-Abina et al., 2003a,b; Howe et al., 2008).

Curiously, similar trials of gene therapy without marrow conditioning in adenosine deaminase-deficient severe combined immune deficiency (ADA-SCID), where there also should have been a significant growth/survival advantage provided by the gene therapy, did not lead to substantial immune reconstitution (Blaese et al., 1995; Kohn et al., 1998). However, when Aiuti and colleagues added a relatively low dose of busulfan marrow conditioning as preparation for receipt of the gene-corrected HSC, the outcome was remarkably improved (Aiuti et al., 2002, 2009). This was a landmark observation that has led to the routine use of busulfan, melphalan, and/or other agents as chemotherapy conditioning for gene therapy. As a general principle, in disorders where there is some selective cell growth or survival advantage expected from the HSC gene transfer therapy, none or modest conditioning can be sufficient to achieve the desired outcome, while for disorders in which no selective advantage is expected from the gene transfer treatment of HSC subablative to marrow ablative levels of chemotherapy may be needed to achieve significant immune reconstitution. Correction of the CGD functional defect would not be expected to provide any proliferative or survival advantage to either HSC or to mature phagocytic cells, and in retrospect, the outcome of the early gene therapy clinical trials for CGD is consistent with that hypothesis.

In 2005, the group in Frankfurt, Germany, led by Manuel Grez, conducted a clinical trial of gene therapy for X-CGD using a vector derived from murine spleen focus-forming virus because of its expected high performance in myeloid lineage (Ott et al., 2006). The two adult patients in this trial were conditioned with a total of 8 mg/kg busulfan before receiving autologous gene-corrected HSC. Early marking of >20% of circulating neutrophils was seen, but rather than observing decrease of early marking to some long-term stable baseline level as might be expected in a disorder where no selective advantage accrued from the gene therapy, they observed a steady increase in myeloid marking over time to a level significantly >50% of neutrophils as would be the case for selective outgrowth of gene-marked cells. Although initially providing significant clinical benefit to these patients in the form of resolution of infections, over time gene silencing was observed with loss of oxidase activity despite sustained high levels of gene marking. Further, the marked outgrowth of myeloid marking was at first oligoclonal for clones with inserts in the *EVI1/MDS1* gene complex and other myeloid specific regulatory growth genes and then almost monoclonal, and both patients developed a monosomy 7 associated myelodysplasia. One patient succumbed to the myelodysplasia while the other was successfully transplanted. In retrospect, it appears that the murine spleen focus forming vector may have a predilection for insertion pattern and myeloid gene activation capability that was responsible for this outcome (Stein et al., 2010).

In late 2006, we initiated a follow up clinical trial of gene therapy for gp91phox-deficient X-CGD using the same amphotropic MFGS-gp91phox vector we had used in the 1998 trial, but now using 10 mg/kg total busulfan conditioning. In fact, the first patient enrolled in the 2007 trial was one of the patients enrolled in the 1998 trial, providing the opportunity to compare outcomes without and with conditioning using the same vector, same target cells (G-CSF mobilized CD34+ HSC), and same *ex vivo* transduction conditions (Kang et al., 2010).

The results of this more recent trial of gene therapy for X-CGD have been published in 2010 (Kang et al., 2010). Also as noted, the first patient (Subject #1) of the 2006 trial was also treated in the 1998 trial, with and without conditioning, respectively, and in both cases received autologous HSC transduced with the exact same lot of vector. In the 1997 trial without conditioning, Subject #1 achieved ≈0.05% oxidase normal neutrophils in his peripheral blood, decreasing from that level over time, with marking dropping to below detectability after 3–4 months. In the 2006 trial, this same patient after receiving 10 mg/kg busulfan conditioning and similar number of similarly transduced autologous HSC achieved a level of 25% of circulating neutrophils oxidase normal early after the gene transfer therapy, 8% at 2 months, and 1.1% at 6 months post-therapy. This patient at almost 5 years post-treatment still has 0.7% of his circulating neutrophils that are oxidase normal, demonstrating the critical difference caused by busulfan conditioning in achieving such remarkably sustained long-term correction.

Although the second and third patients in the 2006 trial also received autologous HSC transduced with the same vector, vector lots of lower titer had to be used because additional vector from the highest titer lot used in Subject #1 was no longer available. Notably, the first clinically produced lot has had the highest titers to date, despite efforts to recapitulate its high titer in subsequent production lots, and demonstrates the vagaries of vector production. We will describe in the following sections the vectors used, as well as the steps and processes needed to proceed from the design of a vector to the actual treatment of patients.

2. Vector Design

The same murine retrovirus MFGS vector backbone was used for both our 1995 trial of gene therapy for p47phox AR-CGD and for the trials in 1998 and 2006 for gp91phox X-CGD. Both the MFG and almost identical MFGS vectors were designed in Richard Mulligan's laboratory from the murine Moloney leukemia virus (MMLV) using the now standard split vector design which removes or inactivates the gag, polymerase and envelope from the transfer vector retaining only the active 5'- and 3'-flanking LTRs for promoter activity and genome insertion potential, plus the essential psi packaging element (Danos and Mulligan, 1988; Krall et al., 1996). MFGS differs from MFG by only a few basepair changes, insertions, or deletions designed to add stop codons that further reduce the potential for translation of any gag sequence from either residual sequence or from any hypothetical recombination event. The Mulligan laboratory designed and selected the vector based on *in vitro* testing demonstrating high-transduction efficiency and more importantly, high transgene expression due to retention of the MMLV splice donor and acceptor sequence (Krall et al., 1996). The 5'LTR sequence is thus able to generate both a full-length viral RNA necessary for the generation of viral particles and a splice modified subgenomic mRNA analogous to the MMLV *env* mRNA, which results in high-level expression of the gp91phox and p47phox transgenes. It is important to note that even to the present time, no other retrovirus or lentivirus vector designed by our lab or any other lab results in as much gp91phox expression per cell as MFGS-gp91phox. It is an amount of gp91phox sufficient to result in full restoration of oxidase function in the gene-corrected X-CGD neutrophils that differentiate in patients from single copy MFGS-gp91phox transduced human CD34+ HSC.

Because transgene is expressed from the LTR of MFGS-gp91phox insert, expression is not tissue specific. Some control of gp91phox expression occurs because coexpression of gp22phox in the same cells is required for stability of gp91phox within the p22phox/gp91phox heterodimeric cytochrome B558 component of the oxidase. To date, there has been no evidence of toxicity from gp91phox expression in nonmyeloid hematopoietic cells. As will be discussed further below, many in the field have focused in the past 10 years on development of lentivirus-based vectors in which 5'LTR function is crippled in the final genomic vector insert by introducing a 3'LTR deletion that ends up in the 5'end of the insert resulting in the inactivation of the 5'LTR (Brenner and Malech, 2003; Dull et al., 1998; Naldini et al., 1996). This self-inactivating design has allowed the introduction of internal promoters that may be of mammalian origin and/or which are designed to have tissue specificity. Myeloid specificity for expression of gp91phox or p47phox transgene might provide an additional safety feature.

For our MFGS-gp91phox and MFGS-p47phox vectors, the transgene itself is an unmanipulated (wild-type) cDNA of the gp91phox *CYBB* gene for the X-CGD trial and the wild-type cDNA of the p47phox *NCF1* gene for the AR-CGD trial (Kang et al., 2010; Malech et al., 1997). No other

SD SA

Figure 7.1 MFGS-gp91phox vector schematic. This vector is very similar to the large class of transfer vectors that have been designed from murine retroviruses. At the ends, the vector retains the long-terminal repeats that contain promoter activities and the sequence required for insertion into the target cell genome. The psi region has been altered by deletions and insertion of stop codons to prevent production of any gag or pol protein sequence but retains critical psi elements necessary for the vector RNA to be packaged in the amphotropic envelope producer line HEK-293-SPA into an infectious, but not replication-competent vector particle. The psi region has also retained splice donor and acceptor sequence that originally were used by the virus to enhance production of envelope protein. The envelope sequence has been replaced by cDNA encoding the gp91phox protein open reading frame. Not shown is that the same exact vector was also used to produce p47phox transgene in our 1995 clinical trial of gene therapy for p47phox-deficient AR-CGD. LTR, long-terminal repeat; ψ, packaging recognition signal; ORF, open reading frame; SD/SA, splice donor/splice acceptor sites. (See Color Insert.)

elements, including insulators, selectable markers, or nonexpressed elements were added to either vector for these studies. Figure 7.1 summarizes these features for MFGS-gp91phox, and MFGS-p47phox has similar features. As will be summarized below, for both vectors we used amphotropic packaging lines, though for the MFGS-p47phox study this was a murine packaging line while for the MFGS-gp91phox vector this was a human HEK-293 packaging line.

3. Vector Production

3.1. Producer cell line description

As the vector itself has no viral protein-encoding sequences, the transfer vector plasmid must be incorporated into a special "packaging" cell line genetically modified to produce the gag–pol and envelope proteins necessary to create an infectious viral particle, but chosen to eliminate the possibility of recombinatorial events leading to a replication competent virus. For example, potential open-reading frames and protein-initiation sites are eliminated except for those needed for the transcription of the therapeutic transgene.

The packaging cell line chosen for the MFGS-p47phox vector was an amphotropic expressing cell line developed initially in Dr. Richard Mulligan's laboratory (Danos and Mulligan, 1988). This line, called psi-CRIP, was created by cotransforming an NIH3T3 cell line with two separate packaging genomes (gag–pol and env) from which psi sequence had been deleted. This psi deletion thereby prevents the encapsidation of retroviral packaging sequence genomes into viral particles. The 3'LTR encoding the polyadenylation and termination sequences was also eliminated in this line, further reducing the possibility of generating viral RNA sequences capable of being transferred to recipient cells. The amphotropic envelope was chosen as it had been shown in many studies to effectively bind to hematopoietic cells. For the preclinical mouse to mouse studies, an ecotropic cell line (psi-CRE) was used as murine cells poorly express the amphotropic receptor and therefore are not efficiently transduced by amphotropic envelope packaged virus.

Titers for this line were remarkably high, in the 10^6 range, and clones were chosen after transfecting with the MFGS-p47phox containing plasmid. The titer of this vector was measured by analyzing transfection of K562 human erythroleukemia cell line using Western blotting because at the time of initial development of that vector no anti-p47phox antibody was available that could be used for flow cytometry analysis (Leto et al., 2007). Since then, antibodies have become available that can detect p47phox in individual cells by flow cytometry and is the method being used currently to develop the current generation of lentivector expressing p47phox in our laboratory.

For reasons we could never fully determine, we could never produce adequately high titers of MFGS-gp91phox from any of the then available murine NIH3T3-based packaging lines. Our clinical work with MFGS-gp91phox had to await development of an amphotropic line using the human HEK 293 cell. The specific HEK-293 producer line that amphotropic envelope-pseudotyped MFGS-gp91phox at very high titer, named 293SPA-MFGS-gp91-155, was developed by Dr. S. Kay Spratt at Somatix before its acquisition by Cell Genesis. Titers of greater than 2×10^7 amphotropic-pseudotyped MFGS-gp91phox could be produced by this line as measured in K562 cells, and four daily transductions of human X-CGD CD34+ HSC could routinely achieve rates of expression of gp91phox in 40–80% of cells. Because much of the transmembrane gp91phox protein remains in intracellular vesicles in neutrophils, flow cytometry measurement of expression of gp91phox in neutrophils is performed by permeablizing with 0.1% saponin and staining with the gp91-specific antibody, 7D5, that recognizes a membrane surface accessible epitope (Yamauchi et al., 2001). The availability of this high-affinity/high-specificity antibody for flow cytometry studies greatly facilitated vector development and optimization of transduction regimens for treatment of X-CGD. This antibody is now readily available as a fluorescence-labeled congener from commercial source (MBL International Corp, Product Code No.: D162-5).

4. Preclinical Testing of Vector Titer and Transgene Function (*In Vitro* Culture, Mouse Knockout (KO) CGD Models, and Mouse Models of Human CD34+ HSC Xenograft)

Cell cultures that model CGD were and are essential tools for preclinical testing of vector development. Although we are able to obtain G-CSF mobilized CD34+ HSC from adult CGD volunteers to use for preclinical development, these stem cells from patients are a precious and limited resource that we reserve for the most part for the final preclinical studies in mouse xenograft models. Dr. Mary Dinauer generated a widely used and robust X-CGD model cell culture line by knocking out the *CYBB* in the single X-chromosome remaining in the female XO aneuploid myeloid cell line PLB-985 (Zhen *et al.*, 1993). This line has been widely used in the study of X-CGD. The PLB-985 line requires treatment with butyrate or other induction agents to complete differentiation to a neutrophil phenotype that expresses the other subunits of the oxidase enzyme to manifest oxidase activity. The human erythroleukemia cell line K562 constitutively expresses p22phox and rac2, but not any of the other phox components. Using MFGS retroviruses expressing oxidase components, we have generated a series of transgenic K562 lines which express all but one of the oxidase components necessary for the cell to constitutively produce oxidase when nonspecifically activated with phorbol myristate acetate (PMA; Leto *et al.*, 2007). When the final oxidase component is expressed after transduction with vector encoding the missing component, the amount of oxidase activity manifested following PMA activation is directly proportional to the transduction efficiency and the amount of transgene expressed. Thus, we have K562 transgenic lines designed to facilitate testing vector transduction efficiency and transgene expression for vectors encoding gp91phox, p47phox, and p67phox. In this way, these K562 lines are used for standardization/quality control to titer and test performance of CGD vector lots and to test the stability of vector preparations.

The mouse knockout model of gp91phox X-lined CGD was developed by Dr. Mary Dinauer at the University of Indiana and that of p47phox AR-CGD by our program at the NIH. Both models are close models of the human disease and are now widely available and used to model pathology and treatment of CGD infections, inflammation, bone marrow transplant, and gene therapy. These models have been particularly useful to demonstrate that gene therapy can provide protection against bacterial or fungal infections (Björgvinsdóttir *et al.*, 1997; Dinauer *et al.*, 2001; Ding *et al.*, 1996; Mardiney *et al.*, 1997).

The experimental design for such studies was similar for both models. Bone marrow was collected from donor CGD KO mice, transduced with the disease-specific vector using various culture conditions, and the cells then infused into irradiated recipient mice. In some cases, the bone marrow was collected after the mouse was treated with 5-Fluoruracil to increase the number of progenitor cells within the marrow and either whole bone marrow or SCA-1-positive selected cells were used for the transplant. The radiation was used to increase the level of engraftment in the mouse but was not intended for use in the clinical trials. Samples from the mice were assayed for the level of correction in the peripheral blood using PCR to assess the presence of the transgene and flow cytometry (for gp91phox) or Western blot (for early studies of p47phox) using a monoclonal or polyclonal antibody directed against either the gp91 or the p47 protein to measure the level expressed on the cell surface or within the cell. Flow cytometry was also used to measure the increase in dihydrorhodamine (DHR) fluorescence in neutrophils as a measure of oxidase activity (Vowells *et al.*, 1996). Additionally, engrafted mice were challenged with infectious agents such as *Aspergillus fumigatus* or *Burkholderia cepacia* to assess the functionality of gene-corrected neutrophils and infection resistance phenotype of the mice after transplantation of gene-corrected stem cells (Björgvinsdóttir *et al.*, 1997; Dinauer *et al.*, 2001; Ding *et al.*, 1996; Mardiney *et al.*, 1997).

For the p47 mice, using nonlethal irradiation of 300–325cGY levels of up to 12% could be seen in the peripheral blood 1 month post-transplant. Over time these levels decreased to a steady state of about 2% at which time the mice were challenged with the *B. cepacia* bacteria. Fifty percent of the gene therapy-treated mice survived, whereas control nontransplanted p47 mice all died indicating that even relatively low levels of correction afforded substantial protection against infection challenge. For the X-CGD mouse gene therapy experiments, the level of marking improved with the titer of the vector, up to 50% in some animals. Those animals receiving the higher titer vector and thus with the better level of marking (vs. 5% for the lower titer vector) tolerated the challenge with *Aspergillus* fungus fairly well. In contrast, the lower marked animals were unable to recover from the *Aspergillus*. The results overall confirmed the proof of principal that use of a retroviral vector to correct CGD cells can provide protection against infection in some proportion to the level of correction achieved, but that even very low levels of genetic correction provided significant benefit.

While CGD cell line models can be of great benefit as a tool to facilitate vector development and while gene correction of CGD KO mice provides critical proof of principle that gene therapy can protect against infection, the final target cells for a clinical trial are the G-CSF mobilized, CD34+ selected HSC from patients with CGD. While these primary HSC can be differentiated to neutrophils in bulk culture or examined in myeloid colony assays to assess vector insert copy number by quantitative PCR, expression of oxidase function by DHR assay in neutrophils differentiating in the culture, or expression of transgene protein by immunofluorescence flow cytometry, such culture systems can only assess out to about 21 days at most after transduction. Ideally, it would be helpful to be able to observe the engraftment potential, extent of transduction in engraftable stem cells, and expression of transgene in neutrophils differentiating from long-term engrafting stem cells. For this type of analysis, we have employed immunodeficient mouse models that accept and support long-term human hematopoietic marrow xenografts. In our development studies until 2 years ago, we used exclusively the NOD–SCID mouse model for such human HSC xenografts which allowed stable engraftment for only 6–8 weeks before the animal had to be sacrificed for examination of the xenograft (Brenner et al., 2003, 2007; Roesler et al., 2002). Animals are prepared to receive the human HSC graft by irradiation with 300–325 cGy. In more recent years, we were able to improve the engraftment in NOD–SCID mice by injecting purified rat antimouse NK cell monoclonal antibody (McDermott et al., 2010). However, in the past 2 years, we have adopted the improved NSG mouse model which more completely lacks mouse NK cells, resulting in a more durable and more easily established human HSC-derived xenograft following either 250 cGy radiation or busulfan 18–20 μg/kg conditioning where the animals can be followed for 16 weeks or longer before analysis of the xenograft. This allows much longer term *in vivo* assessment of the transduction efficiency of longer term repopulating HSC that continuously give rise to human neutrophils in marrow and peripheral blood of these animals (McDermott et al., 2010).

We used the NOD–SCID mouse model to assess the performance of our lots of amphotropic-pseudotyped MFGS-gp91phox vector in transduction of G-CSF mobilized CD34+ HSC from patients with X-CGD in preclinical assessment of that vector. Mobilized CD34+ selected HSC from either patients or normal volunteers were transduced as per the anticipated clinical manner, that is, with about 18-h activation of the cells with cytokines, followed by a daily 6- to 7-h transduction for 4 days. At the end of the last transduction on day 4, about $2–3 \times 10^6$ cultured/transduced cells were injected into each NOD–SCID mouse that had been conditioned the previous day with 300–325 cGy gamma irradiation and for experiments conducted after 2006, also injected with 1 mg rat antimouse NK cell antibody.

At 6–8 weeks after transplant of mobilized peripheral blood human CD34+ HSC, the engrafted human cells comprise about 2–15% of the mouse bone marrow if the rat antimouse NK cell antibody is not used and about double that after we began using that antibody in preparation for the transplant. When the xenograft was derived from healthy human volunteer CD34+ HSC that had not been transduced (or were transduced only with a marker gene such as green fluorescent protein), all human antigen CD13 positive neutrophils arising from the xenograft expressed human gp91phox by flow cytometry, and if these human cells in the marrow were derived from nontransduced CD34+ HSC from X-CGD patients, then none of the human antigen CD13 positive neutrophils expressed human gp91phox by flow cytometry. However, when the xenograft was derived from amphotropic MFGS-gp91phox transduced CD34+ HSC from X-CGD patients (about 50% transduction efficiency in the bulk culture), then 1–3% of the human antigen CD13 positive neutrophils expressed human gp91phox by flow cytometry (Brenner et al., 2003; Roesler et al., 2002). As noted above, the shortcomings of the NOD–SCID xenograft model are the significant number of mobilized, cultured CD34+ HSC that must be injected to achieve useful levels of human cell engraftment in the marrow xenograft even with use of the rat anti-mouse NK cell antibody and the 300–325 cGy radiation, and the limited period of time (6–8 weeks) that the engrafted NOD–SCID mice could be followed before loss of animals from a variety of causes occurred. The NSG mouse model is more robust, allowing higher levels of human cell engraftment with fewer input human CD34+ HSC and a longer time span that the engrafted animals could be studied (routinely >16 weeks) before harvest and study of the xenograft. Thus, many of our studies of new vectors, particularly with our self-inactivating CGD lentivector program, are being assayed now in the NSG mouse model (McDermott et al., 2010).

5. CLINICAL SCALE PRODUCTION

This section will focus on our work for treatment of X-CGD for the 1998 and 2006 studies. A 100 vial master cell bank (MCB) was established for producer line, 293-SPA MFGS gp91–155, at BioReliance, Inc. in Gaithersburg, MD, in 1997 following the then standard safety studies required by the FDA. More specifically, to create this MCB, cells from one vial of the producer cell line, 293-SPA MFGS gp91–155, were thawed and then seeded at a concentration of 3.41×10^4 in a 162-ml flask in a total volume of 40 ml. The cells were passaged four times over 24 days. At the end of 24 days, the cells were then harvested and resuspended in 90% FBS and 10% DMSO at a concentration of 1×10^7 cells/ml and aliquoted into 1-ml cryovials.

Five percent of the vials subsequently underwent testing for post-cryostorage viability. These cells were also used to test for mycoplasma, the presence of adventitial viruses, as well as for sterility. The MCB has been kept in storage at BioReliance. Updated testing needed to be preformed for the follow-up clinical trial in 2007 and again an aliquot of the MCB was supplied to BioReliance for the testing which included HIV 2 and any other new requirements promulgated by the FDA.

The clinical scale production of Lot 1 amphotropic MFGS-gp91 from 293-SPA MFGS gp91–155 was performed at BioReliance with close oversight by our staff. Preclinical runs were performed to establish the production variables that impact on growth and vector production such as initial seeding density, number of media changes, type of media used, and the optimal confluence of cells at the time of harvest. As mentioned, many of these variables were first tested in the laboratory using cells from the master bank to test; however, as scaling up actual production often requires modifications, these parameters were assayed first at a small scale within the production facility before proceeding to the final full scale production. Final production was performed in 175 cm^2 flasks. Master bank cells were grown in standard HEK-293 medium that contained fetal calf serum until the cells reached dense, but not peak, confluence. The 293-SPA MFGS gp91–155 has a tendency to peel from the flask when it reaches highest confluence; so the time of harvest was a compromise between achieving highest confluence, but not so high that the cells peel from the flask. For virus collection, the regular growth medium was removed and replaced by a minimal volume of animal protein-free X-VIVO 10^{TM} medium (BioWhittaker, Inc.) with 1% human serum albumin (the same medium that would be used for culture and transduction of the patient CD34+ HSC). After 12 h, the vector production medium was harvested and the cells allowed to recover in the regular HEK-293 cell medium for 12 h before repeating a harvest in the X-VIVO 10^{TM}. The three harvests were pooled, and after removing portions for required safety testing, the pooled lot was subjected to a filtration step before aliquoting for freeze storage. We found in test runs that attention to how the filtration was conducted could have a major effect on final product vector titer. Use of graded porosity in-line filters was necessary to prevent premature clogging. Further, we found it essential to change filter sets before there was any perceptible slowing of filtrate through the filter. Even in the earliest phases of filter clogging, there was a very significant loss of vector titer, but this could be prevented by early judicious changes of the filter set several times during the filtration step. This has been a critical issue that we find in our experience is not sufficiently appreciated by some vector production facilities.

When planning how much vector supernate to produce, it is important to appreciate that in general 10% of the initial product harvest will be needed for safety testing, and if there are any repeats of the testing, that may consume additional product. It is important to keep updated on FDA requirements for safety testing, quality testing, and composition testing of vector as these are updated periodically. In general, the product will require testing for sterility, mycoplasma, for presence of replication-competent retrovirus (or lentivirus if it is a lentivector), for a variety of adventitial viruses, and specific tests for a number of human viruses (and animal viruses dependent upon the use of animal cells or products at any place in the producer line development or vector production process). Also, generally required is complete sequencing of the vector genome from the final product. Most experienced production facilities can help to advise what testing is necessary, but a critical element is to seek a pre-IND meeting with the FDA before establishment of the MCB or planned vector production to go over the full plan of safety, quality, and composition testing so that final testing includes all that the FDA may require to qualify the final production lot.

6. TARGET CELL

Because CGD is a functional deficiency of phagocytic cells that arise from marrow HSC, this is a disorder which in theory is treatable by introducing a corrective gene into patient autologous HSC. The patient population for our studies is older children and young adults. In this population, the most efficient means of collecting large numbers of CD34+ HSC is by mobilizing these cells to the peripheral blood (Sekhsaria et al., 1996), collecting mononuclear cells by apheresis, and using an immune selection device to select CD34+ HSC from the apheresis product. Using CD34 as the biomarker, patients underwent a standard G-CSF mobilization and apheresis. For our 1997 and 2006 X-CGD gene therapy studies (Kang et al., 2010), the apheresis product was selected using the Isolex 300i anti-CD34 antigen immunomagnetic bead system originally developed by Baxter Healthcare, but subsequently distributed by other entities including the former Nexell Therapeutics, and most recently by Miltenyi Biotec. However, the Isolex is now out of production, and we and most other centers now primarily use Miltenyi Biotec's CliniMACS system for selecting CD34+ HSC from apheresis product or bone marrow for clinical studies. In the United States, use of the CliniMACS must be under IND in an approved clinical trial, with letter of cross-reference to their master file from Miltenyi. Immunoselected CD34+ HSC has been shown in multiple clinical studies by various institutions to have long-term repopulating ability.

As patients with immunodeficiencies, particularly CGD, often do not mobilize well with G-CSF alone (Sekhsaria et al., 1996), it was decided for our studies that selected CD34+ HSC would be collected in advance and cryopreserved prior to treatment, particularly in the setting of patients receiving conditioning before gene therapy. This assures that sufficient autologous CD34+ HSC is available for the transduction before proceeding with treatment in each patient. Cell dose has been shown to be important for adequate engraftment, even in allogeneic settings where there is also a graft versus marrow mediated effect; therefore, ensuring adequate cell numbers prior to administering chemotherapy to patients was paramount. Further, multiple collections were often required to obtain the target cell dose of at least 5×10^6/kg body weight in some patients. The additional use of newer mobilizing agents such as plerixafor (Mozibil®) may make this less of an issue for future gene therapy studies; however, studies need to be done to assess if these agents impact in any way on transducibility of the cell product. Also, if the murine retrovirus transduction efficiency is less than 40% of patient CD34+ HSC in the bulk culture, we find that the great majority of transduced cells are short-term repopulating cells. Both achieving bulk transduction rates higher than 50% and achieving larger cell doses seem to be necessary to ensure reasonable levels of long-term marking with murine retrovirus vectors. These targets may end up being less critical with use of newer lentivirus vectors where higher targeting of the most primitive HSC may occur at more modest levels of total bulk culture transduction efficiencies. In this way, such improved vectors and other manipulations to improve transduction and engraftment of actual stem cells may make these large cell doses unnecessary in the future to achieve long-term clinical benefit.

7. Clinical Transduction

All clinical transduction and culture of cell product for our clinical trials were performed in our NIH Clinical Center Department of Transfusion Medicine, Cell Processing Facility which is a GLP facility that maintains a Master File of procedures and processes with the FDA. Though the process for transduction that we chose for our 293-SPA MFGS gp91–155-derived vector for the X-CGD clinical trial of gene therapy was developed at small scale in the laboratory, there were a number of adaptations that required trial runs at full scale in the Cell Processing Facility to assure translation to the clinic.

Approaches that work at small scale in the lab had to be translated to the clinic with considerations for our Cell Processing Facility technician staff time, and there were also logistical issues relating to the final clinical testing before release. Even more importantly are the requirements to maintain Good Laboratory Practice (GLP) or Good Manufacturing Process (GMP). This includes incorporating use of closed systems for culture and media exchange as much as possible (Malech, 2000). Further, the clinical transduction involves dealing with significantly larger cell numbers and volumes overall as compared to preclinical testing. Other considerations such as nursing and patient care may impact on the transduction process. For our trial, the decision was to infuse the cells immediately after transduction, and hence it was preferable to have the transduction completed on a weekday.

Therefore, for the gene therapy trials, a standard operating procedure (SOP) was developed in concert with the Department of Transfusion Medicine, where the transductions were performed.

In the most recent 2006 trial, the decision was to have the patient cryopreserved CD34+ HSC thawed Sunday night, which would allow the completion of the transduction to be done by Friday morning, therefore, avoiding a weekend infusion.

In order to maintain an appropriate cell concentration, in this case 0.5–2 million CD34's per ml of media and a 90% vector concentration, cells were washed and resuspended in an appropriate volume of media during the day, the vector added at night and the cells transduced overnight for 16h each night starting at 24h after first placing the cells in culture. On the morning of the fourth transduction (Friday if the cells were thawed on Sunday), cells were harvested, washed, and analyzed before release. We used Lifecell X-FOLD™ gas-permeable bags produced by Baxter Healthcare, Inc for this trial. Manufacture of this product has ceased so current and future trials of this type must use alternative gas permeable container systems or return to use of tissue-culture flasks. Before use, the bags were coated with fibronectin fragment CH-296, Retronectin® (Takara Bio Inc.) to enhance transduction. The same Retronectin®-coated bag was used throughout the multiple transductions, with addition of bags to the process as needed to maintain the number of cells at $0.5–2.0 \times 10^6$/ml. A bag size was chosen such that the fluid layer in the bag was less than 0.5-cm deep to maximize interaction of vector and cells with the Retronectin-coated surface.

The media used for culture and transduction of patient autologous CD34+ HSC was X-VIVO 10, a serum-free media produced by the BioWhittaker, Inc. to reduce exposure to animal proteins. The media was supplemented with the following growth factors: stem cell factor, thrombopoietin, and Flt-3 ligand (SCF, TPO, and Flt-3-L) at concentrations of 50 ng/ml and interleukin 3 (IL-3) at a concentration of 5 ng/ml. The growth factors were pharmaceutical grade with respect to manufacturing, but because these are not approved pharmaceutical products, all manufacturing information, composition analysis, activity testing, and necessary sterility, and adventitial agent testing had to be provided to the FDA as part of our IND application process. Protamine was also used at 6 µg/ml to enhance interaction of cells and vector. All cytokines were maintained by the NIH Clinical Center Pharmacy department until needed. Media was added to the culture bags using sterile tubing connections. Vector, which had been stored frozen at −80 °C in bags or bottles depending upon the particular production lot, was rapidly thawed at 37 °C and added via tubing through this closed system. Prior to infusion into the patient, the cells were given four washes out of vector media and resuspended in phosphate-buffered saline containing 1% human serum albumin in a volume of approximately 50cc in a 60-cc syringe. Prior to release of the product for infusion, trypan blue visual cell count for viability and cell number, as well as flow cytometry evaluating cell viability and actual CD34+ cell count was performed. A sterility culture obtained the previous day had to be reported negative at 24h, and a Gram stain of the final product had to be negative. Testing to show the product was negative for endotoxin also had to be completed before release. The product was released only after the above testing was completed and certified. Other tests, including but not limited to mycoplasma and final sterility culture and PCR and anti-gp91phox flow cytometry, and DHR functional testing on neutrophils that arise in the culture as measures of transduction performed on the final product could be "pending" at the time of product release. Samples were also frozen for archiving as well as for RCR testing if needed. Transduction efficiency was not used as a release criteria, as this analysis for the end product could not be performed real time so that the product could be infused before the expiration period set for this type of product.

8. Patient Care and Monitoring

For the initial studies of X-CGD gene therapy done in 1997, transduced autologous patient CD34+ HSCs were administered into the patients without any prior conditioning of the patient. The patient was therefore admitted to the Clinical Center on the day of infusion, the cells were infused through a peripheral line, the patient observed overnight, and then discharged. With these early studies of X-CGD gene therapy without conditioning, there was detectable marking in four of the six treated patients, which declined over time and became undetectable as noted in detail in Section 1. Despite repeated infusions, the marking quickly decreased over a few weeks to very low levels, and for the most part disappeared completely in all patients within a few months after each infusion. Hence for the more recent 2006 trial of gene therapy for X-CGD, the use of conditioning was added to the design. As noted in some detail in Section 1, the use of conditioning was first shown to be effective in ADA-SCID clinical trial, where prior studies for this disease had also shown low level marking which did not persist. In 2004, Aiuti et al. demonstrated clinical benefit after patients received cells after 2 mg/kg of busulfan given daily for 2 days and the cessation of pegylated-ADA (Aiuti et al., 2002, 2009). The myelosuppression provided by the busulfan alone with the increased survival advantage conferred by the therapeutic transgene and promoted by the cessation of the Peg-ADA allowed for enhanced growth of corrected lymphocytes in the ADA-SCID trials.

As the correction of CGD requires myeloid engraftment and the correction of the defect does not affect cell growth or survival, the necessity for higher dose conditioning to obtain adequate engraftment of transduced cells was determined to be necessary. The optimum dose of and type of conditioning to use for CGD were not clear. Most animal studies use total body irradiation; however, presumably only an agent targeted to marrow HSC is needed and lymphocyte immunosupprression is not needed for treatment of CGD. Busulfan is specifically myelosuppressive and with the availability of an intravenous formulation, it is an attractive drug to use as a single agent in chemonaive populations. We based our dosing on our experience with a patient with X-CGD in our regular transplant program who had undergone an allogeneic transplant from a matched sibling ending up with mixed chimerism more than a year after the transplant where the chimerism had the unusual pattern of 100% donor T-cells but less than 3% marrow donor CD34s or donor myeloid engraftment. A "boost" transplant with the same sibling donor was performed using single agent busulfan conditioning of 5 mg/kg for 2 days (10 mg/kg total), allowing him to achieve 100% myeloid engraftment with little to no perturbation of his lymphocyte counts following the conditioning.

Based on this experience, we used the same dosing regimen for our 2006 X-CGD gene therapy study. Further, we noted that Ott et al. from Frankfurt had used 8 mg/kg total dosing for their X-CGD gene therapy trial and it was well tolerated in their study (Ott et al., 2006). Although they obtained very significant engraftment of gene-marked neutrophils, there was a major contribution to that marking by the insertion mutagenesis driven by clonal outgrowth, and overall base levels of true marrow engraftment were not discernable from that trial (Stein et al., 2010). It should be noted that in a number of current and planned trials of gene therapy targeting patient HSC, where there is little or no expected growth advantage conferred by the gene transfer treatment, a number of investigators in Europe have chosen to use doses of busulfan or similar conditioning agents at or near myeloablative levels to achieve maximum engraftment of gene-corrected cells. It is an area of continued debate and discussion whether fully ablative conditioning is necessary to achieve the best outcome in the setting of disorders with no selective advantage from the gene therapy. It is hoped that newer methods may be developed to enhance engraftment that may reduce the need for ablative or any chemotherapy.

One of the reasons we chose to use a subablative rather than fully ablative conditioning regimen in our 2006 trial was that an entry criteria for our 2006 X-CGD gene therapy study was that patients had to have a severe infection that was not responding to conventional therapy, and the patient could not have a potential matched sibling or a potential matched unrelated transplant donor available. In this setting, we felt that ablative conditioning was too risky, but that subablative conditioning would be tolerated. Thus, the intent of this study was to use gene therapy as salvage treatment in the management of infection. Subject #1 in this 2006 X-CGD gene therapy study (who had been treated in the 1997 trial without conditioning) had Staphylococcal liver abscesses affecting more than 80% of liver mass. Because of many previous abdominal and liver infections, he was not a candidate for surgical removal of the abscesses. Subject #2 had year-long Paecilomyces fungus infection of lung, chest wall, and spine slowly progressing despite maximum antifungal therapy. Subject #3 had a year-long *Aspergillus* infection of lung, chest wall, and spine also slowly progressing despite maximum antifungal therapy.

For our most recent 2006 X-CGD study, given the inclusion of busulfan preconditioning and the presence of ongoing infection in our patient population, patients were inpatients for the conditioning and for several weeks following the gene therapy. A central line was used for both administration of the chemotherapy and for the infusion of the cells. The busulfan was given as a 2-h infusion of 5 mg/kg for 2 days, and although the drug was not dose adjusted based on the integrated area under the curve (AUC), the AUC levels were obtained. The cells were infused a minimum of 48 h post the last dose of the busulfan to ensure that the drug was no longer present in the patient's body at the time of cell infusion. The patients' blood counts were monitored daily as the total dose of busulfan (10 mg/kg) given was very myelosuppressive although not completely myeloablative. Patients were discharged when they were no longer transfusion dependent and their ANC was greater than 1000/μl. They were kept on all their prior antibiotics except sulphamethoxazole/trimethoprim due to its marrow suppressive effects and treated during their course of neutropenia with standard antibiotics for neutropenic fever when it occurred as well as transfusions as needed for anemia or thrombocytopenia. After discharge, patients were monitored at regular intervals with assays for transduction, and monitoring for any changes in the CBC or evidence of clonal outgrowth. A bone marrow was performed routinely at 1 year post cell infusion. Patients were also monitored for the ongoing status of their infection.

The overall results from this study are now published (Kang et al., 2010). Subject #1 was treated with the highest titer vector almost 5 years ago in November 2006. The level of marking over the first year as shown by the number of oxidase-positive neutrophils in his circulating blood is shown in Fig. 7.2. High early marking (24% of circulating neutrophils were oxidase normal initially) was followed by a decline in marking over the first 6–8 months and then stabilized at about 1%. Even at almost 5 years post gene therapy, he is maintaining correction of 0.7% of circulating neutrophils fully oxidase normal until the present time. Figure 7.3 shows the actual DHR assay of his peripheral blood neutrophils at 38 months post-treatment. The graphs show that those neutrophils that are oxidase positive are producing as much or more reactive oxidants than the neutrophils from the normal donor control. Not only did this patient fully resolve his massive liver abscess but he has had a dramatic decrease in the extent and severity of infection problems over the past 5 years. Subject #3 was treated more than 3 years ago and he fully resolved his *Aspergillus* lung, chest wall, and spine infection. His early marking was only 4% of circulating neutrophils, but he also continues to have stable low level persistence of oxidase-normal neutrophils in his peripheral blood stable for the past 2 years. Thus, in both of these patients, there appeared to be clinical benefit with resolution of the infection present at treatment in response to the gene therapy treatment.

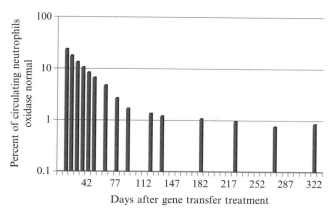

Figure 7.2 Percent of circulating blood neutrophils oxidase positive in Subject #1 of the 2006 trial of gene therapy for X-CGD. Each bar shows the percent of neutrophils (y-axis) that are oxidase positive by dihydrorhodamine (DHR) flow cytometry assay on the indicated day (x-axis) after the injection of transduced autologous CD34+ HSC. The y-axis is a logarithmic scale so the straight line decrease from about 24% at the earliest time point to about 2–3% at 4 months indicates a parabolic arithmetic curve. Stability of marking occurred by 6–8 months at about 0.7–0.8% of circulating neutrophils oxidase positive. Not shown is that to the present time at almost 5 years post gene transfer treatment, the patient retains functional gene marking of 0.7% of circulating neutrophils oxidase normal.

Subject #2 on this trial appeared to have immune-mediated clearance of the marked cells and eventually succumbed to his infection. As a result of this, future patients, including the third patient on the trial was/will be given sirolimus along with the busulfan to try to reduce the potential for development of an immune rejection of marked cells.

9. Future Endeavors

As of this writing in late 2011, the current trial with amphotropic MFGS-gp91phox remains open as salvage therapy for infections unresponsive to conventional therapy in X-CGD patients who do not have a potential transplant donor. However, we have been developing candidate lentivectors for X-CGD ourselves based on the CL20 lentivector construct (Throm et al., 2009; Zhou et al., 2010) as well as working with other investigators developing novel lentivector systems for treatment of X-CGD (Santilli et al., 2011). We believe self-inactivating lentivectors may provide improved targeting of long-term repopulating HSC, while the self-inactivating feature which

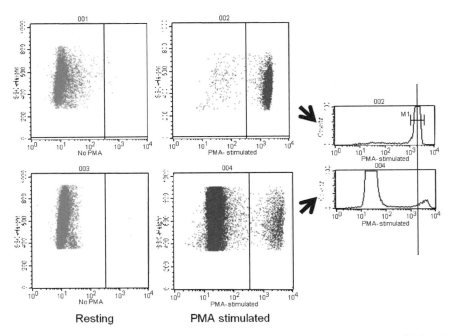

Figure 7.3 Flow cytometry (DHR) assay of oxidase activity in peripheral blood neutrophils from Subject #1 of the 2006 trial of gene therapy for X-CGD at 38 months (more than 3 years) after gene transfer therapy. The upper panels show analysis of neutrophils from a normal control, while the lower panels show analysis of neutrophils from Subject #1. Not shown is that all of the analyses have already been gated (using forward and side scatter) to only include neutrophils from the peripheral blood. The large panels are dot plot format where the *y*-axis is measuring side-scatter and the *x*-axis is measuring DHR fluorescence which is an indicator of the intensity of oxidase activity. The right large dot plot panels show resting neutrophils that have not been activated by phorbol myristate acetate (PMA) and the neutrophils have low DHR fluorescence. Upon PMA stimulation, cells activate the oxidase and electrons flow to the DHR to increase fluorescence. Note that 99% of the activated neutrophils from the normal volunteer control are oxidase positive, and 0.7% of the activated neutrophils from X-CGD Subject #1 are oxidase positive. The dot plots of the activated neutrophil DHR analysis from control and Subject #1 are reformatted (arrows) to histogram plots at the right (*y*-axis is cell number; *x*-axis is DHR fluorescence), in order to line up the graphs to demonstrate that average oxidase activity in the 0.7% of neutrophils oxidase positive in Subject #1 are actually on average producing slightly more oxidants than the oxidase-positive neutrophils from the normal control. This indicates the high per cell efficiency of the MFGS vector in achieving functional correction of oxidase activity. (See Color Insert.)

eliminates LTR activity may improve safety. We are also examining other more distant future options such as the opportunities provided by induced pluripotent stem cells (iPSCs) developed from X-CGD patients (Zou *et al.*, 2011) as well as more precise gene targeting provided by zinc finger nucleases and the TALENs (Wood *et al.*, 2011; Zou *et al.*, 2011).

9.1. Lentivectors

Given the observation of insertional mutagenesis leukemias or myelodysplasias observed in various gene therapy studies using murine retrovirus vectors to treat X-SCID (Hacein-Bey-Abina et al., 2003a,b; Howe et al., 2008), Wiskott Aldrich (Boztug et al., 2010), or CGD (Ott et al., 2006; Stein et al., 2010), the use of self-inactivating lentiviral vectors are being considered not only because of their improved profile for transducing long-term repopulating HSC but also for their improved safety profile as well. Their insertion pattern as compared to those of retroviral vectors would suggest less risk of insertional mutagenic events (Wang et al., 2007). A deterrent to their implementation, however, relates to the difficulties in their production. Stable producer cell lines were previously untenable due to the cell toxicity from the continuous production of some of the lentivector packaging elements. Recently, John Gray and his colleagues have been successful in producing a lentivector packaging line that uses tetracycline repressor elements to prevent expression of toxic packaging component until the line is removed from media containing tetracycline (Throm et al., 2009). Thus, with a stable producer line, it may become more efficient to produce large quantities of lentivector in a GLP facility.

We have thus been pursuing the development of an X-CGD lentivector line built upon the system described for production of a lentivector to treat X-SCID using the CL20 lentivector backbone (Zhou et al., 2010). But a number of issues still need to be resolved. First and foremost is the issue of protein production. The MFGS vector as discussed has the unique characteristic of producing very high levels of gp91phox transgene protein per transgene copy. This is an important consideration for a gene therapy trial for X-CGD as the level of protein expression is essential to achieve full functional correction of oxidase activity, and this is intimately connected to the overall therapeutic benefit. The MFGS-gp91phox insert in our clinical trial achieves full functional correction of oxidase activity in neutrophils arising from HSC with a single copy insert of the vector. Although clinical benefit is expected to accrue from any level of oxidase activity achieved through gene therapy, clinical studies of CGD patients with different levels of residual oxidase activity relating to the type of CGD mutation causing their disease show a direct and highly statistically significant connection between amount of oxidase produced and patient survival (Kuhns et al., 2010).

Self-inactivating lentivectors use an internal promoter to express the transgene, and for safety reasons it is preferable to use a mammalian gene-derived internal promoter rather than a viral promoter (Brenner and Malech, 2003; Dull et al., 1998; Naldini et al., 1996). To date no mammalian internal promoter tested in a lentivector appears to achieve the output of gp91phox as the MFGS-gp91phox vector. One maneuver that has been used to enhance gp91phox protein production in a lentivector is to codon optimize the cDNA. This has been shown to significantly enhance production of gp91phox, and we have included this approach in our vector design plan (Moreno-Carranza et al., 2009). The internal promoter also presents opportunities to make transgene production more myeloid specific, hoping that this will improve production, but also may prevent off target expression. Hybrid promoters may be constructed to achieve both myeloid specificity and excellent transgene expression. Dr. Adrian Thrasher's laboratory has designed a hybrid promoter in which they have fused together the "Cathepsin G" minimal 5′-flanking region (360 bp) to the "c-Fes 5"-flanking region (446 bp) to form a chimeric promoter (Santilli et al., 2011). "Cathepsin G" is a serine protease stored in azurophil granules of neutrophil granulocytes, and "c-Fes" is a proto-oncogene playing a critical role in myeloid development. This is a candidate vector that their group is proposing to bring to the clinic to treat patients with X-CGD in the near future.

9.2. Zinc finger/TALENs

Site-specific correction of genetic mutations could be considered as the ultimate gene therapy and would even be applicable to disorders were there are dominant-negative effects. Although dominant-negative effects are not an issue with gene correction of CGD, the ability to directly correct the specific gene mutation to wild-type sequence would also circumvent the risk of insertional mutagenesis. To date, rapid advance in the use of zinc finger nucleases to enhance targeted recombinational gene correction (Wood et al., 2011) is hampered by the fact that efficient functioning zinc fingers cannot be designed for all genomic targets leading to inefficient targeting. As design algorithms improve, this may become less of a problem. Moreover, the X-linked form of CGD is a result of various mutations throughout the gene, making application of zinc finger nuclease for mutation repair difficult for this disorder, though an approach in which a minigene gp91phox cDNA was placed in such a way as to harness the activity of the *CYBB* promoter might serve to have one targeted approach to treat most X-CGD mutations. On the other hand, most p47phox patients have been found to have a single common mutational deletion at the splice acceptor site of exon 2 (Vázquez et al., 2001), and therefore this would prove to be an attractive target for their treatment. More recently, there has been growing interest in the use of TALENs (Wood et al., 2011) which are designed starting from the proteins responsible for a plant-based recombination repair system. These may be more flexible and efficient, but fewer studies have been done using these and so more research is needed. Ultimately, direct gene correction using zinc finger nuclease or TALEN-type techniques are foreseeable in the near future, eliminating the need for permanent integration of viral vector sequence within the genome of patient-derived cells.

9.3. iPSC

Recently, Yamanaka and Thompson (Takahashi and Yamanaka, 2006; Takahashi *et al.*, 2007; Yu *et al.*, 2007) both described a method that can reprogram somatic cells from mice or humans, respectively, to an embryonic stem cell state and these have been termed induced pluripotent stem cells (iPSCs). Subsequently, numerous labs have been using this technique or a modified version to develop iPS from a variety of cell types and then differentiating these cells to various specific lineages. These labs have also been using them to study specific diseases. Our own lab has used both marrow mesenchymal stem cells and CD34+ HSC to create iPS clones from patients with X-CGD and have successfully corrected these clones using "zinc finger nuclease" directed targeting of a gp91phox minigene to a safe harbor site (Zou *et al.*, 2011). Neutrophils differentiating from X-CGD iPSC lack oxidase activity, while neutrophils differentiated from zinc finger nunclease-targeted gene-corrected X-CGD iPSC were able to express normal NADPH oxidase activity. Current studies are aimed at achieving *CYBB* gene mutation correction or minigene insertion. The iPSC targeting gene-correction technology allows selection of only those clones with the desired genetic change. Autologous corrected neutrophils produced from gene-corrected iPSC could be used to manage severe infection, and as technology is developed in the future for generation of engraftable CD34+ HSC from iPSC, such corrected autologous HSCs may in the more distant future be used to achieve permanent correction of CGD, negating the need for immunosuppression and presumably reduce the risk of rejection associated with donor stem cell transplants. Although labs have successfully produced HSCs from mouse iPSCs that are transplantable into mice (Hanna *et al.*, 2007), to date there has been minimal success in engrafting human iPS-derived HSCs into mouse xenograft models. Therefore, in order to implement this method clinically, a number of obstacles must be overcome.

As noted, in the nearer future, these cells may provide a source of corrected yet autologous neutrophils, which may allow the treatment of an infection without the risk of alloimmunization, thereby preserving the patient's ability to undergo allogeneic transplant without increasing the risk of graft rejection, as is the case after having received third-party granulocytes.

10. CONCLUSION

Gene therapy has now been proven successful in a number of immunodeficiencies. In CGD, although we have not achieved an actual cure, a number of patients have benefitted, specifically for the treatment of various infections. Unfortunately, serious adverse events have also marred some of the success seen in this field.

Implementing a gene therapy trial requires numerous steps as we have described and often the most recent advances cannot be implemented into a trial which has already been planned.

REFERENCES

Aiuti, A., Slavin, S., Aker, M., Ficara, F., Deola, S., Mortellaro, A., Morecki, S., Andolfi, G., Tabucchi, A., Carlucci, F., Marinello, E., Cattaneo, F., et al. (2002). Correction of ADA-SCID by stem cell gene therapy combined with nonmyeloablative conditioning. *Science* **296,** 2410–2413.

Aiuti, A., Cattaneo, F., Galimberti, S., Benninghoff, U., Cassani, B., Callegaro, L., Scaramuzza, S., Andolfi, G., Mirolo, M., Brigida, I., Tabucchi, A., Carlucci, F., et al. (2009). Gene therapy for immunodeficiency due to adenosine deaminase deficiency. *N. Engl. J. Med.* **360,** 447–458.

Björgvinsdóttir, H., Ding, C., Pech, N., Gifford, M. A., Li, L. L., and Dinauer, M. C. (1997). Retroviral-mediated gene transfer of gp91phox into bone marrow cells rescues defect in host defense against Aspergillus fumigatus in murine X-linked chronic granulomatous disease. *Blood* **89,** 41–48.

Blaese, R. M., Culver, K. W., Miller, A. D., Carter, C. S., Fleisher, T., Clerici, M., Shearer, G., Chang, L., Chiang, Y., Tolstoshev, P., Greenblatt, J. J., Rosenberg, S. A., et al. (1995). T lymphocyte-directed gene therapy for ADA-SCID: Initial trial results after 4 years. *Science* **270,** 475–480.

Boztug, K., Schmidt, M., Schwarzer, A., Banerjee, P. P., Díez, I. A., Dewey, R. A., Böhm, M., Nowrouzi, A., Ball, C. R., Glimm, H., Naundorf, S., Kühlcke, K., et al. (2010). Stem-cell gene therapy for the Wiskott-Aldrich syndrome. *N. Engl. J. Med.* **363,** 1918–1927.

Brenner, S., and Malech, H. L. (2003). Current developments in the design of oncoretrovirus and lentivirus vector systems for hematopoietic cell gene therapy. *Biochim. Biophys. Acta* **1640,** 1–24.

Brenner, S., Whiting-Theobald, N. L., Linton, G. F., Holmes, K. L., Anderson-Cohen, M., Kelly, P. F., Vanin, E. F., Pilon, A. M., Bodine, D. M., Horwitz, M. E., and Malech, H. L. (2003). Concentrated RD114-pseudotyped MFGS-gp91phox vector achieves high levels of functional correction of the chronic granulomatous disease oxidase defect in NOD/SCID/beta -microglobulin−/− repopulating mobilized human peripheral blood CD34+ cells. *Blood* **102,** 2789–2797.

Brenner, S., Ryser, M. F., Whiting-Theobald, N. L., Gentsch, M., Linton, G. F., and Malech, H. L. (2007). The late dividing population of gamma-retroviral vector transduced human mobilized peripheral blood progenitor cells contributes most to gene-marked cell engraftment in nonobese diabetic/severe combined immunodeficient mice. *Stem Cells* **25,** 1807–1813.

Chinen, J., Davis, J., De Ravin, S. S., Hay, B. N., Hsu, A. P., Linton, G. F., Naumann, N., Nomicos, E. Y., Silvin, C., Ulrick, J., Whiting-Theobald, N. L., Malech, H. L., et al. (2007). Gene therapy improves immune function in preadolescents with X-linked severe combined immunodeficiency. *Blood* **110,** 67–73.

Danos, O., and Mulligan, R. C. (1988). Safe and efficient generation of recombinant retroviruses with amphotropic and ecotropic host ranges. *Proc. Natl. Acad. Sci. USA* **85,** 6460–6464.

Dinauer, M. C., Gifford, M. A., Pech, N., Li, L. L., and Emshwiller, P. (2001). Variable correction of host defense following gene transfer and bone marrow transplantation in murine X-linked chronic granulomatous disease. *Blood* **97,** 3738–3745.

Ding, C., Kume, A., Björgvinsdóttir, H., Hawley, R. G., Pech, N., and Dinauer, M. C. (1996). High-level reconstitution of respiratory burst activity in a human X-linked chronic granulomatous disease (X-CGD) cell line and correction of murine X-CGD bone marrow cells by retroviral-mediated gene transfer of human gp91phox. *Blood* **88,** 1834–1840.

Dull, T., Zufferey, R., Kelly, M., Mandel, R. J., Nguyen, M., Trono, D., et al. (1998). A third-generation lentivirus vector with a conditional packaging system. *J. Virol.* **72,** 8463–8471.

Gaspar, H. B., Parsley, K. L., Howe, S., King, D., Gilmour, K. C., Sinclair, J., Brouns, G., Schmidt, M., Von Kalle, C., Barington, T., Jakobsen, M. A., Christensen, H. O., et al. (2004). Gene therapy of X-linked severe combined immunodeficiency by use of a pseudotyped gammaretroviral vector. *Lancet* **364,** 2181–2187.

Hacein-Bey-Abina, S., Le Deist, F., Carlier, F., Bouneaud, C., Hue, C., De Villartay, J. P., Thrasher, A. J., Wulffraat, N., Sorensen, R., Dupuis-Girod, S., Fischer, A., Davies, E. G., et al. (2002). Sustained correction of X-linked severe combined immunodeficiency by *ex vivo* gene therapy. *N. Engl. J. Med.* **346,** 1185–1193.

Hacein-Bey-Abina, S., von Kalle, C., Schmidt, M., Le Deist, F., Wulffraat, N., McIntyre, E., Radford, I., Villeval, J. L., Fraser, C. C., Cavazzana-Calvo, M., and Fischer, A. (2003a). A serious adverse event after successful gene therapy for X-linked severe combined immunodeficiency. *N. Engl. J. Med.* **348,** 255–256.

Hacein-Bey-Abina, S., Von Kalle, C., Schmidt, M., McCormack, M. P., Wulffraat, N., Leboulch, P., Lim, A., Osborne, C. S., Pawliuk, R., Morillon, E., Sorensen, R., Forster, A., et al. (2003b). LMO2-associated clonal T cell proliferation in two patients after gene therapy for SCID-X1. *Science* **302,** 415–419.

Hanna, J., Wernig, M., Markoulaki, S., Sun, C. W., Meissner, A., Cassady, J. P., Beard, C., Brambrink, T., Wu, L. C., Townes, T. M., and Jaenisch, R. (2007). Treatment of sickle cell anemia mouse model with iPS cells generated from autologous skin. *Science* **318,** 1920–1923.

Horwitz, M. E., Barrett, A. J., Brown, M. R., Carter, C. S., Childs, R., Gallin, J. I., Holland, S. M., Linton, G. F., Miller, J. A., Leitman, S. F., Read, E. J., and Malech, H. L. (2001). Treatment of chronic granulomatous disease with nonmyeloablative conditioning and a T-cell-depleted hematopoietic allograft. *N. Engl. J. Med.* **344,** 881–888.

Howe, S. J., Mansour, M. R., Schwarzwaelder, K., Bartholomae, C., Hubank, M., Kempski, H., Brugman, M. H., Pike-Overzet, K., Chatters, S. J., de Ridder, D., Gilmour, K. C., Adams, S., et al. (2008). Insertional mutagenesis combined with acquired somatic mutations causes leukemogenesis following gene therapy of SCID-X1 patients. *J. Clin. Invest.* **118,** 3143–3150.

Kang, E. M., and Malech, H. L. (2009). Advances in treatment for chronic granulomatous disease. *Immunol. Res.* **43,** 77–84.

Kang, E. M., Choi, U., Theobald, N., Linton, G., Long Priel, D. A., Kuhns, D., and Malech, H. L. (2010). Retrovirus gene therapy for X-linked chronic granulomatous disease can achieve stable long-term correction of oxidase activity in peripheral blood neutrophils. *Blood* **115,** 783–791.

Kang, E. M., Marciano, B. E., DeRavin, S., Zarember, K. A., Holland, S. M., and Malech, H. L. (2011). Chronic granulomatous disease: Overview and hematopoietic stem cell transplantation. *J. Allergy Clin. Immunol.* **127,** 1319–1326.

Kohn, D. B., Hershfield, M. S., Carbonaro, D., Shigeoka, A., Brooks, J., Smogorzewska, E. M., Barsky, L. W., Chan, R., Burotto, F., Annett, G., Nolta, J. A., Crooks, G., et al. (1998). T lymphocytes with a normal ADA gene accumulate after transplantation of transduced autologous umbilical cord blood CD34+ cells in ADA-deficient SCID neonates. *Nat. Med.* **4,** 775–780.

Krall, W. J., Skelton, D. C., Yu, X. J., Riviere, I., Lehn, P., Mulligan, R. C., and Kohn, D. B. (1996). Increased levels of spliced RNA account for augmented expression from the MFG retroviral vector in hematopoietic cells. *Gene Ther.* **3,** 37–48.

Kuhns, D. B., Alvord, W. G., Heller, T., Feld, J. J., Pike, K. M., Marciano, B. E., Uzel, G., DeRavin, S. S., Priel, D. A., Soule, B. P., Zarember, K. A., Malech, H. L., et al. (2010). Residual NADPH oxidase and survival in chronic granulomatous disease. *N. Engl. J. Med.* **363,** 2600–2610.

Leto, T. L., Lavigne, M. C., Homoyounpour, N., Lekstrom, K., Linton, G., Malech, H. L., and de Mendez, I. (2007). The K-562 cell model for analysis of neutrophil NADPH oxidase function. *Methods Mol. Biol.* **412,** 365–383.

Malech, H. L. (2000). Use of serum-free medium with fibronectin fragment enhanced transduction in a system of gas permeable plastic containers to achieve high levels of retrovirus transduction at clinical scale. *Stem Cells* **18,** 155–156.

Malech, H. L., and Hickstein, D. D. (2007). Genetics, biology and clinical management of myeloid cell primary immune deficiencies: Chronic granulomatous disease and leukocyte adhesion deficiency. *Curr. Opin. Hematol.* **14,** 29–36.

Malech, H. L., Maples, P. B., Whiting-Theobald, N., Linton, G. F., Sekhsaria, S., Vowells, S. J., Li, F., Miller, J. A., DeCarlo, E., Holland, S. M., Leitman, S. F., Carter, C. S., et al. (1997). Prolonged production of NADPH oxidase-corrected granulocytes after gene therapy of chronic granulomatous disease. *Proc. Natl. Acad. Sci. USA* **94,** 12133–12138.

Malech, H. L., Choi, U., and Brenner, S. (2004). Progress toward effective gene therapy for chronic granulomatous disease. *Jpn. J. Infect. Dis.* **57,** S27–S28.

Mardiney, M., 3rd, Jackson, S. H., Spratt, S. K., Li, F., Holland, S. M., and Malech, H. L. (1997). Enhanced host defense after gene transfer in the murine p47phox-deficient model of chronic granulomatous disease. *Blood* **89,** 2268–2275.

Matute, J. D., Arias, A. A., Wright, N. A., Wrobel, I., Waterhouse, C. C., Li, X. J., Marchal, C. C., Stull, N. D., Lewis, D. B., Steele, M., Kellner, J. D., Yu, W., et al. (2009). A new genetic subgroup of chronic granulomatous disease with autosomal recessive mutations in p40 phox and selective defects in neutrophil NADPH oxidase activity. *Blood* **114,** 3309–3315.

McDermott, S. P., Eppert, K., Lechman, E. R., Doedens, M., and Dick, J. E. (2010). Comparison of human cord blood engraftment between immunocompromised mouse strains. *Blood* **116,** 193–200.

Moreno-Carranza, B., Gentsch, M., Stein, S., Schambach, A., Santilli, G., Rudolf, E., Ryser, M. F., Haria, S., Thrasher, A. J., Baum, C., Brenner, S., and Grez, M. (2009). Transgene optimization significantly improves SIN vector titers, gp91phox expression and reconstitution of superoxide production in X-CGD cells. *Gene Ther.* **16,** 111–118.

Naldini, L., Blomer, U., Gallay, P., Ory, D., Mulligan, R., Gage, F. H., et al. (1996). In vivo gene delivery and stable transduction of nondividing cells by a lentiviral vector. *Science* **272,** 263–267.

Ott, M. G., Schmidt, M., Schwarzwaelder, K., Stein, S., Siler, U., Koehl, U., Glimm, H., Kühlcke, K., Schilz, A., Kunkel, H., Naundorf, S., Brinkmann, A., et al. (2006). Correction of X-linked chronic granulomatous disease by gene therapy, augmented by insertional activation of MDS1-EVI1, PRDM16 or SETBP1. *Nat. Med.* **12,** 401–409.

Roesler, J., Brenner, S., Bukovsky, A. A., Whiting-Theobald, N., Dull, T., Kelly, M., Civin, C. I., and Malech, H. L. (2002). Third-generation, self-inactivating gp91(phox) lentivector corrects the oxidase defect in NOD/SCID mouse-repopulating peripheral blood-mobilized CD34+ cells from patients with X-linked chronic granulomatous disease. *Blood* **100,** 4381–4390.

Santilli, G., Almarza, E., Brendel, C., Choi, U., Beilin, C., Blundell, M. P., Haria, S., Parsley, K. L., Kinnon, C., Malech, H. L., Bueren, J. A., Grez, M., et al. (2011). Biochemical correction of X-CGD by a novel chimeric promoter regulating high levels of transgene expression in myeloid cells. *Mol. Ther.* **19,** 122–132.

Segal, B. H., Veys, P., Malech, H., and Cowan, M. J. (2011). Chronic granulomatous disease: Lessons from a rare disorder. *Biol. Blood Marrow Transplant.* **17**(Suppl. 1), S123–S131.

Seger, R. A. (2010). Hematopoietic stem cell transplantation for chronic granulomatous disease. *Immunol. Allergy Clin. North Am.* **30**, 195–208.

Seger, R. A., Gungor, T., Belohradsky, B. H., Blanche, S., Bordigoni, P., Di Bartolomeo, P., Flood, T., Landais, P., Müller, S., Ozsahin, H., Passwell, J. H., Porta, F., *et al.* (2002). Treatment of chronic granulomatous disease with myeloablative conditioning and an unmodified hemopoietic allograft: A survey of the European experience, 1985–2000. *Blood* **100**, 4344–4350.

Sekhsaria, S., Fleisher, T. A., Vowells, S., Brown, M., Miller, J., Gordon, I., Blaese, R. M., Dunbar, C. E., Leitman, S., and Malech, H. L. (1996). Granulocyte colony-stimulating factor recruitment of CD34+ progenitors to peripheral blood: Impaired mobilization in chronic granulomatous disease and adenosine deaminase-deficient severe combined immunodeficiency disease patients. *Blood* **88**, 1104–1112.

Stein, S., Ott, M. G., Schultze-Strasser, S., Jauch, A., Burwinkel, B., Kinner, A., Schmidt, M., Krämer, A., Schwäble, J., Glimm, H., Koehl, U., Preiss, C., *et al.* (2010). Genomic instability and myelodysplasia with monosomy 7 consequent to EVI1 activation after gene therapy for chronic granulomatous disease. *Nat. Med.* **16**, 198–204.

Takahashi, K., and Yamanaka, S. (2006). Induction of pluripotent stem cells from mouse embryonic and adult fibroblast cultures by defined factors. *Cell* **126**, 663–676.

Takahashi, K., Tanabe, K., Ohnuki, M., Narita, M., Ichisaka, T., Tomoda, K., and Yamanaka, S. (2007). Induction of pluripotent stem cells from adult human fibroblasts by defined factors. *Cell* **131**, 861–872.

Thrasher, A. J., Hacein-Bey-Abina, S., Gaspar, H. B., Blanche, S., Davies, E. G., Parsley, K., Gilmour, K., King, D., Howe, S., Sinclair, J., Hue, C., Carlier, F., *et al.* (2005). Failure of SCID-X1 gene therapy in older patients. *Blood* **105**, 4255–4257.

Throm, R. E., Ouma, A. A., Zhou, S., Chandrasekaran, A., Lockey, T., Greene, M., De Ravin, S. S., Moayeri, M., Malech, H. L., Sorrentino, B. P., and Gray, J. T. (2009). Efficient construction of producer cell lines for a SIN lentiviral vector for SCID-X1 gene therapy by concatemeric array transfection. *Blood* **113**, 5104–5110.

Vázquez, N., Lehrnbecher, T., Chen, R., Christensen, B. L., Gallin, J. I., Malech, H., Holland, S., Zhu, S., and Chanock, S. J. (2001). Mutational analysis of patients with p47-phox-deficient chronic granulomatous disease: The significance of recombination events between the p47-phox gene (*NCF1*) and its highly homologous pseudogenes. *Exp. Hematol.* **29**, 234–243.

Vowells, S. J., Fleisher, T. A., Sekhsaria, S., Alling, D. W., Maguire, T. E., and Malech, H. L. (1996). Genotype-dependent variability in flow cytometric evaluation of reduced nicotinamide adenine dinucleotide phosphate oxidase function in patients with chronic granulomatous disease. *J. Pediatr.* **128**, 104–107.

Williams, D. A., Tao, W., Yang, F., Kim, C., Gu, Y., Mansfield, P., Levine, J. E., Petryniak, B., Derrow, C. W., Harris, C., Jia, B., Zheng, Y., *et al.* (2000). Dominant negative mutation of the hematopoietic-specific Rho GTPase, Rac2, is associated with a human phagocyte immunodeficiency. *Blood* **96**, 1646–1654.

Wang, G. P., Ciuffi, A., Leipzig, J., Berry, C. C., and Bushman, F. D. (2007). HIV integration site selection: analysis by massively parallel pyrosequencing reveals association with epigenetic modifications. *Genome Res.* **17**, 1186–1194.

Wood, A. J., Lo, T. W., Zeitler, B., Pickle, C. S., Ralston, E. J., Lee, A. H., Amora, R., Miller, J. C., Leung, E., Meng, X., Zhang, L., Rebar, E. J., *et al.* (2011). Targeted genome editing across species using ZFNs and TALENs. *Science* **333**, 307.

Yamauchi, A., Yu, L., Pötgens, A. J., Kuribayashi, F., Nunoi, H., Kanegasaki, S., Roos, D., Malech, H. L., Dinauer, M. C., and Nakamura, M. (2001). Location of the epitope for 7D5, a monoclonal antibody raised against human flavocytochrome b558, to the extracellular peptide portion of primate gp91phox. *Microbiol. Immunol.* **45,** 249–257.

Yu, J., Vodyanik, M. A., Smuga-Otto, K., Antosiewicz-Bourget, J., Frane, J. L., Tian, S., Nie, J., Jonsdottir, G. A., Ruotti, V., Stewart, R., Slukvin, I. I., and Thomson, J. A. (2007). Induced pluripotent stem cell lines derived from human somatic cells. *Science* **318,** 1917–1920.

Zhen, L., King, A. A., Xiao, Y., Chanock, S. J., Orkin, S. H., and Dinauer, M. C. (1993). Gene targeting of X chromosome-linked chronic granulomatous disease locus in a human myeloid leukemia cell line and rescue by expression of recombinant gp91phox. *Proc. Natl. Acad. Sci. USA* **90,** 9832–9836.

Zhou, S., Mody, D., DeRavin, S. S., Hauer, J., Lu, T., Ma, Z., Hacein-Bey Abina, S., Gray, J. T., Greene, M. R., Cavazzana-Calvo, M., Malech, H. L., and Sorrentino, B. P. (2010). A self-inactivating lentiviral vector for SCID-X1 gene therapy that does not activate LMO2 expression in human T cells. *Blood* **116,** 900–908.

Zou, J., Sweeney, C. L., Chou, B. K., Choi, U., Pan, J., Wang, H., Dowey, S. N., Cheng, L., and Malech, H. L. (2011). Oxidase-deficient neutrophils from X-linked chronic granulomatous disease iPS cells: Functional correction by zinc finger nuclease-mediated safe harbor targeting. *Blood* **117,** 5561–5572.

CHAPTER EIGHT

Alternative Splicing Caused by Lentiviral Integration in the Human Genome

Arianna Moiani* *and* Fulvio Mavilio*,†

Contents

1. Introduction 156
2. Analysis of Aberrantly Spliced Transcripts Generated by the Integration of SIN Lentiviral Vectors in Human Genes 158
   2.1. Cell transduction, cell cloning, and mapping of lentiviral integration sites 158
   2.2. Detection of chimeric transcripts containing proviral and cellular gene sequences 161
   2.3. Estimating the abundance of aberrant splicing events 165
Acknowledgments 168
References 168

Abstract

Gene transfer vectors derived from murine oncoretroviruses or human lentiviruses are widely used in human gene therapy. Integration of these vectors in the human genome may, however, have genotoxic effects, caused by deregulation of gene expression at the transcriptional or posttranscriptional level. In particular, integration of lentiviral vectors within transcribed genes has a significant potential to affect their expression by interfering with splicing and polyadenylation of primary transcripts. Aberrant splicing is caused by the usage of both constitutive and cryptic splice sites located in the retroviral backbone as well as in the gene expression cassettes. We describe a set of simple methods that allow the identification of chimeric transcripts generated by the insertion of a lentiviral vector within genes and the evaluation of their relative abundance. Identification of the splice sites, either constitutive or cryptic, that are frequently used by the cell splicing machinery within a given vector provides a useful resource to attempt recoding of the vector with the objective of reducing its potential genotoxicity in a clinical context.

* Division of Genetics and Cell Biology, Istituto Scientifico H. San Raffaele, Milan, Italy
† Center for Regenerative Medicine, University of Modena and Reggio Emilia, Modena, Italy

1. INTRODUCTION

Retroviral integration is a nonrandom process, whereby the viral RNA genome, reverse transcribed into double-stranded DNA and assembled in preintegration complexes (PICs), associates with the host cell chromatin and integrates through the activity of the viral integrase (Coffin *et al.*, 1997). Large-scale surveys of retroviral integration in murine and human cells uncovered some genomic features systematically and specifically associated with retroviral insertions and revealed that each retrovirus type has a unique, characteristic pattern of integration within mammalian genomes (Bushman *et al.*, 2005). Target site selection depends on both viral and cellular determinants, poorly defined for most retroviruses. The Moloney murine leukemia virus (MLV) and its derived vectors integrate preferentially in transcriptionally active promoters and regulatory elements (Cattoglio *et al.*, 2007, 2010b; Hematti *et al.*, 2004; Mitchell *et al.*, 2004; Wu *et al.*, 2003), suggesting that their PICs are tethered to regions engaged by basal components of the RNA polymerase II transcriptional machinery (Cattoglio *et al.*, 2010b; Felice *et al.*, 2009). On the contrary, the simian (SIV) and human immunodeficiency virus (HIV), and their derived lentiviral vectors, target expressed genes in their transcribed portions, away from regulatory elements (Cattoglio *et al.*, 2010b; Hematti *et al.*, 2004; Mitchell *et al.*, 2004; Schroder *et al.*, 2002; Wang *et al.*, 2007). The host cell factor LEDGF/p75 has a major role in tethering HIV PICs to active chromatin by directly binding the viral integrase (Engelman and Cherepanov, 2008), a major viral determinant of target site selection (Lewinski *et al.*, 2006). Recent evidence indicate that the interactions between PICs and component of the nuclear import machinery may as well play a role in tethering and targeting HIV integration to chromatin (Matreyek and Engelman, 2011; Ocwieja *et al.*, 2011).

Seminal clinical studies have shown that transplantation of stem cells genetically modified by retroviral vectors may cure severe genetic diseases such as immunodeficiencies (Aiuti *et al.*, 2009; Boztug *et al.*, 2010; Hacein-Bey-Abina *et al.*, 2002), skin adhesion defects (Mavilio *et al.*, 2006), and lysosomal storage disorders (Cartier *et al.*, 2009). Some of these studies showed also the genotoxic consequences of retroviral gene transfer technology. Insertional activation of proto-oncogenes by MLV-derived vectors caused T-cell lymphoproliferative disorders in patients undergoing gene therapy for X-linked severe combined immunodeficiency (X-SCID) (Hacein-Bey-Abina *et al.*, 2008; Howe *et al.*, 2008) and premalignant expansion of myeloid progenitors in patients treated for chronic granulomatous disease (Ott *et al.*, 2006; Stein *et al.*, 2010). The strong transcriptional enhancers present in the MLV LTR (long terminal repeat) played a major role in deregulating gene expression. Preclinical studies showed that enhancer-less (self-inactivating, SIN), HIV-derived lentiviral vectors are less likely to cause insertional tumors than MLV-derived vectors (Maruggi et al., 2009; Modlich et al., 2009; Montini et al., 2009, 2006).

Transcriptional gene activation is just one of the genotoxic events that may result from retroviral vector integration. Data obtained both *in vitro* and *ex vivo* suggest that the insertion of retroviral splicing and polyadenylation signals within transcription units may cause posttranscriptional deregulation of gene expression with a certain frequency (Almarza et al., 2011; Cattoglio et al., 2010b; Maruggi et al., 2009). This may include aberrant splicing, premature transcript termination, and the generation of chimeric, read-through transcripts originating from vector-borne promoters (Almarza et al., 2011), a classical cause of insertional oncogenesis (Nilsen et al., 1985). Analysis of gamma-retroviral (MLV) integrations in human T cells before and after infusion in patients showed that proviruses inserted within genes in the same transcriptional orientation are counterselected *in vivo* and suggested that the introduction of splicing and polyadenylation signals causes deregulation of gene expression leading to rapid loss of a substantial number of clones (Cattoglio et al., 2010a). The propensity of lentiviral vectors to integrate into the body of transcribed genes increases the probability of such events compared to MLV-derived vectors. In addition, the deletion of the U3 region in SIN lentiviral vectors results in decreased transcriptional termination and increased generation of chimeric transcripts (Yang et al., 2007). In a clinical context, insertion of a lentiviral vector caused posttranscriptional activation of a truncated proto-oncogene in one patient treated for β-thalassemia, resulting in benign clonal expansion of hematopoietic progenitors (Cavazzana-Calvo et al., 2010). Analyzing the nature and frequency of posttranscriptional genotoxic events in relevant models is therefore crucial to determine the biosafety of clinical gene transfer vectors and to drive intelligent improvement of their design.

Many tests can be used to analyze aberrant splicing caused by the insertion of lentiviral vectors in active genes. High-throughput RNA sequencing of polyclonal cell populations may provide a broad information on the variety of genomic transcripts containing hybrid gene–vector sequences. Analysis of individual clones with identified vector integrations may provide also quantitative estimates of the relative frequency of normal and aberrantly spliced transcripts from individual genes. We describe relatively simple methods to identify chimeric transcripts generated by the insertion of a lentiviral vector within genes and dose their relative abundance without the need of extensive RNA sequencing. Identification of the cryptic or constitutive splicing signals within a certain vector provides a useful resource to attempt recoding of the vector with the objective of reducing its potential genotoxicity in a clinical context.

2. Analysis of Aberrantly Spliced Transcripts Generated by the Integration of SIN Lentiviral Vectors in Human Genes

In this part, we provide optimal conditions and procedures for sequencing and analyzing aberrantly spliced, chimeric transcripts caused by the integration of SIN lentiviral vectors within introns and exons of transcribed genes in human cells. The techniques we adopted include transduction of cell lines with SIN lentiviral vectors, cell cloning, identification of integration sites by linker-mediated polymerase chain reaction (LM-PCR), and amplification and sequencing of transcripts by 5′RACE and exon-specific RT-PCR (Fig. 8.1). The use of cell clones provides the best conditions to analyze individual integration events and transcripts originating from individual genes. Cell clones coming from cultures transduced at high multiplicity of infection (MOI) usually contains several integrated proviruses and allow the study of multiple genes at the same time. Cell lines are reliable, easy to maintain in culture, transduce, and clone, and can be stored and reanalyzed unlimited times. To characterize the splicing induced by vector integration, the use of primary cells is in general not necessary, unless there are particular, cell-specific issues that cannot be addressed in immortal cell lines.

2.1. Cell transduction, cell cloning, and mapping of lentiviral integration sites

Based on our experience, we recommend the use of human cell lines that can be easily transduced at high efficiency and easily cloned by limiting dilution. These include HeLa cells, the T-cell lines JurkaT or SupT1, and the keratinocyte-derived cell line HaCaT. Ideal primary cells are peripheral blood T lymphocytes, which can be transduced and cloned as well at high efficiency but require specific cell culture skills that are not necessarily available in all laboratories. In addition, primary T cells cannot be unlimitedly expanded and are difficult to reexpand and reanalyze once frozen.

Materials

- Cell lines of interest: JurkaT, SupT1, HaCaT, primary T lymphocytes
- VSV-G pseudotyped SIN lentiviral vectors expressing a reporter gene
- Growth medium: The growth medium for JurkaT, SupT1, and HaCaT cells contains DMEM supplemented with 10% FCS, 1% glutamine, and 1% antibiotics (penicillin, streptomycin). T primary lymphocytes are

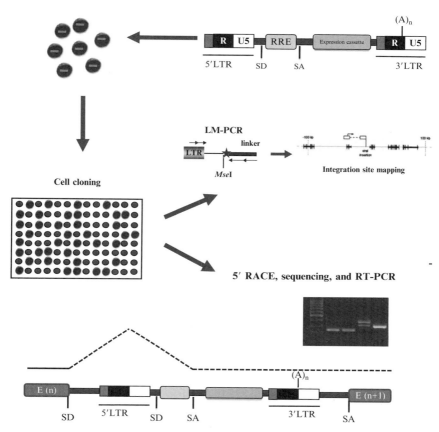

Figure 8.1 The experimental approach. A schematic structure of a SIN lentiviral vector is shown in the top right panel. RRE, rev responsive element; SD, splicing donor site; SA, splicing acceptor site; A_n, polyadenylation signal. 5'LTR and 3'LTR lack the U3 region containing the viral enhancer and promoter. The vector is used to transduce target cells, which are then cloned by limiting dilution. Vector integrations are mapped by linker-mediated (LM) PCR: genomic DNA is digested with a frequently cutting restriction enzyme (*Mse*I), ligated to a linker, amplified with vector- and linker-specific nested primers, and sequenced and aligned to the genome. Chimeric transcripts are amplified by RT-PCR or RACE and analyzed by gel electrophoresis, as showed in the middle panel. The bottom panel shows an example of an aberrant splicing event giving rise to a cellular–proviral chimeric transcript. The blue boxes represent the exons right upstream (E(n)) and downstream (E(n+1)) the provirus with their respective splice donor and acceptor sites (SA/SD). In this example, splicing occurs between the upstream exon SD site and the constitutive proviral SA site. (See Color Insert.)

stimulated in culture with X-VIVO-15 (BioWhittaker, Verviers, Belgium) supplemented with 10% human serum (Cambrex BioScience, Walkersville, MD), 50 U/ml IL-2 (Chiron, Emeryville, CA), and 25 U/ml IL-7 (ImmunoTools, Friesoythe, Germany) in the presence of CD3/CD28 T-cell expander (Dynal, Lake Success, NY).

Other materials

- Trypsin–EDTA solution
- Ca^{2+}- and Mg^{2+}-free PBS (1×)
- Restriction enzymes (*Mse*1, *Sac*1)
- T4 DNA ligase (high concentration)
- 10× PCR buffer
- Taq Platinum (Invitrogen, 5 U/µl)
- dNTP mix, 10 m*M*
- Linker, 10 µ*M*
- 3'LTR specific primer, 10 µ*M*
- 3'LTR specific primer nested, 10 µ*M*
- Linker primer, 10 µ*M*
- Linker nested primer, 10 µ*M*
- TOPO TA cloning kit (Invitrogen)

Disposables

- 1.5-ml (Eppendorf) and 10-ml tubes
- 0.2-ml PCR tubes
- 25- and 75-cm tissue culture flasks

Protocol

Perform cells transduction by overnight incubation with the lentiviral vector preparation in the presence of 4 µg/ml polybrene at low MOI (e.g., 10) to obtain low-copy number of integrated vectors (VCN). If the vector expresses a reporter gene such as green fluorescent protein (GFP), transduction efficiency can be evaluated by analysis of gene expression by cytofluorimetry. Average VCN can be evaluated by Southern blotting, or more easily, by real-time PCR using appropriate standards. Transduced cells may be enriched by fluorescence-activated cell sorting (FACS) and cloned by limiting dilution (0.3 cells/well).

Vector integrations can be mapped in each cell clone by amplifying and sequencing the vector-genome junctions at the 3'LTR end by LM-PCR, according to the following protocol.

1. Extract genomic DNA from transduced clones (5×10^6 cells) using Qiagen QIAamp DNA Blood Mini Kit (see manufacturer's instructions).
2. Take 1 μg of DNA and digest it with 2 μl (20 U) of *Mse*I at 37 °C for 2 h.
3. Digest the DNA with 2 μl of *Sac*I or *Nar*I (20 U) at 37 °C overnight. This second restriction is necessary to prevent amplification of internal 5'LTR fragments.
4. The next day perform several ligation reactions starting from 5 μl of the double-digested DNA, using an *Mse*I double-stranded linker (10 μ*M*), and 2 μl T4 DNA ligase (NEB high concentration), and carry the reactions at 16 °C overnight.
5. Perform LM-PCR using nested primers specific for the linker and the 3'LTR (example of liker and primer sequences are reported in Cattoglio *et al.*, 2007). Run two nested PCRs as follows: initialization step at 95 °C for 2 min, followed by 20 cycles of denaturation at 95 °C for 0.5 min, annealing at 58/60 °C for 0.5 min, and elongation at 72 °C for 1 min, followed by a final elongation step at 72 °C for 2 min.
6. PCR products are shotgun-cloned (TOPO TA cloning kit; Invitrogen) and then sequenced. Sequences between the 3'LTR and the linker primers can be mapped onto the human genome by the UCSC BLAT alignment tool as described in Cattoglio *et al.* (2007).

2.2. Detection of chimeric transcripts containing proviral and cellular gene sequences

Integration downstream to a transcriptionally active cellular promoter may cause read-through transcription of the integrated provirus. If the vector integrates in direct (forward) transcriptional orientation within an intron or exon of a transcribed gene, aberrant splicing may occur by the use of constitutive or cryptic splicing signals contained in the provirus. These events may generate fusion mRNA that can be cloned, sequenced, and analyzed by simple PCR-based techniques. The extent of this phenomenon may vary from negligible use of proviral sequences to monoallelic gene knock out. To analyze the propensity of any given vector to generate alternative splicing, and the splicing signals most commonly recognized by the cellular splicing machinery, we recommend focusing the analysis on integrations inside genes (exonic/intronic) and in direct transcriptional orientation. If a vector contains gene expression cassettes with introns inserted in opposite transcriptional orientation, we recommend analyzing integrations in both orientations with respect to cellular genes. A minimum of 20 independent integrations should be analyzed in order to get statistically significant results (note that cell clones usually harbor more than one integration, so it is in general sufficient to analyze RNA from no more than 20 clones).

2.2.1. Extraction and purification of poly(A)$^+$ RNA

Materials

- Invitrogen Dynabeads Oligo (dT)$_{25}$
- Invitrogen DynaMag^{TM}-2 magnet
- Mixing/rotation device
- Sterile and RNase-free test tubes and pipette tips
- Binding buffer (20 mM Tris–HCl, pH 7.5, 500 mM LiCl, 10 mM EDTA)
- Lysis/binding buffer (100 mM Tris–HCl, pH 7.5, 500 mM LiCl, 10 mM EDTA, 1% LiDS, 5 mM dithiothreitol)
- Washing buffer A (10 mM Tris–HCl, pH 7.5, 0.15 M LiCl, 1 mM EDTA, 0.1% LiDS)
- Washing buffer B (10 mM Tris–HCl, pH 7.5, 0.15 M LiCl, 1 mM EDTA, 10 mM Tris–HCl, pH 7.5)
- Syringe and needle
- Water bath or heating block
- Spectrophotometer

Protocol

Polyadenylated transcripts can be magnetically isolated using Dynabeads Oligo dT (Invitrogen) according to the following protocol.

1. Pellet 5×10^6 cells and resuspend in 1 ml lysis/binding buffer, after 1× PBS washing. Allow complete lysis by pipetting a couple of times and reduce viscosity by passing lysate through a 21-gauge needle using a 1–2-ml syringe.
2. Wash Dynabeads Oligo dT by resupending to obtain a uniform brown suspension and transfer 200 μl of beads to a 1.5-ml tube.
3. Place the tube on a magnet for 1–2 min. The beads will migrate to the side of the tube nearest the magnet. Remove the supernatant with a pipette while the tube remains on the magnet. Remove the tube from the magnet and add 100 μl binding buffer to resuspend the beads. Remove binding buffer and resuspend again the beads in 100 μl binding buffer.
4. Remove binding buffer and mix the cell lysate to the beads; place the tube on a rotating mixer for 3–5 min at room temperature to allow annealing. Place the tube on the magnet for 2 min and remove the supernatant.
5. Wash the beads twice using the magnet to separate the beads from the supernatant, first with washing buffer A (0.5–1 ml) and then with

washing buffer B (0.5–1 ml). Washing is performed at room temperature. Make sure to remove the supernatant completely between washing steps.
6. To elute mRNA from the beads, remove the final washing buffer B and add 10–20 μl of 10 mM Tris–HCl. Incubate at 75–80 °C for 2 min, then place the tube on the magnet, and quickly transfer the supernatant containing the mRNA to a new RNase-free tube.
7. Quantify the yield of mRNA using a spectrophotometer and take 100 ng mRNA to perform cDNA synthesis.

It is not usually necessary to add a DNase treatment step, since the isolation with Oligo dT magnetic beads should have minimized DNA contamination.

2.2.2. 5′RACE and RT-PCR

Materials

- 10× PCR buffer
- 25 mM MgCl$_2$
- 10 mM dNTP mix
- Taq DNA Polymerase
- ROCHE First strand cDNA synthesis kit
- 5× tailing buffer (50 mM Tris–HCl, pH 8.4, 125 mM KCl, 7.5 mM MgCl$_2$)
- 2 mM CTP
- Terminal deoxynucleotidyl transferase
- 5′RACE abridged anchor primer (AAP, 10 μM)
- Universal amplification primer (UAP, 10 μM)
- Abridged universal amplification primer (AUAP, 10 μM)
- DEPC-treated water
- Provirus specific primer (Lenti RT, 1 μM)
- Provirus specific primer nested (Lenti rev, 10 μM)
- TOPO TA cloning kit (Invitrogen)

Protocol

5′RACE PCR and RT-PCR can be performed on the reversely transcribed poly(A)$^+$ RNA to detect chimeric proviral–cellular transcripts. 5′RACE is a technique that facilitates the isolation and characterization of 5′ ends from low-copy messages. A sample protocol for 5′RACE, adapted from the Invitrogen 5′RACE kit, is the following.
1. Synthesize first-strand cDNA starting from 100 ng of poly(A)$^+$ RNA using an internal provirus-specific primer annealing downstream to the vector major splice acceptor signal (usually preceding the internal expression cassette) in reverse transcriptional orientation ("Lenti RT" primer in Fig. 8.2).

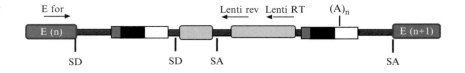
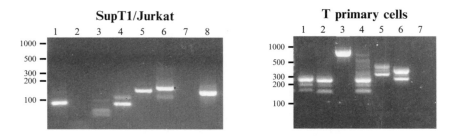
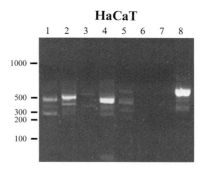

Figure 8.2 Experimental strategy used to detect aberrantly spliced transcripts by 5′RACE and RT-PCR. The upper panel shows a schematic view of a vector inserted in the intron between E(n) and E(n+1) and the position of the primers used in the 5′RACE and RT-PCR. Primers Lenti RT and Lenti rev anneal upstream the major splice acceptor site (SA) of the provirus, usually in the region preceding the expression cassette. The "E for" primer anneals in the exon upstream the integrated provirus. "Lenti RT" is used in the cDNA synthesis reaction. Aberrantly spliced products are detected by 5′RACE and RT-PCR using primer "Lenti rev" in combination with primer "E for." Example of 5′RACE and RT-PCR products detected on a 1% agarose gel electrophoresis and visualized by ethidium bromide staining in SupT1/JurkaT, HaCaT, and T primary lymphocytes is shown in the lower panels. A DNA molecular weight marker is shown at the left of each panel. (See Color Insert.)

2. Perform TdT tailing of cDNA by adding to the reaction DEPC-treated water, 5× tailing buffer, 2 mM dCTP, and 10 μl cDNA, and incubate for 2–3 min at 94 °C. Chill 1 min on ice, add 1 μl TdT, mix gently and incubate for 10 min at 37 °C. Heat-inactivate the TdT for 10 min at 65 °C and place the reaction on ice.

3. PCR amplify dC-tailed cDNA using an AAP primer and a provirus-specific nested primer (Lenti rev primer in Fig. 8.2). Run 35 cycles of PCR as follows: initialization step at 95 °C for 2 min, followed by 35 cycles of denaturation at 95 °C for 0.5 min, annealing at 58 °C for 0.5 min, and elongation at 72 °C for 1 min, followed by a final elongation step at 72 °C for 5 min.
4. Analyze PCR products by 1% agarose gel electrophoresis
5. Clone (TOPO TA cloning kit, Invitrogen) and sequence the PCR products.

If a single PCR round does not generate enough specific products to be detectable by ethidium bromide staining, you can either increase the number of PCR cycles or perform a nested PCR using the AUAP primer and a nested provirus-specific primer.

Chimeric transcripts can also be detected by RT-PCR on cDNA reversely transcribed with the provirus-specific primer ("Lenti RT" in Fig. 8.2). After reverse transcription, PCR amplification of a chimeric transcript is carried out by using an internal, provirus-specific reverse primer (Lenti rev in Fig. 8.2) in combination with a forward primer annealing to an exonic region upstream of the vector integration site ("E for" in Fig. 8.2).

1. Perform the PCR reaction as follows: initialization step at 95 °C for 2 min, followed by 37 cycles of denaturation at 95 °C for 0.5 min, annealing at 58 °C for 0.5 min, elongation at 72 °C for 1 min, followed by a final elongation step at 72 °C for 2 min.
2. Analyze the PCR products by 1% agarose gel electrophoresis.
3. Clone (TOPO TA cloning kit, Invitrogen) and sequence the PCR products.

The sequences resulting from 5'RACE and RT-PCR can be verified on the UCSC Blat alignment tool (2009) and on the known vector sequence. A chimeric transcript would align to the human genome and to the proviral sequence, allowing the identification of the cellular gene splice donor and the provirus splice acceptor site. This analysis allows to map constitutive and cryptic splice signals in the provirus based on their usage by the cell splicing machinery. The relative frequency of splice signal usage provides an estimate of its "strength," and therefore of its genotoxic potential.

2.3. Estimating the abundance of aberrant splicing events

After the identification of aberrant transcripts for a number of genes, it is necessary to analyze their relative abundance compared to wild-type (wt), constitutively spliced transcripts in order to estimate the relevance of the aberrant splicing events in terms of potential perturbation of gene expression. To this aim, we developed a strategy based on semiquantitative RT-PCR amplification of wt and chimeric transcripts. Precise estimation of relative transcript abundance, for example, by quantitative PCR, is usually not necessary. We suggest to rank aberrantly spliced transcripts into four categories (from rare to very abundant) according to their relative level of expression, as estimated by analyzing the PCR products on agarose gels.

Materials

- Invitrogen Dynabeads Oligo (dT)$_{25}$
- Invitrogen DynaMag^{TM}-2 magnet
- ROCHE First strand cDNA synthesis kit
- 5× PCR buffer supplemented with loading dye (Promega)
- Taq Polymerase (Go Taq Promega)
- dNTP mix, 10mM
- Provirus-specific primer, 10μM
- Exon-specific primers, 10μM

Protocol

To amplify wt and chimeric transcripts, poly(A)$^+$ RNA should be reverse transcribed using random hexamer primers. Transcripts are then amplified with specific primers from the reversely transcribed RNA. Oligo (dT) priming is not recommended, since vector insertion can occur at great distance from the polyadenylation signal, beyond the extension capacity of the reverse transcriptase. Exon-specific primers should be designed close to the 3′ end of the exon preceding the vector insertion ("E for" in Fig. 8.3) and close to the 5′ end of the exon immediately downstream ("E rev" in Fig. 8.3). The vector-specific, reverse primer (Lenti rev) should be designed downstream from the splice acceptor site mapped as described in Section 2.2. Based on the predicted wt and chimeric transcript sequence, design primers in order to amplify fragments that can be easily distinguished by their relative mobility on agarose gel electrophoresis. The annealing temperature should ideally be the same for the exon- and the vector-specific primers. The semi-quantitative PCR is performed by amplifying at the same time the wt transcript with primers "E for" and "E rev," and the chimeric transcript with primers "E for" and "Lenti rev."
1. Perform three parallel PCR reactions as follows: initialization step at 95 °C for 2 min, followed by 24, 28, or 33 cycles of denaturation at 95 °C for 0.5 min, annealing at 58 °C for 0.5 min and elongation at 72 °C for 1 min, followed by a final elongation step at 72 °C for 2 min.
2. Run PCR products taken from the different reactions (24, 28, and 33 cycles) on 1–1.5% agarose gels, and visualize the PCR product by ethidium bromide (or equivalent) staining.
3. Identify the number of cycles at which the wt and the chimeric transcript (s) become easily detectable.

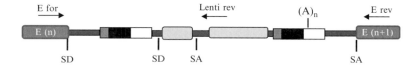

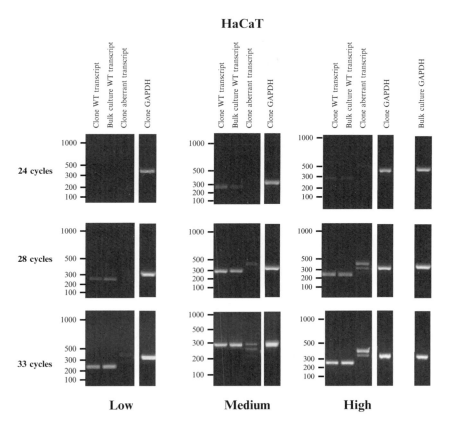

Figure 8.3 Semiquantitative analysis of aberrantly spliced transcripts. cDNA is prepared using random hexamers. Wild-type transcripts are amplified using combination of primer "E for" and "E rev" annealing in the exons immediately upstream and downstream the proviral integration as shown in the upper scheme. PCR products taken from different reactions (24, 28, and 33 cycles) are run on 1.5% agarose gel and visualized by ethidium bromide staining. Each gel panel shows, from left to right, the wild-type transcript obtained in a clone and a bulk culture, the aberrant transcript in the selected clone, and the GAPDH transcript in the selected clone and bulk culture, used for signal normalization. This analysis allows to define four arbitrary classes of relative transcript abundance, that is, High, when chimeric and wild-type transcripts are detected at the same amplification cycle (right panels); Medium, when chimeric transcripts are detected four PCR cycles later than wild-type transcripts (center panels); Low, when chimeric transcripts are detected eight PCR cycles later than wild-type transcript (left panels); and Rare, when aberrant transcripts are not detectable on RNA reverse transcribed by random hexamer primers, although they were identified by RNA reverse transcribed with a vector-specific primer (not shown). (See Color Insert.)

4. Assign chimeric transcripts to one of the following four classes of relative abundance: High, when wt and chimeric transcripts are detected at the same amplification cycle; Medium, when aberrant transcript is detected one step (four PCR cycles) later than wt transcript; Low, when aberrant transcript is detected two steps (eight PCR cycles) later than wt transcript; Rare, when the aberrant transcript is not detectable on RNA reverse transcribed by random hexamer priming, even though it was amplified and sequenced from RNA reverse transcribed by the Lenti rev primer (see Section 2.2).

This analysis allows to estimate the relative abundance of a chimeric, aberrantly spliced transcript compared to the wt, constitutively spliced transcript, independently from the level of transcription of the gene targeted by the vector integration. This in turn allows to estimate the relative use of splicing signals located in the vector compared to the constitutive ones. Frequently used splicing signals are more likely to affect gene expression compared to rarely used ones and should ideally be eliminated by recoding if a vector is destined to a clinical application.

ACKNOWLEDGMENTS

The work described in this chapter has been funded by grants from the European Commission (7th Framework Program, PERSIST Consortium) and the Italian Ministry of University and Research (PRIN 2008).

REFERENCES

Aiuti, A., et al. (2009). Gene therapy for immunodeficiency due to adenosine deaminase deficiency. *N. Engl. J. Med.* **360,** 447–458.
Almarza, D., et al. (2011). Risk assessment in skin gene therapy: Viral-cellular fusion transcripts generated by proviral transcriptional read-through in keratinocytes transduced with self-inactivating lentiviral vectors. *Gene Ther.* **18,** 674–681.
Boztug, K., et al. (2010). Stem-cell gene therapy for the Wiskott-Aldrich syndrome. *N. Engl. J. Med.* **363,** 1918–1927.
Bushman, F., et al. (2005). Genome-wide analysis of retroviral DNA integration. *Nat. Rev. Microbiol.* **3,** 848–858.
Cartier, N., et al. (2009). Hematopoietic stem cell gene therapy with a lentiviral vector in X-linked adrenoleukodystrophy. *Science* **326,** 818–823.
Cattoglio, C., et al. (2007). Hot spots of retroviral integration in human CD34+ hematopoietic cells. *Blood* **110,** 1770–1778.
Cattoglio, C., et al. (2010a). High-definition mapping of retroviral integration sites defines the fate of allogeneic T cells after donor lymphocyte infusion. *PLoS One* **5,** e15688.
Cattoglio, C., et al. (2010b). High-definition mapping of retroviral integration sites identifies active regulatory elements in human multipotent hematopoietic progenitors. *Blood* **116,** 5507–5517.
Cavazzana-Calvo, M., et al. (2010). Transfusion independence and HMGA2 activation after gene therapy of human beta-thalassaemia. *Nature* **467,** 318–322.

Coffin, J. M., et al. (1997). Retroviruses. Cold Spring Harbor Laboratory Press, Cold Spring Harbor, NY.

Engelman, A., and Cherepanov, P. (2008). The lentiviral integrase binding protein LEDGF/p75 and HIV-1 replication. *PLoS Pathog.* **4**, e1000046.

Felice, B., et al. (2009). Transcription factor binding sites are genetic determinants of retroviral integration in the human genome. *PLoS One* **4**, e4571.

Hacein-Bey-Abina, S., et al. (2008). Insertional oncogenesis in 4 patients after retrovirus-mediated gene therapy of SCID-X1. *J. Clin. Invest.* **118**, 3132–3142.

Hacein-Bey-Abina, S., et al. (2002). Sustained correction of X-linked severe combined immunodeficiency by *ex vivo* gene therapy. *N. Engl. J. Med.* **346**, 1185–1193.

Hematti, P., et al. (2004). Distinct genomic integration of MLV and SIV vectors in primate hematopoietic stem and progenitor cells. *PLoS Biol.* **2**, e423.

Howe, S. J., et al. (2008). Insertional mutagenesis combined with acquired somatic mutations causes leukemogenesis following gene therapy of SCID-X1 patients. *J. Clin. Invest.* **118**, 3143–3150.

Lewinski, M. K., et al. (2006). Retroviral DNA integration: Viral and cellular determinants of target-site selection. *PLoS Pathog.* **2**, e60.

Maruggi, G., et al. (2009). Transcriptional enhancers induce insertional gene deregulation independently from the vector type and design. *Mol. Ther.* **17**, 851–856.

Matreyek, K. A., and Engelman, A. (2011). The requirement for nucleoporin NUP153 during human immunodeficiency virus type 1 infection is determined by the viral capsid. *J. Virol.* **85**, 7818–7827.

Mavilio, F., et al. (2006). Correction of junctional epidermolysis bullosa by transplantation of genetically modified epidermal stem cells. *Nat. Med.* **12**, 1397–1402.

Mitchell, R. S., et al. (2004). Retroviral DNA integration: ASLV, HIV, and MLV show distinct target site preferences. *PLoS Biol.* **2**, E234.

Modlich, U., et al. (2009). Insertional transformation of hematopoietic cells by self-inactivating lentiviral and gammaretroviral vectors. *Mol. Ther.* **17**, 1919–1928.

Montini, E., et al. (2009). The genotoxic potential of retroviral vectors is strongly modulated by vector design and integration site selection in a mouse model of HSC gene therapy. *J. Clin. Invest.* **119**, 964–975.

Montini, E., et al. (2006). Hematopoietic stem cell gene transfer in a tumor-prone mouse model uncovers low genotoxicity of lentiviral vector integration. *Nat. Biotechnol.* **24**, 687–696.

Nilsen, T. W., et al. (1985). c-erbB activation in ALV-induced erythroblastosis: Novel RNA processing and promoter insertion result in expression of an amino-truncated EGF receptor. *Cell* **41**, 719–726.

Ocwieja, K. E., et al. (2011). HIV integration targeting: A pathway involving Transportin-3 and the nuclear pore protein RanBP2. *PLoS Pathog.* **7**, e1001313.

Ott, M. G., et al. (2006). Correction of X-linked chronic granulomatous disease by gene therapy, augmented by insertional activation of MDS1-EVI1, PRDM16 or SETBP1. *Nat. Med.* **12**, 401–409.

Schroder, A. R., et al. (2002). HIV-1 integration in the human genome favors active genes and local hotspots. *Cell* **110**, 521–529.

Stein, S., et al. (2010). Genomic instability and myelodysplasia with monosomy 7 consequent to EVI1 activation after gene therapy for chronic granulomatous disease. *Nat. Med.* **16**, 198–204.

Wang, G. P., et al. (2007). HIV integration site selection: Analysis by massively parallel pyrosequencing reveals association with epigenetic modifications. *Genome Res.* **17**, 1186–1194.

Wu, X., et al. (2003). Transcription start regions in the human genome are favored targets for MLV integration. *Science* **300**, 1749–1751.

Yang, Q., et al. (2007). Overlapping enhancer/promoter and transcriptional termination signals in the lentiviral long terminal repeat. *Retrovirology* **4**, 4.

CHAPTER NINE

Genotoxicity Assay for Gene Therapy Vectors in Tumor Prone $Cdkn2a^{-/-}$ Mice

Eugenio Montini *and* Daniela Cesana

Contents

1. Introduction — 172
2. *In Vivo* Genotoxicity Assays Based on Transduction and Transplantation of Tumor-Prone HSPCs — 173
   - 2.1. Isolation of bone marrow-derived lineage depleted cells from $Cdkn2a^{-/-}$ mice — 175
   - 2.2. *Ex vivo* transduction and mouse transplantation procedures — 176
   - 2.3. Assessment of transduction efficiency in *in vitro* cultures and marking levels *in vivo* — 176
   - 2.4. Assessment of engraftment efficiency of transduced $Cdkn2a^{-/-}$ cells in transplanted mice by FACS analysis — 178
   - 2.5. Tumor phenotype analysis, engraftment and vector copy number measurement in tissues — 178
   - 2.6. Linear amplification mediated PCR and genomic integration site analysis — 179
   - 2.7. Quantitative PCR applications to measure of gene expression levels at and nearby vector integration sites — 180
   - 2.8. Survival statistics — 182

References — 183

Abstract

Integrative viral vectors are able to efficiently transduce hematopoietic stem progenitor cells allowing stable transgene expression in the entire hematopoietic system upon transplant in conditioned recipients. For these reasons, integrative vectors based on γ-retroviruses and lentiviruses have been successfully used in gene therapy clinical trials for the treatment of genetic diseases, especially blood disorders. However, in different γ-retroviral-based clinical trials, vector integration into the host cell genome triggered oncogenesis by a mechanism called insertional mutagenesis. Thus, a thorough reassessment of

San Raffaele-Telethon Institute for Gene Therapy, Milan, Italy the safety of available gene transfer systems is a crucial outstanding issue for the whole gene therapy field. Sensitive preclinical models of vector genotoxicity are instrumental to achieve a more detailed understanding of the factors that modulate the risks of insertional mutagenesis. Here, we will describe the methodologies used to address the mutagenesis risk of vector integration using a murine *in vivo* genotoxicity assay based on transduction and transplantation of tumor-prone hematopoietic stem and progenitor cells.

1. INTRODUCTION

Gene therapy vectors, such as those based on replication-defective γ-retroviral or lentiviral vectors (γRVs and LVs, respectively) are able to integrate into the host cellular genome allowing stable and high expression levels in hematopoietic stem and progenitor cells (HSPCs) and their progeny. For this reason, these vectors have been used for the treatment of blood disorders and other genetic diseases (Aiuti *et al.*, 2009, 2002; Cartier *et al.*, 2009; Cavazzana-Calvo *et al.*, 2000; Gaspar *et al.*, 2004). Vector insertions were considered relatively safe because of the replication-defective nature of the vectors and because integrations were thought to be randomly distributed in the genome. Thus, the chance that a limited number of vector integrations could accidentally activate multiple proto-oncogenes within the same cell, in order to promote neoplastic transformation, was considered remote (Baum *et al.*, 2003, 2004; von Kalle *et al.*, 2004). In addition, safety studies performed on wild-type animal models with γRVs did not show genotoxicity related to the vector treatment (Biffi *et al.*, 2004; Dupre *et al.*, 2006; Lo *et al.*, 1999; May *et al.*, 2002; Vigna and Naldini, 2000). The safety profile of γRV was readily reconsidered when four X-linked Severe Combined Immunodeficiency (X-SCID) patients transplanted with γRV transduced HSPCs developed leukemia at 30, 33, 34, and 68 months after treatment (Gansbacher, 2003; Hacein-Bey-Abina *et al.*, 2003, 2008). Further, a patient recruited in an independent γRV based X-SCID trial also developed, a leukemia at 24 months after gene therapy (Howe *et al.*, 2008). These five adverse events have remarkable features in common: all but one of the malignant clones hosted a γRV insertion near to the *LMO-2* proto-oncogene, leading to its overexpression. These data indicate that the leukemogenesis mechanism is likely a consequence of a synergistic interaction between IL2R common γ-chain deficiency and *LMO-2* overexpression. Other insertional mutagenesis driven severe adverse events were observed in two different γRV-based clinical trials: two of three treated patients affected by chronic granulomatous disease (CGD) developed myelodysplastic syndrome (Ott *et al.*, 2006; Stein *et al.*, 2010), and one out of ten treated children affected by Wiskott Aldrich syndrome developed leukemia

(Persons and Baum, 2011). Of note, no adverse effects have been reported in the Adenosine Deaminase Deficiency (ADA)-SCID clinical trials, where more than 17 patients received HSPCs transduced with an ADA-expressing γRV (Aiuti et al., 2002, 2007, 2009). These data suggest that beside vector transduction, other disease-, vector-, or transgene-specific factors may cooperate with insertional gene activation in inducing malignant or premalignant transformation (Baum and Fehse, 2003; Baum et al., 2003, 2004, 2006). The severe adverse events occurred in different clinical trials led the scientific community to reconsider the risks associated with γRV gene transfer in humans, and mandate for a more detailed understanding of the factors that modulate the risks of insertional mutagenesis. In the past years, several preclinical models have been proposed to carefully address the mutagenetic risk of integrative vectors. Here, we will describe how to use an *in vivo* genotoxicity assay to evaluate the safety of integrative vectors based on the transduction and transplantation of tumor-prone murine $Cdkn2a^{-/-}$ HSPCs in wild-type mice (Montini et al., 2006, 2009).

2. *In Vivo* Genotoxicity Assays Based on Transduction and Transplantation of Tumor-Prone HSPCs

The predictive power of safety tests using murine wild-type HSPC transplantation models is limited by the difficulty of inducing cell transformation (Li et al., 2002; Modlich et al., 2005, 2008; Will et al., 2007). Indeed, according to a multiple-hit process, transformation will take time to occur after the first oncogenic hit is provided to cells as additional mutations must be acquired in order to proceed through the *cursus honorum* from premalignant to malignant stage. The lifespan of murine models could be too short to allow this process. Moreover, the number of transplanted cells in a mouse is significantly smaller than the number of cells transplanted in a human patient. A higher number of cells imply receiving a higher number of vector integrations and a consequent higher chance of receiving cells harboring vector integrations targeting oncogenes. Thus, a limited oncogenic potential, apparently neutral in mice, it may pose a significant risk when transducing a high number of cells in a long-lived host.

To increase the sensitivity to vector genotoxicity and to reduce the time required to obtain safety readouts *in vivo*, mouse strains with a predisposition to tumor development can be used. We took advantage of the tumor-prone $Cdkn2a^{-/-}$ mouse model (Lund et al., 2002; Serrano et al., 1996) to perform an *in vivo* comparative analysis of the oncogenic potential of a self-inactivating (SIN) LV and an γRV (Montini et al., 2006) and to dissect the role in genotoxicity of the different features characterizing different vector types and constructs (Montini et al., 2009).

The *Cdkn2a* locus encodes for two proteins, p16INK4a, a regulator of CDK4/6-mediated RB1 phosphorylation, and p19ARF, a modulator of MDM2-mediated degradation of p53 (Sherr, 2004). Loss of *Cdkn2a* results in combined deficiency of the RB1 and P53 pathways, two major regulators of cell proliferation, senescence, and apoptosis. Deficiency of these tumor suppressor pathways does not subvert the cellular growth rate *per se* but enhances the survival and proliferative potential of cells that have acquired an oncogene-activating mutation, and/or are forced to aberrant cell cycle entry. This tumor-prone mouse model bearing germ line inactivation develops a variety of tumors with a well predictable time of onset (Lund et al., 2002; Serrano et al., 1996) and has been an invaluable tool in insertional mutagenesis studies to identify cancer genes (Lund et al., 2002). Moreover, in virtue of its central role in tumor suppression, *Cdkn2a* inactivation may synergize with several types of cancer-promoting lesions allowing the identification of a wide spectrum of oncogenes (Howe et al., 2008; Lund et al., 2002). Thus, the use of *Cdkn2a* deficient cells provides additional practical advantages with respect to wild-type or other tumor-prone mouse models with a restricted number of interacting oncogenes (Kool and Berns, 2009; Mikkers et al., 2002).

In our rationale, genotoxic vector integrations, such as those that activate neighboring proto-oncogenes, would trigger an earlier onset of hematopoietic tumors in transplanted recipients as compared to mice transplanted with mock-treated cells (Fig. 9.1). Using this strategy, it was possible to demonstrate that γRV transduction triggered a dose-dependent acceleration of tumor onset dependent on vector long terminal repeat (LTR) activity and integration at known proto-oncogenes and cell cycle genes,

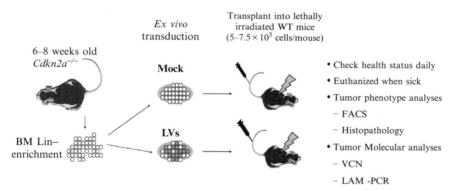

Figure 9.1 Experimental set up. Scheme of the experimental strategy used. (For color version of this figure, the reader is referred to the Web version of this chapter.)

whereas LV transduction, even at high integration load, did not accelerate the tumor onset (Montini et al., 2006). Moreover, we showed that transcriptionally active LTR are major determinants of vector genotoxicity even when reconstituted in LVs and that SIN LTR enhances the safety of γRVs. By comparing the genotoxicity of γRVs and LVs with matched active LTR, we were able to determine that substantially higher LV integration loads are required to approach the same oncogenic risk as γRVs. This difference in facilitating oncogenesis is likely to be explained by the different integration site selection (ISS) of γRVs and LVs (Montini et al., 2006).

The methods used for these *in vivo* studies involve the isolation of bone marrow (BM)-derived lineage depleted cells and *ex vivo* transduction protocols, mouse transplantation procedures, monitoring of mouse health status, phenotype characterization of hematopoietic tumors, application of survival statistics, DNA/RNA isolation for vector copy number measurements, genomic integration and gene expression studies. All the procedures described in this chapter are specifically referred to the safety testing of LVs. Given the proper modifications, the procedures can be translated to any other integrative vector.

2.1. Isolation of bone marrow-derived lineage depleted cells from *Cdkn2a*$^{-/-}$ mice

Six/eight weeks old *Cdkn2a*$^{-/-}$ mice are euthanized by CO_2 inhalation and BM is harvested by flushing femurs and tibias with PBS 2% FBS. FVB or C57/Bl6 *Cdkn2a*$^{-/-}$ mice can be used and are publically available in mouse model repositories such as Jackson laboratories (www.jax.org) or the National Cancer Institute (http://mouse.ncifcrf.gov). Lineage depleted (Lin−) cells are purified by lineage-marker negative selection using specific antibodies and paramagnetic separation on gravity columns. After BM collection, cells are counted in a Burker chamber after a 1:50 dilution (1:10 dilution in acetic acid to lyse erythrocytes, followed by a 1:5 dilution in trypan blue to exclude dead cells). After spinning at 1500 rpm for 5 min at room temperature, cells are resuspended at a concentration of 5×10^7 cells/ml in PBS 2% FBS, 5% rat serum and incubated 15 min at 4°C to block nonspecific binding of antibodies. Cells are then incubated for 15 min at 4°C with an antibody mix containing a combination of rat biotinylated monoclonal antibodies anti-mouse lineage specific markers (CD5, CD11b, CD45R/B220, Gr-1, Neutrophils (7-4), and TER119). After washing with PBS 2% FBS, cells are resuspended at 5×10^7 cells/ml in PBS 2% FBS, incubated with anti-biotin tetrameric antibody complexes for 15 min at 4°C and then magnetic colloid is added for 15 min at 4°C. Cells are then loaded on a primed and equilibrated column in the magnetic field. Flow though, which contains Lin− cells, is harvested and cells are washed, resuspended in StemSpan medium and counted. Murine HSCs are then plated at a density of 1×10^6 cells/ml and cultured in StemSpan SFEMexpansion medium supplemented with penicillin 100 U/ml and streptomycin 100 µg/ml, L-glutamine 2 mM and with a prestimulation cytokine cocktail composed of 100 ng/ml stem cell factor, 100 ng/ml thrombopoietin, 100 ng/ml Flt3-ligand, and 20 ng/ml interleukin-3.

2.2. *Ex vivo* transduction and mouse transplantation procedures

After 24 h of prestimulation with the cytokine cocktail, Lin− cells are split and subjected to LV transduction (10^8 or 10^7 TU/ml). After 12 h, cells are washed and resuspended in the original medium and, if required, reinfected at 48 h using the same culture and transduction conditions. Cells can be kept in culture for 96 h before transplant without severely affecting their repopulating ability. In all the experiments, a portion of cells is maintained in culture in RPMI medium containing 10% FBS and the above mentioned combination of cytokines for 14 days to address the transduction efficiency.

2.3. Assessment of transduction efficiency in *in vitro* cultures and marking levels *in vivo*

The transduction efficiency can be assessed by evaluating the transgene expression levels by fluorescence activated cell sorting (FACS) analysis and/or by quantitative PCR (q-PCR) to evaluate the number of integrated vector copy number (VCN) per cell. These analyses must be performed on cells grown in culture for at least 14 days after transduction, when steady state transgene expression is achieved, and to rule out pseudotransduction (detection of transgene protein associated with the virus particles during vector production and transferred to the target cells during infection). For FACS analysis, transduced cells are washed with PBS and resuspended in PBS 2% FBS containing 5 µg/ml of 7-aminoactinomycin D (7-AAD) staining to exclude dead cells. Untransduced cells are used as negative control to set FACS parameters and gates. At least 10,000–20,000 cells (events) must be scored to obtain reliable readouts.

To calculate the VCN per cell, q-PCR must be performed on genomic DNA of transduced cells. This method allows to measure transduction efficiency independently from transgene expression. After cell lysis, the DNA can be purified on anion exchange resin columns. After elution, genomic DNA is precipitated by adding 0.7 volumes of isopropanol at room temperature and by inverting tubes several times. After formation of the DNA floccule is transferred by winding it onto a glass loop into a 1.5 ml tube containing 1 ml of 70% ethanol. Samples are then centrifuged to remove precipitated salts at 14,000 rpm for 15 min at 4 °C. After supernatant removal, DNA is air-dried for 15 min, resuspended in 100–200 μl TE (10 mM Tris, 1 mM EDTA; pH 8.0), and dissolved on a shaker at 55 °C for 2 h. Genomic DNA can be finally quantified by spectrophotometer analysis and run on a 0.8% agarose gel to check for possible RNA contaminations or degradation of the DNA. DNA is then stored at +4 °C.

q-PCR on genomic DNA is performed using the following thermal cycling conditions: one cycle at 50 °C for 2 min, one cycle at 95 °C for 10 min, 40 cycles at 95 °C for 15 s and 60 °C for 1 min. For each sample, the Sequence Detector System 2.3 software provides an amplification curve constructed by relating the fluorescence signal intensity (Rn) to the cycle number. Cycle threshold (Ct) is defined as the cycle number at which the fluorescence signal is more than 10 of the mean background noise collected from the 3rd to the 15th cycle. For LV sequences, oligonucleotides and probe complementary to the HIV-PBS (primer binding site) can be used. The quantification of the LV sequences in the samples is determined by using as standard curve serial dilutions (200, 100, 50, 25, and 12.5 ng) of genomic DNA extracted from a transgenic mouse containing a known number of LV integrations (copy number of this mouse should be previously defined by Southern blot analysis). It is also possible to set a LV standard curve using the DNA of a cell clone harboring a defined number of LV integrated vector copies.

To determine the VCN per cell, the amount of LV sequences should be normalized to the overall genomic DNA content, which can be assessed either by spectrophotometry or by q-PCR. When assessed by q-PCR, the overall genomic DNA content is measured using the same standard curve used for LV quantification. The overall genomic DNA content can be assessed by using oligonucleotides and probe complementary to the β-*actin* gene or any other nonrepetitive genomic sequence (usually on autosomal chromosomes). Each sample should be run in triplicate in a total volume of 25 μl/reaction, containing 12.5 μl TaqMan Master Mix, 100 ng of sample DNA and one of the following amplification systems at the following concentrations:

- HIV-PBS forward primer: 5-TCTCGACGCAGGACTCG-3′ (300 nM)
- HIV-PBS reverse primer: 5′-TACTGACGCTCTCGCACC-3′ (300 nM)
- HIV-PBSprobe: 5′-(FAM)-ATCTCTCTCCTTCTAGCCTC-(MGB)-3′ (200 nM)

- β-Actin forward primer: 5′-AGAGGGAAATCGTGCGTGAC-3′ (300 nM)
- β-Actin reverse primer: 5′-CAATAGTGATGACCTGGCCGT-3′ (300 nM)
- β-Actinprobe: 5′-(VIC)–AGCTCTCTCGACGCAGGACTCGGC-(MGB)-3′ (200 nM)

VCN was determined by using the following formula:

$$\text{VCN} = (\text{ng of HIV DNA})/(\text{ng of }\beta\text{-actin DNA}) \times \text{VCN standard}(*)$$

(★) The VCN standard is the number of integrated LV copies contained in the genomic DNA of a transgenic mouse or cell clone used as standard.

2.3.1. HSPCs transplantation procedures in wild-type mice

Six weeks old wild-type female mice are lethally irradiated with 11.5 Gy administered in two doses 2 h apart. Four hours after irradiation, mice are heated under an infrared heat lamp and transplanted with vector- and mock-transduced cells (5–7.5×10^5 cells/mouse) by intravenous tail vein injection. All mice are kept in a dedicated pathogen-free animal facility and daily monitored for their health status. Mice must be euthanized when showing signs of severe sickness like hunched posture, labored breath, and ruffled fur.

2.4. Assessment of engraftment efficiency of transduced $Cdkn2a^{-/-}$ cells in transplanted mice by FACS analysis

The engraftment of $Cdkn2a^{-/-}$ HSCP transduced cells in wild-type mice can be evaluated by FACS to estimate the percentage of transgene-expressing cells (direct fluorescence in case of GFP expressing vectors or antibody staining for nonfluorescent transgenes). Six weeks after transplant, peripheral blood from the transplanted mice has to be collected from the tail vein. After blood collection, cells are centrifuged at 1000 rpm for 10 min at room temperature and resuspended in 5 ml/sample of 7% ammonium chloride solution and incubated 20 min on ice to lyse red cells. Cells are then centrifuged at 1200 rpm for 7 min at 4 °C and resuspended in PBS 2% FBS containing 5 μg/ml of 7-AAD to exclude dead cells. Untransduced cells should be used as negative control to set parameters and gates. 10,000–20,000 events should be scored for statistical purposes.

2.5. Tumor phenotype analysis, engraftment and vector copy number measurement in tissues

From the euthanized mice, BM, spleen, liver, thymus, gut, kidney, and lymph nodes are systematically collected. The tumor diagnosis and phenotype must be based on histopathology blinded examination of BM, spleen, thymus, liver, kidney, lung, brain, gut, and lymph nodes, and by FACS analysis of BM, blood, spleen, and thymus. Staining procedures are done following manufacturer instructions depending on the specific antibody used.

q-PCR can be performed on genomic DNA extracted from BM, spleen, thymus, lymph nodes, and liver to evaluate the levels of engraftment, of tumor-infiltrating cells and vector marking in transplanted mice. For DNA extraction, upon harvesting, each tissue is rinsed in PBS, mechanically disrupted, and directly resuspended in 5 ml Buffer G2 containing 0.4 mg/ml RNase A. The amount of tissue used for the extraction depends on the tissue type and the extraction kit used. Samples are then incubated 1 h at 37 °C in a water bath to degrade RNA. Cell lysis is then performed by adding proteinase K to the suspension to a final concentration of 0.4 mg/ml and incubating samples over night at 50 °C in a water bath. The following day samples are vortexed at maximum speed for 10 s and loaded to equilibrated anion exchange resin columns, washed and eluted as described above. The amount of $Cdkn2a^{-/-}$ cells in each tissue can be determined as the ratio between the amount of NeomycinR (specific for transplanted $Cdkn2a^{-/-}$ cells) and β-actin (common to wild-type and $Cdkn2a^{-/-}$ mouse genomic DNA) sequences. Serial dilutions (200, 100, 50, 25, and 12.5 ng) of the DNA extracted from a $Cdkn2a^{-/-}$ mouse are used as standard curve for NeomycinR and β-actin quantification. Tumor VCN is finally determined as the ratio between the amounts of LV sequences and the calculated engraftment levels. PCR cycling conditions are the same as described above.

Sequences of oligonucleotides used:

- Neo forward primer: 5′-CGGTGCCCTGAATGAACTG-3′ (300 nM)
- Neo reverse primer: 5′-CACAGCTGCGCAAGGAA-3′ (300 nM)
- Neo probe: 5′-(VIC)-CAGCGCGGCTATCGT-(MGB)-3′ (200 nM)

2.6. Linear amplification mediated PCR and genomic integration site analysis

High-throughput mapping of vector genomic integration sites in transduced cells allow the identification of the genomic features preferentially targeted by each specific vector used (Hematti et al., 2004; Mitchell et al., 2004; Schroder et al., 2002; Wu et al., 2003). These studies have been instrumental to better understand the virus biology and, from a gene therapy point of view, to highlight if dangerous gene classes, such as those that have a role in cell proliferation, are favored by the virus (Aiuti et al., 2007; Deichmann et al., 2007; Schwarzwaelder et al., 2007). Indeed, virus/vector genomic integration hot spots at specific chromosomal regions can be identified in primary cells (Cattoglio et al., 2007). It is difficult to understand if virus/vector integration hotspots are enriched as a consequence of a selective/proliferative advantage conferred to the cell harboring this type of integration, or if it is the result of an intrinsic genomic bias of virus/vector integration. To partially address this issue, comparison between the vector integration sites in pretransplant cell population kept *in vitro* (considered as an unselected cell population) versus the postengraftment cell population (in which the *in vivo* selection process has already occurred) have been performed. When this approach was applied to preclinical or clinical integration site datasets, it allowed to identify consistent changes in the overall distribution of insertion sites from *in vitro* to *in vivo* conditions

(Aiuti et al., 2007; Deichmann et al., 2007; Schwarzwaelder et al., 2007). Indeed, in X-SCID gene therapy trials the γRV integration pattern in circulating mature lymphocytes is significantly enriched toward multiple different genomic region with respect to the one found *in vitro* in pretransplant cells (Deichmann et al., 2007; Schwarzwaelder et al., 2007). These retroviral integration hot spots preferentially map to growth-regulating genes, highlighting that even without obvious side effects, these vector insertions influence the cell engraftment, clonal proliferation, and survival. In our previous work on the $Cdkn2a^{-/-}$ model this enrichment is further enhanced, as we observed a significant skewing of the integration targeting oncogenes when genotoxic vectors, like those with active LTR, are used (Montini et al., 2009, 2006).

In order to retrieve vector integration sites from vector marked cells, we performed linear amplification mediated (LAM)-PCR. We used 0.5–5 ng of tumor DNA and 10–100 ng of transduced pretransplant DNA, respectively, as template for LAM-PCR (Schmidt et al., 2001, 2007). LAM-PCR is initiated with a 25-cycle linear PCR and restriction digest using Tsp509I, or HpyCHIV4 and ligation of a restriction site compatible linker cassette. The first exponential PCR product was purified via magnetic beads capture and reamplified by PCR with nested oligonucleotides. LAM-PCR amplicons are separated on spreadex gels to address the quality of the PCR and the clonal composition of each sample. PCR products can be shotgun cloned in plasmid and sequenced by Sanger sequencing. Alternatively, the PCR products of the second exponential amplification can be barcoded with a six-nucleotide sequence tag, pooled, and subjected to next-generation sequencing with the 454 GSFlx platform (Roche). For detailed explanation, please refer to Wang et al. (2007, 2008).

Sequences of LAM-PCR primers for LV sequences are the following:

SK-LTR-1: 5′-GAGCTCTCTGGCTAACTAGG-3′ (5′ BIOT)
SK-LTR-2: 5′-GAACCCACTGCTTAAGCCTCA-3′ (5′ BIOT)
SK-LTR 3: 5′-AGCTTGCCTTGAGTGCTTCA-3′
SK-LTR-4: 5′-AGTAGTGTGTGCCCGTCTGT-3′
5′ BIOT: biotinylated oligonucleotide modification at the 5′-end

The obtained sequences are aligned to the mouse genome in order to identify the vector integration sites.

2.7. Quantitative PCR applications to measure of gene expression levels at and nearby vector integration sites

The genomic impact of an integrated LV or γRV provirus could be evaluated by assessing the frequency by which an integration event leads to deregulation of the expression of nearby genes. In the past, several studies have been performed to address this issue.

Coupling integration and gene expression studies in mouse models of insertional mutagenesis with replication-competent γ-retroviruses (aimed at the discovery of novel oncogenes), it was found that malignant cells arising from the screening frequently showed increased expression of the genes targeted by viral integrations, suggesting a putative role of these genes in promoting cell proliferation. In an *in vitro* genotoxicity assay based on the immortalization on primary mouse HSCPs after vector transduction, it has been shown that γRV integrations targeting mainly the complex MECOM locus (also known as MDS1-EVI1), induce its overexpression, and facilitate the immortalization process (Modlich *et al.*, 2006; Zychlinski *et al.*, 2008). These findings were unfortunately also well recapitulated in humans. Proto-oncogenes such as LMO2, in X-SCID, and MECOM, in CGD clinical trials, which are targeted by γRV integrations in tumor cells obtained from the blood of patients that developed severe adverse events, showed also a consistent upregulation. On the other hand, it has been shown that in nonmalignant human γRV-marked T cell clones purified from patients of the thymidine kinase (Ciceri *et al.*, 2009; Recchia *et al.*, 2006) and ADA-SCID clinical trials (Cassani *et al.*, 2009), the perturbation of the expression of the genes surrounding the integration sites occurs 18% and 20% of the times, respectively (Cassani *et al.*, 2009; Recchia *et al.*, 2006). Of note, T cell clones displaying deregulated gene expression did not show increased basal proliferation or aberrant functional responses upon mitogen stimulation, and overall no evidence of clonal selection or preferential survival of transduced T cells were observed in patients in a long follow-up period. These results indicate that perturbation of gene expression in these cases has no major consequences on the biology and function of transplanted cells.

In our murine genotoxicity assay, vector-induced hematopoietic tumors also show deregulation of oncogenes targeted by vector integrations (Montini *et al.*, 2009). The procedures to measure the gene expression levels of genes targeted by vector integrations in tumors arising from $Cdkn2a^{-/-}$ mice are illustrated.

Mouse tumor-infiltrating cells from BM or spleens used for the gene expression analysis are the same described above for the vector integration studies. The cells, resuspended in PBS after red cell lysis performed as described above, are treated with commercially available cell lysis buffers with RNA preservation properties containing chaotropic salts (Montini *et al.*, 2009). Total RNA is then isolated from the crude lysate by using anion exchange column purification kits (Montini *et al.*, 2009). After resuspension in water and spectrophotometric quantification between 100 and 1000 ng of total RNA is subjected to cDNA synthesis is performed using Mo-MLV reverse transcriptase and random hexamers primers. An amount of cDNA equivalent of 100 ng of RNA is thus used as template for real-time q-PCR gene expression assays and fluorescent probes specific for the targeted gene or those surrounding the vector integration sites. The expression level of each gene is calculated using the DDCt method

(Pfaffl, 2001) normalizing to housekeeping control genes (i.e., *Hprt, B2m, Gapdh*). The expression levels normalized to the housekeeping genes must be compared to the expression levels of a calibrator sample and represented as fold change. The choice of the calibrator to measure the relative fold change in expression it will depend on the scientific question the researcher is trying to address. To address the impact on the expression levels of a gene targeted by a vector integration in a tumor cell clone arising from transplanted *Cdkn2a*$^{-/-}$ mice the calibrator should be matched by tumor type (i.e., B or T cell lymphoma or myeloid tumor). The calibrator in this case could be average of the normalized expression levels of different type-matched tumors with no vector integrations or harboring them in different genomic positions. Real-time PCR Miner software (http://www.miner.ewindup.info; Zhao and Fernald, 2005) is used to calculate the mean PCR amplification efficiency for each gene. The qBase software program (http://www.biogazelle.com) was used to measure the relative expression level for each gene (Hellemans *et al.*, 2007).

As calibrators, it is also possible to use mock-transduced samples before cell transformation (i.e., bone marrow); however, this type of calibrators will show differences in gene expression that can be caused by the difference of the sample type rather the effect of vector integration. Therefore, expression levels form control type-matched tumors must be still compared to address the issue. Statistical comparison can be thus performed by the analysis of variance of more than two groups tested.

2.8. Survival statistics

The rationale is that if a vector treatment is genotoxic, then the group of transplanted mice will develop tumors significantly earlier than the group transplanted with mock-transduced cells. The sensitivity of the model allows the comparison of different dosages of the same vector, different vector constructs, and viral origin. For survival analysis, Kaplan Maier survival curves are analyzed by log rank statistics using specific software programs or R-software packages.

Granted that a large number of mice ($n>30$) is tested, it is possible to perform stratifications of cohorts of mice sharing specific attributes. For example, the survival of mice that developed tumors with integrations targeting at least an oncogene was compared to all other mice. Interestingly this subgroup of mice show a strong acceleration of the tumor onset with respects all other subgroups, indicating that these integrations have an impact on survival (Montini *et al.*, 2006). Another type of stratification is based on the comparison of mice that developed tumors with a VCN above or below a certain value. This type of rough stratification allowed to determine that the tumors with high VCN develop earlier with respect marked by few vector integrations (Montini *et al.*, 2009). Because the vectors used are replication defective these data suggest that multiple integrations synergize to accelerate the transition to malignancy.

To analyze the effect of multiple variables such as the VCN in tumors and the vector type/design, statistical modeling with parametric approaches are preferred because smooth continuous estimates of the survivor function are necessary for predictive purposes (Montini et al., 2006). Log-logistic distributions are smooth continuous estimates of the survivor curves function fitting the experimentally generated Kaplan Meier curves. These functions, necessary for predictive risk calculations, allow the simultaneus evaluation of multiple parameters.

The formula describing the log-logistic distribution is: $(\rho^k \times kt^{k-1})/[1+(t\rho)^k]$ in which t is the time, ρ is risk of failure, and k is the log-logistic parameter that determines the effect of the treatment (for further reference, see Bradburn et al., 2003). These types of analyses can be performed with R-software or similar.

REFERENCES

Aiuti, A., Slavin, S., et al. (2002). Correction of ADA-SCID by stem cell gene therapy combined with nonmyeloablative conditioning. Science **296**(5577), 2410–2413.

Aiuti, A., Cassani, B., et al. (2007). Multilineage hematopoietic reconstitution without clonal selection in ADA-SCID patients treated with stem cell gene therapy. J. Clin. Invest. **117**(8), 2233–2240.

Aiuti, A., Cattaneo, F., et al. (2009). Gene therapy for immunodeficiency due to adenosine deaminase deficiency. N. Engl. J. Med. **360**(5), 447–458.

Baum, C., and Fehse, B. (2003). Mutagenesis by retroviral transgene insertion: Risk assessment and potential alternatives. Curr. Opin. Mol. Ther. **5**(5), 458–462.

Baum, C., Dullmann, J., et al. (2003). Side effects of retroviral gene transfer into hematopoietic stem cells. Blood **101**(6), 2099–2114.

Baum, C., von Kalle, C., et al. (2004). Chance or necessity? Insertional mutagenesis in gene therapy and its consequences. Mol. Ther. **9**(1), 5–13.

Baum, C., Kustikova, O., et al. (2006). Mutagenesis and oncogenesis by chromosomal insertion of gene transfer vectors. Hum. Gene Ther. **17**(3), 253–263.

Biffi, A., De Palma, M., et al. (2004). Correction of metachromatic leukodystrophy in the mouse model by transplantation of genetically modified hematopoietic stem cells. J. Clin. Invest. **113**(8), 1118–1129.

Bradburn, M. J., Clark, T. G., et al. (2003). Survival analysis part II: Multivariate data analysis—An introduction to concepts and methods. Br. J. Cancer **89**(3), 431–436.

Cartier, N., Hacein-Bey-Abina, S., et al. (2009). Hematopoietic stem cell gene therapy with a lentiviral vector in X-linked adrenoleukodystrophy. Science **326**(5954), 818–823.

Cassani, B., Montini, E., et al. (2009). Integration of retroviral vectors induces minor changes in the transcriptional activity of T cells from ADA-SCID patients treated with gene therapy. Blood **114**(17), 3546–3556.

Cattoglio, C., Facchini, G., et al. (2007). Hot spots of retroviral integration in human CD34+ hematopoietic cells. Blood **110**(6), 1770–1778.

Cavazzana-Calvo, M., Hacein-Bey, S., et al. (2000). Gene therapy of human severe combined immunodeficiency (SCID)-X1 disease. Science **288**(5466), 669–672.

Ciceri, F., Bonini, C., et al. (2009). Infusion of suicide-gene-engineered donor lymphocytes after family haploidentical haemopoietic stem-cell transplantation for leukaemia (the TK007 trial): A non-randomised phase I-II study. *Lancet Oncol.* **10**(5), 489–500.

Deichmann, A., Hacein-Bey-Abina, S., et al. (2007). Vector integration is nonrandom and clustered and influences the fate of lymphopoiesis in SCID-X1 gene therapy. *J. Clin. Invest.* **117**(8), 2225–2232.

Dupre, L., Marangoni, F., et al. (2006). Efficacy of gene therapy for Wiskott-Aldrich syndrome using a WAS promoter/cDNA-containing lentiviral vector and nonlethal irradiation. *Hum. Gene Ther.* **17**(3), 303–313.

Gansbacher, B. (2003). Report of a second serious adverse event in a clinical trial of gene therapy for X-linked severe combined immune deficiency (X-SCID). Position of the European Society of Gene Therapy (ESGT). *J. Gene Med.* **5**(3), 261–262.

Gaspar, H. B., Parsley, K. L., et al. (2004). Gene therapy of X-linked severe combined immunodeficiency by use of a pseudotyped gammaretroviral vector. *Lancet* **364**(9452), 2181–2187.

Hacein-Bey-Abina, S., von Kalle, C., et al. (2003). A serious adverse event after successful gene therapy for X-linked severe combined immunodeficiency. *N. Engl. J. Med.* **348**(3), 255–256.

Hacein-Bey-Abina, S., Garrigue, A., Wang, G. P., Soulier, J., Lim, A., Morillon, E., Clappier, E., Caccavelli, L., Delabesse, E., Beldjord, K., Asnafi, V., MacIntyre, E., et al. (2008). Insertional oncogenesis in 4 patients after retrovirus-mediated gene therapy of SCID-X1. *J. Clin. Invest.* **118**, 3132–3142.

Hellemans, J., Mortier, G., et al. (2007). qBase relative quantification framework and software for management and automated analysis of real-time quantitative PCR data. *Genome Biol.* **8**(2), R19.

Hematti, P., Hong, B. K., et al. (2004). Distinct genomic integration of MLV and SIV vectors in primate hematopoietic stem and progenitor cells. *PLoS Biol.* **2**(12), e423.

Howe, S. J., Mansour, M. R., et al. (2008). Insertional mutagenesis combined with acquired somatic mutations causes leukemogenesis following gene therapy of SCID-X1 patients. *J. Clin. Invest.* **118**(9), 3143–3150.

Kool, J., and Berns, A. (2009). High-throughput insertional mutagenesis screens in mice to identify oncogenic networks. *Nat. Rev. Cancer* **9**(6), 389–399.

Li, Z., Dullmann, J., et al. (2002). Murine leukemia induced by retroviral gene marking. *Science* **296**(5567), 497.

Lo, M., Bloom, M. L., et al. (1999). Restoration of lymphoid populations in a murine model of X-linked severe combined immunodeficiency by a gene-therapy approach. *Blood* **94**(9), 3027–3036.

Lund, A. H., Turner, G., et al. (2002). Genome-wide retroviral insertional tagging of genes involved in cancer in Cdkn2a-deficient mice. *Nat. Genet.* **32**(1), 160–165.

May, C., Rivella, S., et al. (2002). Successful treatment of murine beta-thalassemia intermedia by transfer of the human beta-globin gene. *Blood* **99**(6), 1902–1908.

Mikkers, H., Allen, J., et al. (2002). High-throughput retroviral tagging to identify components of specific signaling pathways in cancer. *Nat. Genet.* **32**(1), 153–159.

Mitchell, R. S., Beitzel, B. F., et al. (2004). Retroviral DNA integration: ASLV, HIV, and MLV show distinct target site preferences. *PLoS Biol.* **2**(8), E234.

Modlich, U., Kustikova, O. S., et al. (2005). Leukemias following retroviral transfer of multidrug resistance 1 (MDR1) are driven by combinatorial insertional mutagenesis. *Blood* **105**(11), 4235–4246.

Modlich, U., Bohne, J., et al. (2006). Cell-culture assays reveal the importance of retroviral vector design for insertional genotoxicity. *Blood* **108**(8), 2545–2553.

Modlich, U., Schambach, A., et al. (2008). Leukemia induction after a single retroviral vector insertion in Evi1 or Prdm16. *Leukemia* **22**(8), 1519–1528.

Montini, E., Cesana, D., et al. (2006). Hematopoietic stem cell gene transfer in a tumor-prone mouse model uncovers low genotoxicity of lentiviral vector integration. *Nat. Biotechnol.* **24**(6), 687–696.

Montini, E., Cesana, D., et al. (2009). The genotoxic potential of retroviral vectors is strongly modulated by vector design and integration site selection in a mouse model of HSC gene therapy. *J. Clin. Invest.* **119**(4), 964–975.

Ott, M. G., Schmidt, M., et al. (2006). Correction of X-linked chronic granulomatous disease by gene therapy, augmented by insertional activation of MDS1-EVI1, PRDM16 or SETBP1. *Nat. Med.* **12**(4), 401–409.

Persons, D. A., and Baum, C. (2011). Solving the problem of gamma-retroviral vectors containing long terminal repeats. *Mol. Ther.* **19**(2), 229–231.

Pfaffl, M. W. (2001). A new mathematical model for relative quantification in real-time RT-PCR. *Nucleic Acids Res.* **29**(9), e45.

Recchia, A., Bonini, C., et al. (2006). Retroviral vector integration deregulates gene expression but has no consequence on the biology and function of transplanted T cells. *Proc. Natl. Acad. Sci. USA* **103**(5), 1457–1462.

Schmidt, M., Hoffmann, G., et al. (2001). Detection and direct genomic sequencing of multiple rare unknown flanking DNA in highly complex samples. *Hum. Gene Ther.* **12**(7), 743–749.

Schmidt, M., Schwarzwaelder, K., et al. (2007). High-resolution insertion-site analysis by linear amplification-mediated PCR (LAM-PCR). *Nat. Methods* **4**(12), 1051–1057.

Schroder, A. R., Shinn, P., et al. (2002). HIV-1 integration in the human genome favors active genes and local hotspots. *Cell* **110**(4), 521–529.

Schwarzwaelder, K., Howe, S. J., et al. (2007). Gammaretrovirus-mediated correction of SCID-X1 is associated with skewed vector integration site distribution in vivo. *J. Clin. Invest.* **117**(8), 2241–2249.

Serrano, M., Lee, H., et al. (1996). Role of the INK4a locus in tumor suppression and cell mortality. *Cell* **85**(1), 27–37.

Sherr, C. J. (2004). Principles of tumor suppression. *Cell* **116**(2), 235–246.

Stein, S., Ott, M. G., et al. (2010). Genomic instability and myelodysplasia with monosomy 7 consequent to EVI1 activation after gene therapy for chronic granulomatous disease. *Nat. Med.* **16**(2), 198–204.

Vigna, E., and Naldini, L. (2000). Lentiviral vectors: Excellent tools for experimental gene transfer and promising candidates for gene therapy. *J. Gene Med.* **2**(5), 308–316.

von Kalle, C., Fehse, B., et al. (2004). Stem cell clonality and genotoxicity in hematopoietic cells: Gene activation side effects should be avoidable. *Semin. Hematol.* **41**(4), 303–318.

Wang, G. P., Ciuffi, A., et al. (2007). HIV integration site selection: Analysis by massively parallel pyrosequencing reveals association with epigenetic modifications. *Genome Res.* **17**(8), 1186–1194.

Wang, G. P., Garrigue, A., et al. (2008). DNA bar coding and pyrosequencing to analyze adverse events in therapeutic gene transfer. *Nucleic Acids Res.* **36**(9), e49.

Will, E., Bailey, J., et al. (2007). Importance of murine study design for testing toxicity of retroviral vectors in support of phase I trials. *Mol. Ther.* **15**(4), 782–791.

Wu, X., Li, Y., et al. (2003). Transcription start regions in the human genome are favored targets for MLV integration. *Science* **300**(5626), 1749–1751.

Zhao, S., and Fernald, R. D. (2005). Comprehensive algorithm for quantitative real-time polymerase chain reaction. *J. Comput. Biol.* **12**(8), 1047–1064.

Zychlinski, D., Schambach, A., et al. (2008). Physiological promoters reduce the genotoxic risk of integrating gene vectors. *Mol. Ther.* **16**(4), 718–725.

CHAPTER TEN

Lentiviral Hematopoietic Cell Gene Therapy for X-Linked Adrenoleukodystrophy

Nathalie Cartier,[*,†] Salima Hacein-Bey-Abina,[‡,§,¶] Cynthia C. Bartholomae,[∥] Pierre Bougnères,[†] Manfred Schmidt,[∥] Christof Von Kalle,[∥] Alain Fischer,[§] Marina Cavazzana-Calvo,[‡,§,¶] and Patrick Aubourg[*,†]

Contents

1. X-linked Adrenoleukodystrophy	188
2. Hematopoietic Stem Cell Transplantation in X-ALD	189
3. Mechanism of Allogeneic HCT Efficacy in X-ALD	189
4. Lentiviral HSC Gene Therapy in X-ALD	189
5. Design and Production of the Lentiviral Vector	190
6. Transduction of CD34+ Cells from ALD Patients	191
7. Release Testing	192
8. Clinical Protocol	192
9. Patient Biological Follow-up	193
9.1. ALD protein expression in transduced CD34+ cells and PBMCs	193
9.2. Determination of integrated lentiviral vector copy number	194
9.3. Integration site (IS) characterization using 3′-LTR mediated LAM-PCR, sequencing, and data-mining	194
10. Neurological Outcome of the Two Treated Patients	195
References	197

Abstract

X-linked adrenoleukodystrophy (X-ALD) is a severe genetic demyelinating disease caused by a deficiency in ALD protein, an adenosine triphosphate-binding cassette transporter encoded by the ABCD1 gene. When performed

---

[*] INSERM UMR745, University Paris-Descartes, Paris, France
[†] Department of Pediatric Endocrinology and Neurology, Hôpital Bicêtre, Kremlin-Bicêtre, Paris, France
[‡] Department of Biotherapy, Hôpital Necker-Enfants Malades, Paris, France
[§] INSERM UMR768, University Paris-Descartes, Paris, France
[¶] Clinical Investigation Center in Biotherapy, Groupe Hospitalier Universitaire Ouest, Paris, France
[∥] National Center for Tumor Diseases and German Cancer Research Center, Heidelberg, Germany

at an early stage of the disease, allogeneic hematopoietic stem cell transplantation (HCT) can arrest the progression of cerebral demyelinating lesions. To overcome the limitations of allogeneic HCT, hematopoietic stem cell (HSC) gene therapy strategy aiming to perform autologous transplantation of lentivirally corrected cells was developed. We demonstrated the preclinical feasibility of HSC gene therapy for ALD based on the correction of CD34+ cells from X-ALD patients using an HIV1-derived lentiviral vector. These results prompted us to initiate an HSC gene therapy trial in two X-ALD patients who had developed progressive cerebral demyelination, were candidates for allogeneic HCT, but had no HLA-matched donors or cord blood. Autologous CD34+ cells were purified from the peripheral blood after G-CSF stimulation, genetically corrected *ex vivo* with a lentiviral vector encoding wild-type ABCD1 cDNA, and then reinfused into the patients after they had received full myeloablative conditioning. Over 3 years of follow-up, the hematopoiesis remained polyclonal in the two patients treated with 7–14% of granulocytes, monocytes, and T and B lymphocytes expressing the lentivirally encoded ALD protein. There was no evidence of clonal dominance or skewing based on the retrieval of lentiviral insertion repertoire in different hematopoietic lineages by deep sequencing. Cerebral demyelination was arrested 14 and 16 months, respectively, in the two treated patients, without further progression up to the last follow-up, a clinical outcome that is comparable to that observed after allogeneic HCT. Longer follow-up of these two treated patients and HSC gene therapy performed in additional ALD patients are however needed to evaluate the safety and efficacy of lentiviral HSC gene therapy in cerebral forms of X-ALD.

1. X-LINKED ADRENOLEUKODYSTROPHY

X-linked adrenoleukodystrophy (X-ALD) is caused by mutations in the ABCD1 gene that encodes a transporter (ALD protein) localized into the peroxisomal membrane and involved in the metabolism of very-long-chain fatty acids (VLCFA). Deficiency in ALD protein leads to an accumulation of VLCFA in plasma and tissues and progressive demyelination in the central nervous system (CNS). The cerebral form of X-ALD affects boys between 5 and 12 years of age and leads to a vegetative stage or death within 2–5 years (Dubois-Dalcq *et al.*, 1999; Moser *et al.*, 2007). Adult ALD males develop between 20 and 30 years of age a milder form of X-ALD, called adrenomyeloneuropathy, that is characterized by progressive paraplegia due to spinal cord involvement. However, approximately 35% of AMN males are also at risk to develop cerebral demyelination that has the same poor prognosis as in boys. Overall, 65% of ALD males are at risk of developing fatal cerebral demyelination in childhood or adulthood (Dubois-Dalcq *et al.*, 1999; Moser *et al.*, 2007).

2. HEMATOPOIETIC STEM CELL TRANSPLANTATION IN X-ALD

Allogeneic hematopoietic stem cell transplantation (HCT) was shown to result in long-term benefits by arresting the progression of cerebral demyelination in boys with X-ALD. When performed at an early stage of cerebral demyelination, allogeneic HCT is the only therapeutic approach that can stabilize cerebral demyelination in boys with X-ALD (Aubourg et al., 1990; Peters et al., 2004; Shapiro et al., 2000). Similar benefits of allogeneic HCT have been demonstrated in adults with cerebral X-ALD. However, it is not yet known whether allogeneic HCT can prevent or rescue adrenomyeloneuropathy. The absence of biological markers that can predict the evolutivity of cerebral disease is a major limitation to propose in due time allogeneic HCT to X-ALD patients who develop cerebral demyelination.

3. MECHANISM OF ALLOGENEIC HCT EFFICACY IN X-ALD

The long-term benefits of allogeneic HCT in X-ALD are mediated by the replacement of brain microglial cells derived from donor bone marrow myelomonocytic cells (Eglitis and Mezey, 1997; Priller et al., 2001). However, the ALD protein is a transmembrane peroxisomal protein that cannot be secreted. Therefore, in contrast to CNS lysosomal storage disorders like Hurler syndrome, globoid cell leukodystrophy, or metachromatic leukodystrophy, in which normal enzyme produced by donor-derived microglia can be secreted and then recaptured by other CNS cells, this mechanism of cross-correction does not occur in X-ALD. The mechanism by which allogeneic HCT arrests the neuroinflammatory demyelinating process in ALD is still not understood. The conditioning regimen has no effect by itself (Nowaczyk et al., 1997). From 36 ALD patients who received allogeneic HCT in France following the same conditioning regimen, four ALD patients showed a failure or a delay to engraft the allogeneic HCT, and the four patients developed devastating progression of cerebral demyelination. It is hypothesized that allogeneic HCT in X-ALD allows correction of the abnormal function of brain microglia that is deleterious for oligodendrocytes, whether or not this deficiency is directly related to the accumulation of VLCFA.

4. LENTIVIRAL HSC GENE THERAPY IN X-ALD

Overall, 65% of ALD male patients are at risk of developing fatal cerebral demyelination in childhood or adulthood. Allogeneic HCT remains associated with significant morbidity and mortality risks, particularly in adults, and not all X-ALD patients have suitable donors despite the availability of cord blood. Transplantation of autologous HSCs genetically modified to express the missing protein may circumvent the majority of the problems associated with allogeneic HCT. HSC gene therapy could thus be an appropriate therapeutic alternative.

Until the first gene therapy in X-ALD, HSC gene therapy study was shown to provide clinical benefits only in adenosine deaminase deficiency and in severe combined immunodeficiency-X1 and Wiskott–Aldrich syndrome (Aiuti *et al.*, 2002; Cavazzana-Calvo *et al.*, 2000; Gaspar *et al.*, 2004). In these trials, autologous HSCs were genetically corrected using a murine gammaretrovirus taking advantage of the marked selective growth advantage of corrected cells that favors their engraftment and expansion *in vivo*. For diseases like ALD, with no selective growth advantage of corrected cells, lentiviral vectors have generated great therapeutic hope. Lentiviral vectors can transduce non-dividing cells and were shown, *in vitro* and *in vivo* in mice, to allow more efficient gene transfer into HSCs than murine gammaretrovirus vectors (Miyoshi *et al.*, 1999; Naldini *et al.*, 1996) In X-ALD, *in vitro* experiments of ALD gene transfer with lentiviral vector have shown biochemical correction of monocytes/macrophages derived from transduced ALD protein-deficient human CD34+ cells (Benhamida *et al.*, 2003). *In vivo*, the transplantation of lentivirally transduced murine ALD Sca-1+ cells, a functional equivalent of CD34+ cells in humans, into ALD mice resulted in the replacement of 20–25% of brain microglial cells expressing the ALD protein 12 months after transplantation. The fate of lentivirally transduced human ALD CD34+ cells was also investigated in the brain of non-obese diabetic/severe combined immunodeficient (NOD/SCID) mice (Asheuer *et al.*, 2004). *In vivo* expression of ALD protein in human monocytes and macrophages derived from engrafted human stem cells was demonstrated. Human bone marrow-derived cells migrated into the brain of transplanted mice where they differentiated into microglia expressing the human ALD protein.

Based on these pre-clinical data and additional safety data concerning the production and the use of HIV-1 derived lentiviral vector, approval was granted by the French regulatory agency in December 2005 for a HSC gene therapy trial in X-ALD boys with progressive cerebral demyelination but without HLA-matched donor.

5. Design and Production of the Lentiviral Vector

The CG1711 hALD (MND-ALD) vector used for the clinical trial was designed and produced under GMP guidelines by Cell Genesys Inc. (South San Francisco, CA, USA). The CG1711 hALD (MND-ALD)

Figure 10.1 CG1711 hALD (MND-ALD) vector design. The CG1711 hALD (MND-ALD) vector is a self-inactivating (SIN) lentiviral vector in which the U3 region in the 3′LTR was deleted (U3). This vector carries the expression cassette for the human ALD cDNA under the expression of the MND promoter. LTR, long terminal repeat; Ψ+, packaging signal; cPPT/flap, central polypurine tract; RRE, Rev-responsive element; WPRE, Woodchuck hepatitis virus post-transcriptional regulatory element.

vector is a self-inactivating lentiviral vector in which the U3 region in the 3′-LTR was deleted (Fig. 10.1). This vector carries the expression cassette for the human ABCD1 cDNA under the expression of the MND (myeloproliferative sarcoma virus enhancer, negative control region deleted, dl587rev primer binding site substituted) promoter, which was shown to confer stronger and stable expression of the transgene in HSC-derived cells (Halene et al., 1999). Vector was packaged using a third-generation plasmid combination expressing Gag, Pol, and Rev, deleted of the six other HIV-1 genes that are responsible for HIV-1 pathogenesis (vif, vpr, vpu, nef, env, and tat). The lentiviral vector was pseudotyped with the vesicular stomatitis virus (VSV-G) envelope.

Concentration of the lentiviral vector (titered on 293 cells) was 941×10^7 infectious virus particles/ml (infectivity of 5610 I.U./ng p24).

Absence of replication competent lentivirus (RCL) in the vector stock was determined by two assays. The first assay was performed at Cell Genesys Inc. and relied upon the amplification of potential RCL by growth (six passages) of a permissive cell line (C8166) followed by sensitive detection of viral p24 antigen (Escarpe et al., 2003; Farson et al., 2001). The second assay was performed at Genosafe (Evry, France) and aimed at detecting partial *gag–pol* recombinant in the vector stock using the permissive 293G cell line which *trans*-complements the VSV-G envelope and allows amplification of env-defective viral recombinants.

6. TRANSDUCTION OF CD34+ CELLS FROM ALD PATIENTS

Peripheral blood mononuclear cells (PBMCs) were obtained from the patients after stimulation by intravenous injection of granulocyte colony-stimulating factor (G-CSF).

CD34+ cells were positively selected by an immunomagnetic procedure (ClinicMACS, Miltenyi Biotec, Bergisch Gladbach, Germany). Pre-stimulation of CD34+ cells was conducted for 19h at 37 °C in 5% CO_2 in gas-permeable stem cell culture bags (Baxter, France), at a concentration of 2×10^6 cells/ml in *ex vivo* 20 medium (Cambrex Bio Science, Verviers, Belgium) containing protamine sulfate (4μg/ml), stem cell factor, megakaryocyte growth and differentiation factor, Flt3-L (100ng/ml; PeproTech, Nanterre, France), and IL-3 (60ng/ml; PeproTech, Nanterre, France). After washing, CD34+ cells were transferred to containers precoated with the CH296 human fragment of fibronectin (50μg/ml) (Takara Bio Inc., Shiga, Japan) in the same culture medium with cytokines and the lentiviral vector at a final multiplicity of infection (MOI) of 25 for 16h at 37°C in 5% CO_2. This concentration of lentiviral vector was utilized to yield approximately one to two copies of integrated provirus per transduced cell and thus minimize the risk of insertional mutagenesis. After transduction, cells were washed three times and cryopreserved in liquid nitrogen until release testing were completed.

7. Release Testing

RCL assays on transduced CD34+ cells were performed at Genosafe (Evry, France) by coculturing transduced CD34+ cells with the permissive C8166 cell line. Three weeks after co-culture of the transduced CD34+ cells with the C8166-45 cell line, reverse transcriptase-polymerase chain reaction (RT-PCR) and PCR assays were performed to detect gag–pol mRNA and VSV-G DNA. All tests were negative.

8. Clinical Protocol

Patients P1 and P2 were aged of 7.5 and 7 years, respectively, at the time of gene therapy. Both had an older brother who died from cerebral X-ALD before or soon after they were enrolled in the trial. The two patients had progressive cerebral neuroinflammatory demyelinating form of X-ALD determined by brain MRI, were candidates for allogeneic HCT, but had no HLA-matched donor or cord blood. CD34+ cells were selected from PBMCs after G-CSF stimulation and transduced *ex vivo*. A portion of nontransduced CD34+ cells was cryopreserved for rescue transplantation.

A sample of transduced cells (5% of total transduced cells) was used for release testing (including in particular three RCL assays) and analysis of ALD protein expression and vector copy number; remaining cells were cryopreserved. After all release tests were completed, patients received full myeloablative conditioning regimen with cyclophosphamide and busulfan. In the absence of positive selective advantage in ALD, patient's HSCs were ablated to favor engraftment of the gene-corrected HSCs. Cryopreserved transduced CD34+ cells were then thawed and infused (4.6×10^6 to 7.2×10^6 cells/kg, respectively, in P1 and P2).

9. PATIENT BIOLOGICAL FOLLOW-UP

Patients were followed for adverse events and monitoring of bone marrow engraftment and neurological outcome. The procedure was clinically uneventful. Hematopoietic recovery occurred at days 13–15 after transplant and was sustained thereafter. Nearly complete immunological recovery occurred between 9 and 12 months. Bone marrow aspirates were normal at 12 and 24 months after gene therapy and all RCL tests were negative up to the last follow-up.

9.1. ALD protein expression in transduced CD34+ cells and PBMCs

The efficacy of the HSC transduction was assessed by studying the percentage of hematopoietic cells expressing the lentivirally encoded ALD protein using immunohistochemical methods. Both patients had ABCD1 gene mutation that resulted in the absence of detectable ALD protein in their PBMCs. Lentivirally-encoded ALD protein expression was performed in transduced CD34+ cells before reinfusion, then in bone marrow CD34+ cells and monocytes, granulocytes, T and B lymphocytes at different timepoints after gene therapy.

Transduction efficacy of CD34+ cells (analyzed 5 days after transduction) ranged from 33% (P2) to 50% (P1). Between 20% and 33% of PBMCs from the two treated patients expressed the lentivirally encoded ALD protein, 2 months after infusion of the transduced CD34+ cells. The percentage of corrected PBMCs decreased with time but stabilized at 10–13% around 16 months after gene therapy and remained stable up to 36 months after gene therapy.

To evaluate ALD protein expression in the different subpopulations of blood mononuclear cells, granulocytes CD15+), monocytes (CD14+), T lymphocytes (CD3+), and B lymphocytes CD19+) cells were purified by magnetic cell sorting using whole blood purification kits (Miltenyi Biotec). Purity (>99%) was checked on a FACS cell sorter.

In each patient, ALD protein was expressed at similar percentage in granulocytes, monocytes (that have short half-life), and B and T lymphocytes (that have longer half-life) at different timepoints after gene therapy. Thirty-six months after gene therapy, this percentage ranged from 7% to 10% for P1 and 12% to 14% for P2.

Expression of ALD protein in bone marrow CD34+ cells (purity >99%) went from 18–20% (12 months) to 18% (24 months) in patients P1 and P2.

9.1.1. Monoclonal antibodies used for immunofluorescence studies

Monocytes, granulocytes, T and B lymphocytes, and CD34+ cells were purified on microbeads (Miltenyi Biotec) with appropriate monoclonal antibodies from Miltenyi Biotec, Bergisch Gladbach, Germany: anti-CD3 (IgG2a), anti-CD19 (IgG1), anti-CD14 (IgG2a), anti-CD15 (IgM), anti-CD34 (Becton

Dickinson, Le Pont de Claix, France), and anti-ALD protein were used for immunofluorescence and FACS sorting studies. Cells were analyzed on a BD FACSCalibur flow cytometer (Becton Dickinson).

9.2. Determination of integrated lentiviral vector copy number

Quantitative PCR was performed using the ABI Prism 7700 Sequence Detection System (Perkin-Elmer Applied Biosystems, Foster City, CA, USA) and the SYBR® Green PCR Core Reagents kit (Perkin-Elmer Applied Biosystems). For vector copy number, woodchuck posttranscriptional regulatory element and human genomic albumin (*Alb*) sequences (internal control) were simultaneously amplified. Each sample was expressed in terms of its *Alb* content. Quantitative values were obtained from the threshold cycle (Ct) value at which the increase of the signal associated with exponential growth of PCR product was first detected. Results are expressed as N-fold differences in the *WPRE1* sequence copy number relative to the *Alb* gene.

ALD protein expression correlated with the mean number of integrated lentiviral copy per transduced CD34+ cell and PBMC. The mean numbers of integrated lentiviral copy per cell were 0.72 and 0.54 in transduced CD34+ cells from P1 and P2, respectively; it was 0.165 and 0.2 in PBMCs from P1 and P2, respectively, 16 months after gene therapy.

Colony forming assay was performed on CD34+ cells. Vector-derived sequences were present in colony-forming units-granulocyte macrophage (CFU-GM) indicating effective gene transfer into common myeloid progenitors with long-term engraftment capacity.

9.3. Integration site (IS) characterization using 3′-LTR mediated LAM-PCR, sequencing, and data-mining

For 3′-linear amplification-mediated-PCR (LAM-PCR) analyses, 1–1000 ng of DNA served as template for linear PCR using retroviral LTR-specific biotinylated primers (Schmidt *et al.*, 2007). Linear PCR products were separated with paramagnetic beads. Second strand DNA synthesis, restriction digestion (Tsp509I, NlaIII, or HpyCH4IV), and linker ligation were performed and followed by two additional exponential PCR steps.

GS Flx-specific amplification and sequencing primers were added to both ends of the LAM-PCR amplicons for 454 pyrosequencing (GS Flx; Roche Diagnostics). Forty nanograms of purified LAM-PCR products and the following PCR conditions were used: initial denaturation for 120 s at 95 °C; 12 cycles at 95 °C for 45 s, 60 °C for 45 s, and 72 °C for 60 s; final elongation 300 s at 72 °C. LAM-PCR amplicon sequences were trimmed and aligned using BLAT (Kent, 2002).

LAM-PCR on >98% enriched CD14+, CD15+, CD3+, C19+, and bone marrow CD34+ cells revealed a high number of distinct insertion sites (IS), indicating a consistently polyclonal distribution of lentivirally corrected hematopoietic cells.

To evaluate whether identical lentiviral integration sites between lymphoid and myeloid lineages could suggest that primitive hematopoietic progenitors or hematopoietic stem cells have been transduced, the observed number of identical lentiviral integration sites was evaluated using high-throughput 454 pyrosequencing of LAM-PCR amplicons making the null hypothesis that insertions would follow a Poisson distribution. The observed numbers of identical lentiviral amplicons in lymphoid and myeloid cells from P1 ($n=183$) and P2 ($n=114$) were very significantly higher than the values expected by chance alone.

To determine the emergence of individual dominant clones that could result from lentiviral integration, the quantitative contribution of individual clones harboring lentiviral integration to hematopoiesis was determined by ordering the abundance of distinct IS in different hematopoietic cell lineages using high-throughput 454 pyrosequencing of LAM-PCR amplicons. The retrieval frequency of a given lentiviral amplicon by high-throughput sequencing depends on the amount of DNA and therefore the number of cells harboring this IS. This method allows a good estimate of clonal contribution provided that a sufficient amount of DNA is tested and that there is no bias in the amplification of IS by LAM-PCR by choosing an optimized set of restriction enzymes. These studies allowed us to demonstrate that no dominant clone emerged among active hematopoietic clones in the two treated patients up until the last follow-up.

10. NEUROLOGICAL OUTCOME OF THE TWO TREATED PATIENTS

In patient #1, brain MRI showed that progression of cerebral demyelinating lesions was arrested 12–14 months after gene therapy. Up to 36 months after gene therapy, no changes in the extent of cerebral demyelinating lesions have been observed. As is very often observed after allogeneic HCT, a decline of cognitive functions has been observed due to the progression of demyelinating lesions in the frontal white matter during the first 12–14 months post-gene therapy. Before gene therapy, patient #1 had normal verbal IQ (VIQ=104) and performance IQ (PIQ=99). PIQ decreased at 74 and VIQ remained identical (104), 24 months after gene therapy. PIQ has remained stable thereafter, whereas a decrease of VIQ was observed, likely because this patient did not attend school and therefore did not have the opportunity to enrich his vocabulary. Neurologic examination of patient #1 remains normal except for the presence of very moderate spasticity in the right triceps (the only sequelae of right hemiparesis he developed 7 months after gene therapy). This patient is now in a specialized school because of attentional/executive deficits related to white matter lesions in the frontal lobes.

In patient #2, brain MRI showed that progression of the cerebral demyelinating lesions was arrested 16 months after gene therapy. Up to 36 months after gene therapy, no changes in the extent of cerebral demyelinating lesions have been observed. Although the progression of demyelinating lesions defined as the extent of abnormal hypersignal on FLAIR and T2 sequences at brain MRI was arrested, dilatation of the posterior part of the ventricles appeared at month 16 post-gene therapy and progressed up to 30 months after gene therapy. This reflected myelin loss/destruction in the occipital white matter, a finding which is often observed after allogeneic HCT. Patient #2 developed post-gene therapy cuts in the lower parts of his visual field at 16 months, initially without loss of visual acuity. Reduction of his visual fields progressed up to 30 months after gene therapy and was then associated with severe loss of visual acuity, without further aggravation up to 36 months after gene therapy. Patient #2 has no other neurologic deficits. Patient #2 had normal cognitive functions prior to gene therapy (VIQ= 101, PIQ=119, and total IQ=111). Those values remained in the same range (103, 111, and 98) 20 months after gene therapy but declined at 102, 88, and 83, 30 months after gene therapy. The decrease of PIQ (88) and total IQ (83) reflected significant visuospatial, visuoconstructive, visual attention, visual memory deficits as well as severe loss of vision. Despite his visual deficits, patient #2 is performing normally for his age in school, with help for his visual problems.

In this trial, two boys with progressive and lethal cerebral form of X-ALD were successfully treated with lentiviral HSC gene therapy. Long-term follow-up demonstrated stable expression of the lentivirally encoded ALD protein in peripheral blood monocytes, granulocytes, and lymphocytes. The demonstration that identical lentiviral integration sites were present in myeloid and lymphoid cells argues for the transduction of HSCs. However, the limited percentage of corrected hematopoietic cells *in vivo* indicates that improvement of the procedure is clearly needed. This is particularly important for X-ALD to shorten the period during which cerebral demyelination continues to progress after transplant. Lentiviral vector with higher titer and better ratio of infectious/noninfectious particles may help to improve the transduction of HSCs and/or early myeloid progenitors that give rise to brain microglia. Although longer follow-up of the first two treated patients is needed and outcome of two X-ALD patients treated more recently is awaited, HSC-based gene therapy could be a valuable therapeutic option for X-ALD patients with progressive cerebral demyelination as HSC gene therapy abrogates the morbidity and mortality risks associated with conventional allogeneic HCT.

REFERENCES

Aiuti, A., Vai, S., Mortellaro, A., Casorati, G., Ficara, F., Andolfi, G., Ferrari, G., Tabucchi, A., Carlucci, F., Ochs, H. D., Notarangelo, L. D., Roncarolo, M. G., et al. (2002). Immune reconstitution inADA-SCID after PBL gene therapy and discontinuation of enzyme replacement. Nat. Med. **8**, 423–425.

Asheuer, M., Pflumio, F., Benhamida, S., Dubart-Kupperschmitt, A., Fouquet, F., Imai, Y., Aubourg, P., and Cartier, N. (2004). Human CD 34+ cells differentiate into microglia and express recombinant therapeutic protein. Proc. Natl. Acad. Sci. USA **101**, 3557–3562.

Aubourg, P., Blanche, S., Jambaque, I., Rocchiccioli, F., Kalifa, G., Naud-Saudreau, C., Rolland, M. O., Debre, M., Chaussain, J. L., Griscelli, C., Fischer, A., and Bougnères, P. (1990). Reversal of early neurologic and neuroradiologic manifestations of X-linked adrenoleukodystrophy by bone marrow transplantation. N. Engl. J. Med. **322**, 1860–1866.

Benhamida, S., Pflumio, F., Dubart-Kupperschmitt, A., Zhao-Emonet, J. C., Cavazzana-Calvo, M., Rocchiccioli, F., Fichelson, S., Aubourg, P., Charneau, P., and Cartier, N. (2003). Transduced CD34+ cells from adrenoleukodystrophy patients with HIV-derived vector mediate long term engraftment of NOD/SCID mice. Mol. Ther. **7**, 317–324.

Cavazzana-Calvo, M., Hacein-Bey, S., de Saint Basile, G., Gross, F., Yvon, E., Nusbaum, P., Selz, F., Hue, C., Certain, S., Casanova, J. L., Bousso, P., Deist, F. L., et al. (2000). Gene therapy of human severe combined immunodeficiency (SCID)-X1 disease. Science **288**, 669–672.

Dubois-Dalcq, M., Feigenbaum, V., and Aubourg, P. (1999). Neurobiology of X-linked adrenoleukodystrophy, a demyelinating peroxisomal disorder. Trends Neurosci. **22**, 4–122.

Eglitis, M. A., and Mezey, E. (1997). Hematopoietic cells differentiate into both microglia and macroglia in the brains of adult mice. Proc. Natl. Acad. Sci. USA **94**, 4080–4085.

Escarpe, P., Zayek, N., Chin, P., Borellini, F., Zufferey, R., Veres, G., and Kiermer, V. (2003). Development of a sensitive assay for detection of replication competent recombinant lentivirus in large scale HIV-based vector preparations. Mol. Ther. **2**, 332–341.

Farson, D., Witt, R., McGuinness, R., Dull, T., Kelly, M., Song, J., Radeke, R., Bukovsky, A., Consiglio, A., and Naldini, L. (2001). A new generation stable inducible packaging cell line for lentiviral vectors. Hum. Gene Ther. **12**, 981–997.

Gaspar, H. B., Parsley, K. L., Howe, S., King, D., Gilmour, K. C., Sinclair, J., Brouns, G., Schmidt, M., Von Kalle, C., Barington, T., Jakobsen, M. A., Christensen, H. O., et al. (2004). Gene therapy of X-linked severe combined immunodeficiency by use of a pseudotyped gammaretroviral vector. Lancet **364**(9452), 2181–2187, 18–31.

Halene, S., Wang, L., Cooper, R. M., Bockstoce, D. C., Robbins, P. B., and Kohn, D. B. (1999). Improved expression in hematopoietic and lymphoid cells in mice after transplantation of bone marrow transduced with a modified retroviral vector. Blood **94**, 3349–3357.

Kent, W. J. (2002). BLAT, the BLAST-like alignment tool. Genome Res. **12**, 656–664.

Miyoshi, H., Smith, K. A., Mosier, D. E., Verma, I. M., and Torbett, B. E. (1999). Transduction of human CD 34+ cells that mediate long term engraftment of NOD/SCID mice by HIV vectors. Science **283**, 682–686.

Moser, H. W., Mahmood, A., and Raymond, G. V. (2007). X-linked adrenoleukodystrophy. *Nat. Clin. Pract. Neurol.* **3**(3), 140–151.

Naldini, L., Blomer, U., Gallay, P., Ory, D., Mulligan, R., Gage, F. H., *et al.* (1996). In vivo gene delivery and stable transduction of nondividing cells by a lentiviral vector. *Science* **272,** 263–267.

Nowaczyk, M. J., Saunders, E. F., Tein, I., Blaser, S. I., and Clarke, J. T. (1997). Immunoablation does not delay the neurologic progression of X-linked adrenoleukodystrophy. *J. Pediatr.* **131,** 453–455.

Peters, C., Charnas, L. R., Tan, Y., Ziegler, R. S., Shapiro, E. G., De For, T., Grewal, S. S., Orchard, P. J., Abel, S. L., Goldman, A. I., and Ramsay, N. K. (2004). Cerebral X-linked adrenoleukodystrophy: The international hematopoietic cell transplantation experience from 1982 to 1999. *Blood* **10,** 881–888.

Priller, J., Flügel, A., Wehner, T., Boentert, M., Haas, C. A., Prinz, M., Fernández-Klett, F., Prass, K., Bechmann, I., de Boer, B. A., Frotscher, M., Kreutzberg, G. W., *et al.* (2001). Targeting gene-modified hematopoietic cells to the central nervous system: Use of green fluorescent protein uncovers microglial engraftment. *Nat. Med.* **7,** 1356–1361.

Schmidt, M., Schwarzwaelder, K., Bartholomae, C., Zaoui, K., Ball, C., Pilz, I., Braun, S., Glimm, H., and von Kalle, C. (2007). High resolution insertion site analysis by linear amplification mediated PCR (LAMPCR). *Nat. Methods* **4,** 1051.

Shapiro, E., Krivit, W., Lockman, L., Jambaque, I., Peters, C., Cowan, M., Harris, R., Blanche, S., Bordigoni, P., Loes, D., Ziegler, R., Crittenden, M., *et al.* (2000). Long term effect of bone marrow transplantation for childhood onset cerebral X-linked adrenoleukodystrophy. *Lancet* **356,** 713–718.

CHAPTER ELEVEN

Retroviral Replicating Vectors in Cancer

Christopher R. Logg,* Joan M. Robbins,[†] Douglas J. Jolly,[†] Harry E. Gruber,[†] *and* Noriyuki Kasahara*,[‡]

Contents

1. Introduction	200
2. Virus Production by Transient Transfection	202
2.1. Required materials	203
2.2. Transfection procedure	204
2.3. Notes	205
3. Vector Copy Number Assay for Titer Determination and Biodistribution Studies	205
3.1. Preparation of template DNA	206
3.2. qPCR assay for determination of RRV titer and copy number	208
3.3. Illustrative results	211
4. Development of Novel RRV Using Molecular Evolution	211
4.1. Design and construction of prototype RRV for adaptation	212
4.2. Virus production	213
4.3. Molecular evolution and natural selection	213
4.4. Identification and cloning of mutations	214
4.5. Illustrative example	215
5. MTS Assay of RRV-Mediated Cell Killing *In Vitro*	215
5.1. Required materials	217
5.2. Assay procedures	217
6. *In Vivo* Glioma Model for Testing RRV-Mediated Gene Therapy	218
6.1. Required materials	219
6.2. Study procedures	222
6.3. Notes	225
Acknowledgments	225
References	225

* Department of Medicine, University of California, Los Angeles, California, USA
[†] Tocagen Inc., San Diego, California, USA
[‡] Department of Molecular and Medical Pharmacology, University of California, Los Angeles, California, USA

Abstract

The use of replication-competent viruses for the treatment of cancer is an emerging technology that shows significant promise. Among the various different types of viruses currently being developed as oncolytic agents, retroviral replicating vectors (RRVs) possess unique characteristics that allow highly efficient, non-lytic, and tumor-selective gene transfer. By retaining all of the elements necessary for viral replication, RRVs are capable of transmitting genes via exponential *in situ* amplification. Their replication-competence also provides a powerful means whereby novel and useful RRV variants can be generated using natural selection. Their stringent requirement for cell division in order to achieve productive infection, and their preferential replication in cells with defective innate immunity, confer a considerable degree of natural specificity for tumors. Furthermore, their ability to integrate stably into the genome of cancer cells, without immediate cytolysis, contributes to long-lasting therapeutic efficacy. Thus, RRVs show much promise as therapeutic agents for cancer and are currently being tested in the clinic. Here we describe experimental methods for their production and quantitation, for adaptive evolution and natural selection to develop novel or improved RRV, and for *in vitro* and *in vivo* assessment of the therapeutic efficacy of RRVs carrying prodrug activator genes for treatment of cancer.

1. INTRODUCTION

The idea that replication-competent viruses might be used for the treatment of cancer originated more than a century ago, with the first documented case report in 1904 of leukemia remission after influenza infection (Dock, 1904). In 1922, it was reported that vaccinia virus can inhibit rodent tumors (Levaditi and Nicolau, 1922), and during its heyday in the 1950s–1970s, the field of oncolytic virotherapy witnessed numerous clinical trials testing a variety of different viruses administered to patients with advanced cancers (Hammill et al., 2010; Kelly and Russell, 2007). However, in most cases, tumor regression achieved by viral oncolysis was found to be transient, typically followed by immune clearance of the virus, and subsequent tumor recurrence. Faced with such invariably disappointing results, these clinical trials were largely abandoned with the advent of modern chemotherapy and radiation therapy. For many years, the field of oncolytic virotherapy was largely forgotten and has virtually been ignored in recent chronicles recounting the development of modern cancer treatment (Davis, 2007; Mukherjee, 2010; Olson, 1989).

Fast forward to the mid-1980s, advances in molecular biology provided the tools to engineer viruses into efficient gene transfer vectors, spawning the field of gene therapy. However, in clinical trials of gene therapy conducted over the ensuing decade, conventional non-replicating vectors generally showed disappointing levels of gene transfer and inadequate therapeutic effectiveness (Friedmann, 1996; McCormick, 2001; Orkin and Motulsky, 1995). Accordingly, in recent years, there has been renewed interest in the use of tumor-selectively replicating forms of viruses and viral vectors for the treatment of cancer. Various types of replicating viruses, including adenovirus, herpesvirus, reovirus, poliovirus, rhabdovirus, paramyxoviruses, and vaccinia virus, have entered clinical trials (for recent general reviews, see Donnelly et al., 2011; Eager and Nemunaitis, 2011). While all other virus species being investigated for this purpose are inherently cytolytic, retroviral replicating vectors (RRVs), based on simple gamma retroviruses such as murine leukemia virus (MLV), possess properties that render them uniquely well suited for use in cancer therapy as a non-lytic tumor-selective gene transfer agent.

As with MLV itself, MLV-derived RRVs have a strict requirement for cell division, at least in part because their nucleocapsids lack nuclear localization signals for active transport across intact nuclear membranes in quiescent cells, and infection is innately restricted to cells that are mitotically active (Lewis and Emerman, 1994; Roe et al., 1993; Seamon et al., 2002). As most normal cells in the body are quiescent, RRV-mediated gene transfer *in vivo* is largely restricted to the rapidly growing cells of malignant tissue. Furthermore, intrinsic restriction factors activated by innate immunity such as APOBEC-3G, TRIM-5α, tetherin, etc. (Douville and Hiscott, 2010; Jolly, 2011; Oliveira et al., 2010; Takeuchi and Matano, 2008), as well as humoral and cellular adaptive immune responses including neutralizing antibodies and cytotoxic T lymphocytes (Biasi et al., 2011; Hein et al., 1995; Kende et al., 1981) are active against retrovirus infection of normal cells. In contrast, signaling pathways activating innate immunity are frequently mutated or lost in cancer cells (Critchley-Thorne et al., 2009), and the tumor microenvironment is immunosuppressive to adaptive immunity (Flavell et al., 2010; Zitvogel et al., 2006), thereby forming a niche for preferential replication of RRVs *in vivo*. Since retroviruses permanently integrate their reverse-transcribed cDNA into the cancer cell genome and newly formed virions bud off from the surface of infected cells without causing cytolysis, RRVs can spread relatively stealthily and achieve widespread gene transfer throughout tumors without harming cells or triggering a strong antiviral immune response that could impair or prevent viral propagation. In contrast, oncolytic viruses kill cells as a natural part of their replication cycle, resulting in piecemeal destruction of the microenvironment that was shielding them from immune clearance.

The cell-killing function of RRVs lies in the protein encoded by the transgene, and to date most studies evaluating these vectors for cancer therapy have employed prodrug activator ("suicide") genes, which generally encode metabolic enzymes that can convert an inactive substrate into an active chemotoxin (Kirn et al., 2002). As noted above, RRVs can stealthily mediate efficient delivery and stable expression of prodrug activator genes throughout the tumor, and subsequently, simultaneous *en masse* killing of infected cells can be effected at the desired time through administration of the corresponding prodrug. This strategy thus permits a substantially greater amplification of viral replication and spread as compared with other oncolytic viruses, as each cell upon infection becomes a stable virus-producing cell itself, rather than undergoing immediate lysis. Based on highly promising studies conducted by multiple groups in a variety of experimental cancer models (Dalba et al., 2005; Hiraoka et al., 2006, 2007; Hlavaty et al., 2011; Kikuchi et al., 2007a,b; Logg et al., 2001b; Metzl et al., 2006; Solly et al., 2003; Tai et al., 2005, 2010; Wang et al., 2003, 2006), a clinical trial of an RRV encoding the yeast cytosine deaminase (CD) prodrug activator gene for treatment of brain cancer has been initiated and is currently underway in the United States (www.clinicaltrials.gov, NCT01156584).

RRVs are relatively easy to design and construct, due to the small size of their genomes and the breadth of knowledge of MLV biology that already exists, and the production of moderately high titer virus preparations for laboratory use is rapid and simple. The ability of RRV to greatly amplify itself *in situ* furthermore largely abrogates the need for extensively concentrated or massive doses of virus. These vectors are encoded on a single plasmid harboring a full-length vector provirus. No accessory plasmids are required for virus production as is the case with standard defective vector systems. Initial production of most RRV is driven in transfected cells by the cytomegalovirus (CMV) promoter, which is present within the 5′ LTR in place of the MLV U3 sequence. In 293T human embryonic kidney cells, which are used for virus production, expression from the CMV promoter is higher than from the enhancer/promoter of the MLV U3 region. As described in further detail below, upon infection with virus produced by transfection, the CMV sequence is replaced with the wild-type MLV enhancer promoter of the U3 region.

Earlier versions of RRVs, developed in the 1980s, contained the transgene sequences within the long terminal repeat (LTR) (Lobel et al., 1985; Reik et al., 1985; Stuhlmann et al., 1989). These earlier vectors, however, exhibited a strong tendency to rapidly delete the inserted transgene sequences upon replication. We therefore developed a different RRV design, in which the transgene is located immediately downstream of the *env* stop codon (Logg et al., 2001a,b). We and subsequently others (Metzl et al., 2006; Paar et al., 2007) have found that this configuration affords much greater genomic stability, and allows replication kinetics comparable to that of wild-type MLV.

2. Virus Production by Transient Transfection

To produce virus, we routinely perform calcium phosphate transfection of 293T human embryonic kidney cells in 10-cm cell culture dishes. The transfection protocol below is for transfection of one 10-cm dish. If more dishes will be transfected or if culture vessels of different size are used, the components can be scaled in proportion to total surface area. A low-endotoxin Maxi-prep kit (e.g., an Invitrogen HiPure or Qiagen Endo-Free kit) is used to generate plasmid preparations of adequate purity for high transfection efficiency. We have found that the *Escherichia coli* strains DH-5α, DH10B, and derivatives such as NEB-5α (NEB) and TOP10 (Invitrogen) serve as good hosts for propagating RRV plasmids, reliably providing high yields of high-quality plasmid without recombination.

2.1. Required materials

2.1.1. Cell culture

- SV40 T antigen-transformed human embryonic kidney (293T) cells: 293T cells may be obtained from the American Type Culture Collection (ATCC; catalog no. CRL-11,268).
- Culture medium: Dulbecco's modified Eagle's medium (DMEM) supplemented with 10% fetal calf serum, 100 U/ml penicillin and 100 μg/ml streptomycin.
- The 293T cells are maintained in the above culture medium at 37 °C and 5% CO_2 in a humidified incubator, and are split every 3–4 days at a ratio between 1:4 and 1:8. During propagation, cells should not be allowed to reach full confluency at any point as this may reduce transfection efficiency. Frozen aliquots of early passage stocks should be maintained and thawed periodically, as optimal transfection efficiencies are obtained with cells that have been passaged fewer than 40 times.

2.1.2. Reagents and solutions

- HEPES-buffered saline (HBS) solution: Prepare 2× HBS stock by adding the following to 400 ml dH_2O: 50 ml 1 *M* HEPES, 28 ml 5 *M* NaCl and 1.5 ml 0.5 *M* Na_2HPO_4. After adjusting the pH to 7.05–7.10 with 5 *M* and 1 *M* NaOH, bring the total volume to 500 ml with dH_2O.
- Calcium chloride solution: Prepare 2.5 *M* $CaCl_2$ by bringing 3.68 g of cell culture grade reagent (dihydrate form, Sigma cat# C7902) to 10 ml with dH_2O.
- Sodium butyrate solution: Prepare 0.5 *M* sodium butyrate (NaB; Sigma cat# B5887) by bringing 551 mg to 10 ml with dH_2O.
- The HBS, $CaCl_2$, and NaB solutions should all be sterile-filtered with 0.2 μ*M* filters, aliquoted and stored at −20 °C. The solutions are stable at this temperature for several months.
- Poly-L-lysine solution (to enhance cell adhesion to culture dishes): 0.01% poly-L-Lysine solution is obtained from Sigma (cat# P4832). Coating of culture dishes with poly-L-lysine allows the cell monolayer to grow to high density without detaching.

2.2. Transfection procedure

2.2.1. One day before transfection

1. All procedures should be performed with aseptic technique in a biosafety hood. Coat one 10-cm cell culture dish with poly-L-lysine solution. This is carried out by pipetting 3–4 ml of 0.01% poly-L-lysine solution onto the dish, distributing it over the entire surface and then removing all excess liquid. Allow dish to air dry for 10–15 min. If the dish will not be used right away, it can be wrapped in Saran wrap and stored at 4 °C.
2. The day before the cells are to be transfected, seed $2-4 \times 10^6$ 293T cells on the coated dish in 10 ml growth media. The number of cells plated should be such that, at the time of transfection, cell confluency is on the order of 50–75%. (*Note*: different sublines of 293T cells, and even the same subline after prolonged culture, may show different growth rates; hence for each batch of cells, various initial plating densities should be checked to determine what level of confluency results in optimal transfection efficiencies.)

2.2.2. The day of transfection

1. At least 2 h before the transfection is performed, replace the medium on the cells with 10 ml of fresh medium.
2. Cell transfection is performed using the following procedure:
    a. Bring 23 μg of the RRV plasmid to 450 μl with dH$_2$O. Sterile filter by spinning in a 0.22-μm Costar Spin-X centrifuge tube filter at $15,000 \times g$ for 2 min at room temperature.
    b. Add 50 μl 2.5 M CaCl$_2$ to the plasmid solution and mix by gently flicking the tube.
    c. Pipet 500 μl of 2× HBS into a 5-ml polystyrene tube. Add the 500 μl DNA/CaCl$_2$ solution dropwise to the HBS. Briefly vortex on medium speed and let sit at room temperature for 5 min.
    d. Distribute the 1 ml of precipitate dropwise over the cells.
    e. Gently rock the dish back-and-forth and side-to-side and return to the incubator.

2.2.3. One day after transfection

1. Replace the medium with 10 ml of fresh medium containing 10 mM NaB.
2. 6–8 h later, completely aspirate the NaB-containing medium and gently replace with 5 ml fresh medium containing no NaB.

2.2.4. Two days after transfection

1. 24 h after the media change, collect the supernatant from cells and filter through a 0.45-μm surfactant-free cellulose acetate syringe filter.
2. Aliquot volumes of 0.2–1.0 ml directly into sterile microfuge tubes. If the virus will be used within 1 day, it may be kept at 4 °C with negligible loss in titer. If the virus will not be used within 1 day, place immediately at −80 °C for storage. Smaller volumes should be avoided for freezing as they can lead to an accelerated decrease in titer during storage.

2.3. Notes

- If the RRV encodes a fluorescent reporter transgene, transfection efficiency can be readily assessed by flow cytometry. With an RRV encoding green fluorescent protein (GFP), the percentage of GFP-positive 293T cells at 2 days post-transfection is typically 60–80%.
- As the limit of efficient packaging capacity for RRVs is on the order of up to 1.5 kb of inserted transgene sequences in addition to the full-length retroviral genome, it may be difficult to incorporate both a marker gene and a therapeutic gene into the same vector. In this case, RRV titers can be determined by molecular methods, as described in the following section.

3. VECTOR COPY NUMBER ASSAY FOR TITER DETERMINATION AND BIODISTRIBUTION STUDIES

In many cases, RRVs carrying therapeutic genes cannot be readily quantitated using reporter assays. An alternative approach is to determine the number of RRV copies stably integrated into cells or tissues of interest. A method of determining copy number employing quantitative PCR (qPCR) analysis of cellular DNA using fluorogenic 5′ nuclease ("TaqMan") chemistry is described here. This assay can be used both for determination of biological titer of virus preparations and for evaluating RRV biodistribution in animals.

Advantages of this method include: (A) its ability to detect RRV containing any transgene. As the assay specifically detects MLV sequence within the RRV it thus can be used for vectors carrying any insert; (B) its measurement of biological titer, which may better correlate with functionality, rather than physical particle titer (as with any biological titer method, different cell lines can give widely different functional titers even when exposed to the same number of physical particles); (C) the ability to measure copy number regardless of vector expression levels in transduced cells, which may be influenced by the strength of viral promoter activity in a particular cell type; (D) the ability to determine titer without testing of a dilution series. In contrast, methods based on determination of the number of marker-expressing cells require low (<20%) transduction levels to preclude multiple infection of individual cells.

Additionally, this assay can distinguish between and quantitate contaminating vector plasmid versus *bona fide* reverse-transcribed virus in genomic DNA from transduced cells. This specificity relies on the mechanism whereby the retroviral LTR at each end of the proviral genome are reconstituted during each replication cycle. The LTR consists of the U3, R, and U5 regions, and during reverse transcription, the U3 region of the 3′ LTR serves as template for reconstitution of the 5′ LTR. Within most RRV vector plasmids, however, the 5′ LTR contains the CMV promoter in place of U3, where it serves to drive robust initial transcription during virus production in transfected cells. As the start site of transcription is at the 5′ border of the R region, the upstream CMV sequence is not included in the viral genomic transcript that is packaged into virions (Fig. 11.1A), and in cells infected with these virions, the MLV U3 in the 3′ LTR is copied over to the 5′ LTR. Thus, a qPCR primer pair that selectively amplifies the sequence from the CMV promoter to the packaging signal (ψ) will be specific for vector plasmid, whereas an MLV U3-specific primer paired with the ψ primer will detect the reverse-transcribed virus (Fig. 11.1A). Using genomic DNA from uninfected cells spiked with serial dilutions of vector plasmid DNA containing either the U3 or CMV sequence in the 5′ LTR, we found that reactions with oligonuclotides designed according to this strategy are highly specific for their respective targets and that the quantitation is linear over 7 orders of magnitude, from 50 to 5×10^8 copies per reaction, with correlation coefficients of >0.99 (not shown).

3.1. Preparation of template DNA

3.1.1. For vector titer determination

1. Plate cells of interest in 6-well tissue culture plates. Plate at a density such that 72h later the cultures will have just reached or be nearing confluence.
2. The following day, determine cell numbers in two of the wells using a hemacytometer. Add polybrene to a final concentration of 4μg/ml to the remaining wells. Add 100μl of the RRV of interest and gently swirl the plate to mix. For virus samples of very low titer (<10^4 TU/ml), this volume may be increased to up to half that of the culture medium to ensure that the quantitation is within the linear range of the assay.
3. 24h after infection, replace the medium with an equal volume of fresh medium containing 50μM AZT to block secondary replication cycles.
4. 48h after infection, harvest the cells for extraction of genomic DNA as indicated below.

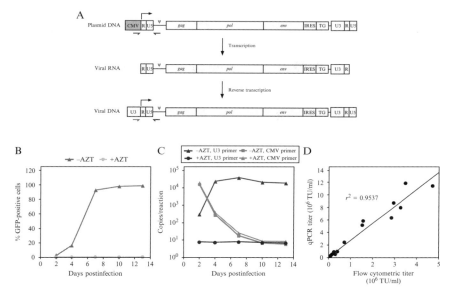

Figure 11.1 qPCR assay for determining provirus copy number and biological titer of RRV. (A) Strategy for detection of virus versus RRV plasmid contaminants. The stepwise changes in the RRV genome that occur through transcription of the plasmid and reverse transcription of the viral RNA are shown. Colored arrows indicate the binding location of primers that specifically amplify either RRV plasmid or virus. The use of a CMV-specific forward primer (red) will detect RRV plasmid in infected cells, either carried over from transfection or from another source, whereas a U3-specific primer (green) will detect only genuine reverse-transcribed RRV genomes. (B) RRV-mediated transmission of GFP expression following infection of cultured cells at low MOI with virus produced by transient transfection. In control cultures, AZT was included in the medium from the time of infection. GFP was detected by flow cytometry at 2, 4, 7, 10, and 13 days postinfection. (C) PCR quantitation of virus versus plasmid copies in genomic DNA from the same cultures as in panel B. The reactions employed a forward primer specific either for MLV U3 or for the CMV promoter. Note that the copy numbers per reaction below 50 are outside of the linear range of the assay. (D) Correlation between titer as determined by copy number determination by qPCR versus analysis of GFP expression by flow cytometry. (See Color Insert.)

3.1.2. For vector biodistribution studies

At the desired time point following administration of RRV, animals are sacrificed and tissue samples are harvested for isolation of genomic DNA. Tissues may be placed at $-80\,^\circ$C for storage if genomic DNA will not be immediately extracted. Ten to 25 mg of each tissue is processed with a column-based kit as indicated below. The expected yield is typically in the range of 3–30 μg DNA, and yield can vary by tissue.

3.1.3. Genomic DNA extraction

1. Commercially available kits, such as the PureLink Genomic DNA Mini Kit (Invitrogen) or DNeasy Blood & Tissue Kit with the optional Rnase A treatment (Qiagen), are used according to the manufacturer's guidelines for genomic DNA extraction. The resulting DNA is high molecular weight (20–50 kb), but not so high as to make pipetting difficult, and is free of contaminating RNA that might confound spectrophotometric quantitation.
2. The concentration of each purified genomic DNA sample should be determined by spectroscopic measurement of absorbance at 260 nm, and run on a 0.5% agarose gel stained with ethidium bromide to confirm that the size is within the expected range. No low molecular weight smears, which may indicate RNA contamination or DNA degradation, should be present.
3. For use in qPCR reactions, all genomic DNA samples are normalized to 10 ng/µl in the same Tris–EDTA buffer supplied in the kit for elution.

3.2. qPCR assay for determination of RRV titer and copy number

3.2.1. Materials required

- Real-time PCR instrument (e.g., Applied Biosystems 7900HT or Bio-Rad MyiQ2)
- qPCR primers at 100 µM: U3 forward primer 5′-AGCCCACAACCC-CTCACTC-3′; CMV forward primer 5′-GGTGGGAGGTCTATA-TAAGCAGAG-3′; reverse primer, 5′-TCTCCCGATCCCGGACGA-3′
- qPCR probe, at 100 µM: 5′-CCCCAAATGAAAGACCCCCGCT-GACG-3′. The probe is labeled with a 5′ reporter and a 3′ quencher suitable for the qPCR instrument to be used. Common labels are FAM, HEX, or TET at the 5′ end with TAMRA or BHQ1 at the 3′ end
- 2× qPCR master mix (e.g., Applied Biosystems Taqman Gene Expression Master Mix or Bio-Rad iQ Super Mix)
- Optical 96-well plates
- Optical adhesive film
- Centrifuge with swinging bucket rotor for microtiter plates

3.2.2. Prepare standard curve

To determine absolute RRV copy numbers in gDNA samples, a standard curve is generated with RRV plasmid containing a wild-type 5′ LTR. Plasmids pAZE-GFP (Logg et al., 2001a) or pAZ3-GFP (Kimura et al., 2010) can be used for this purpose. The number of molecules in

1 ng of any given plasmid can be conveniently calculated with the following formula:

$$\#\text{molecules(ng)} = \frac{9.17 \times 10^{11}}{\text{bp in plasmid}}$$

One nanogram of the 12,255-bp pAZE-GFP or 12,136-bp pAZ3-GFP plasmid thus contains approximately 74.8 million or 75.6 million copies, respectively. To derive a standard curve, the plasmid DNA is serially diluted in H_2O to produce five separate 10-fold dilutions covering a range of 10–100,000 plasmid copies per μl. Each dilution will be amplified in triplicate, using 5 μl of diluted plasmid per reaction. The standard curve reactions are set up as summarized for the experimental samples in Table 11.1, but the volume of water is reduced by 5 μl per reaction to accommodate the volume of plasmid.

3.2.3. Perform qPCR reactions

1. Prepare two master mixes in accordance with Table 11.2 and the total number of wells to be used. One is for the experimental samples and the other is for the standard curve. To allow for loss during pipetting, prepare 10% more than will be required for all wells.
2. Add 5 μl of template gDNA to each well. For the standard curve wells, use gDNA from uninfected cells.
3. Add 5 μl of diluted plasmid to the standard curve wells.
4. Add either 25 or 20 μl master mix to the experimental or standard curve wells, respectively.

Table 11.1 Preparation of reagents for qPCR analysis

Component	Volume per 30-μl reaction	Final concentration
2× qPCR master mix	15 μl	1×
Forward primer	0.15 μl	500 nM
Reverse primer	0.15 μl	500 nM
Probe	0.06 μl	200 nM
Water	To 25 μl (to 20 μl for std. curve)	–

Table 11.2 Time and temperature parameters for qPCR amplification

Initial denaturation and enzyme activation	1 cycle	95 °C for 10 min
Amplification	40 cycles	95 °C for 15 s
		60 °C for 1 min

5. Seal the plate with a sheet of optical adhesive film.
6. Spin the plate at 2000 rpm for 2 min in a swinging bucket rotor.
7. Run the plate in the qPCR instrument under the conditions specified in Table 11.2.

3.2.4. Analysis of results

1. Using the qPCR instrument's software, obtain threshold cycle (Ct) values for each sample. The Ct is defined as the fractional cycle number at which fluorescence crosses a threshold set above background but sufficiently low to be within the exponential region of the amplification curve.
2. Export the results to Microsoft Excel.
3. For the standard curve reactions, plot the log of plasmid copy number versus Ct.
4. Right-click on a data point on the plot and select "Add trendline."
5. Under Trendline Options, select "linear" and check "Display equation on chart." The equation will be of the form $y=mx+b$, where m is the slope and b is the y-intercept.
6. For each experimental sample, calculate the log of the copy number per well using the numbers for the y-intercept and slope obtained in the previous step using the following formula:

$$\log(\text{copy number}) = \frac{Ct - y\text{-intercept}}{\text{slope}}$$

7. For the experimental samples, determine the copy number per well by taking the antilog of the log copy number. In Excel, this is calculated with the function $=10^X$, where X is log copy number.
8. Calculate the copy number per cell by dividing the copy number with the number of cell equivalents represented by the 50 ng of gDNA present in the reaction. For normal human or mouse cells, 50 ng corresponds to approximately 7140 or 7460 cells, respectively. The mass of gDNA per cell for other species may be obtained at www.genomesize.com.
9. For calculation of titer (in transducing units per ml) use the following formula:

$$\text{Titer} = \text{Copy number per cell} \times \text{total cells} \times \text{dilution factor}$$

where "total cells" is the number of cells in the culture at the time of infection. The "dilution factor" is the ratio of the final culture volume to the volume of virus preparation added.

3.3. Illustrative results

We used this assay to monitor viral spread and to measure the level of plasmid carried over in cells exposed to RRV produced by transient transfection. PC-3 cells were infected at low multiplicity of infection (MOI) with RRV expressing GFP. The reverse transcriptase inhibitor 3′-azido-3′-deoxythymidine (AZT) was included in the medium of some cultures to $50\,\mu M$ starting from the time of infection to inhibit replication, while others were given no AZT. At 2, 4, 7, 10, and 13 days postinfection, the cells were passaged and aliquots were taken for flow cytometric detection of GFP and for copy number determination by qPCR. Flow cytometry showed that in the absence of AZT, the virus spread with the expected sigmoidal kinetics, while inclusion of AZT had effectively prevented infection (Fig. 11.1B). Copy number determinations were carried out with either the plasmid- or virus-specific primer/probe set. The results revealed that a substantial amount of plasmid DNA, as detected by the plasmid-specific oligo set, was indeed carried over from transfection, but by day 10 had dropped to undetectable levels (Fig. 11.1C). The virus-specific set showed that the cultures exposed to virus but not AZT underwent a burst of replication in the first 4 days after infection, followed by a peak and plateau, while those that contained AZT exhibited vastly lower, relatively constant signals comparable to that obtained with genomic DNA from naïve cells.

Measurement of the titers of a large number of independent RRV preparations using both this copy number assay and flow cytometric detection of GFP revealed a very strong correlation between the two methods ($r^2=0.9537$, $P<0.0001$) (Fig. 11.1D). Titers as determined by qPCR, however, were roughly two- to four-fold higher than those determined by flow cytometry. This difference may at least in part be explained by the low level of GFP expressed from RRV integrated at some loci and the relatively low detection sensitivity of GFP fluorescence.

4. DEVELOPMENT OF NOVEL RRV USING MOLECULAR EVOLUTION

The ability of RRVs to replicate provides a powerful tool for vector development. Within a spreading virus population, genetic variants with characteristics providing a replicative advantage will eventually predominate. Such variant RRVs can then be cloned, genetically stabilized, and employed as novel vectors. The genetic diversity in a population can either be experimentally introduced, by the construction of libraries of virus mutants, or it can occur naturally during replication. Retroviruses can exhibit a high rate of spontaneous mutation, in large part due to errors by reverse transcriptase. The rate of base misincorporation per nucleotide for retroviruses is estimated to be about a millionfold higher than the rate for eukaryotic cells (Preston and Dougherty, 1996).

We and others have taken advantage of the propensity of retroviruses to naturally undergo genetic diversification to evolve novel RRVs with useful properties. The general approach is to passage a poorly replicating, or potentially even replication-defective RRV on susceptible cells, so to allow the random appearance of mutations that facilitate replication. This strategy has been employed in the isolation of more efficiently replicating (Barsov and Hughes, 1996) and less cytotoxic (Barsov et al., 2001) mutants of a Rous sarcoma virus-based RRV containing the *env* gene of 4070A MLV; to improve the replication efficiency of a doxycycline-dependent human immunodeficiency virus-based RRV (Marzio et al., 2001); and to generate robustly replicating MLV-GALV chimeras (Logg et al., 2007).

An advantage of this evolutionary approach over rational methods of vector development is that knowledge of the mechanism behind any given deficiency in replication is unnecessary. Furthermore, little experimental intervention is required: viral replication itself mediates the selection, and spontaneous mutations provide the genetic diversity. The guidelines provided here for adapting RRVs assume that a poorly- or non-replicating "prototype" RRV, designed to possess some novel property, is in hand.

4.1. Design and construction of prototype RRV for adaptation

The prototype vector might be constructed to contain, for example, a gene or *cis*-acting element from another viral strain or species, exogenous regulatory elements, a targeting moiety, etc. For the vector to be a viable candidate for molecular evolution, it should contain all of the elements necessary for viral replication, including genes for production of Gag, Pol, and Env proteins and sequences for expression, reverse transcription and integration of the viral RNA. In order for the vector to be a viable candidate for molecular evolution, it should be capable of producing at least small amounts of transduction-competent particles upon transfection.

The presence of an easily assayed marker transgene in the RRV will greatly facilitate the titration and monitoring of replication of RRVs during attempts to evolve mutants. GFP is perhaps best suited for this purpose as it can be rapidly quantitated on a cell-by-cell basis by flow cytometry, and the Emerald and EGFP mutants are among the most sensitive fluorescent reporters when detected with a flow cytometer containing the common 488-nm argon laser. We therefore recommend that prototype RRVs be constructed to contain one of these two genes if possible.

4.2. Virus production

1. Virus is prepared by transfection with the prototype RRV plasmid as detailed in Section 2. As a control, a GFP-encoding RRV that is known to replicate in the cells to be used for infection should be prepared in parallel.
2. After harvest of virus, the transfected 293T cells are analyzed by flow cytometry to assess transfection efficiency and to confirm that expression from the vector plasmid is occurring.
3. The preparation should be titered on the cells of interest by flow cytometric detection of GFP, as described previously (Logg et al., 2002). Polybrene (hexadimethrine bromide; Sigma) should be included in the culture medium to 4 µg/ml during titer determinations and all other infections to maximize the available effective titer of prototype virus.

4.3. Molecular evolution and natural selection

1. To initiate replication, the cells of interest are infected with the prototype and control vectors at two or more MOIs between 0.05 and 0.5. If an MOI as high as 0.5 is not possible due to low titer, the maximum amount of virus supernatant the cells will tolerate (as a fraction of total media volume) should be used. While virus doses at the higher end of this range will provide greater genetic diversity in the initial infection, those at the lower end may be sufficient and will leave a greater number of cells for infection during secondary and later replication cycles.
2. The kinetics of vector spread is monitored by analyzing the cells by flow cytometry every 2–3 days thereafter for 1–2 weeks or until all of the cells are GFP-positive. Polybrene may also be included in the media throughout the culturing of the infected cells to enhance the effective titer of any virus produced from them and thereby stimulate the generation of genetic diversity. Adaptation is indicated by a change in replication kinetics.
3. If a poorly replicating prototype RRV exhibits no sign of horizontal infection during the 10–20 days following initial inoculation, the procedure may be repeated with a higher MOI to improve the likelihood that an adaptive mutation will occur during the initial transduction. Alternatively, the use of different cells as hosts may improve the ability of the virus to undergo initial replication and adaptation.
4. Once evidence of adaptation is observed in cultures infected with the prototype vector, the virus should be passaged through cultures of naïve cells at low MOI (0.01–0.05) to confirm the improved phenotype and to further purify the adapted vector or vectors. The change in kinetics may not be readily apparent until such infection of naïve cells is performed with the passaged virus.
5. In preparation for cloning, two to three additional passages at low MOI are performed to further refine the population. During each passage the virus should be given enough time to infect >90% of the cells.

4.4. Identification and cloning of mutations

The passaged vector is screened for replication-enabling or -enhancing mutations by cloning and functional testing of PCR fragments amplified from integrated provirus in the cells used in the final virus passage.

1. PCR primers are designed to flank restriction sites in the prototype RRV plasmid so as to facilitate reinsertion of the amplified sequence into the prototype RRV plasmid for testing and subsequent sequencing.
2. Genomic DNA is prepared from cell lines infected with the passaged virus according to the procedures described in Section 3.1.
3. Subregions of the provirus of 2–4 kb are amplified from genomic DNA for this purpose using a very high-fidelity polymerase such as Phusion (NEB) or PfuUltra II Fusion (Agilent).
4. The amplification products are fractionated by agarose electrophoresis, gel extracted, cut at the selected restriction sites and cloned back into the parental prototype RRV plasmid for testing. The pool of recloned RRV plasmids are transformed into a suitable bacterial strain such as DH-5α or DH10B, or a derivative thereof, as recommended above.
5. As the cells may include a mixture of proviruses containing both adaptive and incidental mutations, or no mutations, several plasmid clones are picked and prepped from bacteria. Miniprep-scale amounts of plasmid are normally sufficient for the screening process.
6. Virus is prepared by transfection of 293T cells as detailed in Section 2, but scaled down for transfection of single wells of 6-well or 12-well plates with minipreps of each plasmid clone.
7. Virus preparations are individually harvested, and tested for titer and replication on the same cell type previously used for evolutionary adaptation.
8. The replication kinetics of each clone is compared to that of the prototype and control RRVs by infection at low MOI followed by monitoring of GFP expression by flow cytometry. Improved kinetics in any of the clones relative to the prototype vector will indicate the presence of adaptive mutations. If no improvement is detected, a different region of the vector should be amplified and tested using the same procedure until one is observed.
9. After clones exhibiting improved replication are obtained, the entire inserted fragment is sequenced. Since the insert may also contain mutations unrelated to the adaptation, multiple independent clones exhibiting the phenotype should be sequenced. The existence of particular mutations in multiple clones will suggest an adaptive role. If necessary, the importance of these mutations can be confirmed by testing them in isolation after introduction into the prototype vector plasmid.

4.5. Illustrative example

Figure 11.2 shows the results of infections with an impaired prototype vector, GZAP-GFP, and an efficiently replicating control, AZE-GFP. The two are identical MLV-based RRV, except that the former contains the *env* gene of GALV and the latter contains the *env* gene of MLV 4070A. The GALV Env protein utilizes the Pit-1 receptor to mediate viral entry into cells whereas the 4070A Env utilizes Pit-2. While AZE-GFP spread efficiently and without delay, GZAP-GFP exhibited a lag phase of about 9 days before rapid replication commenced (Fig. 11.2A). Transfer of the supernatants from the infected cultures to naïve cells demonstrated that during the initial passage, GZAP-GFP had acquired the ability to spread with kinetics similar to that of AZE-GFP (Fig. 11.2B). Cloning and testing of the passaged GZAP-GFP revealed that it had acquired a mutation in the MLV 3′ splice site, causing an increase in the level of spliced viral RNA and thereby greatly improving replication.

5. MTS Assay of RRV-Mediated Cell Killing In Vitro

In this section a procedure is presented for assessing of the capacity of RRVs encoding a prodrug activator ("suicide") gene to kill tumor cells in culture upon addition of prodrug. The procedure can be adapted for use in

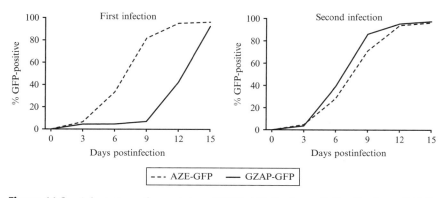

Figure 11.2 Adaptation of a prototype RRV. (A) (Left panel) Replication of AZE-GFP, an efficiently replicating amphotropic MLV RRV, and GZAP-GFP, a replication-impaired prototype MLV-GALV hybrid RRV, following initial infection of cultured cells at equal MOI. The virus used for these infections were generated by transient transfections with the corresponding plasmids and replication in infected cells was monitored by flow cytometry. (B) (Right panel) Infection of fresh cells with virus taken from Day 15 of the initial infection, demonstrating greatly improved kinetics for GZAP-GFP.

evaluating any prodrug activator gene/prodrug combination. Cytotoxicity is determined using a colorimetric assay that quantitates the reducing potential of cells, which is a measure of their metabolic function. Upon addition to culture medium, the chromogenic substrate MTS (3-(4,5-dimethylthiazol-2-yl)-5(3-carboxymethonyphenol)-2-(4-sulfophenyl)-2H-tetrazolium) is reduced by metabolically active cells into a product having a peak absorbance of 490 nm. The absorbance at this wavelength is directly proportional to the number of viable cells present.

In the example provided, U-87 MG malignant glioma cells are infected with an RRV expressing the yeast CD gene, which converts the non-toxic prodrug 5-Fluorocytosine (5-FC) to the classic chemotherapy drug 5-Fluorouracil (5-FU) (Kievit et al., 1999). Naïve or infected RRV-infected cancer cells are exposed to various concentrations of 5-FC to determine the IC50, the prodrug concentration at which cell viability is reduced by 50%. IC50 values are useful when comparing the potency of different prodrugs, different vectors carrying the same suicide gene, or different cell lines subjected to the same vector/prodrug treatment. An example kill curve from treatment of infected and uninfected U-87 cells exposed to several concentrations of 5-FC is given in Fig. 11.3. Statistical software was used to fit curves to the collected data points. As illustrated, the IC50 represents the concentration at which the curve reaches the midpoint between the upper and lower asymptotes (Fig. 11.3).

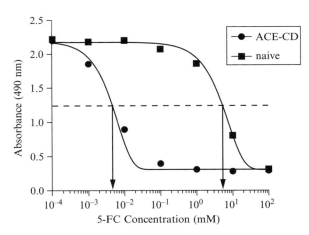

Figure 11.3 Illustration of IC50 determination by nonlinear regression analysis of MTS assay results. U-87 cells, either naive or infected with ACE-CD, were exposed to a series of 5-FC concentrations and cell viability was subsequently determined by the MTS assay. Sigmoidal dose–response curves were fit to each data set using Prism software. As shown, the IC50 values correspond to the concentration at which the curves cross the midpoint between the baseline and the maximal response, which is indicated by the dashed line.

5.1. Required materials

- Flat-bottomed 96-well tissue culture plates
- U-87 human glioma cells (U-87 MG, ATCC catalog no. HTB-14)
- Growth media for U-87 cells: DMEM supplemented with 10% fetal bovine serum (FBS)
- Titered preparation of RRV virus encoding CD (e.g., ACE-CD)
- 5-FC (Sigma-Aldrich, catalog no. F7129): prepare 10 mg/ml aqueous stock solution and sterile-filter through a 0.2-μm filter. Alternatively, pre-sterilized stock solutions are also commercially available (e.g., InvivoGen Inc., catalog no. sud-5 fc).
- MTS reagent (e.g., Promega CellTiter 96 AQueous One Solution Cell Proliferation Assay)
- Spectrophotometric plate reader capable of reading absorbance at 490 nm (e.g., Spectra Max 190 Plate Reader)
- GraphPad Prism 5 software

5.2. Assay procedures

1. Generate U-87 cells fully infected with ACE-CD: expose the cells to virus at an MOI of 1 and allow the infection to proceed for 1 week.
2. Optional: complete infection by can be verified by exposure of the cells to ACE-GFP at an MOI of 1, followed by flow cytometric analysis 3–4 days later. Interference mediated by the first virus should prevent any cells from becoming GFP-positive.
3. Plate the infected cells and naïve control cells in 96-well plates at a density of 2000 cells/well in growth medium containing 5-FC concentrations of 0, 0.5, 5, 50, 500, 5,000, and 50,000 μM. The final volume in each well should be 200 μl. For determination of background signal generated from the growth medium itself, set up "no-cell" wells containing 200 μl of medium alone. All samples should be in triplicate, and great care should be taken to plate the same number of cells in each well as consistently as possible (e.g., periodically invert and remix cell suspension between plating each row of wells). In order to avoid confounding effects of medium evaporation on assay results (Patel et al., 2005), either use only the inner 32 wells of the plate and fill the outer and any unused wells with 200 μl of sterile water, or use plates specifically designed to eliminate evaporation (e.g., Thermo Scientific Nunc Edge plates).
4. Incubate the plate at 37 °C in a humidified, 5% CO_2 atm. for 7 days. Note that the number of days necessary to reach a maximal therapeutic index will depend on the vector, suicide gene, prodrug, and cells used and will need to be experimentally determined for other systems. Note that it is often difficult to maintain viability of cells in untreated control wells for longer than 8–10 days or so.

5. Thaw the MTS reagent and add 40 μl to each well. Return the plate to the incubator and leave for 2 h.
6. Measure absorbance at 490 nm. If readings are below 1, absorbance unit for wells containing no 5-FC, incubate for another 30–60 min and repeat the measurement. Absorbances can be measured up to 4 h after addition of the MTS solution. To obtain the background-corrected absorbances, subtract the average reading from the three no-cell wells from all of the other readings.
7. Enter data into GraphPad Prism to plot the kill curve and determine the IC50:
   a. Create an XY data table with triplicate data values. Enter the seven tested 5-FC concentrations into the X column. Enter the corrected absorbances from the infected cells into column A and those from the naïve cells in column B.
   b. Log transform the concentrations: click the "Analyze" button and select "Transform" from the list of analyses. In the Transform dialog box, check the "Transform X values using" box, select "X=log(X)."
   c. Fit a four-parameter dose–response curve to the transformed data: from the data table, click "Analyze" and select "Nonlinear regression (curve fit)," then from the list of equations, under "Dose–response–Inhibition," choose "log(inhibitor) vs. response–Variable slope (four parameters)," click "OK."
   d. A new table will appear containing the IC50 values along with other statistics from the regression analysis.

6. *In Vivo* Glioma Model for Testing RRV-Mediated Gene Therapy

Glioblastoma multiforme (GBM) is the most common primary brain tumor in adults, and due to its dismal prognosis and the lack of effective conventional therapeutic options, this disease offers a risk:benefit ratio that justifies its use as a target for testing various experimental treatment strategies, particularly gene therapy (Aghi and Chiocca, 2006; Dent *et al.*, 2008; Rainov and Heidecke, 2011). Furthermore, lessons learned in treating primary brain tumors such as GBM can also be applied to CNS metastasis of systemic cancers, which occur in 10–30% of patients, and similarly carry a dire prognosis (Aragon-Ching and Zujewski, 2007). In fact, much of the preclinical testing, which supported the entry of RRV into clinical trials for treatment of patients with recurrent GBM, was performed in subcutaneous and intracranial brain tumor models (Hlavaty *et al.*, 2011; Ostertag *et al.*, 2012; Tai *et al.*, 2005; Wang *et al.*, 2003, 2006).

Subcutaneous models of glioma provide a larger tumor mass and allow examination of a longer time course of replication without rapid lethality, representing a tumor size that (although still small) is closer to what would be encountered clinically. Procedures for establishment of subcutaneous tumors and testing RRV in these models have been previously described in detail (Logg and Kasahara, 2004).

Accordingly, here we will focus on the use of intracranial models of GBM. Intracranial tumor models are, of course, smaller in mass, but are more rapidly fatal, since death is induced through the same mechanisms of increased intracranial pressure and loss of neuroregulation that maintains respiratory function as seen in malignant gliomas in human patients. Intracranial GBM xenografts in immunodeficient hosts can be used to model intratumoral virus replication in human cells in the absence of confounding effects from the immune system, while intracranial syngeneic tumors in immunocompetent hosts can model viral replication and potential antiviral immune responses in the context of the immunosuppressive tumor microenvironment placed within the semiprivileged immunological milieu of the CNS.

In this illustrative example, intracranial tumor models are established by stereotactic intracerebral injection of U-87 MG human glioma cells in athymic nude mice (Tai et al., 2005; Wang et al., 2003). Other intracranial tumor models can also be used, for example Tu-2449 murine gliomas in syngeneic B6C3F1 mice (Hlavaty et al., 2011; Ostertag et al., 2012) or RG2 gliomas in syngeneic Fischer rats (Wang et al., 2006). Subsequently, animals receive intratumoral injections of RRV through the same stereotactic procedure. After allowing virus spread to proceed for 2–3 weeks, treatment groups then receive systemic injections of the non-toxic prodrug 5-FC, administered as a single or twice daily intraperitoneal (IP) injection, while control groups receive phosphate-buffered saline (PBS), for 7–8 consecutive days. Subsequently, these 7–8-day cycles of prodrug administration are repeated at 2–3-week intervals to achieve long-term survival. In previously published studies (Tai et al., 2005), survival could be maintained in the majority of treated animals receiving both vector and prodrug for the entire duration of the experiment, i.e. >135 days, as compared to control groups receiving vector only, prodrug only, or no treatment, which all showed a median survival time on the order of 35 days ($P<0.0001$) (Fig. 11.4).

6.1. Required materials

6.1.1. Animals

- Athymic nude (nu/nu) laboratory mice (Mus musculus), 6–9 weeks old
- Note: Animals selected for use in this study should be as uniform in age and weight as possible. Each animal should be identified by ear tag or other standard method of identification. The animals should be housed

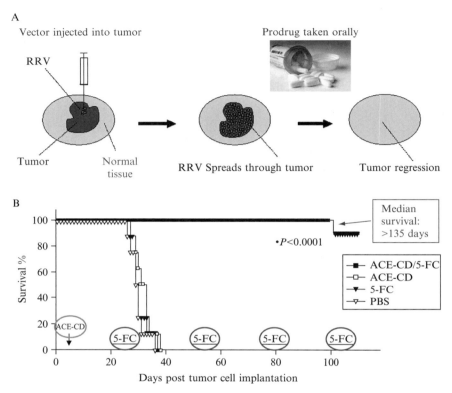

Figure 11.4 (A) The concept of Controlled Active Gene Transfer: Vector spreads through tumor but not normal tissue, hence the prodrug is activated into an anti-cancer drug only in the infected tumor cells. In the case of human clinical trials currently being conducted by Tocagen Inc., the RRV contains the cytosine deaminase (CD) gene from yeast. This enzyme converts the prodrug 5-FC, an FDA-approved anti-fungal compound that can be taken as an oral pill, into the anti-cancer drug 5-FU directly within the RRV-infected cancer cells. (B) RRV-mediated prodrug activator gene therapy can achieve long-term survival in a U-87 intracranial glioma model. PBS: control group injected with phosphate-buffered saline instead of vector and prodrug; 5-FC: control group injected with PBS instead of RRV, followed by systemic prodrug administration; ACE-CD: control group injected with RRV expressing CD gene, but without receiving prodrug afterward; ACE-CD/5-FC: treatment group receiving both RRV and prodrug. In the treatment group, only a single injection of vector was performed (ACE-CD, circled with downward arrow), followed by multiple cycles of prodrug (5-FC, circled with bars indicating duration of each cycle). (See Color Insert.)

individually in microisolator cages so they do not disturb each other's wounds. The room in which the animals are kept should be well ventilated (>10 air changes per hour) with 100% fresh air (no air recirculation), and maintained with a 12-h light/12-h dark cycle and room temperatures between 18 and 26 °C. All cages, equipment, bedding, and water for immunodeficient animals should be autoclaved prior to use, and rodent chow should be irradiated. All study animals should be acclimatized to their designated housing for 3–7 days prior to the first day of experimentation.

6.1.2. Therapeutic and analgesic agents

- Ketamine/Xylazine: anesthetic/analgesic combination, 100 mg/kg Ketamine + 10 mg/kg Xylazine via single IP injection, 1 h prior to surgical procedures
- Bacitricin Zinc-Polymyxin S04 ointment: antibiotic for topical use, to cover wounds, once after surgery and as needed thereafter
- Sulfamethoxazole/Trimethoprim: antibiotic, per os (PO), 1 mg/40 ml in drinking water, for 5 days following intracranial procedures
- Betadine (povidone-iodine): topical use, applied once over the surgical site at the beginning of surgery to achieve sterile field
- Carprofen: non-steroidal anti-inflammatory drug (NSAID), 5 mg/ml, single subcutaneous (SC) injection daily for 5 days following intracranial procedures
- Buprenex (buprenorphine): analgesic, 0.01–0.05 mg/kg SC or IP, once following cranial procedures; also to manage pain if any adverse effects are seen
- Sodium pentobarbital: 100 mg/kg IP, for euthanasia
- PBS solution: sterile-filtered, given IP as needed for fluid replacement and as control vehicle for prodrug administration
- 5-FC (Sigma-Aldrich, catalog no. F7129): 10 mg/ml stock solution, sterile-filtered

6.1.2. Virus and cells

- RRV carrying GFP marker gene (e.g., ACE-GFP)
- RRV carrying yeast CD prodrug activator gene (e.g., ACE-CD)
- Human glioma cell line U-87 (for tumor establishment). Cell lines are cultured in DMEM supplemented with 10% FBS and 1% penicillin-streptomycin, and maintained in a humidified atmosphere with 5% CO_2.

6.1.3. Surgical equipment

- Stereotaxic mounting frame with micromanipulator, for mice (e.g., Narishige Instruments catalog no. SR-5M or SR-6M; or Stoelting Co. Lab Standard catalog no. 51,600 or 51,603)
- Cordless surgical drill (e.g., Ideal Micro-Drill with 1–1.2 mm burr; or Dremel MiniMite cordless drill with 1/64 in. drill bit)
- Electric shaver (e.g., Oster Mini-Clipper)
- Electrical heating pad

- Surgical scalpels (#10 blade), forceps, scissors
- Serum separator tube (BD Microtainer, #365,956) for blood collection
- Tissue-Tek Optimal Cutting Temperature (OCT) compound and tissue sample cassettes, for embedding frozen sections
- Nunc cryovials for tissue collection and freezing
- Hamilton syringe, 700-series, 26s-gauge (e.g., catalog no. 20,734 or 20,779)
- Calibrated stereotaxic injector apparatus (e.g., Stoelting Co. Stereotaxic Injector with universal adaptor for Hamilton syringes)
- For IP and SC injections: 1–1 ml Luer-lock syringe with 20–25 gauge, 5/8 to 1 in. length needle (e.g., Becton-Dickinson BD Eclipse series)

6.2. Study procedures

6.2.1. Stereotactic injection procedure for intracranial tumor establishment and intratumoral vector administration

1. Mice are anesthetized with ketamine/xylazine and mounted into a stereotaxic frame with blunt ear bars.
2. Once properly positioned and immobilized, the fur overlaying the scalp is shaved and betadine is swabbed to prepare the surgical site.
3. The animal is placed on a heating pad and a scalpel is used under sterile conditions to make a midline incision through the skin. Retraction of the skin and reflection of the fascia at the incision site allows for visualization of the skull.
4. A small burr hole is drilled in the skull, and a Hamilton syringe containing the tumor cells is positioned at the appropriate intracranial *stereotaxic* coordinates. The coordinates that we typically employ are AP=3.0 mm, ML=2.0 mm, DV=3.0 mm (from midline, lambda, and dura, respectively). Exact coordinates for animals at different ages and weights should be determined in a pilot experiment by injecting dye and determining its location.
5. The tumor cell suspension (2×10^5 U-87 cells in 1–5 µl PBS) is injected into the brain. The injection should be performed very slowly over an interval of 5–10 min, using a calibrated injector assembly which allows precise control of injection speed and volume (e.g., screw-type plunger controller gradiated in 0.01 mm increments, one complete revolution advances the plunger of the syringe 0.5 mm), to minimize injection pressure and equilibrate with tissue pressure. If the plunger is advanced too rapidly, the injected cell suspension will immediately leak out via the route of least resistance, that is, up the needle track.
6. Five minutes after completing the injection, the injector assembly is slowly raised out of the skull (again, rapid removal will result in a suction effect that causes the tumor cell suspension to leak out via the injection track) and the skin is closed with suture or surgical staples.

7. The animals are monitored during anesthesia recovery. Analgesics should be administered once before the end of the procedure, and again every 12–24 h for up to 3 days. Antibiotics should be administered topically and in the drinking water for 5 days after surgery, and the animals should be monitored on a daily basis and weighed weekly after each procedure.
8. For intratumoral injection of virus vectors (RRV at titers of up to 10^6 total transduction units (TU) in 1–5 µl total volume), the stereotactic injection procedure is repeated at the same coordinates. Control animals are injected with PBS instead of vector solution. In the intracranial U-87 model, vector injection is generally performed 5–7 days after glioma cell inoculation, in order to allow an adequate time interval for virus spread before control animals succumb to intracranial tumor growth (starting at about 4 weeks post-tumor establishment).
9. The time-frames given here and below are appropriate for the U-87 xenograft model and are usually slightly different for other brain tumor models such as the B6C3F1/Tu2449 model, where the tumors grow more rapidly.

6.2.2. Prodrug administration

1. After allowing adequate time for viral spread (generally 2–3 weeks post-intratumoral vector injection, which corresponds to 3–4 weeks post-tumor establishment for the U-87 model), the prodrug (5-FC, 500 mg/kg) is administered systemically via ip injection. Control animals should receive IP injections of PBS instead of prodrug.
2. Mice should be manually restrained, but need not be anesthetized for IP injections. The abdominal area is swabbed with Betadine or ethanol, and injection into the abdomen is performed with care taken to avoid the bladder.
3. Prodrug administration is performed daily for a period of 7–8 consecutive days. Depending on the tumor model, if an adequate therapeutic effect is not achieved in this time interval, twice daily IP injections of 5-FC at the same dose may also be performed.
4. Prodrug administration is stopped after 7–8 days and the animals are monitored daily. This 7- to 8-day cycle of prodrug or vehicle administration should be repeated at intervals of every 2–3 weeks. Of course, the exact duration and scheduling of these prodrug cycles to achieve the optimum therapeutic effect will depend on the particular tumor model used.

6.2.3. In-life observations and measurements

1. Animals should be observed within their cages at least once daily throughout the acclimation and study period.
2. Starting Day-3 to -5 and continuing throughout the study, each animal should be observed for changes in general appearance and behavior. In

particular, after intracranial procedures, if any one or combination of adverse symptoms (bleeding, paresis, hunching, inactivity, or not feeding or grooming or weight loss >15% of baseline body weight) is observed and no sign of improvement is shown within a 2-day period, the animal should be humanely euthanized.
3. Body weights should be measured prior to group assignment and weekly thereafter until termination.

6.2.4. Tissue collection

1. Animals should be fully anesthesized prior to initiation of terminal procedures and tissue collection. Urine is collected prior to blood collection via direct bladder tap, then transferred to a sterile 1.5 ml tube and kept refrigerated.
2. Blood samples are also collected under anesthesia, by cardiac puncture. Half of the blood is transferred into a serum separator tube, and the collected serum is frozen at $-80\,^\circ\mathrm{C}$.
3. Euthanasia can be performed by sodium pentobarbital overdose, followed by cervical dislocation. Only sterile single-use instruments, draping, and collection tubes or containers should be used for all dissections and tissue collection, which should be conducted according to a protocol designed to minimize the possibility of cross-contamination by harvesting specimens in a fixed order: for example, Skin→Axillary lymph node→Lung →Esophagus→Heart→Intestine→Kidney→Liver→Spleen→Ovary→ Femur dissection/Bone Marrow flush→Spinal cord→Brain. Representative samples from each organ are harvested using sterile, single-use biopsy punches or scalpels, placed in sterile cryovials, and flash frozen in liquid nitrogen and stored at $-80\,^\circ\mathrm{C}$ until analysis for vector biodistribution by qPCR as described in Section 3. Dissection of the femur to obtain bone marrow is performed from the distal portion of the thigh in order to prevent contamination from intraperitoneal organs. Tissues for immunohistochemistry are preserved by embedding in OCT media and freezing.

6.2.5. Statistical analysis

1. According to a statistical power analysis, $n=14$ animals per treatment group would be required to achieve a significant result at $P<0.05$ using Fisher's exact test, assuming 0% survival of the reference group and 50% survival with the treatment of interest at the same time of follow-up.
2. Survival curves are constructed by use of the Kaplan–Meier method and compared by the log-rank test, with P values of <0.05 considered statistically significant.

6.3. Notes

1. Based on the time course of cell killing observed in MTS *in vitro* assays over a period of 8 days (above), the majority of cell killing should occur within the first 4–5 days after exposure to prodrug, followed by asymptotic reduction in infected cells during the remaining few days. However, as 5-FU produced intracellularly in the infected tumor cells (by 5-FC prodrug conversion) generally acts through interfering with DNA synthesis, it appears that not all the infected cells are dividing within this time interval and hence a small fraction of infected cells remain without being killed.
2. *In vivo*, further daily administration of prodrug beyond this time interval may result in complete loss of virus-infected tumor cells, which could result in a shorter duration of survival (Wang *et al.*, 2003), as these cells seem to act as a reservoir for reinfection of recurrent tumor growth.
3. Accordingly, we have found 7–8-day cycles of daily prodrug administration can be interspersed with periods of no treatment for up to 3–4 weeks to allow reinfection during tumor recurrence (Tai *et al.*, 2005). Thus, a single injection of RRV carrying a prodrug activator gene can achieve long-lasting therapeutic benefit with repeated cycles of non-toxic prodrug administration, without incurring the systemic adverse side effects associated with conventional chemotherapy.

ACKNOWLEDGMENTS

This work was supported in part by Tocagen Inc., NIH grants R01 CA121258 and U01 NS059821 (to N.K.), and a pilot project grant from the UCLA CURE Digestive Disease Research Center (to C.R.L.). We would like to acknowledge the UCLA Vector Core & Shared Resource facility for their assistance with vector construction and characterization. Tocagen has also received support for this work from Accelerate Brain Cancer Cure (ABC^2) Foundation, Washington DC.

REFERENCES

Aghi, M., and Chiocca, E. A. (2006). Gene therapy for glioblastoma. *Neurosurg. Focus* **20**, E18.

Aragon-Ching, J. B., and Zujewski, J. A. (2007). CNS metastasis: An old problem in a new guise. *Clin. Cancer Res.* **13**, 1644–1647.

Barsov, E. V., and Hughes, S. H. (1996). Gene transfer into mammalian cells by a Rous sarcoma virus-based retroviral vector with the host range of the amphotropic murine leukemia virus. *J. Virol.* **70**, 3922–3929.

Barsov, E. V., *et al.* (2001). Adaptation of chimeric retroviruses *in vitro* and *in vivo*: Isolation of avian retroviral vectors with extended host range. *J. Virol.* **75**, 4973–4983.

Biasi, G., et al. (2011). Immune response to Moloney-murine leukemia virus-induced antigens in bone marrow. *Immunol. Lett.* **138,** 79–85.

Critchley-Thorne, R. J., et al. (2009). Impaired interferon signaling is a common immune defect in human cancer. *Proc. Natl. Acad. Sci. USA* **106,** 9010–9015.

Dalba, C., et al. (2005). Beyond oncolytic virotherapy: Replication-competent retrovirus vectors for selective and stable transduction of tumors. *Curr. Gene Ther.* **5,** 655–667.

Davis, D. (2007). The Secret History of the War on Cancer. Basic Books, New York, NY.

Dent, P., et al. (2008). Searching for a cure: Gene therapy for glioblastoma. *Cancer Biol. Ther.* **7,** 1335–1340.

Dock, G. (1904). Influence of complicating diseases upon leukaemia. *J. Med. Sci.* **127,** 563–592.

Donnelly, O. G., et al. (2011). Recent clinical experience with oncolytic viruses. *Curr. Pharm. Biotechnol.*

Douville, R. N., and Hiscott, J. (2010). The interface between the innate interferon response and expression of host retroviral restriction factors. *Cytokine* **52,** 108–115.

Eager, R. M., and Nemunaitis, J. (2011). Clinical development directions in oncolytic viral therapy. *Cancer Gene Ther.* **18,** 305–317.

Flavell, R. A., et al. (2010). The polarization of immune cells in the tumour environment by TGFbeta. *Nat. Rev. Immunol.* **10,** 554–567.

Friedmann, T. (1996). The maturation of human gene therapy. *Acta Paediatr.* **85,** 1261–1265.

Hammill, A. M., et al. (2010). Oncolytic virotherapy reaches adolescence. *Pediatr. Blood Cancer* **55,** 1253–1263.

Hein, A., et al. (1995). Effects of adoptive immune transfers on murine leukemia virus-infection of rats. *Virology* **211,** 408–417.

Hiraoka, K., et al. (2006). Tumor-selective gene expression in a hepatic metastasis model after locoregional delivery of a replication-competent retrovirus vector. *Clin. Cancer Res.* **12,** 7108–7116.

Hiraoka, K., et al. (2007). Therapeutic efficacy of replication-competent retrovirus vector-mediated suicide gene therapy in a multifocal colorectal cancer metastasis model. *Cancer Res.* **67,** 5345–5353.

Hlavaty, J., et al. (2011). Comparative evaluation of preclinical *in vivo* models for the assessment of replicating retroviral vectors for the treatment of glioblastoma. *J. Neurooncol* **102,** 59–69.

Jolly, C. (2011). Cell-to-cell transmission of retroviruses: Innate immunity and interferon-induced restriction factors. *Virology* **411,** 251–259.

Kelly, E., and Russell, S. J. (2007). History of oncolytic viruses: Genesis to genetic engineering. *Mol. Ther.* **15,** 651–659.

Kende, M., et al. (1981). Naturally occurring humoral immunity to endogenous xenotropic and amphotropic type-C virus in the mouse. *Int. J. Cancer* **27,** 235–242.

Kievit, E., et al. (1999). Superiority of yeast over bacterial cytosine deaminase for enzyme/prodrug gene therapy in colon cancer xenografts. *Cancer Res.* **59,** 1417–1421.

Kikuchi, E., et al. (2007a). Delivery of replication-competent retrovirus expressing *Escherichia coli* purine nucleoside phosphorylase increases the metabolism of the prodrug, fludarabine phosphate and suppresses the growth of bladder tumor xenografts. *Cancer Gene Ther.* **14,** 279–286.

Kikuchi, E., et al. (2007b). Highly efficient gene delivery for bladder cancers by intravesically administered replication-competent retroviral vectors. *Clin. Cancer Res.* **13,** 4511–4518.

Kimura, T., et al. (2010). Optimization of enzyme-substrate pairing for bioluminescence imaging of gene transfer using Renilla and Gaussia luciferases. *J. Gene Med.* **12,** 528–537.

Kirn, D., et al. (2002). The emerging fields of suicide gene therapy and virotherapy. *Trends Mol. Med.* **8,** S68–S73.

Levaditi, C., and Nicolau, S. (1922). Affinite du virus herpetique pour les neoplasmes epitheliaux. *Compt. Rend. Soc. Biol.* **87,** 498–500.

Lewis, P. F., and Emerman, M. (1994). Passage through mitosis is required for oncoretroviruses but not for the human immunodeficiency virus. *J. Virol.* **68,** 510–516.

Lobel, L. I., et al. (1985). Construction and recovery of viable retroviral genomes carrying a bacterial suppressor transfer RNA gene. *Science* **228,** 329–332.

Logg, C. R., and Kasahara, N. (2004). Retrovirus-mediated gene transfer to tumors: Utilizing the replicative power of viruses to achieve highly efficient tumor transduction in vivo. *Methods Mol. Biol.* **246,** 499–525.

Logg, C. R., et al. (2001a). Genomic stability of murine leukemia viruses containing insertions at the Env-3′ untranslated region boundary. *J. Virol.* **75,** 6989–6998.

Logg, C. R., et al. (2001b). A uniquely stable replication-competent retrovirus vector achieves efficient gene delivery in vitro and in solid tumors. *Hum. Gene Ther.* **12,** 921–932.

Logg, C. R., et al. (2002). Tissue-specific transcriptional targeting of a replication-competent retroviral vector. *J. Virol.* **76,** 12783–12791.

Logg, C. R., et al. (2007). Adaptive evolution of a tagged chimeric gammaretrovirus: Identification of novel cis-acting elements that modulate splicing. *J. Mol. Biol.* **369,** 1214–1229.

Marzio, G., et al. (2001). In vitro evolution of a highly replicating, doxycycline-dependent HIV for applications in vaccine studies. *Proc. Natl. Acad. Sci. USA* **98,** 6342–6347.

McCormick, F. (2001). Cancer gene therapy: Fringe or cutting edge? *Nat. Rev. Cancer* **1,** 130–141.

Metzl, C., et al. (2006). Tissue- and tumor-specific targeting of murine leukemia virus-based replication-competent retroviral vectors. *J. Virol.* **80,** 7070–7078.

Mukherjee, S. (2010). The Emperor of All Maladies: A Biography of Cancer. Scribner, New York, NY.

Oliveira, N. M., et al. (2010). A novel envelope mediated post entry restriction of murine leukaemia virus in human cells is Ref1/TRIM5alpha independent. *Retrovirology* **7,** 81.

Olson, J. S. (1989). The History of Cancer: An Annotated Bibliography. Greenwood Press, Westport, CT.

Orkin, S. H., and Motulsky, A. G. (1995). Report and Recommendations of the Panel to Assess the NIH Investment in Research on Gene Therapy. National Institutes of Health, Bethesda, MD.

Ostertag, D., et al. (2012). Brain Tumor Eradication and Prolonged Survival from Intratumoral Conversion of 5-Fluorocytosine to 5-Fluorouracil Using a Nonlytic Retroviral Replicating Vector. *Neuro Oncol.* 2011 Nov 9. [Epub ahead of print].

Paar, M., et al. (2007). Effects of viral strain, transgene position, and target cell type on replication kinetics, genomic stability, and transgene expression of replication-competent murine leukemia virus-based vectors. *J. Virol.* **81,** 6973–6983.

Patel, M. I., et al. (2005). A Pitfall of the 3-(4,5-dimethylthiazol-2-yl)-5(3-carboxymethoxyphenol)-2-(4-sulfophenyl)-2 H-tetrazolium (MTS) assay due to evaporation in wells on the edge of a 96 well plate. *Biotechnol. Lett.* **27,** 805–808.

Preston, B. D., and Dougherty, J. P. (1996). Mechanisms of retroviral mutation. *Trends Microbiol.* **4,** 16–21.

Rainov, N. G., and Heidecke, V. (2011). Clinical development of experimental virus-mediated gene therapy for malignant glioma. *Anticancer Agents Med Chem.* **11,** 739–747.

Reik, W., et al. (1985). Replication-competent Moloney murine leukemia virus carrying a bacterial suppressor tRNA gene: Selective cloning of proviral and flanking host sequences. *Proc. Natl. Acad. Sci. USA* **82,** 1141–1145.

Roe, T., et al. (1993). Integration of murine leukemia virus DNA depends on mitosis. *EMBO J.* **12,** 2099–2108.

Seamon, J. A., et al. (2002). Inserting a nuclear targeting signal into a replication-competent Moloney murine leukemia virus affects viral export and is not sufficient for cell cycle-independent infection. *J. Virol.* **76,** 8475–8484.

Solly, S. K., et al. (2003). Replicative retroviral vectors for cancer gene therapy. *Cancer Gene Ther.* **10,** 30–39.

Stuhlmann, H., et al. (1989). Construction and properties of replication-competent murine retroviral vectors encoding methotrexate resistance. *Mol. Cell. Biol.* **9,** 100–108.

Tai, C. K., et al. (2005). Single-shot, multicycle suicide gene therapy by replication-competent retrovirus vectors achieves long-term survival benefit in experimental glioma. *Mol. Ther.* **12,** 842–851.

Tai, C. K., et al. (2010). Enhanced efficiency of prodrug activation therapy by tumor-selective replicating retrovirus vectors armed with the *Escherichia coli* purine nucleoside phosphorylase gene. *Cancer Gene Ther.* **17,** 614–623.

Takeuchi, H., and Matano, T. (2008). Host factors involved in resistance to retroviral infection. *Microbiol. Immunol.* **52,** 318–325.

Wang, W. J., et al. (2003). Highly efficient and tumor-restricted gene transfer to malignant gliomas by replication-competent retroviral vectors. *Hum. Gene Ther.* **14,** 117–127.

Wang, W., et al. (2006). Use of replication-competent retroviral vectors in an immunocompetent intracranial glioma model. *Neurosurg. Focus* **20,** E25.

Zitvogel, L., et al. (2006). Cancer despite immunosurveillance: Immunoselection and immunosubversion. *Nat. Rev. Immunol.* **6,** 715–727.

CHAPTER TWELVE

Adeno-Associated Virus Vectorology, Manufacturing, and Clinical Applications

Joshua C. Grieger* *and* R. Jude Samulski*,†

Contents

1. Adeno-Associated Virus Biology 230
2. AAV Vectorology 231
   2.1. AAV serotypes and strategies for targeted vector generation 235
   2.2. Capsid library generation for directed evolution of AAV capsids *in vivo* 237
3. rAAV Manufacturing Methods 238
4. Recent Clinical Trials Utilizing rAAV Vectors 240
   4.1. Leber's congenital amaurosis clinical trials 242
   4.2. Hemophilia B 243
5. Conclusion 245
Acknowledgments 246
References 246

Abstract

Adeno-associated virus (AAV) has emerged as an attractive vector for gene therapy. The benefits of using AAV for gene therapy include long-term gene expression, the inability to autonomously replicate without a helper virus, transduction of dividing and nondividing cells, and the lack of pathogenicity from wild-type infections. A number of Phase I and Phase II clinical trials utilizing AAV have been carried out worldwide (Aucoin *et al.*, 2008; Mueller and Flotte, 2008). A number of challenges have been identified based upon data generated from these clinical trials. These challenges include (1) large scale manufacturing technologies in accordance with current Good Manufacturing Practices (cGMP), (2) tissue specific tropism of AAV vectors, (3) high-quality/high potency recombinant AAV vectors (rAAV), and (4) immune response to AAV capsids and transgene. In this chapter, we will provide an overview of AAV biology, AAV vectorology, rAAV manufacturing, and the current status on the latest rAAV clinical trials.

* Gene Therapy Center, University of North Carolina, Chapel Hill, North Carolina, USA
† Department of Pharmacology, University of North Carolina, Chapel Hill, North Carolina, USA

1. ADENO-ASSOCIATED VIRUS BIOLOGY

Adeno-associated virus (AAV) is a member of the parvovirus family. Parvoviruses are among the smallest of the DNA animal viruses with a virion of approximately 25 nm in diameter composed entirely of protein and DNA. AAV has been classified as a dependovirus because it requires coinfection with helper viruses such as adenovirus (Ad; Atchison et al., 1965), herpes simplex virus (HSV; Georg-Fries et al., 1984), vaccinia virus (Schlehofer et al., 1986), and human papillomavirus (Ogston et al., 2000; Walz et al., 1997) for productive infection. The AAV2 genome is a linear, single-stranded DNA molecule containing 4679 nucleotides (Srivastava et al., 1983). The wild-type (wt) AAV genome is made up of two genes that encode four replication proteins, three capsid proteins, and an assembly activating protein and is flanked on either side by 145 bp inverted terminal repeats (Lusby et al., 1980; Sonntag et al., 2010; Srivastava et al., 1983), as shown in Fig. 12.1. The larger replication proteins, Rep 78 and 68, are splice variants originating from the p5 promoter. Rep 68

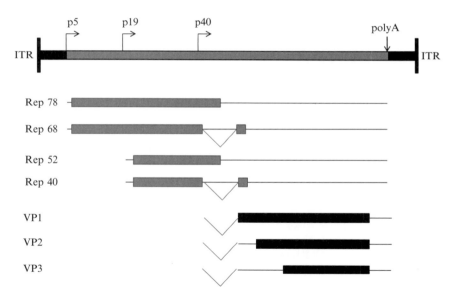

Figure 12.1 AAV2 genome organization. The horizontal arrows indicate the three transcriptional promoters (p5, p19, and p40). The black solid lines represent the transcripts and the introns are shown by the angled lines. The polyadenylation signal is common to all transcripts. The first open reading frame encodes the four replication proteins arising from the p5 and p19 promoters and alternative splicing. The second open reading frame utilizes the p40 promoter and encodes the three capsid proteins from two transcripts. VP1 is initiated from the first capsid transcript and VP2 and VP3 are initiated at two different codon sites from the second capsid transcript. (For color version of this figure, the reader is referred to the Web version of this chapter.)

and 78 are multifunctional and play a role in almost every aspect of the life cycle of AAV such as transcription, viral DNA replication, and site-specific integration into human chromosome 19. The small replication proteins, Rep 40 and 52, have been characterized as important for DNA packaging into the preformed viral capsid within the nucleus (King *et al.*, 2001). Capsid structures have been solved for eight serotypes of AAV: AAV1 (Miller *et al.*, 2006), AAV2 (Kronenberg *et al.*, 2001; Xie *et al.*, 2002), AAV4 (Kaludov *et al.*, 2003), AAV5 (Walters *et al.*, 2004), AAV6 (Xie *et al.*, 2008), AAV7 (Quesada *et al.*, 2007), AAV8 (Nam *et al.*, 2007), and AAV9 (Mitchell *et al.*, 2009). It has been estimated that the three capsid proteins combine to form the capsid in a ratio of VP1:VP2:VP3 of 1:1:10 (Kronenberg *et al.*, 2001). The capsid proteins are produced from the same open reading frame but utilize different translational start sites. Each capsid contains 50 copies of VP3 and 5 copies each of VP1 and VP2. However, it has been shown that it is possible to generate VP3 only, VP3–VP1 only, and VP3–VP2 only capsids suggesting that AAV virions can be assembled with various ratios of capsid subunits. Variation of these capsid subunits in a virion will most likely impact infectivity of the virus (specifically low incorporation of VP1).

Latent infection with AAV is common in the human population. However, no pathogenesis has ever been linked with AAV (Berns *et al.*, 1975). When AAV encounters a host cell, it enters the cell via phagocytosis through receptor-mediated endocytosis (Bartlett *et al.*, 2000). The capsid must then escape from the endosome and be transported into the nucleus, where uncoating occurs (Johnson and Samulski, 2009). The unique N-termini of VP1 and VP2 both contain nuclear localization signals required for the nuclear transport of the capsid (Grieger *et al.*, 2007). In addition, VP1 contains a phospholipase domain that is required for endosomal escape of the virion (Rabinowitz and Samulski, 2000). Virions lacking VP1 are not infectious. After the capsid releases the genome, the genome must then be converted to double stranded form by Rep and cellular DNA synthesis machinery (Ferrari *et al.*, 1996). In the absence of a helper virus, wtAAV DNA can be retained in circular episomal form or can be integrated into the chromosome at the AAVS1 integration site (Cheung *et al.*, 1980; Kotin *et al.*, 1990; Samulski *et al.*, 1991). The episomal and integrated DNA are both found in ITR (Inverted Terminal Repeat) mediated concatemerized forms. In the presence of a helper virus or cellular stress, AAV transcription and DNA replication are reactivated, and AAV completes its replication cycle.

2. AAV Vectorology

AAV has developed into one of the most important and safest viral gene delivery vectors in the field of gene therapy. The simplicity of the AAV genome allows for the straightforward design of rAAV vectors to deliver transgenes of interest. The inverted terminal repeats flanking either side of the genome are the only cis-acting elements necessary for genome replication, integration, and packaging into the capsid. rAAV can therefore be produced by replacing the replication and capsid genes with a promoter and therapeutic gene of interest (vector DNA). The *rep* and *cap* genes are expressed in trans from a different plasmid lacking ITRs. The separation of these genes from the vector plasmid DNA is critical for circumventing the formation of wtAAV.

Ad and HSV have been shown to be necessary for rAAV replication and production. In regard to the Ad helper virus, the E1a, E1b, E2a, E4orf6, and VA RNA genes were determined to supply the helper functions necessary for rAAV production (Xiao et al., 1998). Infection of Ad into producer cells to generate rAAV was effective in producing rAAV, but a consequence was that it also produced Ad particles. A significant improvement in the evolution of rAAV production was the introduction of the triple plasmid transfection method (Xiao et al., 1998). This method used an AAV serotype specific *rep* and *cap* plasmid as well as the vector DNA plasmid but eliminated the use of Ad infection by supplying the essential Ad genes on a third plasmid (pXX6). Supplying the Ad helper genes on the pXX6 plasmid eliminated Ad production in the transfected cells yielding only rAAV vector. Multi-plasmid transient transfection of adherent HEK293 cells remains the most widely used method for rAAV production (Grimm et al., 1998; Matsushita et al., 1998; Xiao et al., 1998).

In addition to the development of the Ad helper plasmid, two additional improvements were made on the AAV helper plasmid. The first was establishing that reduced levels of Rep protein expression lead to increased levels of Cap protein production. This, in turn, leads to increased rAAV production per cell (5–10-fold increase over the AAV/Ad production system). This was achieved by mutating the strong ATG start site of Rep78 to a weak ACG codon start site (Li et al., 1997). The second improvement was cross-packaging a single AAV2 vector genome (containing ITRs from serotype 2) into multiple AAV serotype capsids (Rabinowitz et al., 2002). This was achieved by replacing the AAV2 *rep* sequence downstream of the p19 promoter with the respective serotype *rep* sequence. The AAV helper plasmid series was named pXR1-5, which enabled replication and packaging of an AAV2 ITR recombinant genome into serotypes 1–5. This has now expanded to other AAV serotypes and chimeric capsids. The advantage of using this system is that the same vector genome is packaged within different capsids making the only variable the AAV serotype capsid. Due to equivalent genomes being packaged within various serotype capsids, the cross-packaging system allows for direct comparison of *in vivo* tropism in animal models (Rabinowitz et al., 2002; Zincarelli et al., 2008).

Transduction efficiencies for rAAV vectors generally range from 20 to several thousand vector genome-containing particles per transducing unit. This depends greatly on the cell type since a cell line permissive to all AAV serotypes has not been identified or standardized. Since cell lines and infectivity assays have not been established or standardized for AAV, transduction assays can only be used as a guide or measuring tool to qualitatively determine whether a rAAV serotype prep falls within a specific acceptance criteria set by the investigator. As discussed thoroughly in Aucoin *et al.*, comparison of transduction or infectivity data between publications should be done with vigilance due to the variety of assays utilized to quantitate infectivity (Aucoin *et al.*, 2008). Investigations into the rate limiting steps for transduction (post entry into the cell) highlighted the importance of converting the single-stranded DNA vector genome into double-stranded DNA prior to gene expression (Ferrari *et al.*, 1996; Fisher *et al.*, 1996) Studies have revealed additional bottlenecks that include rAAV transport to the nucleus and/or uncoating from the capsid (Duan *et al.*, 2000; Johnson and Samulski, 2009; Sipo *et al.*, 2007; Thomas *et al.*, 2004). There is also a period of genome instability after dsDNA conversion that leads to a significant loss of gene expression (Wang *et al.*, 2007). A combination of these limiting steps is likely to contribute to the dose required to achieve transduction of the cell and therapeutic benefit in humans. Regardless of these bottlenecks, rAAV genomes that reach the nucleus still require the synthesis or recruitment of a complementary strand to achieve gene expression. This is the critical step that was effectively bypassed by work conducted by McCarty *et al.* through the use of self-complementary AAV (scAAV) vectors.

To produce the scAAV vectors efficiently, McCarty *et al.* deleted the terminal resolution site from one rAAV ITR, preventing the initiation of replication at the mutated end (McCarty *et al.*, 2003). These constructs generate single-stranded, inverted repeat genomes with a wt ITR at each end and a mutated ITR in the middle (Fig. 12.2). It is believed that after uncoating, the vector genome folds through intramolecular base pairing initiating within the mutant ITR. The intramolecular base pairing then proceeds through the vector genome to form a double-stranded or self-complementary genome. Due to the design of the self-complementary genome, the vector genome particle to transducing particle ratio was decreased dramatically to levels of 20:1 and below. The kinetics for the onset of gene expression was also increased. Interestingly, cells that generally did not express transgenes from single-stranded genomes expressed transgenes from sc genomes delivered by the same serotype capsid, suggesting that a number of cells *in vivo* are infected by ssAAV but do not transduce these cells due to the inability to generate the complementary second strand.

As previously described, there are numerous advantages to using AAV for gene delivery. However, one of the main disadvantages of using AAV as

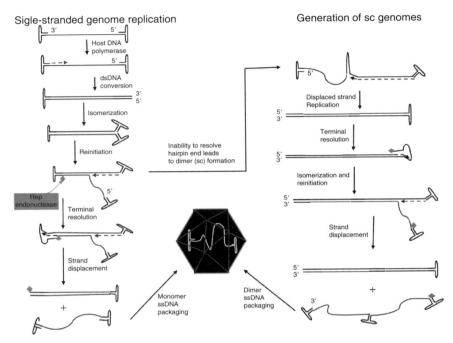

Figure 12.2 AAV replication cycles. Replication cycles for the generation of (left) single-stranded genomes and (right) self-complementary genomes. (See Color Insert.)

a vector is its limited packaging capacity. The packaging capacity of AAV has been studied by a number of laboratories with each resulting in varying packaging limitations (Allocca et al., 2008; Dong et al., 1996; Grieger and Samulski, 2005; Hermonat et al., 1997; Wu et al., 2010). As a result of these studies, the vector cassette including the ITRs should not exceed 4.7–5.0 kb in length or 4.4–4.7 kb of unique transgene sequence between the ITRs. While ssAAV vectors generally deliver approximately 4.4 kb of unique transgene sequence, scAAV will be able to carry only half of that because the unique transgene sequence is in duplex form in the scAAV genome. With 500–1000 bp for transcription elements, proteins of 40–55 kDa can be encoded within the sc genome. Because space is even a larger issue with scAAV vectors, optimization of transcriptional and posttranscriptional regulatory elements and codon optimization of the transgene will likely provide an increase in levels of transgene expression (Wu et al., 2008). Various groups have tried to expand upon the packaging capacity of AAV utilizing a split vector approach (using two rAAV) and homologous recombination (Duan et al., 2001; Halbert et al., 2002).

2.1. AAV serotypes and strategies for targeted vector generation

An essential requirement of gene therapy is the ability to target specific tissues for therapeutic gene delivery and to detarget those tissues that are unnecessary for correction of a given genetic disease. Production of an efficient targeted delivery system overcomes several limitations, including deleterious effects arising from the introduction of genetic material into the incorrect cell type, large therapeutic doses, and inefficient transduction of the tissue or cell type being targeted. To attempt to produce a targeted virus, extensive research has been performed on both the biology of individual natural serotypes as well as the construction of laboratory-derived serotypes that maximize gene delivery to intended tissue while evading nontargeted tissue types. As increasing numbers of naturally occurring serotypes from different primate and other animal species have been discovered (Bantel-Schaal and zur Hausen, 1984; Gao et al., 2002; Grimm and Kay, 2003; Grimm et al., 2003; Mori et al., 2004; Rutledge et al., 1998; Schmidt et al., 2006), the known diversity of AAV's tropism has broadened (Blankinship et al., 2004; Chao et al., 2000; Gregorevic et al., 2004; Mori et al., 2008; Pacak et al., 2006; Zincarelli et al., 2008).

Zincarelli et al. conducted a thorough *in vivo* mouse study utilizing AAV serotypes 1–9 packaging the luciferase reporter gene (Zincarelli et al., 2008). The rAAV vectors were injected via the tail vein for systemic delivery to look at tropism and expression profiles. It is evident from their study that tropism varies among the different serotypes as well as gene expression kinetics. The overall level of amino acid identity in the capsid protein of AAV serotypes 1–9 is ~45%, with the most divergent serotypes being AAV4 and AAV5 (Chiorini et al., 1997, 1999; Gao et al., 2002, 2004; Rutledge et al., 1998). The variability between serotypes is not evenly distributed throughout the capsid protein sequence but is concentrated in the looped out domains that are displayed on the capsid surface (Gao et al., 2003). It has not been conclusively determined whether the variable domains are responsible for the differences in the expression levels and duration of expression seen in *in vivo* comparative studies of AAV serotype gene expression. Interestingly, a recent study conducted by Keiser et al. concluded that each serotype may utilize distinct mechanisms for intracellular trafficking, and that postnuclear events play an important role in determining the efficiency of transduction. Therefore, overcoming postnuclear barriers that limit uncoating and/or promote virion degradation may enhance the efficiency of AAV serotypes in specific tissues (Keiser et al., 2011).

Harnessing the AAV crystal structure and cryo-EM data along with the vast amount of serotype specific information in the literature, and the availability of AAV serotype plasmids it is now possible to (1) select natural

AAV serotype/s for specific disease types, (2) rationally design a capsid with specific characteristics, and (3) generate capsid libraries to be used in directed evolution studies. One of the first successful rationally designed capsids was generated by Dr. Bowles using an AAV2 capsid. This new AAV2 variant, designated AAV2.5, contains five AAV1 amino acids. The choice and locations of the five amino acid modifications were determined by comparative analysis of capsid sequences of high muscle tropic serotypes combined with structural information. The criteria for selection of amino acid candidates were: (1) they must differ from AAV2 and be similar to the high muscle tropic serotypes, (2) they must be located in a structurally variable region on the capsid surface, or (3) they must be located in an AAV2 antigenic region that is recognized by an antibody. AAV2.5 combines the improved muscle transduction properties of AAV1 with reduced antigenic cross-reactivity against antibodies directed at both parental serotypes, while keeping the receptor binding properties of AAV2. The AAV2.5 vector successfully delivered the minidystrophin gene into the muscle of six patients with muscular dystrophy in a Phase I clinical trial with no adverse effects. The results of this rationally designed AAV2 capsid variant have laid the foundation that AAV vectors can be customized to improve efficacy and change immunological profiles for clinical applications. Another example of a rationally designed AAV capsid that has shown success was developed by Asokan *et al.* (2010). Asokan *et al.* swapped the heparin binding residues (585–590) from AAV2 with an AAV8 surface loop (585–590). AAV8 is known for its systemic transduction profile (Zincarelli *et al.*, 2008). These mutations were made based on the alignment of the receptor footprint of the serotypes. It was hypothesized that the swap would ablate the binding of AAV2 to its natural receptor (detargeting the virus from the liver), in addition to conferring some of the systemic delivery properties of AAV8. The new construct was termed AAV2i8. AAV2i8 exhibited a 40-fold decrease in liver gene expression compared to AAV2 upon IV injection into C57BL/6 mice. Additionally, high expression levels were observed in all types of muscle cells, including cardiac, skeletal, facial, and diaphragm with low levels of expression observed in brain, lung, and spleen. This illustrated that AAV2i8 has a tropism unique from either of its parental serotypes. A potential explanation for part of AAV2i8 widespread muscle transduction is that it exhibited markedly reduced blood clearance—persisting for well over 48 h in circulation. A final set of rationally designed vectors was engineered by Petrs-Silva *et al.* (2009) and Zhong *et al.* (2008b). These studies were the first to significantly demonstrate that site-directed mutagenesis of the capsid can also be used to improve transduction efficiency while maintaining the AAV serotypes' natural tropism. It had been previously established that EGFR-PTK phosphorylation of tyrosines in the viral capsid proteins negatively affects transduction levels, targeting the capsid for proteasomal degradation and thereby preventing nuclear localization of the vector (Zhong et al., 2007, 2008a). Zhong et al. hypothesized that mutating each of the tyrosines that undergo phosphorylation to phenylalanine would improve transduction by allowing the virus to escape degradation. Seven solvent-exposed residues were chosen for mutagenesis. It was determined that mutants Y444F and Y730F had the greatest impact on transduction efficiency, with up to a 29-fold gene expression increase observed in mouse hepatocytes after IV administration of these vectors to C57BL/6 mice. Similar increased transduction efficiencies (>20-fold over wt capsids) were also observed by Petrs-Silva et al., who used tyrosine to phenylalanine mutations on the capsids of AAV2, AAV8, and AAV9 to examine changes in transduction efficiency upon intravitreal delivery of vector to the retina.

2.2. Capsid library generation for directed evolution of AAV capsids *in vivo*

Directed evolution is the most efficient way to create novel AAV vectors that display enhanced tropism for specific cell types while detargeting them from undesirable cell types and achieve efficient transduction. Directed evolution allows the investigator to select a capsid that displays the characteristics needed for their application. This approach involves fragmenting the *cap* genes of multiple AAV serotypes, reannealing the fragments in a primerless PCR reaction, and creating full-length chimeric *cap* genes in a second PCR reaction. This is an expansion of the DNA shuffling techniques developed by Stemmer (Crameri et al., 1998; Stemmer, 1994). An *in vitro* directed evolution approach was conducted by Li et al. using chimeric viral libraries (serotypes 1–6, 8, and 9) comprising the chimeric capsid encapsidating the corresponding *cap* gene and AAV2 *rep* gene (Li et al., 2008). This group evolved a chimeric variant (AAV1829) that could transduce several melanoma cell lines and evade AAV2 neutralizing antibodies by shuffling the *cap* genes from AAV1-9 and selecting on Chinese hamster ovary cells.

The *in vitro* directed evolution approach has moved to an *in vivo* setting providing a more physiologically relevant methodology for selecting chimeric capsids with specific properties. The directed evolution approach in an *in vivo* setting removes any bias associated with cultured cells and produces variants with properties that can achieve therapeutic goals. This process involves systemic administration of a chimeric viral library, resection of the target tissue, and PCR amplification of the sequestered vector genomes. These vector genomes are then used in the next cycle of selection. Using an AAV1-9 shuffled library, Yang et al. evolved a heart tropic vector (AAVM41) with similar transduction levels as AAV9 but enhanced tropism (and detargeting from other tissue types) after systemic delivery (Yang et al., 2009). Using a combined technique of compromising the blood brain barrier with Kanic acid-induced seizure followed by the application of an AAV1-9 shuffled library, Gray *et al.* evolved a vector (clone 83) that could efficiently transduce neurons and oligodendrocytes while being detargeted from the liver and most other organs when administered systemically (Gray *et al.*, 2010). One caveat to the directed evolution approaches is that results in cell lines and animal models may not be directly comparable to results in large animals and humans. However, *in vivo* directed evolution should be and will continue to be explored, as this methodology could yield AAV variants that are beneficial to human applications. The directed evolution approach has also been explored as a way to isolate novel, immune response-escaping vectors. This approach is extensively reviewed in Mitchell *et al.* (2010). The long-term potential of the directed evolution approach and AAV vectors in general will only be realized when the action of this reagent in humans is understood completely.

3. rAAV Manufacturing Methods

AAV has shown great promise as a gene therapy vector in multiple aspects of preclinical and clinical applications. Many developments including new serotypes as well as self complementary vectors are now entering the clinic. With these ongoing vector developments in addition to new methods such as directed evolution, continued effort has been focused on scalable manufacturing processes that can efficiently generate high titer, highly pure, and potent quantities of rAAV vectors. Advances and novel approaches in rAAV production have been made recently that have allowed some laboratories to move away from production using adherent HEK293 cells and move toward scalable technologies. Three of these novel approaches are (1) development of HEK293 cell lines that can grow in serum-free suspension conditions, (2) the baculovirus expression vector system (BEVS) utilizing Spodoptera frugiperda (Sf9) cells, and (3) the dual rHSV infection system of suspension baby hamster kidney (BHK) cells. Each of these systems shows promise as the next generation scalable production platform for GMP manufacturing of rAAV vectors for clinical applications.

Recently, infection-based technologies using the BEVS and rHSV vectors have been developed and are currently being modified to support scalable production of rAAV vectors. The BEVS utilizes insect cells (SF9) grown in serum-free suspension conditions at 27 °C. The production of rAAV in insect cells does not require helper virus functions from Ad or HSV, for example, but does require infection of one to four baculoviruses (number depends on the constructs and strategy used) carrying the AAV *rep* and *cap* genes as well as the TR transgene cassette (Aslanidi *et al.*, 2009;

Cecchini et al., 2008; Kohlbrenner et al., 2005; Negrete and Kotin, 2008; Negrete et al., 2007; Urabe et al., 2002, 2006). A number of studies have been conducted using the BEVS to generate rAAV (Aucoin et al., 2008). Yields from these studies ranged from 5×10^3 to 5×10^4 vector genomes (vg)/cell. The advantages of using BEVS are (1) it is considered an animal component free scalable system, (2) utilizes infection instead of transfection of plasmids (although both are considered transient in nature, it is believed that infection is more consistent from batch to batch), and (3) it is already in use at the commercial scale for protein manufacturing. However, the disadvantages for its use to generate rAAV are (1) it can take up to 2 months to generate and fully characterize the baculoviruses needed to infect the insect cells, (2) the poor stability of the baculoviruses (characterized by the loss of recombinant gene products such as AAV Rep) after multiple passages of recombinant baculovirus (Kohlbrenner et al., 2005), (3) rAAV are less infectious than those generated from HEK293 cells (Kohlbrenner et al., 2005; Merten et al., 2005; Urabe et al., 2006), (4) capsid subunit ratios are different when compared to HEK293 cell produced capsids (VP1 and VP2 are incorporated less efficiently in BEVS with serotypes other than AAV2), and (5) potential differences in posttranslational modifications of the capsid (insect cells vs. mammalian cells). Much effort is being placed on resolving these disadvantages. Chen et al. inserted an artificial intron, containing the insect cell polyhedron promoter, into the AAV *rep* and *cap* coding sequences (Chen, 2008). This design maintained expression of Rep and Cap while extending the stability of the baculoviruses. When the VP1 protein level was increased by various modifications such as phospholipase A2 domain swapping or optimization of the start codon of VP1 and its surrounding nucleotide sequence, the infectivity of the vectors was improved (Kohlbrenner et al., 2005; Merten et al., 2005; Urabe et al., 2006).

Current rHSV infection systems used to generate rAAV are based on infection only and do not depend on a proviral cell line as previous generations of this system. The cis- and trans-acting elements necessary for rAAV replication and packaging are packaged within the rHSV vectors (Hwang et al., 2003), and the rAAV is produced via a single infection step using two rHSV vectors. One rHSV vector contains the AAV *rep* and *cap* genes, and the second rHSV vector contains the ITR transgene cassette. This system and a similar system generated by Kang et al. generated 1×10^5 vg/cell at a scale of 1×10^9 cells (Kang et al., 2009). The most recent advancement using this technology was developed by Thomas et al. (2009). This technology is a scalable version of the rHSV coinfection process in which an adherent HEK293 cell culture was replaced with a suspension culture of the BHK cell line. Briefly, this technology reduced the amount of input rHSV, increased cell density at time of infection, decreased time of harvest, and generated 7×10^4 to 1.1×10^5 vg/cell. Similar to the BEVS, a significant amount of time and effort is spent generating and characterizing the rHSV vectors needed to carry out the infection to generate rAAV at all scales. In addition, the production of high-titer, infectious rHSV stocks is challenging because engineering the rHSV vectors to be replication incompetent (making the rHSV vectors safe for use) reduces the overall yield and HSV particles can be easily inactivated since they are sensitive to production and processing conditions.

A few recent studies have reported rAAV production from mammalian HEK293 cells adapted to grow in serum-free suspension media (Durocher et al., 2007; Hildinger et al., 2007; Park et al., 2006). These studies demonstrated that approximately 1.4×10^4 and 3×10^4 vg/cell, respectively, were generated using their optimized serum-free suspension HEK293 cell production systems. Common with these studies is the fact that the yield of vector continues to be the impediment and significantly below the vg/cell generated via transfection of adherent HEK293 cells. Grieger et al. recently developed a HEK293 cell line that grows and transfects significantly well in an optimized serum-free suspension media (manuscript submitted for publication). The HEK293 cell line was expanded from a clinical master cell bank clone that was selected for high transfection efficiency and high rAAV vector production/cell. Utilizing the triple transfection method, the suspension HEK293 cell line generates greater than 1×10^5 rAAV vg/cell and greater than 1×10^{14} total rAAV vg/L of cell culture. rAAV vg/cell achieved by Grieger et al.'s cell line is similar to vg/cell yields obtained using their adherent HEK293 cells. With the exception of the HEK293 cell line, this user-friendly production system is considered an animal component-free/antibiotic-free system and was successfully scaled to wave bioreactors for the manufacture of preclinical and clinical rAAV vectors.

Currently, a number of potential rAAV manufacturing platforms exist with each having specific challenges to overcome to be designated the platform technology for large scale manufacturing of clinical and commercial grade rAAV vectors. Advances in rAAV biology/vectorology will be essential in assisting with the development of these technologies. As newly designed rAAV vectors are generated with increased targeting/detargeting as well as increased infectivity, manufacturing demands decrease. For example, vector capsids containing the tyrosine mutants have shown >20-fold increase in transduction (Petrs-Silva et al., 2009; Zhong et al., 2008b), which could translate to a 20-fold decrease in both patient doses and rAAV vector manufacturing.

4. Recent Clinical Trials Utilizing rAAV Vectors

rAAV vectors possess a number of characteristics that make them suitable for clinical gene therapy, including being based upon a virus for which there is no known pathology and a natural predisposition to persist in

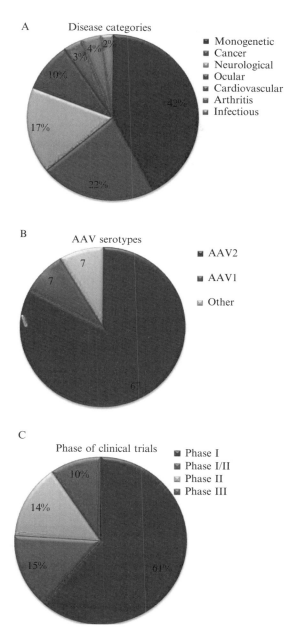

Figure 12.3 Clinical trials utilizing rAAV. The percentage of clinical trials using rAAV (A) in different categories of disease, (B) in different phases, and (C) using capsids of different AAV serotypes. Other serotypes include AAV8, AAV2.5, AAV6, and AAVrh.10. Clinical trial data were compiled from the Gene Therapy Clinical Trials Worldwide Database (http://www.wiley.com/legacy/wileychi/genmed/clinical/). (See Color Insert.)

human cells. Based on this and a wide range of preclinical studies in mice, rabbits, dogs, and nonhuman primates, a growing number of clinical trials have been initiated with these vectors. Approximately 80 clinical trials using rAAV are currently open, have been completed, or are being reviewed. Statistics encompassing most of these trials are depicted in Fig. 12.3. The earliest clinical trials were conducted with rAAV2 vectors to treat cystic fibrosis and hemophilia B. Lately, the number of diseases being tested for treatment utilizing AAV has greatly expanded and the newest clinical trials are utilizing serotypes other than AAV2. All of these trials utilize a genome with AAV2 ITRs transencapsidated into the various serotype capsids. The increase in the number of capsid isolates available for use as vectors as well as the development of rationally designed and directed evolution vectors has now provided the potential to broaden the application of AAV-based gene therapy to other target tissues. In the subsequent paragraphs, we give an overview of the latest promising rAAV Phase I clinical trials (the rAAV clinical history is reviewed extensively in Mitchell *et al.*, 2010; Mueller and Flotte, 2008). Before moving on to these recent Phase I clinical trials, it is very important to note that Phase II clinical trials were completed for congestive heart failure (Celladon) and Parkinson's disease (Neurologix and Ceregene) using rAAV vectors.

4.1. Leber's congenital amaurosis clinical trials

Leber's congenital amaurosis (LCA) is a group of inherited disorders involving retinal degeneration with severe vision loss noted in early infancy. The condition is usually identified through behaviors, including abnormal roving-eye movements (nystagmus). The diagnosis is confirmed by both abnormal electroretinographic responses and pupillary light reflexes (Aleman *et al.*, 2004; Lorenz *et al.*, 2000; Perrault *et al.*, 1999; Simonelli *et al.*, 2007). Most patients with LCA have severe visual impairment throughout childhood with vision deteriorating over time and total blindness by the third or fourth decade of life (Perrault *et al.*, 1999). There is no treatment for LCA. The LCA2 form of the disease is associated with mutations in *RPE65*, which encodes a protein necessary for the isomerohydrolase activity of the retinal pigment epithelium. This activity produces 11-*cis*-retinal from all-*trans*-retinyl esters. In the absence of 11-*cis*-retinal, the natural ligand and chromophore of the opsins of rod and cone photoreceptors, the opsins cannot capture light and transduce it into electrical responses to initiate vision (Gu *et al.*, 1997; Perrault *et al.*, 1999; Redmond *et al.*, 2005).

A few groups are currently conducting Phase I clinical trials delivering rAAV2-RPE65 vector (Cideciyan *et al.*, 2009; Hauswirth *et al.*, 2008; Maguire *et al.*, 2008, 2009; Simonelli *et al.*, 2010). rAAV2-RPE65 is a replication deficient AAV vector containing *RPE65* cDNA. rAAV2-RPE65 that was injected behind the retina of animal models of LCA2 resulted in rapid development of visual function (Bennicelli et al., 2008). The work conducted by Bennett et al. has documented long-term, sustained (>7.5 years, with ongoing observation) restoration of visual function in a canine model of LCA2 after a single subretinal injection of rAAV2-RPE65 (Acland et al., 2001, 2005; Bennicelli et al., 2008). This and additional safety and efficacy data has provided the basis for their Phase I trial of gene therapy in human patients with LCA2.

The initial LCA2 Phase I clinical trial conducted by Bennet et al. consisted of three patients between the ages of 19 and 26 to receive the low dose of rAAV2-RPE65. The eye with the worst function was identified for delivery of the AAV2-RPE65 vector. They reported that as little as 2 weeks after surgery, all three reported having improved vision in dimly lit environments. Each of the three eyes that received injection became more effective in driving the pupillary response becoming at least three times as sensitive to light as it had been at baseline surpassing that of the non-injected eye. Additionally, significant improvement in navigation through an obstacle course was noted. Both the safety and the efficacy noted at early timepoints persisted through 1.5 years after injection in the three LCA2 patients enrolled in the low dose cohort of the trial (Simonelli et al., 2010). The data from the 1.5 year follow-up led to some significant conclusions: (1) transgene expression resulting from AAV delivery is stable over time, (2) 4efficacy of gene transfer does not seem to be affected by the progressive degenerative nature of LCA2, and (3) AAV2-mediated gene transfer to the human retina does not elicit cytotoxic T-lymphocyte responses to AAV capsids.

An additional Phase I clinical trial was conducted to assess the retinal and visual function in 12 patients (aged 8–44 years) given one subretinal injection of rAAV-RPE65 at low $(1.5 \times 10^{10}$ vg), medium $(4.8 \times 10^{10}$ vg), or high dose $(1.5 \times 10^{11}$ vg) for up to 2 years (Maguire et al., 2009). All 12 patients given AAV2-hRPE65v2 in one eye showed improvement in retinal function and the gene therapy seemed to be safe at all doses. Overall, the results of the tests show that the greatest improvement in visual function with subretinal gene therapy will occur in young individuals. The visual recovery in the children confirms their hypothesis that efficacy will be improved if treatment is applied before retinal degeneration has progressed.

4.2. Hemophilia B

Hemophilia B, an X-linked bleeding disorder, is suited for gene replacement approaches. This is because the disease is attributable to the lack of a single gene product, clotting Factor IX (FIX) and because the therapeutic goal is modest as 1% of physiological levels would improve the severe bleeding phenotype. Correction of the bleeding diathesis after a single administration of AAV2 into either muscle or liver has been consistently observed in murine (Herzog et al., 1997; Nathwani et al., 2001; Wang et al., 2000) and canine (Herzog et al., 1999; Mount et al., 2002; Snyder et al., 1999; Wang et al., 2000) models of hemophilia without significant toxicity. These preclinical studies have supported clinical evaluation of AAV2 vectors in more than 100 patients with hemophilia B (Manno et al., 2003, 2006), cystic fibrosis (Flotte et al., 1996; Wagner et al., 1998, 1999a,b), Canavan disease (Janson et al., 2002), limb-girdle muscular dystrophy (Stedman et al., 2000), alpha-1-antitrypsin deficiency(Flotte et al., 2004), and LCA (Cideciyan et al., 2009; Hauswirth et al., 2008; Maguire et al., 2008, 2009; Simonelli et al., 2010). These early studies suggest that AAV2 vectors are safe in humans. To date, three Phase I clinical trials have been conducted for the treatment of hemophilia B with rAAV2 mediated gene therapy. The clinical trials conducted so far have attempted to transduce either the muscle or the liver in order to facilitate the production of the FIX protein. The first report on a hemophilia clinical trial with rAAV2 was with three patients conducted by Avigen, Inc (Fabb and Dickson, 2000). This trial demonstrated that (1) the treatment was safe with no evidence of germline transduction by the vector and (2) there was a small response by two of the patients. A Phase I/II, dose escalation trial with seven patients, demonstrated that treatment was safe with little to no toxicity. However, it was determined that effective therapeutic levels of FIX were achieved with the highest dose of vector, but the levels returned to baseline 8 weeks posttreatment due to a cell-mediated immune response (Manno et al., 2006). Overall, these trials demonstrated promising therapeutic value that is limited by immune responses, which were not predicted by animal model studies (Snyder et al., 1997). Other studies have examined the possibility of transducing the skeletal muscle through intramuscular injection. An early report with three patients determined that treatment was safe with no evidence for antibody production to FIX or of transmission of the vector. However, this study demonstrated no evidence of transduction or changes in the FIX serum levels (Kay et al., 2000). In a follow-up study, it was determined that this expression lasted for at least 10 months, with a biopsy from one patient maintaining expression after almost 4 years (Jiang et al., 2006). This was the first evidence of successful, long-term expression of a transgene delivered by rAAV in humans. Overall, the trials with hemophilia and AAV2 suggest that clinical long-term transduction is possible but greatly depends on the route of delivery, vector dose, and immune response.

Based on previous clinical trials, additional research was conducted on optimizing AAV for FIX delivery. A couple studies utilized self-complementary vectors and a codon optimized FIX gene for delivery into animal models (Nathwani et al., 2006; Wu et al., 2008). However, in addition to FIX codon optimization, Wu et al. systematically screened transcriptional regulatory elements to produce constitutive and liver-specific scAAV FIX expression cassettes that increased expression 4–20-fold when compared to a single-stranded rAAV vector that was used in a recent clinical trial (Wu *et al.*, 2008).

One currently open trial (initiated in August 2009) focused on the use of a rAAV vector pseudotyped with capsid protein of an alternative serotype, serotype 8. Although serotypes AAV2, AAV5, and AAV8 were generally similar in their ability to target and transduce rhesus hepatocytes (Davidoff *et al.*, 2005), it was found that the prevalence of immunity to AAV8, resulting from prior wt viral infection, is lower in humans than for the other two serotypes (Gao *et al.*, 2002; Parks *et al.*, 1970). This result is important because it has been shown that serotype-specific neutralizing antibodies preclude successful hepatocyte transduction (Davidoff *et al.*, 2005; Hurlbut *et al.*, 2010). Clearance of AAV8 from the systemic circulation in macaques was also significantly quicker than with other serotypes (Nathwani *et al.*, 2007), and AAV8 was shown to have reduced heparin binding, which reduced cytotoxic capsid-specific T cell activation (Vandenberghe *et al.*, 2006). In addition to utilizing AAV8, Nathwani *et al.* (2006) adopted the use of self-complementary genomes that provided enhanced human FIX expression in murine and nonhuman primate liver. Three dose levels were evaluated. The scAAV8 vector was infused through the peripheral vein without side effects. At the time the abstract was published (ASH meeting 2010), the longest follow-up was 5 months for the first patient infused with the lowest dose. The patient's FIX levels increased from <1% to between 1.5% and 2% within 2 weeks and was maintained beyond 5 months. Additional patients have now been infused with increasing doses scAAV8 vector. These early data are very encouraging and suggest that low doses of scAAV8 vector mediate therapeutic levels of FIX for several months without illiciting an immunological response that was identified in previous trials.

5. Conclusion

The intricate biology of AAV as a nonpathogenic human virus has lead to its development into a promising clinical viral vector. As a better understanding of AAV biology becomes available, we have and will continue to see a direct impact on vector development, vector manufacturing, clinical applications, and clinical efficacy. AAV vectors have an excellent safety record with data accumulated from thousands of animal studies and hundreds of human patients. The long-term potential of this vector will only be realized when the action of this reagent in humans is understood completely.

ACKNOWLEDGMENTS

This work was supported in part by US National Institutes of Health Wellstone-1 U54 AR056953, AFM (French Muscular Dystrophy Association) 14753/DDT12010, and North Carolina Biotechnology Center (NCBC) 2007-CFG-8012.

REFERENCES

Acland, G. M., Aguirre, G. D., Ray, J., Zhang, Q., Aleman, T. S., Cideciyan, A. V., Pearce-Kelling, S. E., Anand, V., Zeng, Y., Maguire, A. M., Jacobson, S. G., Hauswirth, W. W., et al. (2001). Gene therapy restores vision in a canine model of childhood blindness. *Nat. Genet.* **28**, 92–95.

Acland, G. M., Aguirre, G. D., Bennett, J., Aleman, T. S., Cideciyan, A. V., Bennicelli, J., Dejneka, N. S., Pearce-Kelling, S. E., Maguire, A. M., Palczewski, K., Hauswirth, W. W., and Jacobson, S. G. (2005). Long-term restoration of rod and cone vision by single dose rAAV-mediated gene transfer to the retina in a canine model of childhood blindness. *Mol. Ther.* **12**, 1072–1082.

Aleman, T. S., Jacobson, S. G., Chico, J. D., Scott, M. L., Cheung, A. Y., Windsor, E. A., Furushima, M., Redmond, T. M., Bennett, J., Palczewski, K., and Cideciyan, A. V. (2004). Impairment of the transient pupillary light reflex in Rpe65(−/−) mice and humans with leber congenital amaurosis. *Invest. Ophthalmol. Vis. Sci.* **45**, 1259–1271.

Allocca, M., Doria, M., Petrillo, M., Colella, P., Garcia-Hoyos, M., Gibbs, D., Kim, S. R., Maguire, A., Rex, T. S., Di Vicino, U., Cutillo, L., Sparrow, J. R., et al. (2008). Serotype-dependent packaging of large genes in adeno-associated viral vectors results in effective gene delivery in mice. *J. Clin. Invest.* **118**, 1955–1964.

Aslanidi, G., Lamb, K., and Zolotukhin, S. (2009). An inducible system for highly efficient production of recombinant adeno-associated virus (rAAV) vectors in insect Sf9 cells. *Proc. Natl. Acad. Sci. USA* **106**, 5059–5064.

Asokan, A., Conway, J. C., Phillips, J. L., Li, C., Hegge, J., Sinnott, R., Yadav, S., DiPrimio, N., Nam, H. J., Agbandje-McKenna, M., McPhee, S., Wolff, J., et al. (2010). Reengineering a receptor footprint of adeno-associated virus enables selective and systemic gene transfer to muscle. *Nat. Biotechnol.* **28**, 79–82.

Atchison, R. W., Casto, B. C., and Hammon, W. M. (1965). Adenovirus-associated defective virus particles. *Science* **149**, 754–756.

Aucoin, M. G., Perrier, M., and Kamen, A. A. (2008). Critical assessment of current adeno-associated viral vector production and quantification methods. *Biotechnol. Adv.* **26**, 73–88.

Bantel-Schaal, U., and zur Hausen, H. (1984). Characterization of the DNA of a defective human parvovirus isolated from a genital site. *Virology* **134**, 52–63.

Bartlett, J. S., Wilcher, R., and Samulski, R. J. (2000). Infectious entry pathway of adeno-associated virus and adeno-associated virus vectors. *J. Virol.* **74**, 2777–2785.

Bennicelli, J., Wright, J. F., Komaromy, A., Jacobs, J. B., Hauck, B., Zelenaia, O., Mingozzi, F., Hui, D., Chung, D., Rex, T. S., Wei, Z., Qu, G., et al. (2008). Reversal of blindness in animal models of leber congenital amaurosis using optimized AAV2-mediated gene transfer. *Mol. Ther.* **16**, 458–465.

Berns, K. I., Pinkerton, T. C., Thomas, G. F., and Hoggan, M. D. (1975). Detection of adeno-associated virus (AAV)-specific nucleotide sequences in DNA isolated from latently infected Detroit 6 cells. *Virology* **68**, 556–560.

Blankinship, M. J., Gregorevic, P., Allen, J. M., Harper, S. Q., Harper, H., Halbert, C. L., Miller, A. D., and Chamberlain, J. S. (2004). Efficient transduction of skeletal muscle using vectors based on adeno-associated virus serotype 6. *Mol. Ther.* **10**, 671–678.

Cecchini, S., Negrete, A., and Kotin, R. M. (2008). Toward exascale production of recombinant adeno-associated virus for gene transfer applications. *Gene Ther.* **15**, 823–830.

Chao, H., Liu, Y., Rabinowitz, J., Li, C., Samulski, R. J., and Walsh, C. E. (2000). Several log increase in therapeutic transgene delivery by distinct adeno-associated viral serotype vectors. *Mol. Ther.* **2**, 619–623.

Chen, H. (2008). Intron splicing-mediated expression of AAV Rep and Cap genes and production of AAV vectors in insect cells. *Mol. Ther.* **16**, 924–930.

Cheung, A. K., Hoggan, M. D., Hauswirth, W. W., and Berns, K. I. (1980). Integration of the adeno-associated virus genome into cellular DNA in latently infected human detroit 6 cells. *J. Virol.* **33**, 739–748.

Chiorini, J. A., Yang, L., Liu, Y., Safer, B., and Kotin, R. M. (1997). Cloning of adeno-associated virus type 4 (AAV4) and generation of recombinant AAV4 particles. *J. Virol.* **71**, 6823–6833.

Chiorini, J. A., Kim, F., Yang, L., and Kotin, R. M. (1999). Cloning and characterization of adeno-associated virus type 5. *J. Virol.* **73**, 1309–1319.

Cideciyan, A. V., Hauswirth, W. W., Aleman, T. S., Kaushal, S., Schwartz, S. B., Boye, S. L., Windsor, E. A., Conlon, T. J., Sumaroka, A., Pang, J. J., Roman, A. J., Byrne, B. J., et al. (2009). Human RPE65 gene therapy for Leber congenital amaurosis: Persistence of early visual improvements and safety at 1 year. *Hum. Gene Ther.* **20**, 999–1004.

Crameri, A., Raillard, S. A., Bermudez, E., and Stemmer, W. P. (1998). DNA shuffling of a family of genes from diverse species accelerates directed evolution. *Nature* **391**, 288–291.

Davidoff, A. M., Gray, J. T., Ng, C. Y., Zhang, Y., Zhou, J., Spence, Y., Bakar, Y., and Nathwani, A. C. (2005). Comparison of the ability of adeno-associated viral vectors pseudotyped with serotype 2, 5, and 8 capsid proteins to mediate efficient transduction of the liver in murine and nonhuman primate models. *Mol. Ther.* **11**, 875–888.

Dong, J. Y., Fan, P. D., and Frizzell, R. A. (1996). Quantitative analysis of the packaging capacity of recombinant adeno-associated virus. *Hum. Gene Ther.* **7**, 2101–2112.

Duan, D., Yue, Y., Yan, Z., Yang, J., and Engelhardt, J. F. (2000). Endosomal processing limits gene transfer to polarized airway epithelia by adeno-associated virus. *J. Clin. Invest.* **105**, 1573–1587.

Duan, D., Yue, Y., and Engelhardt, J. F. (2001). Expanding AAV packaging capacity with trans-splicing or overlapping vectors: A quantitative comparison. *Mol. Ther.* **4**, 383–391.

Durocher, Y., Pham, P. L., St-Laurent, G., Jacob, D., Cass, B., Chahal, P., Lau, C. J., Nalbantoglu, J., and Kamen, A. (2007). Scalable serum-free production of recombinant adeno-associated virus type 2 by transfection of 293 suspension cells. *J. Virol. Methods* **144**, 32–40.

Fabb, S. A., and Dickson, J. G. (2000). Technology evaluation: AAV factor IX gene therapy. *Curr. Opin. Mol. Ther.* **2**, 601–606.

Ferrari, F. K., Samulski, T., Shenk, T., and Samulski, R. J. (1996). Second-strand synthesis is a rate-limiting step for efficient transduction by recombinant adeno-associated virus vectors. *J. Virol.* **70**, 3227–3234.

Fisher, K. J., Gao, G. P., Weitzman, M. D., DeMatteo, R., Burda, J. F., and Wilson, J. M. (1996). Transduction with recombinant adeno-associated virus for gene therapy is limited by leading-strand synthesis. *J. Virol.* **70**, 520–532.

Flotte, T., Carter, B., Conrad, C., Guggino, W., Reynolds, T., Rosenstein, B., Taylor, G., Walden, S., and Wetzel, R. (1996). A phase I study of an adeno-associated virus-CFTR gene vector in adult CF patients with mild lung disease. *Hum. Gene Ther.* **7**, 1145–1159.

Flotte, T. R., Brantly, M. L., Spencer, L. T., Byrne, B. J., Spencer, C. T., Baker, D. J., and Humphries, M. (2004). Phase I trial of intramuscular injection of a recombinant adeno-associated virus alpha 1-antitrypsin (rAAV2-CB-hAAT) gene vector to AAT-deficient adults. *Hum. Gene Ther.* **15**, 93–128.

Gao, G. P., Alvira, M. R., Wang, L., Calcedo, R., Johnston, J., and Wilson, J. M. (2002). Novel adeno-associated viruses from rhesus monkeys as vectors for human gene therapy. *Proc. Natl. Acad. Sci. USA* **99,** 11854–11859.

Gao, G., Alvira, M. R., Somanathan, S., Lu, Y., Vandenberghe, L. H., Rux, J. J., Calcedo, R., Sanmiguel, J., Abbas, Z., and Wilson, J. M. (2003). Adeno-associated viruses undergo substantial evolution in primates during natural infections. *Proc. Natl. Acad. Sci. USA* **100,** 6081–6086.

Gao, G., Vandenberghe, L. H., Alvira, M. R., Lu, Y., Calcedo, R., Zhou, X., and Wilson, J. M. (2004). Clades of adeno-associated viruses are widely disseminated in human tissues. *J. Virol.* **78,** 6381–6388.

Georg-Fries, B., Biederlack, S., Wolf, J., and zur Hausen, H. (1984). Analysis of proteins, helper dependence, and seroepidemiology of a new human parvovirus. *Virology* **134,** 64–71.

Gray, S. J., Blake, B. L., Criswell, H. E., Nicolson, S. C., Samulski, R. J., and McCown, T. J. (2010). Directed evolution of a novel adeno-associated virus (AAV) vector that crosses the seizure-compromised blood–brain barrier (BBB). *Mol. Ther.* **18,** 570–578.

Gregorevic, P., Blankinship, M. J., Allen, J. M., Crawford, R. W., Meuse, L., Miller, D. G., Russell, D. W., and Chamberlain, J. S. (2004). Systemic delivery of genes to striated muscles using adeno-associated viral vectors. *Nat. Med.* **10,** 828–834.

Grieger, J. C., and Samulski, R. J. (2005). Packaging capacity of adeno-associated virus serotypes: Impact of larger genomes on infectivity and postentry steps. *J. Virol.* **79,** 9933–9944.

Grieger, J. C., Johnson, J. S., Gurda-Whitaker, B., Agbandje-McKenna, M., and Samulski, R. J. (2007). Surface-exposed adeno-associated virus VP1-NLS capsid fusion protein rescues infectivity of noninfectious wild-type VP2/VP3 and VP3-only capsids but not that of fivefold pore mutant virions. *J. Virol.* **81,** 7833–7843.

Grimm, D., and Kay, M. A. (2003). From virus evolution to vector revolution: Use of naturally occurring serotypes of adeno-associated virus (AAV) as novel vectors for human gene therapy. *Curr. Gene Ther.* **3,** 281–304.

Grimm, D., Kern, A., Rittner, K., and Kleinschmidt, J. A. (1998). Novel tools for production and purification of recombinant adenoassociated virus vectors. *Hum. Gene Ther.* **9,** 2745–2760.

Grimm, D., Kay, M. A., and Kleinschmidt, J. A. (2003). Helper virus-free, optically controllable, and two-plasmid-based production of adeno-associated virus vectors of serotypes 1 to 6. *Mol. Ther.* **7,** 839–850.

Gu, S. M., Thompson, D. A., Srikumari, C. R., Lorenz, B., Finckh, U., Nicoletti, A., Murthy, K. R., Rathmann, M., Kumaramanickavel, G., Denton, M. J., and Gal, A. (1997). Mutations in RPE65 cause autosomal recessive childhood-onset severe retinal dystrophy. *Nat. Genet.* **17,** 194–197.

Halbert, C. L., Allen, J. M., and Miller, A. D. (2002). Efficient mouse airway transduction following recombination between AAV vectors carrying parts of a larger gene. *Nat. Biotechnol.* **20,** 697–701.

Hauswirth, W. W., Aleman, T. S., Kaushal, S., Cideciyan, A. V., Schwartz, S. B., Wang, L., Conlon, T. J., Boye, S. L., Flotte, T. R., Byrne, B. J., and Jacobson, S. G. (2008). Treatment of leber congenital amaurosis due to RPE65 mutations by ocular subretinal injection of adeno-associated virus gene vector: Short-term results of a phase I trial. *Hum. Gene Ther.* **19,** 979–990.

Hermonat, P. L., Quirk, J. G., Bishop, B. M., and Han, L. (1997). The packaging capacity of adeno-associated virus (AAV) and the potential for wild-type-plus AAV gene therapy vectors. *FEBS Lett.* **407,** 78–84.

Herzog, R. W., Hagstrom, J. N., Kung, S. H., Tai, S. J., Wilson, J. M., Fisher, K. J., and High, K. A. (1997). Stable gene transfer and expression of human blood coagulation factor IX after intramuscular injection of recombinant adeno-associated virus. *Proc. Natl. Acad. Sci. USA* **94,** 5804–5809.

Herzog, R. W., Yang, E. Y., Couto, L. B., Hagstrom, J. N., Elwell, D., Fields, P. A., Burton, M., Bellinger, D. A., Read, M. S., Brinkhous, K. M., Podsakoff, G. M., Nichols, T. C., et al. (1999). Long-term correction of canine hemophilia B by gene transfer of blood coagulation factor IX mediated by adeno-associated viral vector. *Nat. Med.* **5,** 56–63.

Hildinger, M., Baldi, L., Stettler, M., and Wurm, F. M. (2007). High-titer, serum-free production of adeno-associated virus vectors by polyethyleneimine-mediated plasmid transfection in mammalian suspension cells. *Biotechnol. Lett.* **29,** 1713–1721.

Hurlbut, G. D., Ziegler, R. J., Nietupski, J. B., Foley, J. W., Woodworth, L. A., Meyers, E., Bercury, S. D., Pande, N. N., Souza, D. W., Bree, M. P., Lukason, M. J., Marshall, J., et al. (2010). Preexisting immunity and low expression in primates highlight translational challenges for liver-directed AAV8-mediated gene therapy. *Mol. Ther.* **18,** 1983–1994.

Hwang, k.K., Mandell, T., Kintner, H., Zolotukhin, S., Snyder, R., and Byrne, B. J. (2003). High titer recombinant adeno-associated virus production using replication deficient herpes simplex viruses type 1. *Mol. Ther.* **7,** S14–S15.

Janson, C., McPhee, S., Bilaniuk, L., Haselgrove, J., Testaiuti, M., Freese, A., Wang, D. J., Shera, D., Hurh, P., Rupin, J., Saslow, E., Goldfarb, O., et al. (2002). Clinical protocol. Gene therapy of canavan disease: AAV-2 vector for neurosurgical delivery of aspartoacylase gene (ASPA) to the human brain. *Hum. Gene Ther.* **13,** 1391–1412.

Jiang, H., Pierce, G. F., Ozelo, M. C., de Paula, E. V., Vargas, J. A., Smith, P., Sommer, J., Luk, A., Manno, C. S., High, K. A., and Arruda, V. R. (2006). Evidence of multiyear factor IX expression by AAV-mediated gene transfer to skeletal muscle in an individual with severe hemophilia B. *Mol. Ther.* **14,** 452–455.

Johnson, J. S., and Samulski, R. J. (2009). Enhancement of adeno-associated virus infection by mobilizing capsids into and out of the nucleolus. *J. Virol.* **83,** 2632–2644.

Kaludov, N., Padron, E., Govindasamy, L., McKenna, R., Chiorini, J. A., and Agbandje-McKenna, M. (2003). Production, purification and preliminary X-ray crystallographic studies of adeno-associated virus serotype 4. *Virology* **306,** 1–6.

Kang, W., Wang, L., Harrell, H., Liu, J., Thomas, D. L., Mayfield, T. L., Scotti, M. M., Ye, G. J., Veres, G., and Knop, D. R. (2009). An efficient rHSV-based complementation system for the production of multiple rAAV vector serotypes. *Gene Ther.* **16,** 229–239.

Kay, M. A., Manno, C. S., Ragni, M. V., Larson, P. J., Couto, L. B., McClelland, A., Glader, B., Chew, A. J., Tai, S. J., Herzog, R. W., Arruda, V., Johnson, F., et al. (2000). Evidence for gene transfer and expression of factor IX in haemophilia B patients treated with an AAV vector. *Nat. Genet.* **24,** 257–261.

Keiser, N. W., Yan, Z., Zhang, Y., Lei-Butters, D. C., and Engelhardt, J. F. (2011). Unique characteristics of AAV1, 2, and 5 viral entry, intracellular trafficking, and nuclear import define transduction efficiency in HeLa cells. *Hum. Gene Ther. Nov.* **22**(11), 1433–1444.

King, J. A., Dubielzig, R., Grimm, D., and Kleinschmidt, J. A. (2001). DNA helicase-mediated packaging of adeno-associated virus type 2 genomes into preformed capsids. *EMBO J.* **20,** 3282–3291.

Kohlbrenner, E., Aslanidi, G., Nash, K., Shklyaev, S., Campbell-Thompson, M., Byrne, B. J., Snyder, R. O., Muzyczka, N., Warrington, K. H., Jr., and Zolotukhin, S. (2005). Successful production of pseudotyped rAAV vectors using a modified baculovirus expression system. *Mol. Ther.* **12,** 1217–1225.

Kotin, R. M., Siniscalco, M., Samulski, R. J., Zhu, X. D., Hunter, L., Laughlin, C. A., McLaughlin, S., Muzyczka, N., Rocchi, M., and Berns, K. I. (1990). Site-specific integration by adeno-associated virus. *Proc. Natl. Acad. Sci. USA* **87,** 2211–2215.

Kronenberg, S., Kleinschmidt, J. A., and Bottcher, B. (2001). Electron cryo-microscopy and image reconstruction of adeno-associated virus type 2 empty capsids. *EMBO Rep.* **2,** 997–1002.

Li, J., Samulski, R. J., and Xiao, X. (1997). Role for highly regulated rep gene expression in adeno-associated virus vector production. *J. Virol.* **71,** 5236–5243.

Li, W., Asokan, A., Wu, Z., Van Dyke, T., DiPrimio, N., Johnson, J. S., Govindaswamy, L., Agbandje-McKenna, M., Leichtle, S., Redmond, D. E., Jr., McCown, T. J., Petermann, K. B., *et al.* (2008). Engineering and selection of shuffled AAV genomes: A new strategy for producing targeted biological nanoparticles. *Mol. Ther.* **16,** 1252–1260.

Lorenz, B., Gyurus, P., Preising, M., Bremser, D., Gu, S., Andrassi, M., Gerth, C., and Gal, A. (2000). Early-onset severe rod-cone dystrophy in young children with RPE65 mutations. *Invest. Ophthalmol. Vis. Sci.* **41,** 2735–2742.

Lusby, E., Fife, K. H., and Berns, K. I. (1980). Nucleotide sequence of the inverted terminal repetition in adeno-associated virus DNA. *J. Virol.* **34,** 402–409.

Maguire, A. M., Simonelli, F., Pierce, E. A., Pugh, E. N., Jr., Mingozzi, F., Bennicelli, J., Banfi, S., Marshall, K. A., Testa, F., Surace, E. M., Rossi, S., Lyubarsky, A., *et al.* (2008). Safety and efficacy of gene transfer for Leber's congenital amaurosis. *N. Engl. J. Med.* **358,** 2240–2248.

Maguire, A. M., High, K. A., Auricchio, A., Wright, J. F., Pierce, E. A., Testa, F., Mingozzi, F., Bennicelli, J. L., Ying, G. S., Rossi, S., Fulton, A., Marshall, K. A., *et al.* (2009). Age-dependent effects of RPE65 gene therapy for Leber's congenital amaurosis: A phase 1 dose-escalation trial. *Lancet* **374,** 1597–1605.

Manno, C. S., Chew, A. J., Hutchison, S., Larson, P. J., Herzog, R. W., Arruda, V. R., Tai, S. J., Ragni, M. V., Thompson, A., Ozelo, M., Couto, L. B., Leonard, D. G., *et al.* (2003). AAV-mediated factor IX gene transfer to skeletal muscle in patients with severe hemophilia B. *Blood* **101,** 2963–2972.

Manno, C. S., Pierce, G. F., Arruda, V. R., Glader, B., Ragni, M., Rasko, J. J., Ozelo, M. C., Hoots, K., Blatt, P., Konkle, B., Dake, M., Kaye, R., *et al.* (2006). Successful transduction of liver in hemophilia by AAV-Factor IX and limitations imposed by the host immune response. *Nat. Med.* **12,** 342–347.

Matsushita, T., Elliger, S., Elliger, C., Podsakoff, G., Villarreal, L., Kurtzman, G. J., Iwaki, Y., and Colosi, P. (1998). Adeno-associated virus vectors can be efficiently produced without helper virus. *Gene Ther.* **5,** 938–945.

McCarty, D. M., Fu, H., Monahan, P. E., Toulson, C. E., Naik, P., and Samulski, R. J. (2003). Adeno-associated virus terminal repeat (TR) mutant generates self-complementary vectors to overcome the rate-limiting step to transduction in vivo. *Gene Ther.* **10,** 2112–2118.

Merten, O. W., Geny-Fiamma, C., and Douar, A. M. (2005). Current issues in adeno-associated viral vector production. *Gene Ther.* **12**(Suppl. 1), S51–S61.

Miller, E. B., Gurda-Whitaker, B., Govindasamy, L., McKenna, R., Zolotukhin, S., Muzyczka, N., and Agbandje-McKenna, M. (2006). Production, purification and preliminary X-ray crystallographic studies of adeno-associated virus serotype 1. *Acta Crystallogr. Sect. F Struct. Biol. Cryst. Commun.* **62,** 1271–1274.

Mitchell, M., Nam, H. J., Carter, A., McCall, A., Rence, C., Bennett, A., Gurda, B., McKenna, R., Porter, M., Sakai, Y., Byrne, B. J., Muzyczka, N., *et al.* (2009). Production, purification and preliminary X-ray crystallographic studies of adeno-associated virus serotype 9. *Acta Crystallogr. Sect. F Struct. Biol. Cryst. Commun.* **65,** 715–718.

Mitchell, A. M., Nicolson, S. C., Warischalk, J. K., and Samulski, R. J. (2010). AAV's anatomy: Roadmap for optimizing vectors for translational success. *Curr. Gene Ther.* **10,** 319–340.

Mori, S., Wang, L., Takeuchi, T., and Kanda, T. (2004). Two novel adeno-associated viruses from cynomolgus monkey: Pseudotyping characterization of capsid protein. *Virology* **330,** 375–383.

Mori, S., Takeuchi, T., Enomoto, Y., Kondo, K., Sato, K., Ono, F., Sata, T., and Kanda, T. (2008). Tissue distribution of cynomolgus adeno-associated viruses AAV10, AAV11, and AAVcy.7 in naturally infected monkeys. *Arch. Virol.* **153,** 375–380.

Mount, J. D., Herzog, R. W., Tillson, D. M., Goodman, S. A., Robinson, N., McCleland, M. L., Bellinger, D., Nichols, T. C., Arruda, V. R., Lothrop, C. D., Jr., and High, K. A. (2002). Sustained phenotypic correction of hemophilia B dogs with a factor IX null mutation by liver-directed gene therapy. *Blood* **99,** 2670–2676.

Mueller, C., and Flotte, T. R. (2008). Clinical gene therapy using recombinant adeno-associated virus vectors. *Gene Ther.* **15,** 858–863.

Nam, H. J., Lane, M. D., Padron, E., Gurda, B., McKenna, R., Kohlbrenner, E., Aslanidi, G., Byrne, B., Muzyczka, N., Zolotukhin, S., and Agbandje-McKenna, M. (2007). Structure of adeno-associated virus serotype 8, a gene therapy vector. *J. Virol.* **81,** 12260–12271.

Nathwani, A. C., Davidoff, A., Hanawa, H., Zhou, J. F., Vanin, E. F., and Nienhuis, A. W. (2001). Factors influencing in vivo transduction by recombinant adeno-associated viral vectors expressing the human factor IX cDNA. *Blood* **97,** 1258–1265.

Nathwani, A. C., Gray, J. T., Ng, C. Y., Zhou, J., Spence, Y., Waddington, S. N., Tuddenham, E. G., Kemball-Cook, G., McIntosh, J., Boon-Spijker, M., Mertens, K., and Davidoff, A. M. (2006). Self-complementary adeno-associated virus vectors containing a novel liver-specific human factor IX expression cassette enable highly efficient transduction of murine and nonhuman primate liver. *Blood* **107,** 2653–2661.

Nathwani, A. C., Gray, J. T., McIntosh, J., Ng, C. Y., Zhou, J., Spence, Y., Cochrane, M., Gray, E., Tuddenham, E. G., and Davidoff, A. M. (2007). Safe and efficient transduction of the liver after peripheral vein infusion of self-complementary AAV vector results in stable therapeutic expression of human FIX in nonhuman primates. *Blood* **109,** 1414–1421.

Negrete, A., and Kotin, R. M. (2008). Large-scale production of recombinant adeno-associated viral vectors. *Methods Mol. Biol.* **433,** 79–96.

Negrete, A., Yang, L. C., Mendez, A. F., Levy, J. R., and Kotin, R. M. (2007). Economized large-scale production of high yield of rAAV for gene therapy applications exploiting baculovirus expression system. *J. Gene Med.* **9,** 938–948.

Ogston, P., Raj, K., and Beard, P. (2000). Productive replication of adeno-associated virus can occur in human papillomavirus type 16 (HPV-16) episome-containing keratinocytes and is augmented by the HPV-16 E2 protein. *J. Virol.* **74,** 3494–3504.

Pacak, C. A., Mah, C. S., Thattaliyath, B. D., Conlon, T. J., Lewis, M. A., Cloutier, D. E., Zolotukhin, I., Tarantal, A. F., and Byrne, B. J. (2006). Recombinant adeno-associated virus serotype 9 leads to preferential cardiac transduction in vivo. *Circ. Res.* **99,** e3–e9.

Park, J. Y., Lim, B. P., Lee, K., Kim, Y. G., and Jo, E. C. (2006). Scalable production of adeno-associated virus type 2 vectors via suspension transfection. *Biotechnol. Bioeng.* **94,** 416–430.

Parks, W. P., Boucher, D. W., Melnick, J. L., Taber, L. H., and Yow, M. D. (1970). Seroepidemiological and ecological studies of the adenovirus-associated satellite viruses. *Infect. Immun.* **2,** 716–722.

Perrault, I., Rozet, J. M., Gerber, S., Ghazi, I., Leowski, C., Ducroq, D., Souied, E., Dufier, J. L., Munnich, A., and Kaplan, J. (1999). Leber congenital amaurosis. *Mol. Genet. Metab.* **68,** 200–208.

Petrs-Silva, H., Dinculescu, A., Li, Q., Min, S. H., Chiodo, V., Pang, J. J., Zhong, L., Zolotukhin, S., Srivastava, A., Lewin, A. S., and Hauswirth, W. W. (2009). High-efficiency transduction of the mouse retina by tyrosine-mutant AAV serotype vectors. *Mol. Ther.* **17,** 463–471.

Quesada, O., Gurda, B., Govindasamy, L., McKenna, R., Kohlbrenner, E., Aslanidi, G., Zolotukhin, S., Muzyczka, N., and Agbandje-McKenna, M. (2007). Production, purification and preliminary X-ray crystallographic studies of adeno-associated virus serotype 7. *Acta Crystallogr. Sect. F Struct. Biol. Cryst. Commun.* **63,** 1073–1076.

Rabinowitz, J. E., and Samulski, R. J. (2000). Building a better vector: The manipulation of AAV virions. *Virology* **278**, 301–308.

Rabinowitz, J. E., Rolling, F., Li, C., Conrath, H., Xiao, W., Xiao, X., and Samulski, R. J. (2002). Cross-packaging of a single adeno-associated virus (AAV) type 2 vector genome into multiple AAV serotypes enables transduction with broad specificity. *J. Virol.* **76**, 791–801.

Redmond, T. M., Poliakov, E., Yu, S., Tsai, J. Y., Lu, Z., and Gentleman, S. (2005). Mutation of key residues of RPE65 abolishes its enzymatic role as isomerohydrolase in the visual cycle. *Proc. Natl. Acad. Sci. USA* **102**, 13658–13663.

Rutledge, E. A., Halbert, C. L., and Russell, D. W. (1998). Infectious clones and vectors derived from adeno-associated virus (AAV) serotypes other than AAV type 2. *J. Virol.* **72**, 309–319.

Samulski, R. J., Zhu, X., Xiao, X., Brook, J. D., Housman, D. E., Epstein, N., and Hunter, L. A. (1991). Targeted integration of adeno-associated virus (AAV) into human chromosome 19. *EMBO J.* **10**, 3941–3950.

Schlehofer, J. R., Ehrbar, M., and zur Hausen, H. (1986). Vaccinia virus, herpes simplex virus, and carcinogens induce DNA amplification in a human cell line and support replication of a helpervirus dependent parvovirus. *Virology* **152**, 110–117.

Schmidt, M., Grot, E., Cervenka, P., Wainer, S., Buck, C., and Chiorini, J. A. (2006). Identification and characterization of novel adeno-associated virus isolates in ATCC virus stocks. *J. Virol.* **80**, 5082–5085.

Simonelli, F., Ziviello, C., Testa, F., Rossi, S., Fazzi, E., Bianchi, P. E., Fossarello, M., Signorini, S., Bertone, C., Galantuomo, S., Brancati, F., Valente, E. M., et al. (2007). Clinical and molecular genetics of Leber's congenital amaurosis: A multicenter study of Italian patients. *Invest. Ophthalmol. Vis. Sci.* **48**, 4284–4290.

Simonelli, F., Maguire, A. M., Testa, F., Pierce, E. A., Mingozzi, F., Bennicelli, J. L., Rossi, S., Marshall, K., Banfi, S., Surace, E. M., Sun, J., Redmond, T. M., et al. (2010). Gene therapy for Leber's congenital amaurosis is safe and effective through 1.5 years after vector administration. *Mol. Ther.* **18**, 643–650.

Sipo, I., Fechner, H., Pinkert, S., Suckau, L., Wang, X., Weger, S., and Poller, W. (2007). Differential internalization and nuclear uncoating of self-complementary adeno-associated virus pseudotype vectors as determinants of cardiac cell transduction. *Gene Ther.* **14**, 1319–1329.

Snyder, R. O., Miao, C. H., Patijn, G. A., Spratt, S. K., Danos, O., Nagy, D., Gown, A. M., Winther, B., Meuse, L., Cohen, L. K., Thompson, A. R., and Kay, M. A. (1997). Persistent and therapeutic concentrations of human factor IX in mice after hepatic gene transfer of recombinant AAV vectors. *Nat. Genet.* **16**, 270–276.

Snyder, R. O., Miao, C., Meuse, L., Tubb, J., Donahue, B. A., Lin, H. F., Stafford, D. W., Patel, S., Thompson, A. R., Nichols, T., Read, M. S., Bellinger, D. A., et al. (1999). Correction of hemophilia B in canine and murine models using recombinant adeno-associated viral vectors. *Nat. Med.* **5**, 64–70.

Sonntag, F., Schmidt, K., and Kleinschmidt, J. A. (2010). A viral assembly factor promotes AAV2 capsid formation in the nucleolus. *Proc. Natl. Acad. Sci. USA* **107**(22), 10220–10225. Epub 2010 May 17. PMID: 20479244.

Srivastava, A., Lusby, E. W., and Berns, K. I. (1983). Nucleotide sequence and organization of the adeno-associated virus 2 genome. *J. Virol.* **45**, 555–564.

Stedman, H., Wilson, J. M., Finke, R., Kleckner, A. L., and Mendell, J. (2000). Phase I clinical trial utilizing gene therapy for limb girdle muscular dystrophy: Alpha-, beta-, gamma-, or delta-sarcoglycan gene delivered with intramuscular instillations of adeno-associated vectors. *Hum. Gene Ther.* **11**, 777–790.

Stemmer, W. P. (1994). DNA shuffling by random fragmentation and reassembly: In vitro recombination for molecular evolution. *Proc. Natl. Acad. Sci. USA* **91**, 10747–10751.

Thomas, C. E., Storm, T. A., Huang, Z., and Kay, M. A. (2004). Rapid uncoating of vector genomes is the key to efficient liver transduction with pseudotyped adeno-associated virus vectors. *J. Virol.* **78,** 3110–3122.

Thomas, D. L., Wang, L., Niamke, J., Liu, J., Kang, W., Scotti, M. M., Ye, G. J., Veres, G., and Knop, D. R. (2009). Scalable recombinant adeno-associated virus production using recombinant herpes simplex virus type 1 coinfection of suspension-adapted mammalian cells. *Hum. Gene Ther.* **20,** 861–870.

Urabe, M., Ding, C., and Kotin, R. M. (2002). Insect cells as a factory to produce adeno-associated virus type 2 vectors. *Hum. Gene Ther.* **13,** 1935–1943.

Urabe, M., Nakakura, T., Xin, K. Q., Obara, Y., Mizukami, H., Kume, A., Kotin, R. M., and Ozawa, K. (2006). Scalable generation of high-titer recombinant adeno-associated virus type 5 in insect cells. *J. Virol.* **80,** 1874–1885.

Vandenberghe, L. H., Wang, L., Somanathan, S., Zhi, Y., Figueredo, J., Calcedo, R., Sanmiguel, J., Desai, R. A., Chen, C. S., Johnston, J., Grant, R. L., Gao, G., et al. (2006). Heparin binding directs activation of T cells against adeno-associated virus serotype 2 capsid. *Nat. Med.* **12,** 967–971.

Wagner, J. A., Reynolds, T., Moran, M. L., Moss, R. B., Wine, J. J., Flotte, T. R., and Gardner, P. (1998). Efficient and persistent gene transfer of AAV-CFTR in maxillary sinus. *Lancet* **351,** 1702–1703.

Wagner, J. A., Messner, A. H., Moran, M. L., Daifuku, R., Kouyama, K., Desch, J. K., Manley, S., Norbash, A. M., Conrad, C. K., Friborg, S., Reynolds, T., Guggino, W. B., et al. (1999a). Safety and biological efficacy of an adeno-associated virus vector-cystic fibrosis transmembrane regulator (AAV-CFTR) in the cystic fibrosis maxillary sinus. *Laryngoscope* **109,** 266–274.

Wagner, J. A., Nepomuceno, I. B., Shah, N., Messner, A. H., Moran, M. L., Norbash, A. M., Moss, R. B., Wine, J. J., and Gardner, P. (1999b). Maxillary sinusitis as a surrogate model for CF gene therapy clinical trials in patients with antrostomies. *J. Gene Med.* **1,** 13–21.

Walters, R. W., Agbandje-McKenna, M., Bowman, V. D., Moninger, T. O., Olson, N. H., Seiler, M., Chiorini, J. A., Baker, T. S., and Zabner, J. (2004). Structure of adeno-associated virus serotype 5. *J. Virol.* **78,** 3361–3371.

Walz, C., Deprez, A., Dupressoir, T., Durst, M., Rabreau, M., and Schlehofer, J. R. (1997). Interaction of human papillomavirus type 16 and adeno-associated virus type 2 co-infecting human cervical epithelium. *J. Gen. Virol.* **78**(Pt 6), 1441–1452.

Wang, L., Nichols, T. C., Read, M. S., Bellinger, D. A., and Verma, I. M. (2000). Sustained expression of therapeutic level of factor IX in hemophilia B dogs by AAV-mediated gene therapy in liver. *Mol. Ther.* **1,** 154–158.

Wang, J., Xie, J., Lu, H., Chen, L., Hauck, B., Samulski, R. J., and Xiao, W. (2007). Existence of transient functional double-stranded DNA intermediates during recombinant AAV transduction. *Proc. Natl. Acad. Sci. USA* **104,** 13104–13109.

Wu, Z., Sun, J., Zhang, T., Yin, C., Yin, F., Van Dyke, T., Samulski, R. J., and Monahan, P. E. (2008). Optimization of self-complementary AAV vectors for liver-directed expression results in sustained correction of hemophilia B at low vector dose. *Mol. Ther.* **16,** 280–289.

Wu, Z., Yang, H., and Colosi, P. (2010). Effect of genome size on AAV vector packaging. *Mol. Ther.* **18,** 80–86.

Xiao, X., Li, J., and Samulski, R. J. (1998). Production of high-titer recombinant adeno-associated virus vectors in the absence of helper adenovirus. *J. Virol.* **72,** 2224–2232.

Xie, Q., Bu, W., Bhatia, S., Hare, J., Somasundaram, T., Azzi, A., and Chapman, M. S. (2002). The atomic structure of adeno-associated virus (AAV-2), a vector for human gene therapy. *Proc. Natl. Acad. Sci. USA* **99,** 10405–10410.

Xie, Q., Ongley, H. M., Hare, J., and Chapman, M. S. (2008). Crystallization and preliminary X-ray structural studies of adeno-associated virus serotype 6. *Acta Crystallogr. Sect. F Struct. Biol. Cryst. Commun.* **64,** 1074–1078.

Yang, L., Jiang, J., Drouin, L. M., Agbandje-McKenna, M., Chen, C., Qiao, C., Pu, D., Hu, X., Wang, D. Z., Li, J., and Xiao, X. (2009). A myocardium tropic adeno-associated virus (AAV) evolved by DNA shuffling and in vivo selection. *Proc. Natl. Acad. Sci. USA* **106,** 3946–3951.

Zhong, L., Zhao, W., Wu, J., Li, B., Zolotukhin, S., Govindasamy, L., Agbandje-McKenna, M., and Srivastava, A. (2007). A dual role of EGFR protein tyrosine kinase signaling in ubiquitination of AAV2 capsids and viral second-strand DNA synthesis. *Mol. Ther.* **15,** 1323–1330.

Zhong, L., Li, B., Jayandharan, G., Mah, C. S., Govindasamy, L., Agbandje-McKenna, M., Herzog, R. W., Weigel-Van Aken, K. A., Hobbs, J. A., Zolotukhin, S., Muzyczka, N., and Srivastava, A. (2008a). Tyrosine-phosphorylation of AAV2 vectors and its consequences on viral intracellular trafficking and transgene expression. *Virology* **381,** 194–202.

Zhong, L., Li, B., Mah, C. S., Govindasamy, L., Agbandje-McKenna, M., Cooper, M., Herzog, R. W., Zolotukhin, I., Warrington, K. H., Jr., Weigel-Van Aken, K. A., Hobbs, J. A., Zolotukhin, S., et al. (2008b). Next generation of adeno-associated virus 2 vectors: Point mutations in tyrosines lead to high-efficiency transduction at lower doses. *Proc. Natl. Acad. Sci. USA* **105,** 7827–7832.

Zincarelli, C., Soltys, S., Rengo, G., and Rabinowitz, J. E. (2008). Analysis of AAV serotypes 1–9 mediated gene expression and tropism in mice after systemic injection. *Mol. Ther.* **16,** 1073–1080.

CHAPTER THIRTEEN

Gene Delivery to the Retina: From Mouse to Man

Jean Bennett,[*,†] Daniel C. Chung,[*,†] and Albert Maguire[*,†]

Contents

1. Introduction	256
2. Nucleic Acid Delivery to the Outer Retina in the Mouse	257
2.1. Surgical procedures in the mouse	258
2.2. Determination of retinal health/transduction outcome post surgery	260
3. Nucleic Acid Delivery to the Outer Retina in the Dog	261
3.1. Surgical procedures in the dog	261
3.2. Determination of transduction outcome	264
4. Nucleic Acid Delivery to the Outer Retina in the Non-Human Primate	265
4.1. Surgical procedures in the non-human primate	265
4.2. Determination of transduction outcome	267
5. Nucleic Acid Delivery to the Outer Retina in the Human	267
5.1. Surgical procedures in the human	267
5.2. Determination of transduction outcome	270
Acknowledgments	270
References	270

Abstract

With the recent progress in identifying disease-causing genes in humans and in animal models, there are more and more opportunities for using retinal gene transfer to learn more about retinal physiology and also to develop therapies for blinding disorders. Success in preclinical studies for one form of inherited blindness have led to testing in human clinical trials. This paves the way to consider a number of other retinal diseases as ultimate gene therapy targets in human studies. The information presented here is designed to assist scientists and clinicians to use gene transfer to probe the biology of the retina and/or to move appropriate gene-based treatment studies from the bench to the clinic.

[*] F.M. Kirby Center for Molecular Ophthalmology, Scheie Eye Institute, University of Pennsylvania, Philadelphia, Pennsylvania, USA
[†] The Center for Cellular and Molecular Therapeutics, The Children's Hospital of Philadelphia, Philadelphia, Pennsylvania, USA

1. Introduction

Because of its ease of access, its benign immunologic response to gene transfer, and the ability to perform noninvasive functional and structural studies, the mammalian eye is an ideal target for gene transfer. Gene transfer has been used to learn about varied topics such as the development and differentiation of the retina and the nature of the suppressive immune response in this tissue. Gene augmentation strategies whereby a wild type copy of a gene is delivered have also been used successfully in proof-of-concept gene therapy studies in small and large animal models of more than a dozen different conditions (Acland et al., 2001, 2005; Alexander et al., 2007; Ali et al., 2000; Allocca et al., 2008, 2011; Andrieu-Soler et al., 2007; Batten et al., 2005; Bennett et al., 1996; Bennicelli et al., 2008; Boye et al., 2010; Cai et al., 2009, 2010; Carvalho et al., 2011; Conley and Naash, 2010; Dejneka et al., 2004; Gargiulo et al., 2009; Georgiadis et al., 2010; Ho et al., 2002; Janssen et al., 2008; Kjellstrom et al., 2007; Kong et al., 2008; Kumar-Singh and Chamberlain, 1996; Mancuso et al., 2009; Mao et al., 2011; Michalakis et al., 2010; Mihelec et al., 2011; Min et al., 2005; Narfstrom et al., 2003a,b,c; Pang et al., 2006, 2008, 2010a,b, 2011; Pawlyk et al., 2005; Sarra et al., 2001; Simons et al., 2011; Sun et al., 2011; Surace et al., 2005; Takahashi et al., 1999; Tan et al., 2009; Williams et al., 2006; Zeng et al., 2004; Zou et al., 2011). In addition, there has been success with strategies aimed at rescuing disease due to toxic gain-of-function mutations (Chadderton et al., 2009; LaVail et al., 2000; Lewin et al., 1998; Millington-Ward et al., 2011; Mussolino et al., 2011; O'Reilly et al., 2007; Palfi et al., 2006; Tam et al., 2008, 2010).

The expansion of the "toolkit" of vectors that can be used to deliver nucleic acids to retinal cells in recent years will enhance these opportunities. There are now a number of promising physicochemical (nonviral) reagents as well as recombinant virus vectors available for studies of the retina. New recombinant viral vectors contain modifications of capsids, envelopes, and surface proteins designed to achieve the desired transduction parameters. A large number of these will continue to be useful to evaluate a variety of biochemical, cell biological, developmental, immunologic, physiologic, and therapeutic parameters involving wild-type and mutant retinal proteins. Some of them may also be useful in generating new animal models via somatic gene transfer.

At this point in time, the vast majority of vectors that have been tested have a very limited ability to diffuse across tissue interfaces. Thus, to deliver nucleic acids to the outer retina (photoreceptors and retinal pigment epithelium (RPE) cells), it is necessary to carry out a subretinal injection. Expertise in delivering vectors to the outer retina is not widely available, particularly for large animal models and humans. Therefore, we describe the procedures for carrying out subretinal injections in small and large animal models and ultimately in humans. All of the studies described in this chapter assume that the investigator obtains the appropriate institutional and federal approvals (including rDNA approvals) before carrying out such studies. They also assume that the investigator has an appropriately trained assistant. The investigator should also be sure to use the minimum number of animals/subjects to obtain statistically significant results, all instruments that come in contact with the eye should be sterile, investigators should have the appropriate qualifications, rigorous safety studies in large animal models should be carried out before testing retinal gene transfer in humans, informed consent should be obtained from human subjects enrolling in clinical trials, and animals/subjects should be appropriately anesthetized and be given the appropriate postoperative care.

2. Nucleic Acid Delivery to the Outer Retina in the Mouse

In this section, we provide procedures for delivering nucleic acids to the outer retina of the mouse (Bennett *et al.*, 1996; Liang *et al.*, 2000). The surgical approach to subretinal injection depends mainly on the size of the eye. In the mouse, the relative volume of the vitreous space occupied by the crystalline lens is large (Fig. 13.1). Thus, it is difficult to introduce a cannula using an anterior approach as there is a high likelihood of damaging the lens and causing damage that will lead to a cataract and/or inflammation. Therefore, a posterior approach is used, in which an incision is made across the posterior sclera and choroid, and a cannula is placed in the subretinal space. An assistant pushes the plunger of the injection syringe when signaled

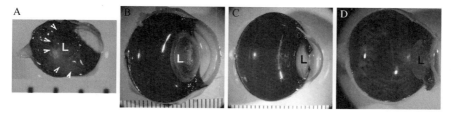

Figure 13.1 The space occupied by the lens relative to the vitreous cavity differs across species and dictates the surgical approach. In the mouse (A), the lens (L) occupies the majority of the cavity and so injections are usually carried out with a trans-choroidal approach. In larger animals ((B) dog, (C) monkey) and (D) humans, the lens is much smaller relative to the rest of the eye and so injections can be carried out from an anterior approach under direction visualization. Arrowheads in (A) indicate the borders of the lens. Distance between each bar in (A)–(C) is 1 mm.

by the surgeon. The injection is not performed by direct visualization as the pupil is rotated away from the surgeon in order to expose the posterior part of the globe. Accuracy of the injection should be assessed after the procedure.

2.1. Surgical procedures in the mouse

2.1.1. Required materials

Devices and materials of the injection apparatus

Dissecting microscope (15×) (e.g., Nikon SMZU microscope, Optical Apparatus, Ardmore, PA)
Heating pad
Fiber optic light source
10 μl Hamilton syringe with 33-gauge blunt needle (e.g., Hamilton #801RN and Hamilton #79633, Baxter Scientific Products, Edison, NJ)

Additional materials

Indirect ophthalmoscope with 78 or 90D lens; ideally equipped with a blue filter (e.g., Keeler Vintage)
Vector of interest, stored on wet ice, purified, and in a sterile container (e.g., sterile 1.5 ml Eppendorf tube)
Vannas iridotomy scissors (e.g., catalog #RS-5610, Roboz Surgical, Rockville, MD)
Jeweler's forceps (e.g., Dumont #5, Roboz Surgical)
Hand rest (e.g., foam block, towel)

Other reagents

Dilating agents: 1.0% (w/v) tropicamide, topical (Alcon, Fort Worth, TX)
Analgesic: for example, Meloxicam 5 mg/kg SC once at the time of surgery; acetaminophen (1.6 mg/ml) in drinking water for 3 days prior to and 2–3 days following surgery
Anesthetic: (subject to approval of Institutional Animal Care and Use Committee and to controlled substance approvals) Ketamine/Xylazine (100 mg/kg/10 mg/kg IM) or isoflurane 1–5% (decreased following initial dose depending on the degree of sedation); topical—proparacaine HCl 0.5%
Povidone-iodine 5% (Betadine 5%, Escalon Ophthalmics, Skillman, NJ)
Wratten 47B gelatin excitation filter (Kodak, Rochester, NY)
PredG ointment (prednisolone acetate–gentamicin, 0.3%/0.6%, Allergan Pharmaceuticals, Irvine, CA)
Phosphate-buffered saline (PBS)
Disinfectant: 10% bleach (i.e., 0.525 g sodium hypochlorite/100 ml) or cidex

Disposables

Personal protective apparel (gloves, mask, gowns, booties)
Surgical tape
30-gauge 0.5-in. needles

2.1.2. Injection

Starting 3 days prior to injection, the animal is treated with acetaminophen in drinking water. Just prior to surgery, the animal is anesthetized and the pupils are dilated with 0.5% tropicamide. The anesthetized animal is positioned under the microscope with surgical tape. All procedures are done aseptically using sterile instruments, surgical fields, and solutions. One drop of betadine solution is placed in the fornix and ocular adnexa. A conjunctival incision is made parallel to the base of the cornea, just anterior to the equator of the eye. The eye is rotated by grasping the conjunctival flap near the cornea. The topical anesthetic, ophthetic (proparacaine HCl 0.5%), is applied (1 drop OU). A sclerotomy is made by advancing a 27-gauge needle so that the bevel just enters the vitreous cavity. A 33-gauge needle connected to a Hamilton syringe is inserted through the sclerotomy 2mm posterior to the temporal limbus. The cannula is then advanced tangential to the curvature of the globe to the subretinal space in the posterior pole. One microliter of transfection solution (e.g., purified recombinant adeno-associated virus) is injected into the subretinal space with the assistant pushing the plunger. (There is a Hamilton syringe under development that can be operated by a single person, if an assistant is unavailable.) The needle/plunger is held in place for 5s following injection in order to minimize reflux from the injection site. A successful subretinal injection raises a dome-shaped retinal detachment (bleb, Fig. 13.2), apparent immediately after injection when viewed with an

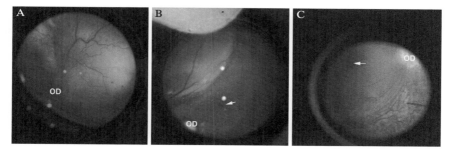

Figure 13.2 Appearance of the "bleb" immediately following subretinal injection in (A) dog, (B) non-human primate (NHP), and (C) human. OD, optic disc; arrow indicates the fovea in the NHP and the human. Panel (C) was taken from an intraoperative video recording. (See Color Insert.)

indirect ophthalmoscope. The detachment covers only a small fraction (~1/5) of the retina. The solution is not drained but is resorbed within a few hours by the retina.

Each retina undergoes only one injection. A localized retinal detachment indicates a successful injection. With practice, an investigator can expect to achieve accurate subretinal injections in ~80% of the animals.

The eyes are dressed with PredG ointment immediately following the surgery so that the cornea does not dessicate. The animals are attended and are kept warm through recovery on a heating pad. Animals are provided appropriate postoperative analgesia (meloxicam and acetaminophen). Animals are monitored for potential (although rare) postoperative complications including infection. The investigators will be prepared to treat a rodent or terminate the experiment for any particular animals that appears to be in distress. In our experience, the infection rate and mortality rate for this procedure is low.

2.2. Determination of retinal health/transduction outcome post surgery

Indirect ophthalmoscopy is useful for assessing retinal health serially over time and can be carried out without anesthesia. Expression of certain reporter genes (e.g., green fluorescent protein (GFP)) can be appreciated through indirect ophthalmoscopy if the ophthalmoscope is equipped with a cobalt blue filter (Fig. 13.2; Bennett et al., 1997). If not, a Wratten filter can be taped over the light source in order to excite the GFP at the appropriate wavelength. Additional imaging will depend upon availability of specialized equipment. The following noninvasive imaging protocols are provided as examples. In all of these, the retinas are first dilated with tropicamide: (1) animals injected with vectors carrying lucifererase can be evaluated with an *In Vivo* Imaging System (IVIS; XENOGEN, Caliper Life Sciences, Hopkinton, MA) after i.p. injection of the luciferase substrate, luciferin (150 mg/kg); (2) the retina can be evaluated through optical coherence tomography (OCT) in order to determine the thicknesses of the various retinal cell layers. Animals are positioned in front of the OCT probe (e.g., Spectral Domain Imaging System, Bioptigen, Inc., Durham, NC) and the retina is brought into focus. Images are collected; (3) fundus photographs may be obtained if an appropriate camera is available (e.g., Micron III fundus camera, Phoenix Research Laboratories, Inc., Pleasanton, CA). For this, the animals are held under the photographic lens so that the retina can be viewed through the fundus camera. Photographs are taken. The procedure is quick (generally 1 min per eye for a trained investigator) and does not require anesthesia. Sterile eye lubricant (saline or Artificial Tears) can be applied to the corneas topically as needed. After any of these procedures, the corneas can be dressed with 1 cm of PredG ointment to alleviate dryness.

The investigator should make sure that the animals do not become hypothermic during any imaging procedures incorporating anesthesia as there is a tendency for this to induce a (temporary) cataract which will limit visualization of the retina.

3. Nucleic Acid Delivery to the Outer Retina in the Dog

3.1. Surgical procedures in the dog

In this section, we provide procedures for delivering nucleic acids to the outer retina of the dog. The same procedures would be effective in other similar-sized animals (e.g., cats, rabbits). In these large animals, the relative volume of the vitreous space occupied by the crystalline lens is small (Fig. 13.1). This allows instruments to be introduced under direct visualization with an anterior approach (Acland *et al.*, 2001, 2005; Amado *et al.*, 2010; Bennicelli *et al.*, 2008). A transchoroidal approach is possible, but it requires manipulations through the choriocapillaris layer that do not allow direct visualization. It is advisable to avoid introduction of potentially immunogenic material in this region of high blood flow (vascularity). Further, the injection procedure cannot be visualized using a transchoroidal approach. In contrast, an anterior approach allows one to assess accuracy of the injection during and immediately after the procedure. There is no need in the dog to carry out a vitrectomy prior to performing the injection as long as an equivalent amount of fluid (as that to be injected) is removed from the anterior segment. The injection is achieved by delivery through a small retinotomy with a 39-gauge cannula. As with the mouse, a localized retinal detachment is generated, raising a "bleb" (Fig. 13.2). Most, if not all the volume, of injected material is trapped between the outer retina and RPE in this scenario. Thus, cells expressing the transgene are in this region of the retina (Fig. 13.3). There is negligible escape of material back through the retinotomy site into the vitreous as evidenced by the fact that the size of the bleb does not change once it is formed. It is possible to use an even smaller diameter (i.e., 41-gauge) cannula for subretinal injections, but it is more difficult to maneuver this flexible device through the semisolid vitreous.

3.1.1. Required materials

Devices and materials of the microinjection apparatus

Operating microscope (e.g., Zeiss operating microscope)
Heating pad

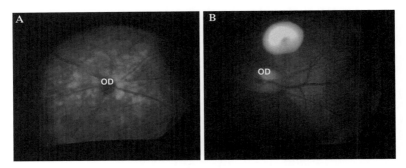

Figure 13.3 Green fluorescent protein (GFP) is visible through illumination with blue light with an ophthalmoscope in animals after subretinal injection of a recombinant adeno-associated virus vector delivering the gene encoding GFP. The GFP is below the layer of the inner retinal vessels, which appear dark. (A) Mouse; (B) non-human primate; OD, optic disc. (For interpretation of the references to color in this figure legend, the reader is referred to the web version of this chapter.)

Innorex 25/39-gauge subretinal injection syringe (Surmodics, Edina, MN). This unit is not FDA approved, but contains all of the elements needed for subretinal injection in 1 unit. Alternatively, the reagents listed below for human surgery (Section 5.1.1) can be used.
Hand rest (e.g., foam block, towel)

Additional materials

Indirect ophthalmoscope with 20 or 28D lens; ideally equipped with a blue filter
Vector of interest, stored on wet ice, purified, and in a sterile container (e.g., sterile 1.5 ml Eppendorf tube)

Other reagents

Sedation: acepromazine, 0.1–0.3 mg/kg IM then atropine 0.03 mg/kg IM
Anesthetic (subject to approval of Institutional Animal Care and Use Committee and to controlled substance approvals): Induction (after aseptic preparation of the site of venipuncture) will be with propofol (2 mg/kg IV). This will be followed by intubation and ventilation with 2.5% isofluorane.
Povidone-iodine 5% (Betadine 5%, Escalon Ophthalmics)
Eye drops:
1.0% (w/v) tropicamide (Alcon)
Mydfin (phenylephrine hydrochloride) 2.5%
Flurbiprofen, 0.3% ophthalmic solution
Ciprofloxacin, 0.3% ophthalmic solution
Analgesic: topical—proparacaine HCl 0.5%

Artificial Tears: Systane lubricant, Alcon
PredG ointment (prednisolone acetate–gentamicin, 0.3%/0.6%, Allergan Pharmaceuticals)
Gonak/Goniosol solution (hydrocellulose; Akorn, Inc., Lake Forest, IL)
Wescote scissors
Bishop forceps
Steven's scissors
Hemostat
Lid speculum
2.5% fluorescein in PBS
Wratten 47B gelatin excitation filter (Kodak)
Phosphate-buffered saline
Disinfectant: 10% bleach (i.e., 0.525 g sodium hypochlorite/100 ml)

Disposables

Corneal contact lens (e.g., flat Machemer lens)
Surgical tape
Hi Temperature Ophthalmic Cautery
1 cc syringe with 27-gauge needle (e.g., tuberculin (tb) syringe with removable needle)
30-gauge needle
5 cc syringe with 1⅝-in. 25-gauge needle

3.1.2. Injection

All procedures are done aseptically using sterile instruments, surgical fields, and solutions. Anesthesia in the dog is achieved after pre-anesthesia with pre-acepromazine, 0.1–0.3 mg/kg IM and then atropine, 0.03 mg/kg IM. Induction (after aseptic preparation of the site of venipuncture) will be with propofol. This will be followed by intubation and ventilation with 2.5% isofluorane. This will be decreased to maintain anesthesia. It is estimated that the dog will be intubated a maximum of 20 min. Although isoflurane anesthesia is our preferred method, it is possible to perform the procedure using IV ketamine and diazepam.

Just prior to surgery (i.e., during the time that acepromazine is administered), the pupils are dilated with mydriacyl (tropicamide) 1% and mydfin (phenylephrine hydrochloride) 2.5%, and flurbiprofen, proparacaine, and ciprofloxacin drops are applied to the cornea. After the animal has been anesthetized, it is positioned chin down in the operating field and one drop of betadine solution is placed in the fornix and ocular adnexa. Five cc of sterile saline or local anesthetic solution is injected retro-orbitally with a 25-gauge needle to proptose the eye and prevent the nicatating membrane from covering the surgical field. Proptosis will prevent Bell's phenomenon, which prevents visualization in the dog, due to rotation of the eye from the straight ahead position. The fluid injected retro-orbitally prior to the procedure is resorbed within 1 h.

A speculum is used to hold the eyelids open. A drop of proparacaine may be added for topical anesthesia. A radial conjunctival incision is performed to expose the sclera incision site. This site is lightly treated with Cautery for hemostasis. The anterior chamber is tapped with a 30-gauge needle to remove ~0.1 µl of aqueous humor. It is advisable to bend the tip of the 30-gauge needle 30° in order to avoid damaging the lens or iris. The anterior chamber tap is carried out to make space for the volume of vector solution that will be injected under the retina. This fluid can be stored frozen and used later for other studies. A sclerotomy is then made by penetration of a 27-gauge needle into the geometric center of the globe. The site is ~3 mm posterior to the limbus in a quadrant accessible to the dominant (injecting) hand.

The cornea is kept moist through application of sterile PBS. The contact lens is coupled to the cornea with Goniosol. Visualization is achieved via coaxial illumination through an operating microscope. A Surmodics 30-gauge subretinal injection syringe is inserted through the sclerotomy site. Alternative devices may be used (e.g., the devices that are used in human surgery). The cannula is then advanced through the vitreous. The speculum can be removed at this point since the corneal contact lens holds the eyelids open. The cannula is advanced to the point where the retina is indented and draped over the tip. It may be easiest for visualization to perform the injection in the tapetal (superior) portion of the retina rather than the nontapetal region. Under microscopic control, 25–400 µl of the vector (which may contain 2.5% fluorescein for visualization) is injected into the subretinal space when the assistant is told to push the plunger on the syringe. This raises a dome-shaped retinal detachment (bleb; Fig. 13.2). The plunger is held in place for 5 s following injection in order to overcome any impedance through the small gauge cannula which delays egress of the fluid. The solution is resorbed within a few hours by the retina. The retinas can be examined with indirect ophthalmoscopy to verify location of the retinal detachment immediately after the injection has taken place. Photographs may be taken with a fundus camera. A subconjunctival injection of 0.15 ml of kenalog solution (40 mg/ml) is delivered. The cornea is dressed with PredG ointment and the animal is awoken.

3.2. Determination of transduction outcome

Similar to studies in mice, indirect ophthalmoscopy is useful for assessing canine retinal health serially over time after gene transfer. Ophthalmoscopy can be carried out (after dilating the pupil) without anesthesia. If the subretinal injection has been targeted for the tapetal retina, there will likely be a change in the reflectivity of this layer in the region that had been injected. This is likely due to injection-induced alterations in the crystalline structure of molecules in the tapetum and forms a convenient marker for identifying the region of retina that has been exposed to the vector. Additional imaging can include OCT and fundus photography. Again, these procedures can be carried out without anesthesia. Sterile eye lubricant (saline or Artificial Tears) can be applied to the corneas topically as needed. After any of these procedures, the corneas can be dressed with 1 cm of PredG ointment to alleviate dryness.

4. NUCLEIC ACID DELIVERY TO THE OUTER RETINA IN THE NON-HUMAN PRIMATE

4.1. Surgical procedures in the non-human primate

In this section, we provide procedures for delivering nucleic acids to the outer retina of the non-human primate (NHP). The procedure is similar to that used for dogs (Section 3) except that care is taken to avoid damage to the fovea. The fovea is a structure that is present only in primates (NHP and humans, Fig. 13.3). As with the dog, the instruments to be introduced under direct visualization with an anterior approach (and without vitrectomy) and accuracy of the injection can be assessed during and immediately after the procedure (Amado *et al.*, 2010; Bennett *et al.*, 1999; Lebherz *et al.*, 2005a,b; Vandenberghe *et al.*, 2011).

4.1.1. Required materials

Devices and materials of the injection

Operating microscope (e.g., Zeiss operating microscope)
Heating pad
Innorx 30-gauge subretinal injection syringe (Surmodics). This unit is not FDA approved, but contains all of the elements needed for subretinal injection in one unit. Alternatively, the devices listed below for human surgery (Section 5.1.1) can be used.

Additional materials

Indirect ophthalmoscope with 23 D lens; ideally equipped with a blue filter
Vector of interest, stored on wet ice, purified, and in a sterile container (e.g., sterile 1.5 ml Eppendorf tube)
Hand rest (e.g., foam block, towel)

Other reagents

Anesthetic (subject to approval of Institutional Animal Care and Use Committee and to controlled substance approvals): Induction (after aseptic

preparation of the site of venipuncture) will be with propofol. This will be followed by intubation and ventilation with 2.5% isofluorane. As an alternative, Dexmedetomidine (0.05–0.1 mg/kg IM) may be used.
Povidone-iodine 5% (Betadine 5%, Escalon Ophthalmics)
Eye drops:
1.0% (w/v) tropicamide (Alcon)
Mydfin (phenylephrine hydrochloride) 2.5%
Flurbiprofen, 0.3% ophthalmic solution
Ciprofloxacin, 0.3% ophthalmic solution
Analgesic: topical—proparacaine HCl 0.5%
Artificial Tears: Systane lubricant, Alcon
PredG ointment (prednisolone acetate–gentamicin, 0.3%/0.6%, Allergan Pharmaceuticals)
Gonak/Goniosol solution (hydrocellulose; Akorn, Inc.)
Analgesic: Flunixin meglumine (1.0 mg/kg IM once a day)
Wescote scissors
Bishop forceps
Steven's scissors
Hemostat
Phosphate-buffered saline
Lid speculum
Disinfectant: 10% bleach (i.e., 0.525 g sodium hypochlorite/100 ml)
Mold to immobilize the head (e.g., foam pillow with space carved out to rest the back of the head)
2.5% fluorescein in PBS

Disposables

Corneal contact lens (e.g., flat Machemer lens)
Specialized protective apparel for the investigators (face shields, biohazard suits)
Hi Temperature Cautery
1 cc syringe with 27-gauge needle (e.g., tb syringe with removable needle)
30-gauge needle
5 cc syringe with 1.5-in. 25-gauge needle

4.1.2. Injection

All procedures are done aseptically using sterile instruments, surgical fields, and solutions. Additional care should be taken to minimize the possibility of body contact with fluids from the NHP due to the dangers imposed by Simian B virus. For example, face shields and biohazard suits should be worn by the investigators. Anesthesia in the NHP can be achieved with Dexmedetomidine.

Retinal Gene Delivery

The pupils are dilated with mydriacyl (tropicamide) 1% and mydfin (phenylephrine hydrochloride) 2.5%, and flurbiprofen, proparacaine, and ciprofloxacin drops are applied to the cornea. The animal is positioned in the foam pillow so that its chin is facing up in the operating field. One drop of betadine solution is placed in the fornix and ocular adnexa. A drop of proparacaine is used for topical anesthesia. The remainder of the surgical procedure is carried out similarly as described above, for the dog (Section 3.1.2). After the subretinal injection has been completed, flunixin meglumine (1.0 mg/kg IM once a day) is delivered for analgesia. The cornea is dressed with PredG ointment and the animal is awoken.

4.2. Determination of transduction outcome

Similar to studies in dogs, indirect ophthalmoscopy is useful for assessing retinal health serially over time in the NHP. Ophthalmoscopy (and other imaging procedures) in the NHP should be carried out under anesthesia due to the challenges of having the animal cooperate. Additional imaging can include OCT and fundus photography. Pupil dilation will be required prior to all of these procedures. Sterile eye lubricant (saline or Artificial Tears) can be applied to the corneas topically as needed. After any of these procedures, the corneas can be dressed with 1 cm of PredG ointment to alleviate dryness.

5. NUCLEIC ACID DELIVERY TO THE OUTER RETINA IN THE HUMAN

5.1. Surgical procedures in the human

In this section, we provide procedures for delivering nucleic acids to the outer retina in human subjects (Maguire *et al.*, 2008, 2009). The procedure is similar to that used for NHP (Section 4; Fig. 13.2) except that special procedures are implemented to provide the surgeon with more control and to ensure the maximum amount with respect to the surgical procedure. These procedures include a three-port approach with a full vitrectomy and protection of the fovea during the procedure mediated by the dense liquid, Perfluoron. There is a wealth of experience with three port pars plana vitrectomy in human retinal surgery and the details of this procedure are not described here. The injection approach has been developed for other human subretinal applications. As with the other large animals, accuracy of the injection can be assessed both during (Fig. 13.2) and immediately after the procedure. Further, additional maneuvers such as fluid–gas exchange and laserpexy can be employed with the pars plana approach in order to manipulate the subretinal bleb or to manage potential complications.

5.1.1. Required materials

Devices and materials for the injection

Operating microscope (e.g., Zeiss operating microscope)
Good manufacturing processes (GMP) vector, stored in a cooler and delivered to the operating room in a sterile vessel (e.g., capped syringe)

Other reagents

Anesthetic (subject to the surgeon's preferences and IRB approval): For example, induction (after aseptic preparation of the site of venipuncture) can be with propofol. This can be followed by intubation and ventilation with 2.5% isofluorane.
Indirect ophthalmoscope with 23D lens
Povidone-iodine 5% (Betadine 5%, Escalon Ophthalmics)
Eye drops:
1.0% (w/v) tropicamide (Alcon)
Mydfin (phenylephrine hydrochloride) 2.5%
Flurbiprofen, 0.3% ophthalmic solution
Ciprofloxacin, 0.3% ophthalmic solution
Analgesic: topical—proparacaine HCl 0.5%
Artificial Tears: Systane lubricant, Alcon
PredG ointment (prednisolone acetate–gentamicin, 0.3%/0.6%, Allergan Pharmaceuticals)
Gonak/Goniosol solution (hydrocellulose; Akorn, Inc.)
Wescote scissors
Bishop forceps
Steven's scissors
Hemostat
Phosphate-buffered saline
Lid speculum
Perfluorooctane liquid (Perfluoron, Alcon)

Disposables

Corneal contact lens (e.g., flat Machemer lens)
Surgical gowns, masks, gloves, drapes, etc.
Hi Temperature Cautery
Suture

5.1.2. Injection

All procedures are done using standard surgical procedures (i.e., sterile instruments, surgical fields, supplies, and solutions). It may be advisable to provide a short course of oral steroids just prior to and following the procedure to minimize inflammation due to the surgical procedure (Maguire et al., 2008, 2009).

General anesthesia or monitored sedation anesthetic technique can be used according to the surgeon's/patient's preferences. The pupil is dilated with mydriacyl (tropicamide) 1% and mydfin (phenylephrine hydrochloride) 2.5%, and flurbiprofen, proparacaine, and ciprofloxacin drops are applied to the cornea. The subject is positioned supine in the operating field. One drop of betadine solution is placed in the fornix and ocular adnexa.

The eye is stabilized with a retrobulbar injection of 4.0 ml marcaine (0.25%) and 1.0 ml triamcinolone acetonide (40 mg/ml). Standard techniques for vitreo-retinal surgery are used. The surgery can be recorded intraoperatively with a video device, if that is available. A three port pars plana vitrectomy is performed with removal of posterior cortical vitreous. If epiretinal membranes are observed, they should be removed (Maguire et al., 2009).

The fovea should be buttressed from hydrodynamic stress during injection with perfluorooctane (PFO) liquid (Perfluoron, Alcon), which is heavier than water. A 21-guage silicone-tipped cannula is used to inject a small amount (~0.2 ml) of PFO liquid over the macula. This will coalesce into a small bubble over the fovea. The subretinal injection cannula is placed no closer than 3000 μm from the foveal center. Injection in eyes with advanced degeneration is typically easier if the tip is placed in the vicinity of the papillomacular bundle between the fovea and disc, as the retina is thicker in this area. Even in eyes with advanced retinal degeneration, the retina in this area is usually thick enough to allow for successful placement of the cannula tip and the injection into the subretinal space. A Bausch & Lomb Storz 39-gauge translocation cannula (San Dimas, CA) is used. The cannula tip is positioned so as to avoid direct injury to retinal arterioles. After lowering infusion pressure, the surgeon asks the assistant to inject a small amount of material to be sure that the injection is going subretinally. When subretinal delivery is observed, the remainder of the injection solution is injected into the subretinal space. Under microscopic control, 100–400 μl of the GMP grade vector is injected into the subretinal space when the assistant is told to push the plunger on the syringe. This creates a localized dome-shaped retinal detachment. The cannula/plunger is held in place for 5 s after the injection to minimize reflux of the injection material. The PFO liquid is then aspirated. A 50% fluid–air exchange is then performed prior to closure of incisions, carefully avoiding draining through the retinotomy created for the subretinal injection. Air exchange is principally done to compartmentalize the subretinal injection so that the vector does not come into contact with anterior uveal structures and remains central to the area of the retina–RPE.

The subject is recovered from anesthesia but is positioned in the postoperative period to orient the eye so that the desired area of retinal exposure is in the most dependent position while the subretinal injection fluid is resorbed and the retina reattaches. The sclera incisions are closed.

5.2. Determination of transduction outcome

As in the other species, indirect ophthalmoscopy is useful for assessing retinal health serially over time in humans. It is often difficult to appreciate the borders of the original injection site following subretinal injection in the human. Usually the retinotomy site can be identified as a small atrophic region. Additional imaging can include OCT, autofluorescence imaging, fundus photography, and adaptive optics scanning laser ophthalmoscopy, depending upon the data desired, equipment availability and expertise of the personnel. Pupil dilation will be required prior to all of these procedures. Sterile eye lubricant (Artificial Tears) can be applied to the corneas topically as needed. After any of these procedures, the corneas can be dressed with 1 cm of PredG ointment to alleviate dryness.

ACKNOWLEDGMENTS

We are grateful for support from NIH (Pioneer Award 1DP1OD008267-01 and 1R24EY019861-01A1), the University of Pennsylvania, Institute for Translational Medicine and Experimental Therapeutics (ITMAT), Choroideremia Research Foundation, Foundation Fighting Blindness, the Center for Cellular and Molecular Therapeutics at The Children's Hospital of Philadelphia, Research to Prevent Blindness, the Grousbeck Family Foundation, Paul and Evanina Mackall Foundation Trust, and the F.M. Kirby Center for Molecular Ophthalmology.

REFERENCES

Acland, G. M., Aguirre, G. D., Ray, J., et al. (2001). Gene therapy restores vision in a canine model of childhood blindness. *Nat. Genet.* **28**, 92–95.

Acland, G. M., Aguirre, G. D., Bennett, J., et al. (2005). Long-term restoration of rod and cone vision by single dose rAAV-mediated gene transfer to the retina in a canine model of childhood blindness. *Mol. Ther.* **12**, 1072–1082.

Alexander, J. J., Umino, Y., Everhart, D., et al. (2007). Restoration of cone vision in a mouse model of achromatopsia. *Nat. Med.* **13**, 685–687.

Ali, R., Sarra, G.-M., Stephens, C., et al. (2000). Restoration of photoreceptor ultrastructure and function in retinal degeneration slow mice by gene therapy. *Nat. Genet.* **25**, 306–310.

Allocca, M., Doria, M., Petrillo, M., et al. (2008). Serotype-dependent packaging of large genes in adeno-associated viral vectors results in effective gene delivery in mice. *J. Clin. Invest.* **118**, 1955–1964.

Allocca, M., Manfredi, A., Iodice, C., et al. (2011). AAV-mediated gene replacement either alone or in combination with physical and pharmacological agents results in partial and transient protection from photoreceptor degeneration associated with {beta}PDE deficiency. *Invest. Ophthalmol. Vis. Sci.* **52**(8), 5713–5719.

Amado, D., Mingozzi, F., Hui, D., et al. (2010). Safety and efficacy of subretinal re-administration of an AAV2 vector in large animal models: Implications for studies in humans. *Sci. Transl. Med.* **2**, 21ra16.

Andrieu-Soler, C., Halhal, M., Boatright, J., et al. (2007). Single-stranded oligonucleotide-mediated in vivo gene repair in the rd1 retina. *Mol. Vis.* **13**, 682–706.

Batten, M. L., Imanishi, Y., Tu, D. C., et al. (2005). Pharmacological and rAAV gene therapy rescue of visual functions in a blind mouse model of Leber congenital amaurosis. *PLoS Med.* **2**, e333.

Bennett, J., Tanabe, T., Sun, D., et al. (1996). Photoreceptor cell rescue in retinal degeneration (*rd*) mice by in vivo gene therapy. *Nat. Med.* **2**, 649–654.

Bennett, J., Duan, D., Engelhardt, J. F., et al. (1997). Real-time, noninvasive in vivo assessment of adeno-associated virus-mediated retinal transduction. *Invest. Ophthalmol. Vis. Sci.* **38**, 2857–2863.

Bennett, J., Maguire, A. M., Cideciyan, A. V., et al. (1999). Stable transgene expression in rod photoreceptors after recombinant adeno-associated virus-mediated gene transfer to monkey retina. *Proc. Natl. Acad. Sci. USA* **96**, 9920–9925.

Bennicelli, J., Wright, J. F., Komaromy, A., et al. (2008). Reversal of blindness in animal models of leber congenital amaurosis using optimized AAV2-mediated gene transfer. *Mol. Ther.* **16**, 458–465.

Boye, S. E., Boye, S. L., Pang, J., et al. (2010). Functional and behavioral restoration of vision by gene therapy in the guanylate cyclase-1 (GC1) knockout mouse. *PLoS One* **5**, e11306.

Cai, X., Nash, Z., Conley, S. M., et al. (2009). A partial structural and functional rescue of a retinitis pigmentosa model with compacted DNA nanoparticles. *PLoS One* **4**, e5290.

Cai, X., Conley, S. M., and Naash, M. I. (2010). Gene therapy in the retinal degeneration slow model of retinitis pigmentosa. *Adv. Exp. Med. Biol.* **664**, 611–619.

Carvalho, L. S., Xu, J., Pearson, R. A., et al. (2011). Long-term and age-dependent restoration of visual function in a mouse model of CNGB3-associated achromatopsia following gene therapy. *Hum. Mol. Genet.* **20**(16), 3161–3175.

Chadderton, N., Millington-Ward, S., Palfi, A., et al. (2009). Improved retinal function in a mouse model of dominant retinitis pigmentosa following AAV-delivered gene therapy. *Mol. Ther.* **17**, 593–599.

Conley, S. M., and Naash, M. I. (2010). Nanoparticles for retinal gene therapy. *Prog. Retin. Eye Res.* **29**, 376–397.

Dejneka, N., Surace, E., Aleman, T., et al. (2004). Fetal virus-mediated delivery of the human RPE65 gene rescues vision in a murine model of congenital retinal blindness. *Mol. Ther.* **9**, 182–188.

Gargiulo, A., Bonetti, C., Montefusco, S., et al. (2009). AAV-mediated tyrosinase gene transfer restores melanogenesis and retinal function in a model of oculo-cutaneous albinism type I (OCA1). *Mol. Ther.* **17**, 1347–1354.

Georgiadis, A., Tschernutter, M., Bainbridge, J. W., et al. (2010). AAV-mediated knock-down of peripherin-2 in vivo using miRNA-based hairpins. *Gene Ther.* **17**, 486–493.

Ho, T. T., Maguire, A. M., Aguirre, G. D., et al. (2002). Phenotypic rescue after adeno-associated virus-mediated delivery of 4-sulfatase to the retinal pigment epithelium of feline mucopolysaccharidosis VI. *J. Gene Med.* **4**, 613–621.

Janssen, A., Min, S. H., Molday, L. L., et al. (2008). Effect of late-stage therapy on disease progression in AAV-mediated rescue of photoreceptor cells in the retinoschisin-deficient mouse. *Mol. Ther.* **16**, 1010–1017.

Kjellstrom, S., Bush, R. A., Zeng, Y., et al. (2007). Retinoschisin gene therapy and natural history in the Rs1h-KO mouse: Long-term rescue from retinal degeneration. *Invest. Ophthalmol. Vis. Sci.* **48**, 3837–3845.

Kong, J., Kim, S. R., Binley, K., et al. (2008). Correction of the disease phenotype in the mouse model of Stargardt disease by lentiviral gene therapy. *Gene Ther.* **15**, 1311–1320.

Kumar-Singh, R., and Chamberlain, J. S. (1996). Encapsidated adenovirus minichromosomes allow delivery and expression of a 14 kb dystrophin cDNA to muscle cells. *Hum. Mol. Genet.* **5**, 913–921.

LaVail, M. M., Yasumura, D., Matthes, M. T., et al. (2000). Ribozyme rescue of photoreceptor cells in P23H transgenic rats: Long-term survival and late stage therapy. *Proc. Natl. Acad. Sci. USA* **97**, 11488–11493.

Lebherz, C., Auricchio, A., Maguire, A. M., et al. (2005a). Long-term inducible gene expression in the eye via adeno-associated virus gene transfer in nonhuman primates. *Hum. Gene Ther.* **16**, 178–186.

Lebherz, C., Maguire, A., Auricchio, A., et al. (2005b). Non-human primate models for retinal neovascularization using AAV2-mediated overexpression of vascular endothelial growth factor. *Diabetes* **54**, 1141–1149.

Lewin, A. S., Drenser, K. A., Hauswirth, W. W., et al. (1998). Ribozyme rescue of photoreceptor cells in a transgenic rat model of autosomal dominant retinitis pigmentosa. *Nat. Med.* **4**, 967–971.

Liang, F.-Q., Anand, V., Maguire, A. M., et al. (2000). Intraocular delivery of recombinant virus. In "Methods in Molecular Medicine: Ocular Molecular Biology Protocols," (P. E. Rakoczy, ed.), pp. 125–139. Humana Press Inc, Totowa, NJ.

Maguire, A. M., Simonelli, F., Pierce, E. A., et al. (2008). Safety and efficacy of gene transfer for Leber's congenital amaurosis. *N. Engl. J. Med.* **358**, 2240–2248.

Maguire, A. M., High, K. A., Auricchio, A., et al. (2009). Age-dependent effects of RPE65 gene therapy for Leber's congenital amaurosis: A phase 1 dose-escalation trial. *Lancet* **374**, 1597–1605.

Mancuso, K., Hauswirth, W. W., Li, Q., et al. (2009). Gene therapy for red-green colour blindness in adult primates. *Nature* **461**(7265), 784–787.

Mao, H., James, T., Jr., Schwein, A., et al. (2011). AAV delivery of wild-type rhodopsin preserves retinal function in a mouse model of autosomal dominant retinitis pigmentosa. *Hum. Gene Ther.* **22**, 567–575.

Michalakis, S., Muhlfriedel, R., Tanimoto, N., et al. (2010). Restoration of cone vision in the CNGA3−/− mouse model of congenital complete lack of cone photoreceptor function. *Mol. Ther.* **18**, 2057–2063.

Mihelec, M., Pearson, R. A., Robbie, S. J., et al. (2011). Long-term preservation of cones and improvement in visual function following gene therapy in a mouse model of Leber congenital amaurosis (LCA) caused by GC1 deficiency. *Hum. Gene Ther.* **20**(16), 3161–3175.

Millington-Ward, S., Chadderton, N., O'Reilly, M., et al. (2011). Suppression and replacement gene therapy for autosomal dominant disease in a murine model of dominant retinitis pigmentosa. *Mol. Ther.* **19**, 642–649.

Min, S. H., Molday, L. L., Seeliger, M. W., et al. (2005). Prolonged recovery of retinal structure/function after gene therapy in an Rs1h-deficient mouse model of x-linked juvenile retinoschisis. *Mol. Ther.* **12**, 644–651.

Mussolino, C., Sanges, D., Marrocco, E., et al. (2011). Zinc-finger-based transcriptional repression of rhodopsin in a model of dominant retinitis pigmentosa. *EMBO Mol. Med.* **3**, 118–128.

Narfstrom, K., Bragadottir, R., Redmond, T. M., et al. (2003a). Functional and structural evaluation after AAV.RPE65 gene transfer in the canine model of Leber's congenital amaurosis. *Adv. Exp. Med. Biol.* **533**, 423–430.

Narfstrom, K., Katz, M. L., Bragadottir, R., et al. (2003b). Functional and structural recovery of the retina after gene therapy in the RPE65 null mutation dog. *Invest. Ophthalmol. Vis. Sci.* **44,** 1663–1672.

Narfstrom, K., Katz, M. L., Ford, M., et al. (2003c). In vivo gene therapy in young and adult RPE65−/− dogs produces long-term visual improvement. *J. Hered.* **94,** 31–37.

O'Reilly, M., Palfi, A., Chadderton, N., et al. (2007). RNA interference-mediated suppression and replacement of human rhodopsin in vivo. *Am. J. Hum. Genet.* **81,** 127–135.

Palfi, A., Ader, M., Kiang, A. S., et al. (2006). RNAi-based suppression and replacement of rds-peripherin in retinal organotypic culture. *Hum. Mutat.* **27,** 260–268.

Pang, J. J., Chang, B., Kumar, A., et al. (2006). Gene therapy restores vision-dependent behavior as well as retinal structure and function in a mouse model of RPE65 Leber congenital amaurosis. *Mol. Ther.* **13,** 565–572.

Pang, J. J., Boye, S. L., Kumar, A., et al. (2008). AAV-mediated gene therapy for retinal degeneration in the rd10 mouse containing a recessive PDEbeta mutation. *Invest. Ophthalmol. Vis. Sci.* **49,** 4278–4283.

Pang, J., Boye, S. E., Lei, B., et al. (2010a). Self-complementary AAV-mediated gene therapy restores cone function and prevents cone degeneration in two models of Rpe65 deficiency. *Gene Ther.* **17,** 815–826.

Pang, J. J., Alexander, J., Lei, B., et al. (2010b). Achromatopsia as a potential candidate for gene therapy. *Adv. Exp. Med. Biol.* **664,** 639–646.

Pang, J. J., Dai, X., Boye, S. E., et al. (2011). Long-term retinal function and structure rescue using capsid mutant AAV8 vector in the rd10 mouse, a model of recessive retinitis pigmentosa. *Mol. Ther.* **19,** 234–242.

Pawlyk, B. S., Smith, A. J., Buch, P. K., et al. (2005). Gene replacement therapy rescues photoreceptor degeneration in a murine model of Leber congenital amaurosis lacking RPGRIP. *Invest. Ophthalmol. Vis. Sci.* **46,** 3039–3045.

Sarra, G.-M., Stephens, C., de Alwis, M., et al. (2001). Gene replacement therapy in the retinal degeneration slow (*rds*) mouse: The effect on retinal degeneration following partial transduction of the retina. *Hum. Mol. Genet.* **10,** 2353–2361.

Simons, D. L., Boye, S. L., Hauswirth, W. W., et al. (2011). Gene therapy prevents photoreceptor death and preserves retinal function in a Bardet-Biedl syndrome mouse model. *Proc. Natl. Acad. Sci. USA* **108,** 6276–6281.

Sun, X., Pawlyk, B., Xu, X., et al. (2011). Gene therapy with a promoter targeting both rods and cones rescues retinal degeneration caused by AIPL1 mutations. *Gene Ther.* **17,** 117–131.

Surace, E. M., Domenici, L., Cortese, K., et al. (2005). Amelioration of both functional and morphological abnormalities in the retina of a mouse model of ocular albinism following AAV-mediated gene transfer. *Mol. Ther.* **12,** 652–658.

Takahashi, M., Miyoshi, H., Verma, I. M., et al. (1999). Rescue from photoreceptor degeneration in the rd mouse by human immunodeficiency virus vector-mediated gene transfer. *J. Virol.* **73,** 7812–7816.

Tam, L. C., Kiang, A. S., Kennan, A., et al. (2008). Therapeutic benefit derived from RNAi-mediated ablation of IMPDH1 transcripts in a murine model of autosomal dominant retinitis pigmentosa (RP10). *Hum. Mol. Genet.* **17,** 2084–2100.

Tam, L. C., Kiang, A. S., Chadderton, N., et al. (2010). Protection of Photoreceptors in a Mouse Model of RP10. *Adv. Exp. Med. Biol.* **664,** 559–565.

Tan, M., Smith, A., Pawlyk, B., et al. (2009). Gene therapy for retinitis pigmentosa and Leber congenital amaurosis caused by defects in AIPL1: Effective rescue of mouse models of partial and complete Aipl1 deficiency. *Hum. Mol. Genet.* **18,** 2099–2114.

Vandenberghe, L., Bell, P., Maguire, A., et al. (2011). Dosage thresholds for AAV2 and AAV8 photoreceptor gene therapy in monkey. *Sci. Transl. Med.* **3,** 88ra54.

Williams, M. L., Coleman, J. E., Haire, S. E., et al. (2006). Lentiviral expression of retinal guanylate cyclase-1 (RetGC1) restores vision in an avian model of childhood blindness. *PLoS Med.* **3,** e201.

Zeng, Y., Takada, Y., Kjellstrom, S., et al. (2004). RS-1 gene delivery to an adult Rs1h knockout mouse model restores ERG b-wave with reversal of the electronegative waveform of X-linked retinoschisis. *Invest. Ophthalmol. Vis. Sci.* **45,** 3279–3285.

Zou, J., Luo, L., Shen, Z., et al. (2011). Whirlin replacement restores the formation of the USH2 protein complex in whirlin knockout photoreceptors. *Invest. Ophthalmol. Vis. Sci.* **52,** 2343–2351.

CHAPTER FOURTEEN

Generation of Hairpin-Based RNAi Vectors for Biological and Therapeutic Application

Ryan L. Boudreau* *and* Beverly L. Davidson*,[†],[‡]

Contents

1. Introduction	276
2. Selecting Candidate siRNA Sequences	278
2.1. Retrieve sequences for your desired target transcript(s)	278
2.2. Select 22-nt long target sites	278
3. Cloning shRNA Expression Cassettes	280
3.1. Notes on siRNA sequence selection for shRNA application	280
3.2. Oligo design and cloning protocol for shRNA-tailed PCR	281
3.3. Materials for shRNA cloning	285
4. Cloning Artificial miRNA Expression Vectors	285
4.1. Considerations for designing artificial miRNAs	285
4.2. Oligo design and cloning protocol for artificial miRNAs	286
4.3. Materials for artificial miRNA cloning	291
5. Screening RNAi Vectors for Silencing Efficacy *In Vitro*	291
5.1. Screening against a cotransfected reporter constructs	292
5.2. Screening against endogenously expressed targets	293
6. Integration into Viral Vectors	293
7. Summary	294
References	294

Abstract

RNA interference (RNAi) is a natural process of gene silencing mediated by small RNAs. Shortly after the discovery of the RNAi mechanism, scientists devised various methods of delivering small interfering RNAs (siRNAs) capable of co-opting the endogenous RNAi machinery and suppressing target gene expression based on sequence complementarity. RNAi has since become a powerful tool to study gene function and is being investigated as a potential

---

* Department of Internal Medicine, University of Iowa, Iowa City, Iowa, USA
[†] Department of Neurology, University of Iowa, Iowa City, Iowa, USA
[‡] Department of Molecular Physiology & Biophysics, University of Iowa, Iowa City, Iowa, USA

therapeutic approach to treat a vast array of human diseases (e.g., cancer, viral infections, and dominant genetic disorders). Among the available RNAi vectors are hairpin-based expression platforms (short-hairpin RNAs and artificial micro-RNAs) designed to mimic endogenously expressed inhibitory RNAs. These RNAi vectors are capable of achieving long-term potent gene silencing *in vitro* and *in vivo*. Here, we describe methods to design and generate these hairpin-based vectors and briefly review considerations for downstream applications.

1. INTRODUCTION

RNA interference (RNAi) is an evolutionarily conserved cellular process involved in gene regulation and innate defense (Krol et al., 2010). RNAi directs sequence-specific gene silencing by double-stranded RNA (dsRNA) which is processed into functional small inhibitory RNAs (~21 nt) (Provost et al., 2002). In nature, small RNAs known as microRNAs (miRNAs) are the key mediators of this gene regulatory process (Krol et al., 2010). Mature miRNAs (~19–25 nts) are excised from stem-loop regions within larger primary miRNA transcripts (pri-miRNAs). A cascade of cleavage reactions catalyzed by the ribonucleases, Drosha–DGCR8 and Dicer (Gregory et al., 2004; Han et al., 2004; Lee et al., 2003), releases the miRNA duplex. A single strand (the antisense "guide" strand) subsequently enters the RNA-induced silencing complex (RISC) (Khvorova et al., 2003; Schwarz et al., 2003), thus producing a functional complex capable of base pairing with target transcripts and reducing their expression by various mechanisms. Perfect or near perfect binding along the length of the small RNA induces target transcript cleavage, whereas imperfect complementarity (typically to transcript 3′-UTRs) causes the canonical miRNA-based repression mechanism resulting in translational repression and mRNA destabilization (Guo et al., 2010; Lewis et al., 2005).

RNAi was initially discovered as a perplexing gene silencing observation in plants and worms (Ecker and Davis, 1986; Fire et al., 1998; Napoli et al., 1990). Over the past decade, our increased understanding of miRNA biogenesis and gene silencing mechanisms has facilitated the development of various strategies for co-opting the cellular RNAi machinery to direct specific silencing of virtually any gene. These RNAi-based technologies have become invaluable molecular tools to study gene function and are being investigated as therapeutic reagents for many human diseases. The potential to artificially induce gene silencing depends on our ability to design inhibitory RNAs that properly engage the RNAi machinery and to introduce them into target cells or tissues. The central RNAi effectors, known as small interfering RNAs (siRNAs), are designed to mimic mature miRNA duplexes, but with the guide strand exhibiting perfect complementarity to the intended target transcript to trigger the potent cleavage-based silencing mechanism. siRNAs are typically synthesized *in vitro* using modified bases for improved stability, specificity, and reduced immunostimulatory properties (Behlke, 2008). Upon entering the cell by endosomal uptake and escape or by electroporation, siRNAs engage the RNAi pathway at the Dicer-to-RISC stage. Alternatively, siRNAs can be incorporated into expression-based systems by embedding the sequences into stem-loop structures designed to mimic pre-miRNAs (short-hairpin RNAs or shRNAs) or pri-miRNAs (artificial miRNAs) (Paul *et al.*, 2002; Zeng *et al.*, 2002). shRNAs are generally expressed from vector systems, most often with strong, constitutive Pol III promoters (e.g., U6 and H1) (Paul *et al.*, 2002; Sui *et al.*, 2002). In order to mimic pre-miRNAs, shRNAs are transcribed as sense and antisense sequences connected by a loop of unpaired nucleotides. Following transcription, shRNAs are exported from the nucleus by Exportin-5 and, once in the cytoplasm, are cleaved by Dicer to generate functional siRNAs (Lund *et al.*, 2004; Provost *et al.*, 2002; Yi *et al.*, 2003). More recently, scientists have embedded siRNA sequences into molecular scaffolds designed to mimic pri-miRNAs which enter the RNAi pathway upstream of Drosha-DGCR8 (Chung *et al.*, 2006; Zeng *et al.*, 2002). These artificial miRNAs more naturally resemble endogenous RNAi substrates, thus improving the efficacy and accuracy of downstream processing events (Boden *et al.*, 2004; Silva *et al.*, 2005). Furthermore, since miRNA hairpins can be embedded within larger transcripts, this approach is more amenable to Pol II-based expression systems, providing enhanced potential for tissue-specific and inducible gene silencing. Expression-based RNAi vectors afford unique opportunities for employment of viral-based delivery systems, stable long-term gene suppression, and finer control of spatiotemporal silencing, among other related advantages associated with transgenic approaches (McJunkin *et al.*, 2011; Shin *et al.*, 2006).

With an array of RNAi triggers and delivery modalities available, researchers must decide which combination will yield a suitable inhibitory RNA dose to achieve potent and highly specific silencing in their experimental setting. Dose optimization remains a crucial consideration given the potential for RNAi treatments to induce cellular toxicity. High levels of exogenously supplied RNAi substrates may disrupt cellular function by saturating the RNAi machinery, thus interfering with natural miRNA-mediated gene regulation (Boudreau *et al.*, 2009a; Castanotto *et al.*, 2007; Grimm *et al.*, 2006). Also, artificial inhibitory RNAs have the potential to bind to and regulate unintended mRNA targets, an effect known as off-target gene silencing (Chi *et al.*, 2003; Jackson *et al.*, 2003; Semizarov *et al.*, 2003). Off-targeting primarily occurs when the seed region (nucleotides 2–8 of the small RNA) pairs with 3′-UTR sequences of unintended mRNAs and directs translational repression and destabilization of those transcripts, similar to the canonical miRNA-based silencing mechanism (Birmingham *et al.*, 2006;

Jackson et al., 2006). Together, these adverse events may have severe consequences; for example, Grimm et al. reported that high-level shRNA expression, from strong Pol III promoters, in mouse liver induced fatality (Grimm et al., 2006). Additional work from our laboratory identified that artificial miRNAs may have lower toxicity potentials. In comparison studies, we found that shRNAs are more potent but induce toxicity *in vitro* and *in vivo*, whereas artificial miRNAs are expressed at tolerably lower levels yet maintain potent gene silencing capacities (Boudreau et al., 2008, 2009a; McBride et al., 2008). Together, these results and others underscore the need to consider and monitor dosing in RNAi experiments. Controlling the dose of synthetic siRNAs is rather uncomplicated. By contrast, inhibitory RNA levels produced by expression-based systems are influenced by many factors (e.g., vector platforms, delivery modalities, promoter selection, hairpin structure, and availability of RNAi pathway components) which are likely to be unique to each experimental setting. Researchers should consider these factors in choosing an appropriate RNAi approach for their purposes.

Here, we provide methods to generate and screen hairpin-based RNAi vectors and discuss the downstream utility and considerations for these vectors in biological and therapeutic applications.

2. Selecting Candidate siRNA Sequences

2.1. Retrieve sequences for your desired target transcript(s)

Choose your target gene(s) of interest and retrieve the relevant mRNA sequence(s) from NCBI, Ensembl, UCSC Genome Browser, or other available databases. We leave it to the reader's responsibility to further investigate target transcripts for variants which may result from RNA processing events (e.g., alternative splicing, alternative polyadenylation, and RNA editing, among others) that may influence the ability to target certain regions within the transcript. In general, we target the coding region; however, targeting the 5′- and 3′-UTR sequences is possible. Careful consideration for the target sequence with regard to the project objectives is important before proceeding with design and screening of inhibitory RNA sequences. For instance, the reader should consider whether allele- or splice-isoform-specific silencing is desirable or whether sequence conservation is important, allowing the RNAi vectors to be tested in multiple species.

2.2. Select 22-nt long target sites

Identifying potent and highly specific siRNA sequences is not trivial. Numerous empirical evaluations of large-scale siRNA knockdown data have allowed researchers to establish several siRNA design guidelines

(Khvorova et al., 2003; Matveeva et al., 2007). For example, one key consideration is that siRNA sequences be selected or manipulated to promote accurate loading of the antisense guide strand into RISC, leaving the sense strand to be degraded (Leuschner et al., 2006; Matranga et al., 2005). Furthermore, GC-content and positional nucleotide preferences also influence siRNA efficacy. Given the multifaceted nature of designing optimal siRNAs, we direct the reader to additional literature on the subject (Birmingham et al., 2007; Davidson and McCray, 2011; Jackson and Linsley, 2010). Also, there are numerous publicly available siRNA design tools online. It is important to note that siRNA design rules serve more as guidelines, and that sequences adhering to them may not silence and vice versa. To date, no algorithm guarantees silencing efficacy, and most recommend the user pick three to four candidates for screening. Here, we will describe a basic strategy for siRNA target site selection which incorporates the most important criteria for promoting efficacy, in addition to certain rules that are specific to the design of hairpin-based RNAi expression vectors.

We identify 22-nt target sites within the target transcript that adhere to four criteria: (1) high propensity to primarily load the antisense guide strand into RISC, (2) GC content between 20% and 70%, (3) void of restriction enzyme sites relevant to downstream applications (e.g., cloning RNAi expression cassettes into viral vector systems), (4) lacking a stretch of four continuous A or T nucleotides (i.e., AAAA or TTTT). The latter prevents premature transcription termination from Pol III promoters which typically terminate at stretches of four to six T's. Strand biasing is determined by the thermodynamic stabilities present at the ends of the siRNA duplex. To achieve faithful loading of the antisense strand, the duplex must be designed such that there is strong G–C base pairing present at the $5'$-end of the sense (passenger) strand and weak A/G-U base paring at the opposing terminus (Khvorova et al., 2003; Schwarz et al., 2003). The RISC complex selects the strand with the weakest $5'$-end thermodynamic stability, in this case, the antisense strand (Fig. 14.1). Hence, we select target sequences that have G or C nucleotides at positions 3 and 4 (*note*: positions 1 and 2 correspond to the dinucleotide $3'$-overhang of the antisense strand) and A, T, or C nucleotides at positions 20, 21, and 22. The C is allowed at the latter positions because we can destabilize the resulting G–C base pairs by converting them to G–U base pairs (G and U weakly pair in RNA) without altering the antisense strand sequence. Next, the GC content of the 22-nt site is calculated by dividing the number of G or C nucleotides by 22. For instance, the target site shown in Fig. 14.1 has 12 G or C nucleotides, resulting in a GC content of 55% which is within the acceptable range. Finally, we avoid stretches of A's and T's and relevant restriction enzyme sites for reasons mentioned above. *Note*: the user must also avoid creating these elements upon incorporating the siRNA into shRNA or artificial miRNA scaffolds.

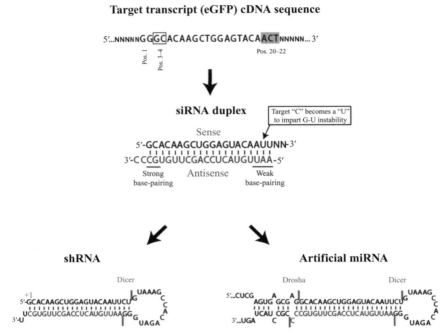

Figure 14.1 siRNA sequence selection for hairpin-based vectors. Schematic depicting the design of an eGFP-targeted siRNA sequence which satisfies the rules outlined in Section 2.2. siRNA sequences may be embedded into shRNA or artificial miRNA scaffolds for expression. (See Color Insert.)

3. Cloning shRNA Expression Cassettes

3.1. Notes on siRNA sequence selection for shRNA application

The criteria defined above are applicable with the following considerations. siRNAs designed for embedding into a shRNAs transcribed from the Pol III U6 promoter should contain a G nucleotide at position 3 to allow for proper expression and strand biasing of the resulting siRNA. U6-driven transcripts initiate at the +1-G position located at the 3′-end of the U6 promoter. By contrast, the H1 promoter (another Pol III promoter) can initiate with any nucleotide at the +1 position. Pol III-based transcription terminates at a stretch of four to six T's, resulting in two to three U's being incorporated at the 3′-end of the shRNA (Fig. 14.1). This provides the short 3′-overhang on the hairpin which is optimal for downstream processing events (e.g., nuclear export and subsequent recognition and cleavage by

Dicer) (Vermeulen et al., 2005; Zeng and Cullen, 2004). Thus, the presence of A or G nucleotides, which are both capable of base pairing with U, at positions 1 and 2 of the target sequence may promote better silencing efficacy. However, there is considerable empirical evidence supporting that shRNAs may be effective silencers even without base pairing at these terminal positions.

3.2. Oligo design and cloning protocol for shRNA-tailed PCR

The shRNA cassette will be designed for transcription by the mouse U6 promoter as sense, loop, and antisense sequences followed by the Pol III terminator. This results in a hairpin transcript that has a short 3'-overhang necessary for downstream processing by the RNAi pathway machinery (Fig. 14.1, shRNA). Positioning the antisense region at the 5'-end (i.e., antisense–loop–sense) is also possible but not described here. Steps 1–6 are outlined in Fig. 14.2 and describe the incorporation of a siRNA (targeting eGFP as the example) into a shRNA expression vector.

Step 1. Layout the shRNA DNA template.

The shRNA DNA template consists of the following from 5' to 3': (1) the sense sequence consisting of nucleotides 3–22 of the target site, (2) the 19-nt loop sequence (5'-CTGTAAAGCCACA-GATGGG-3') which is partially derived from the naturally occurring human miR-30 transcript, and (3) the antisense sequence which is the reverse complement of the target site (positions 3–22).

Step 2. Convert C's to T's at the sense 3'-end to impart duplex instability.

As mentioned above, proper loading of the antisense strand requires weak base pairing at its 5'-end in the siRNA duplex and strong base pairing opposite. Target sequence selection allows C's to be in 3'-end of the sense strand (positions 19–22 of the target site) because the C's, which will be paired with G's on the antisense strand, can be converted to U's, which pair much more weakly to G's. Thus, only the sense strand is manipulated to promote loading of the antisense strand. In the example, the terminal three nucleotides of sense sequence is converted from ACT to ATT.

Note: at this point, it is good practice to ensure that the hairpin sequence is void of stretches of four or more T's and any restriction enzyme sites (e.g., *Xho*I and *Eco*RI) relevant to downstream applications.

Step 3. Convert all T's to U's and fold RNA using UNAfold.

Convert the shRNA DNA template into an RNA transcript by replacing the T nucleotides with U's and use the UNAfold RNA folding tool (available at http://www.idtdna.com/Scitools/Applications/unafold/) to confirm that the shRNA transcript

Step 1. Layout the shRNA DNA template.

Step 2. Convert C's to T's at sense 3'-end to impart duplex instabililty.

Step 3. Convert all T's to U's and fold RNA using UNAfold.

Step 4. Add U6 promoter, terminator, and XhoI sequences to shRNA DNA template.

Step 5. Reverse complement the sequence and order DNA oligos (forward and reverse).

U6-shRNA reverse primer
5'-AAAACTCGAGAAAAAAGCACAAGCTGGAGTACAACTCCCATCTGTGGCTTTACAGAATTGTACTCCAGCTTGTGCAAACAAGGCTTTTCTCCAAGGG-3'

U6 forward primer
5'-CGACGCCGCCATCTCTAG-3'

Step 6. PCR amplify U6-shRNA expression cassette, TOPO clone, and screen.

Figure 14.2 shRNA cloning scheme. The details for designing shRNAs and performing tailed PCR to generate the mU6-driven shRNA expression cassette are described in the text (Section 3.2, Steps 1–6). Shown here is the example of incorporating the eGFP-targeted siRNA into the shRNA expression platform.

forms the desired secondary structure. *Note*: select "RNA" as the input "nucleotide type" at the user interface and leave the remaining parameters set to default. UNAfold generally predicts several possible secondary structures for each input; however, the correct version of the shRNA should be at the top as the most stable structure. In this case, the fold should resemble the shRNA depicted in Fig. 14.1, but lacking the 3'-UU overhang resulting from Pol III termination.

Step 4. Add U6 promoter, terminator, and *Xho*I sequences to the shRNA DNA template.

To generate the shRNA expression cassette, a tailed PCR strategy will be used to amplify the mouse U6 promoter sequence while tailing on the necessary shRNA components and a restriction enzyme site for identifying full-length PCR products. The 3′-end of the U6 promoter sequence (5′-CCCTTGGAGAAAAGCCTTGTTT-3′; minus its 3′G) is placed upstream of the shRNA sequence. This will serve as the priming site during the initial PCR cycles. Next, place the Pol III terminator sequence (six T's) followed by an *Xho*I restriction enzyme site (5′-CTCGAG-3′) downstream of the shRNA sequence.

Step 5. Reverse complement the sequence and order DNA oligos (forward and reverse).

To easily reverse complement DNA sequences, there are several tools available online (e.g., http://www.bioinformatics.org/sms/rev_comp.html). We typically add four A nucleotides to the 5′-end of the reverse-complemented oligo before ordering. The resulting shRNA-tailed reverse oligos can be ordered at reasonable cost from a variety of commercial vendors (e.g., Integrated DNA Technologies or Sigma-Aldrich) using the smallest synthesis scale available and standard desalting purification. A forward primer corresponding to the 5′-end of the mouse U6 promoter (5′-CGACGCCGCCATCTCTAG-3′) should also be ordered. Upon receipt, the oligos can be reconstituted in purified water. In general, we make a 100 μM stock for storage and a 20-μM working stock for use in PCR.

Step 6. PCR amplify the U6-shRNA expression cassette, TOPO clone and screen.

Step 6.1. Prepare and run the PCR as follows:

Reaction conditions

5 µl 10× reaction buffer with $MgCl_2$
1 µl mU6 forward primer (20 μM)
1 µl shRNA-tailed reverse primer (20 μM)
1 µl dNTPs (10 mM)
1 µl of mU6 containing plasmid (Tb:mU6; 10 ng/µl)
0.5 µl enzyme mix (i.e., polymerase)
40.5 µl dH_2O

Amplify in a thermocycler

94 °C for 3 min (initial denaturation)
30 cycles of 94 °C for 30 s, 54 °C for 30 s, and 72 °C for 30 s
72 °C for 5 min (final extension)

Step 6.2. Confirm amplification of the desired PCR product size. Run 5 μl of the PCR on a 1% agarose gel to verify the presence of a single band (~400 bp) in size.

Step 6.3. Perform TOPO cloning reaction as per the manufacturer's protocol by mixing 4 μl of the PCR product with 1 μl of salt solution and 1 μl of pCR®4-TOPO TA cloning vector, and incubating the reaction for 10 min at room temperature.

Step 6.4. Transform 2–3 μl of the TOPO cloning reaction into chemically competent *Escherichia coli* using standard procedures and plate onto ampicillin-selective plates for incubation overnight at 37°C. *Note*: This selects against the undesired template Tb:mU6 plasmid which is kanamycin resistant, allowing for positive selection of the desired pCR®4-TOPO plasmids which confer ampicillin resistance and kanamycin resistance.

Step 6.5. Pick 6–12 colonies and inoculate each into 3 ml of LB growth medium containing 50 μg/ml ampicillin. Grow overnight in a shaker at 37°C.

Step 6.6. Isolate plasmid DNA using standard procedures or a commercial miniprep kit. We recommend using the Qiaprep Miniprep kit (Qiagen) which yields plasmid DNA of sufficient quality for cell culture applications.

Step 6.7. Screen for positive full-length shRNA clones by digesting the plasmids with *Xho*I and performing gel electrophoresis. The presence of a band at ~400 bp indicates that the U6-shRNA cassette is in reverse orientation, while a ~50-bp band indicates forward insertion.

Step 6.8. Sequence positive clones using an automated sequencer that employs Sanger-based fluorescent chemistry. The cycling parameters may need to be optimized to sequence through the hairpin structure. For example, we previously found that adding DMSO (to 4%, v/v) to the sequencing reaction and using a longer initial denaturation step produced more reliable and consistent sequencing through the shRNAs. M13 forward or reverse primers can be used for sequencing TOPO cloning insertions.

The PCR for generating the shRNA expression cassettes is similar to a standard PCR which requires a DNA template (in this case, a plasmid containing the mouse U6 promoter; TOPObluntII: mU6 (Tb:mU6)), forward and reverse primers, dNTPs, reaction buffers, and DNA polymerase. Here, we use the Expand High-Fidelity PCR System (Taq-based, available from Roche); however, in our experience, shRNA-tailed PCRs have also proven successful with a variety of proof-reading polymerases (e.g., Accuprime Pfx, Invitrogen or Phusion, New England Biolabs). With these enzymes, the blunt-ended PCR products can be sub-cloned into pCR®-Blunt-II TOPO, requiring removal of the kanamycin-resistant template plasmid (e.g., by DpnI digestion) prior to TOPO cloning.

3.3. Materials for shRNA cloning

Tb:mU6 plasmid (PCR template)
DNA oligos (mU6 forward and shRNA-tailed reverse primers)
Expand High Fidelity Polymerase and buffers (Roche)
Restriction enzymes and buffers (*Xho*I)
Gel electrophoresis equipment and reagents
pCR®4-TOPO TA cloning kit (Invitrogen)
Chemical competent bacterial cells
Ampicillin-containing LB agar growth plates
LB growth media
Qiaprep Miniprep kit (Qiagen)

4. CLONING ARTIFICIAL miRNA EXPRESSION VECTORS

4.1. Considerations for designing artificial miRNAs

Artificial miRNAs, or miRNA shuttles, are designed to mimic naturally occurring pri-miRNAs, for which the Drosha-DGCR8 and Dicer cleavage sites have been mapped and experimentally validated. With this information, the identity of the small RNA duplex which is processed from the initial stem-loop transcript is known. For artificial miRNAs, this region is replaced by siRNA duplexes, thus creating a miRNA-based hairpin which serves to shuttle siRNA sequences into the RNAi pathway. An important consideration for designing artificial miRNAs is to maintain the structural and sequence recognition motifs required for appropriate processing. Drosha-DGCR8 binds to regions of single-stranded nature located at the base of the pri-miRNA stem loop (Han *et al.*, 2004; Zeng and Cullen, 2005). Thus, including 50–100nts of the flanking sequences (5′ and 3′) native to the pri-miRNA will help to ensure that the stem-loop base folds properly to promote cleavage at the intended site. Artificial miRNAs have been generated using a number of naturally occurring pri-miRNAs as scaffolds for siRNA sequences (Chung *et al.*, 2006; Tsou *et al.*, 2011; Zeng *et al.*, 2002). Here, we will describe a method to generate artificial miRNAs based on the natural human miR-30 pri-miRNA transcript. For ease of cloning, we only include a minimal amount of natural miR-30 flanking sequences, while additional 5′- and 3′-sequences are derived from the mouse U6 expression vector. We have characterized these vectors for appropriate expression and processing using small transcript northern blot, small RNA RT-PCR and RACE analyses (Boudreau et al., 2008; Chen et al., 2005). We recommend that newly designed miRNA shuttles be similarly defined with these techniques, since simply switching promoters, restriction enzyme sites, or expression contexts (e.g., embedding the miRNA-based stem loop in the 3'-UTR or intron of a reporter gene) can alter the pri-miRNA structure and subsequent processing.

4.2. Oligo design and cloning protocol for artificial miRNAs

The following steps are outlined in Fig. 14.3.

Step 1. Start with the artificial miRNA backbone sequence.

The miR-30-based artificial miRNA backbone consists of constant regions (5'- and 3'-flanking and loop sequences) that are partially derived from the human miR-30 pri-miRNA. The backbone is engineered with flanking restriction enzyme sites for cloning purposes. The 5'-XhoI and 3'-SpeI sites were chosen since their respective GAG and ACT nucleotides correspond to natural miR-30 stem-loop sequences. The distal (relative to the loop) single nucleotide bulge pictured in Fig. 14.1 is formed by the 5'- and 3'-flanks, while the proximal bulge consists of the 3'-end of the antisense strand; thus, the opposing nucleotide must be manipulated to maintain the bulge for proper Drosha-DGCR8 processing (see Step 3 below).

Step 2. Insert sense and antisense sequences.

Insert the sense sequence (i.e., target site positions 1–22) immediately after the 5'-flank sequence and the antisense sequence (i.e., the reverse complement of the target site) directly after the loop sequence.

Step 3. Manipulate the 5'-bulge nucleotide to induce a mismatch.

The bulge is natural to the human miR-30 pri-miRNA transcript, and thus, we aim to maintain this structure to promote proper downstream cleavage by Drosha-DGCR8. In the example, the G nucleotide in the sense strand bulge position (5') is converted to an A nucleotide which will create a mismatch with the opposing C bulge nucleotide in the antisense region (3'). In generating artificial miRNAs harboring different siRNA sequences, the user should alter the 5'-bulge nucleotide so that it does not pair with the opposing 3'-bulge nucleotide. This can be done following these simple guidelines:

A: if the 3'-bulge site is an A, the 5'-bulge nucleotide can be changed from T to either C, G, or A.

C: if the 3'-bulge site is a C, the 5'-bulge nucleotide can be changed from G to either A, T, or C.

Step 1. Start with artificial miRNA backbone sequence.

Step 2. Insert sense and antisense sequences.

Step 3. Manipulate 5' bulge nucleotide to induce a mismatch.

Step 4. Convert C's to T's at sense 3' end to impart duplex instability.

Step 5. Convert all T's to U's and fold RNA using UNAfold

Step 6. Design and order DNA oligos.

Step 7. Anneal and polymerase extend oligos.

Step 8. PCR purification, digestion and cloning to pTOPObluntII-U6

Figure 14.3 Artificial miRNA cloning scheme. The details for designing artificial miRNAs and cloning them into the mU6 expression vector are described in the text (Section 4.2, Steps 1–8). Shown here is the example of incorporating the eGFP-targeted siRNA into an artificial miRNA scaffold.

      G: if the 3'-bulge site is a G, the 5'-bulge nucleotide can be changed from C to either A or G.

      T: if the 3' bulge site is a T, the 5'-bulge nucleotide can be changed from A to either C or T.

Step 4. Convert C's to T's at the sense 3'-end to impart duplex instability. See Step 2 in shRNA cloning protocol (Section 3.2).

Note: at this point, it is good practice to ensure the hairpin sequence is void of stretches of four or more T's and any restriction enzyme sites (e.g., internal *Xho*I, *Spe*I, and *Eco*RI sites) relevant to downstream applications.

Step 5. Convert all T's to U's and fold RNA using UNAfold.

See Step 3 in shRNA cloning protocol (Section 3.2). In this case, the fold should resemble the artificial miRNA depicted in Fig. 14.1.

Step 6. Design and order DNA oligos.

The artificial miRNA DNA template is made by annealing two oligos that overlap in the loop region and performing a polymerase extension reaction to create the double-stranded DNA cassette. Oligo 1 consists of the *Xho*I site, 5'-flank, sense sequence, and loop in forward orientation, while Oligo 2 is the reverse complement of the loop, antisense, 3'-flank and *Spe*I site. We add four A nucleotides to the 5'-end of each oligo; thus, Oligo 1 will begin with 5'-AAAACTCGAG...-3' and Oligo 2 with 5'-AAAAACTAGT...-3'. These additional bases are required for efficient restriction enzyme digestion near the ends of the DNA cassette following the polymerase extension. The resulting DNA oligos can be ordered at reasonable cost from a variety of commercial vendors (e.g., Integrated DNA Technologies or Sigma-Aldrich) using the smallest synthesis scale available and standard desalting purification. Upon receipt, the oligos can be reconstituted in purified water to make a 100-μM stock.

Step 7. Anneal and polymerase extend Oligo 1 and Oligo 2.

Reaction conditions

5 µl 10× reaction buffer with $MgCl_2$
1 µl Oligo 1 (100 μM)
1 µl Oligo 2 (100 μM)
1 µl dNTPs (10 mM)
0.5 µl enzyme mix (i.e., polymerase)
41.5 µl dH_2O
Incubate the reaction in a thermocycler:
94 °C for 2 min (denaturation)
54 °C for 1 min (annealing)
72 °C for 15 min (extension)

Step 8. PCR purification, digestion, and cloning into Tb:mU6 plasmid

Step 8.1. Purify the extended product using a PCR Purification kit (QIAquick; Qiagen) as per the manufacturer's instructions and elute in 30 µl water.

Step 8.2. Digest the product with *Xho*I and *Spe*I by adding the following to the 30 µl eluate:

4 µl 10× restriction enzyme buffer (NEB2; New England Biolabs)
4 µl 10× BSA (10 mg/ml)
1 µl SpeI (10 units/µl)
1 µl XhoI (20 units/µl)
Incubate reaction at 37°C for 4 h to overnight.

Step 8.3. Digest 3–4 µg of the Tb:mU6 expression plasmid with XhoI and XbaI. This plasmid contains the mouse U6 promoter followed by a multiple cloning site (MCS) and a Pol III termination signal (six T's) (Fig. 14.4). XbaI- (vector) and SpeI (artificial miRNA)-cleaved sites produce compatible sticky ends for ligation. Although there is a SpeI site in the Tb:mU6 MCS, we have found in prior studies that ligation to the

```
                                         5′ ....U6 forward primer....▶
5′-...XbaI/NheI ligation_GAATTCGACGCCGCCATCTCTAGGCCCGCGCCGG
                         EcoRI
CCCCCTCGCACAGACTTGTGGGAGAAGCTCGGCTACTCCCCTGCCCCGGTTAATTTGCA

TATAATATTTCCTAGTAACTATAGAGGCTTAATGTGCGATAAAAGACAGATAATCTGTT

CTTTTTAATACTAGCTACATTTTACATGATAGGCTTGGATTTCTATAAGAGATACAAAT

ACTAAATTATTATTTTAAAAAACAGCACAAAAGGAAACTCACCCTAACTGTAAAGTAAT
                                 shRNA reverse primer
                          ◀.........................5′
TGTGTGTTTTGAGACTATAAATATCCCTTGGAGAAAAGCCTTGTTTGCGTTTAGTGAAC
                        XhoI                                  EcoRI
CGTCAGATGGTACCGTTTAAACTCGAGGTCGACGGTATCGATAAGCTTGATATCGAATT
                                    XbaI
CCTGCAGCCCGGGGATCCACTAGTTCTAGAGCGGCCGCCACAGCGGGAGATCCAGAC

ATGATAAGATACATTTTTTGAATTC_BglII/BamHI ligation...-3′
                 EcoRI
```

Figure 14.4 TOPObluntII:mU6 (Tb:mU6) plasmid details. Tb:mU6 plasmid is ~4 kb and contains a EcoRI–mU6 promoter–multiple cloning site–TTTTTT terminator–EcoRI cassette which was cloned into the TOPObluntII vector in reverse orientation using the indicated restriction sites (XbaI and BamHI were vector-derived, while NheI and BglII originated from the insert; all sites were destroyed upon ligation). The mU6 +1 transcription start site and Pol III terminator (TTTTTT) are shown in bold. The multiple cloning site (MCS) is underlined and contains KpnI–PmeI–XhoI–SalI–ClaI–HindIII–EcoRV–EcoRI–XmaI–SmaI–BamHI–SpeI–XbaI–NotI restriction enzyme sites. Note: cloning an artificial miRNA into the XhoI and XbaI sites removes the internal EcoRI (in the MCS), and thus, positive clones can be screened for by EcoRI digestion. The relevant primer binding sites for the shRNA-tailed PCR are also shown.

XbaI site produces a stem loop which is more efficiently processed, yielding higher antisense RNA levels and more potent gene silencing (Boudreau et al., 2008).

Step 8.4. Gel purify the digested fragments. We typically run the digested artificial miRNA inserts on a 2% agarose gel and excise the ~100-bp band. The digested Tb:mU6 is run on a 1% agarose gel and the ~4-kb fragment is excised. Gel extraction can be performed by various means; for example, we simply subject the gel slices to ultracentrifugation in a Spin-X column (Corning Incorporated). Precipitation may be necessary to concentrate the DNA fragments for ligation.

Step 8.5. Perform ligation and bacterial transformation using standard protocols. We ligate 6 ng of insert to 50 ng of vector and incubate at room temperature for 1 h before transformation. The Tb:mU6 plasmid is kanamycin resistant, and thus, transformed bacteria should be grown on LB agar plates containing kanamycin. Because the Tb:mU6 vector is digested with two noncohesive enzymes (XhoI and XbaI), the likelihood of intramolecular vector ligation is minimal, and few, if any, kanamycin-resistant colonies grow on "vector-only" control plates. If necessary, the vector can be treated with alkaline phosphatase after restriction enzyme digestion to further minimize the potential for background colonies. We typically pick four to six colonies per artificial miRNA construct and grow each of them overnight in 3 ml liquid LB cultures containing kanamycin. The following day, minipreps are performed using the Qiaprep kit which yields plasmid DNA of sufficient quality for cell culture applications.

Step 8.6. Screen for positive clones by EcoRI digestion and gel electrophoresis. Tb:mU6 vector-only plasmid will yield ~3.5 kb, ~380 bp, and ~80 bp fragments indicating a negative clone; we recommend digesting the parental Tb:mU6 vector alongside the potential artificial miRNA clones to serve as a reference. The positive constructs with successful insertion of an artificial miRNA will yield ~3.5 kb and ~500 bp bands.

Step 8.7. Sequence positive clones using the M13 reverse primer. Refer to the information regarding sequencing of shRNAs (Section 3.2—Step 6.8).

4.3. Materials for artificial miRNA cloning

Tb:mU6 plasmid (expression vector)
Overlapping DNA oligos for artificial miRNA
Expand high-Fidelity Polymerase and buffers (Roche)
Restriction enzymes and buffers (*Xho*I, *Spe*I, *Xba*I, and *Eco*RI)
Gel electrophoresis equipment and reagents
Gel extraction kit or Spin-X columns (Corning)
DNA ligase and buffer
Chemical competent bacterial cells
Kanamycin-containing LB agar growth plates
Qiaprep MIniprep kit (Qiagen)

5. SCREENING RNAi VECTORS FOR SILENCING EFFICACY *IN VITRO*

Although we select siRNA sequences based on the most significant determinants of gene silencing efficacy (i.e., strand biasing and GC content), not all sequences will be functional. Thus, we typically generate several constructs, each with unique sequences, for a given target gene. These constructs must then be screened for gene silencing efficacy to identify those which mediate silencing levels suited to the researchers' needs. Gene silencing can be assessed at the mRNA level by performing real-time quantitative PCR (preferred) and/or the protein level via Western blot analysis (dependent upon protein half-life) following delivery of the RNAi vectors into cells. For practical purposes, RNAi efficacy screens are generally carried out in highly transfectable cultured cell lines. With this approach, the target mRNA may be expressed endogenously or from cotransfected plasmids (Boudreau *et al.*, 2008; Paul *et al.*, 2002; Sui *et al.*, 2002). Screening against exogenously supplied targets (e.g., cotransfected target-reporter expression plasmids) in highly transfectable cell lines is particularly useful if the target is only expressed in poorly transfectable cells (e.g., primary neurons). In addition, silencing of coexpressed targets is more efficient since most cells receive both RNAi and target expression plasmids due to the nature of transfection, and target expression levels can be readily controlled. Conversely, silencing of endogenous mRNAs may be limited by transfection efficiency, particularly when the treated population is analyzed as a whole. For larger-scale screens, the use of target-reporter fusions (e.g., luciferase or fluorescent) may expedite the process of narrowing candidates (Mousses *et al.*, 2003). Silencing of the natural target can be subsequently confirmed with other functional, biological, and biochemical assays.

5.1. Screening against a cotransfected reporter constructs

For screening RNAi expression vectors *in vitro*, we often use HEK293 cells because they are easy to culture and highly transfectable with liposome-based reagents (e.g., Lipofectamine 2000, Invitrogen). Prior studies validate that siRNA, shRNA, and artificial miRNA vectors are capable of effective gene silencing in HEK293 cells, supporting that the necessary RNAi machinery is present. In our example, we test the eGFP-targeted artificial miRNA (miGFP) for its ability to silence GFP expression from a cotransfected reporter plasmid (Fig. 14.5). For this experiment, we seeded 200,000 cells per well in a 24-well culture plate. The next day, cells were cotransfected in triplicate with 75 ng of GFP expression plasmid along with one of the following plasmids: (1) Tb:mU6, serving as the promoter-only control; (2) Tb:mU6-miSCA1, an artificial miRNA targeting ataxin-1, providing a nontargeted RNAi control; and (3) Tb:mU6-miGFP, the on-target RNAi construct in this study. Twenty-four to forty-eight hours posttransfection, we evaluated GFP levels by fluorescent microscopy and observed significant knockdown in miGFP-treated cells, relative to the controls.

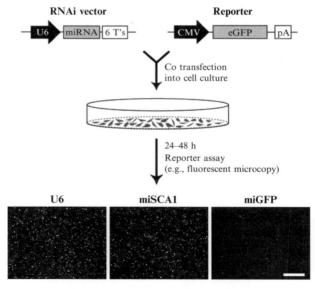

Figure 14.5 Example silencing experiment targeting an eGFP reporter. Plasmids expressing U6-driven artificial miRNAs or a CMV-driven eGFP reporter were cotransfected into HEK293 cells with a ratio of four RNAi:one eGFP (300 ng:75 ng). After 48 h, fluorescence microscopy was performed to evaluate eGFP levels. Representative photomicrographs of eGFP autofluorescence show evident silencing of eGFP expression in miGFP-treated cells, relative to the controls (U6 promoter-only and miSCA1, a nontargeted RNAi control). Scale bar=1 mm.

Additional information and recommendations for screening by cotransfection:

1. A variety of reporter options are available. The user may clone their target transcript into fluorescent or luciferase (psiCheck2, Promega) reporter systems, or as an epitope-tagged version for Western blotting.
2. We recommend master mixing the target reporter prior to distributing among the various treatments to ensure consistency across treatments. In our experience with 24-well plates, we typically use 25–100 ng of fluorescent reporter plasmids and 5–20 ng of luciferase-based reporters.
3. A dose response should be performed to further evaluate the potency of the RNAi constructs. This is important when multiple RNAi constructs show maximal silencing in a given experiment, supporting the need to test lower doses to identify the most efficacious sequence. In our example, we tested a RNAi:reporter ratio of 4:1. We recommend testing a 1:1 ratio as well. In the low-dose condition, supplementing empty Tb:mU6 plasmid to balance total RNAi plasmid amounts, relative to the high-dose condition, is a good practice.
4. Results from reporter assays are generally translatable to the natural transcripts; however, we advise that target silencing be confirmed in the intended experimental setting before downstream effects are queried or long-term studies are initiated.

5.2. Screening against endogenously expressed targets

Silencing of natural targets can be directly assessed if the target gene is expressed in a highly transfectable cell line (e.g., HEK293, HeLa, or NIH3T3, among others) of the relevant species. Previous studies from our laboratory have demonstrated silencing of endogenously expressed transcripts (e.g., encoding huntingtin or ataxin-1) in HEK293s and C2C12 cells (Boudreau et al., 2008, 2009b; McBride et al., 2008). In these experiments, 200–400 ng of RNAi expression plasmids were transfected into cells grown in 24-well plates, and gene silencing was evaluated 24–48 h later by QPCR or Western blot analyses measuring mRNA and protein levels, respectively. The methods for these experiments are not described here but can be found in the relevant references.

6. INTEGRATION INTO VIRAL VECTORS

Upon identifying effective hairpins, the RNAi expression cassettes can be easily subcloned into various viral vector systems for downstream applications. The University of Iowa Gene Transfer Vector Cora facility offers an assortment of plasmids for production of recombinant adenovirus, adeno-associated virus (AAV) and lentivirus (FIV). The mU6-driven artificial miRNA and shRNA expression cassettes described above are flanked by *Eco*RI sites and can be easily cloned into compatible sites (*Eco*RI or *Mfe*I) in each of these viral vector platforms. AAV vectors have become a powerful tool for *in vivo* application and, with several capsids available, offer a broad capacity to transduce many different cell types and tissues. To date, AAV-based RNAi vectors have been successful in achieving gene silencing in a variety of tissues including muscle, liver, and brain, among others (Davidson and McCray, 2011). As a note of caution, AAV-shRNA vectors have been shown to cause toxicity in mouse liver and brain due to high-level shRNA expression in conjunction with such a robust delivery platform (Boudreau et al., 2009a; Grimm et al., 2006; Martin et al., 2011; McBride et al., 2008). Thus, we recommend expressing artificial miRNAs, which are expressed at lower levels relative to shRNAs, from AAVs to achieve more tolerable RNAi expression levels *in vivo*. High-level shRNA expression is more advantageous for low-copy applications, such as generating stable cell lines with integrating lentiviral vectors, where single-copy insertions are often desired. In this instance, shRNAs may be better suited than miRNA-based vectors, where low-level expression may preclude sufficient silencing. These examples highlight the importance of considering the balance of efficacy and toxicity when selecting the most suitable RNAi expression strategy and delivery platform.

7. SUMMARY

RNAi triggers expressed from vector-based systems provide important tools for experimental biology and over the past decade have become a focus for therapeutic development. This chapter provides the reader with a stepwise protocol for generating shRNA- or artificial miRNA-based systems for use in gene silencing experiments. In addition, we describe methods for introducing these RNAi expression systems into viral vectors for *in vivo* applications, providing a powerful approach to query gene function or validate drug targets.

REFERENCES

Behlke, M. A. (2008). Chemical modification of siRNAs for in vivo use. *Oligonucleotides* **18**(4), 305–319.
Birmingham, A., Anderson, E. M., et al. (2006). 3′ UTR seed matches, but not overall identity, are associated with RNAi off-targets. *Nat. Methods* **3**(3), 199–204.

Birmingham, A., Anderson, E., et al. (2007). A protocol for designing siRNAs with high functionality and specificity. *Nat. Protoc.* **2**(9), 2068–2078.

Boden, D., Pusch, O., et al. (2004). Enhanced gene silencing of HIV-1 specific siRNA using microRNA designed hairpins. *Nucleic Acids Res.* **32**(3), 1154–1158.

Boudreau, R. L., Mas Monteys, A., et al. (2008). Minimizing variables among hairpin-based RNAi vectors reveals the potency of shRNAs. *RNA* **14**, 1834–1844.

Boudreau, R. L., Martins, I., et al. (2009a). Artificial microRNAs as siRNA shuttles: Improved safety as compared to shRNAs in vitro and in vivo. *Mol. Ther.* **17**(1), 169–175.

Boudreau, R. L., McBride, J. L., et al. (2009b). Nonallele-specific silencing of mutant and wild-type huntingtin demonstrates therapeutic efficacy in Huntington's disease mice. *Mol. Ther.* **17**(6), 1053–1063.

Castanotto, D., Sakurai, K., et al. (2007). Combinatorial delivery of small interfering RNAs reduces RNAi efficacy by selective incorporation into RISC. *Nucleic Acids Res.* **35**, 5154–5164.

Chen, C., Ridzon, D. A., et al. (2005). Real-time quantification of microRNAs by stem-loop RT-PCR. *Nucleic Acids Res.* **33**(20), e179.

Chi, J. T., Chang, H. Y., et al. (2003). Genomewide view of gene silencing by small interfering RNAs. *Proc. Natl. Acad. Sci. USA* **100**(11), 6343–6346.

Chung, K. H., Hart, C. C., et al. (2006). Polycistronic RNA polymerase II expression vectors for RNA interference based on BIC/miR-155. *Nucleic Acids Res.* **34**(7), e53.

Davidson, B. L., and McCray, P. B., Jr. (2011). Current prospects for RNA interference-based therapies. *Nat. Rev. Genet.* **12**(5), 329–340.

Ecker, J. R., and Davis, R. W. (1986). Inhibition of gene expression in plant cells by expression of antisense RNA. *Proc. Natl. Acad. Sci. USA* **83**(15), 5372–5376.

Fire, A., Xu, S. Q., et al. (1998). Potent and specific genetic interference by double-stranded RNA in Caenorhabditis elegans. *Nature* **391**, 806–811.

Gregory, R. I., Yan, K. P., et al. (2004). The Microprocessor complex mediates the genesis of microRNAs. *Nature* **432**(7014), 235–240.

Grimm, D., Streetz, K. L., et al. (2006). Fatality in mice due to oversaturation of cellular microRNA/short hairpin RNA pathways. *Nature* **441**(7092), 537–541.

Guo, H., Ingolia, N. T., et al. (2010). Mammalian microRNAs predominantly act to decrease target mRNA levels. *Nature* **466**(7308), 835–840.

Han, J., Lee, Y., et al. (2004). The Drosha-DGCR8 complex in primary microRNA processing. *Genes Dev.* **18**(24), 3016–3027.

Jackson, A. L., and Linsley, P. S. (2010). Recognizing and avoiding siRNA off-target effects for target identification and therapeutic application. *Nat. Rev. Drug Discov.* **9**(1), 57–67.

Jackson, A. L., Bartz, S. R., et al. (2003). Expression profiling reveals off-target gene regulation by RNAi. *Nat. Biotechnol.* **21**(6), 635–637.

Jackson, A. L., Burchard, J., et al. (2006). Widespread siRNA "off-target" transcript silencing mediated by seed region sequence complementarity. *RNA* **12**(7), 1179–1187.

Khvorova, A., Reynolds, A., et al. (2003). Functional siRNAs and miRNAs Exhibit Strand Bias. *Cell* **115**(2), 209–216.

Krol, J., Loedige, I., et al. (2010). The widespread regulation of microRNA biogenesis, function and decay. *Nat. Rev. Genet.* **11**(9), 597–610.

Lee, Y., Ahn, C., et al. (2003). The nuclear RNase III Drosha initiates microRNA processing. *Nature* **425**(6956), 415–419.

Leuschner, P. J., Ameres, S. L., et al. (2006). Cleavage of the siRNA passenger strand during RISC assembly in human cells. *EMBO Rep.* **7**(3), 314–320.

Lewis, B. P., Burge, C. B., et al. (2005). Conserved seed pairing, often flanked by adenosines, indicates that thousands of human genes are microRNA targets. *Cell* **120**(1), 15–20.

Lund, E., Guttinger, S., et al. (2004). Nuclear export of microRNA precursors. *Science* **303**(5654), 95–98.

Martin, J. N., Wolken, N., et al. (2011). Lethal toxicity caused by expression of shRNA in the mouse striatum: Implications for therapeutic design. *Gene Ther.* **18**(7), 666–673.

Matranga, C., Tomari, Y., et al. (2005). Passenger-strand cleavage facilitates assembly of siRNA into Ago2-containing RNAi enzyme complexes. *Cell* **123**(4), 607–620.

Matveeva, O., Nechipurenko, Y., et al. (2007). Comparison of approaches for rational siRNA design leading to a new efficient and transparent method. *Nucleic Acids Res.* **35**(8), e63.

McBride, J. L., Boudreau, R. L., et al. (2008). Artificial miRNAs mitigate shRNA-mediated toxicity in the brain: Implications for the therapeutic development of RNAi. *Proc. Natl. Acad. Sci. USA* **105**(15), 5868–5873.

McJunkin, K., Mazurek, A., et al. (2011). Reversible suppression of an essential gene in adult mice using transgenic RNA interference. *Proc. Natl. Acad. Sci. USA* **108**(17), 7113–7118.

Mousses, S., Caplen, N. J., et al. (2003). RNAi microarray analysis in cultured mammalian cells. *Genome Res.* **13**(10), 2341–2347.

Napoli, C., Lemieux, C., et al. (1990). Introduction of a chimeric chalcone synthase Gene into Petunia results in reversible co-suppression of homologous genes in trans. *Plant Cell* **2**(4), 279–289.

Paul, C. P., Good, P. D., et al. (2002). Effective expression of small interfering RNA in human cells. *Nat. Biotechnol.* **20**(5), 505–508.

Provost, P., Dishart, D., et al. (2002). Ribonuclease activity and RNA binding of recombinant human Dicer. *EMBO J.* **21**(21), 5864–5874.

Schwarz, D. S., Hutvagner, G., et al. (2003). Asymmetry in the assembly of the RNAi enzyme complex. *Cell* **115**(2), 199–208.

Semizarov, D., Frost, L., et al. (2003). Specificity of short interfering RNA determined through gene expression signatures. *Proc. Natl. Acad. Sci. USA* **100**(11), 6347–6352.

Shin, K. J., Wall, E. A., et al. (2006). A single lentiviral vector platform for microRNA-based conditional RNA interference and coordinated transgene expression. *Proc. Natl. Acad. Sci. USA* **103**(37), 13759–13764.

Silva, J. M., Li, M. Z., et al. (2005). Second-generation shRNA libraries covering the mouse and human genomes. *Nat. Genet.* **37**(11), 1281–1288.

Sui, G., Soohoo, C., et al. (2002). A DNA vector-based RNAi technology to suppress gene expression in mammalian cells. *Proc. Natl. Acad. Sci. USA* **99**(8), 5515–5520.

Tsou, W. L., Soong, B. W., et al. (2011). Splice isoform-specific suppression of the Ca(V)2.1 variant underlying spinocerebellar ataxia type 6. *Neurobiol. Dis.* **43**(3), 533–542.

Vermeulen, A., Behlen, L., et al. (2005). The contributions of dsRNA structure to Dicer specificity and efficiency. *RNA* **11**(5), 674–682.

Yi, R., Qin, Y., et al. (2003). Exportin-5 mediates the nuclear export of pre-microRNAs and short hairpin RNAs. *Genes Dev.* **17**(24), 3011–3016.

Zeng, Y., and Cullen, B. R. (2004). Structural requirements for pre-microRNA binding and nuclear export by Exportin 5. *Nucleic Acids Res.* **32**(16), 4776–4785.

Zeng, Y., and Cullen, B. R. (2005). Efficient processing of primary microRNA hairpins by Drosha requires flanking non-structured RNA sequences. *J. Biol. Chem.* **280**(30), 27595–27603.

Zeng, Y., Wagner, E. J., et al. (2002). Both natural and designed micro RNAs can inhibit the expression of cognate mRNAs when expressed in human cells. *Mol. Cell* **9**(6), 1327–1333.

CHAPTER FIFTEEN

Recombinant Adeno-Associated Viral Vector Reference Standards

Philippe Moullier*,†,‡ and Richard O. Snyder*,†,§

Contents

1. Introduction	298
2. Utility of Reference Standards	298
3. Volunteer Working Groups	300
4. The AAV2 Reference Standard Material	301
5. The AAV8 Reference Standard Material	302
6. Methods Used to Characterize the AAV RSMs	303
6.1. Assays for protein purity and identity	304
6.2. Assays for rAAV vector genome titer	304
6.3. Assay for infectious rAAV titer and the particle to infectivity ratio	305
6.4. ELISA for determination of intact AAV virions	305
6.5. Assay for transgene expression	306
7. Conclusions	308
Acknowledgments	308
References	309

Abstract

Reference standard materials (RSMs) exist for a variety of biologics including vaccines but are not readily available for gene therapy vectors. To date, a recombinant adeno-associated virus serotype 2 RSM (rAAV2 RSM) has been produced and characterized and was made available to the scientific community in 2010. In addition, a rAAV8 RSM has been produced and will be characterized in the coming months. The use of these reference materials by members of the gene therapy field facilitates the calibration of individual laboratory vector-specific internal standards and the eventual comparison of preclinical and clinical data based on common dosage units. Normalization of data to

---

* INSERM UMR 649, CHU Hôtel Dieu, Nantes, France
† Department of Molecular Genetics and Microbiology, College of Medicine, University of Florida, Gainesville, Florida, USA
‡ Genethon, Evry, France
§ Center of Excellence for Regenerative Health Biotechnology, University of Florida, Gainesville, Florida, USA

determine therapeutic dose ranges of rAAV vectors for each particular tissue target and disease indication is important information that can enhance the safety and protection of patients.

1. INTRODUCTION

Recombinant adeno-associated viral (rAAV) vectors are increasingly being used in the clinic (Gene Therapy Clinical Trials Worldwide, June 2011), and new vectors that incorporate the latest science are being developed to target a wide array of diseases and tissues. As rAAV vector technology has improved, the safety and efficacy of rAAV-mediated gene transfer in animal models has been demonstrated, and the vectors have been shown to persist and to be safe in humans (Brantly et al., 2009; Crystal et al., 2004; During et al., 2001; Flotte et al., 2003, 2004; Janson et al., 2002; Manno et al., 2003, 2006; Stedman et al., 2000; Wagner et al., 1999). Transient correction of hemophilia B following hepatic artery administration of a rAAV2–hFIX vector (High, 2007; Manno et al., 2006) has been reported, and transient transgene expression was detected in patients treated for LPL deficiency (Mingozzi et al., 2009). Immune responses against the AAV capsid (Brantly et al., 2009; High, 2007; Manno et al., 2006; Mingozzi et al., 2009) and transgene product (Mendell et al., 2010) have been observed in humans. Long-term improvements have been reported for the inherited blindness Leber's congenital amaurosis (Bainbridge et al., 2008; Cideciyan et al., 2009; Maguire et al., 2008) and hemophilia B via intravenous infusion (Ponder, 2011). rAAV vectors have the ability to be maintained in a variety of tissues and express therapeutic transgenes long term. In non-human primate skeletal muscle, the vector persists as monomeric and concatemeric episomal circles that assemble into a chromatin structure (Penaud-Budloo et al., 2008) that resists DNA methylation (Leger et al., 2011). In the non-human primate liver, rAAV is maintained as episomal circles (Sun et al., 2010), but rare integration events have been observed in murine liver (Donsante et al., 2007; Nakai et al., 2003). The stable maintenance, longevity of transgene expression, and excellent safety profile have made rAAV vectors highly utilized for gene transfer and are demonstrating long-term efficacy in humans.

2. UTILITY OF REFERENCE STANDARDS

Reference standards for a variety of biologics including therapeutic proteins (such as clotting factors) and vaccines are available from the US Pharmacopeia (http://www.usp.org/referenceStandards/), European

Pharmacopeia (http://www.edqm.eu/en/Ph-Eur-Reference-Standards-627.html), Japanese Pharmacopeia (http://www.sjp.jp/hyojun/html/frm031.php?lang=e), National Institute of Standards and Technology (NIST; http://www.nist.gov/srm/index.cfm), National Institute for Biological Standards and Control (NIBSC; http://www.nibsc.ac.uk/products/reference_standards.aspx), and the World Health Organization (WHO; http://www.who.int/biologicals/en/) with the purpose of providing research laboratories and hospital-based clinical testing laboratories with standards that are practical and recognized as legitimate by the scientific and medical communities. These standards are prepared and characterized for the specific attributes that are evaluated when used in testing and do not necessarily need to be pure, provided that impurities do not interfere with the performance of the assays.

At an International Conference on Harmonisation (ICH) Gene Therapy Workshop sponsored by the Pharmaceutical Research and Manufacturers of America (PhRMA) on September 9, 2002, discussions took place about the importance of establishing vector reference standards for all vector systems. A major point made by Dr. Stephanie Simek of the US Food and Drug Administration (FDA) Center for Biologics Evaluation and Research (CBER) was to use reference standards to validate each laboratory's own product-specific reference standard and test methods. Comparisons of pharmacodynamic, pharmacokinetic, toxicology, and efficacy data from preclinical studies and clinical trials performed by laboratories using different rAAV vector–transgene combinations requires equivalent titer units to calculate a consensus dosage unit (Flotte et al., 2002). Until recently, there has been a lack of standardization and inability to compare titer values between preclinical and clinical studies for vectors made and tested in different laboratories. Standardization is an issue because different assays or variations in protocols for the same assay are often used by individual laboratories to measure the same vector attribute. In response to this need, highly characterized reference standard materials (RSMs) of rAAV have been generated to facilitate these comparisons. An RSM allows researchers to normalize their titer values to the units of measure of the RSM analyte, thus allowing each laboratory to state their titers in units that can be indexed to titers of vectors used in other studies. The RSMs are used primarily to calibrate the internal standards and analytical methods used by individual laboratories to interrelate the doses used in different nonclinical and clinical studies. Further, efficacy and toxicology data can be reported in the literature or to the regulatory authorities with a relationship to an RSM, making the data useful as a guide for dosing in animals and humans.

The acknowledgment that RSMs benefit the field of gene therapy is not new and was first addressed for adenoviral vectors. The AAV reference standard efforts were established to address the lack of normalization of doses administered to animals and humans (Flotte et al., 2002). The AAV

RSMs are available to all members of the research community in a form suitable for nonclinical and clinical data support, and are shipped to investigators along with documentation that includes a manufacturing and characterization summary and profile of the RSM.

3. Volunteer Working Groups

In response to the unfortunate death of a patient during an adenovirus gene therapy trial in 1999 (Carmen, 2001), the Adenoviral Reference Material Working Group (ARMWG) was established in 2000 to oversee the production, characterization, and distribution of an adenovirus type 5 (Ad5) reference standard material (ARM) for the purpose of normalizing titers and doses of Ad5-based gene therapy vectors (Hutchins, 2002; Hutchins et al., 2000). The adenovirus community worked cooperatively to generate and characterize the ARM, and many groups donated time, reagents, and space to the effort. The ARM was made available in 2002 as ATCC VR-1516™ from the American Tissue Type Collection (ATCC, Manassas, VA).

Discussions about the utility of a rAAV vector reference standard began in May 1999 at a joint US FDA/National Institutes of Health (NIH) workshop, but the effort gained momentum starting in 2002 when the FDA and the NIH Recombinant DNA Advisory Committee (RAC) together with the support of the National Gene Vector Laboratory (NGVL, a program of the NIH National Center for Research Resources (NCRR)), encouraged scientists within the AAV community to form an AAV2 Reference Standard Working Group (AAV2RSWG; Snyder and Flotte, 2002) to produce and characterize an AAV serotype 2 RSM since AAV2 was the most widely used vector at the time. The AAV2RSWG was established as a volunteer organization comprised of industry and academic members from nine countries and the International Society for BioProcess Technology (ISBioTech; http://www.isbiotech.org/), with representatives from FDA and NIH who were available for consultation. Organizationally, the AAV2RSWG established four committees to facilitate the effort: Manufacturing, Quality Control, and Donations that were coordinated by the Executive Committee. The AAV2RSWG held several meetings during the past 9 years to plan and coordinate activities. ISBioTech generously offered time and energy toward soliciting donations, posting information on the ISBioTech web site, coordinating meetings and conference calls, and participating directly in the effort. To supplement the NIH NGVL funding, the AAV2RSWG drafted requests for proposals for donations, and generous donations of raw materials and services were provided by several vendors.

In Europe, an effort was established in 2008 to generate an AAV8 RSM (Moullier and Snyder, 2008). Members of the AAV8RSWG have been assembled including industry and academic members from 10 countries.

Again, ISBioTech is participating and the US FDA is available for consultation. There are several AAV8RSWG member laboratories who also participated in the AAV2RSWG and this has facilitated an accelerated effort and maintained continuity. The major difference of the AAV8 RSM was the ability to obtain sufficient funding, resulting in a much shortened timeline. For the AAV2RSWG, the cumbersome solicitation of donations of time, materials, and laboratory space (totaling $414,500) contributed to a very lengthy process to generate the AAV2 RSM. In 2010 a recombinant AAV2 RSM was completed, and in 2009–2010 a rAAV8 RSMRSM was produced and will be characterized in 2011–2012 by volunteer laboratories prior to being available to the community in 2012. To date, 15 laboratories have volunteered to characterize the AAV8 RSM.

4. THE AAV2 REFERENCE STANDARD MATERIAL

One of the early discussions of the AAV2RSWG was focused on whether the AAV2 RSM should be a well-characterized research-grade material or one that is manufactured under current Good Manufacturing Practices (cGMP). Most participants agreed that producing the RSM under cGMP conditions would be ideal, however, the cost for cGMP was a major concern, and the discussion resulted in a decision to produce a highly characterized AAV2 RSM in a research vector core (non-GMP) under controlled and documented conditions. Also, there was much discussion on the choice of the vector genome, where the working group entertained generating an AAV2 RSM based on wild-type AAV2 or a recombinant vector harboring a random stuffer fragment, a therapeutic transgene, or a marker gene; ultimately the AAV2RSWG settled on a vector expressing a marker gene (green fluorescent protein, GFP). The rAAV2 RSM carries a single-stranded DNA vector genome derived from the vector plasmid pTR-UF-11 (Burger et al., 2004). Self-complementary AAV vector genomes (McCarty et al., 2003) incorporate a new technology capable of expressing the vector transgene at an accelerated rate and achieving higher transduction efficiency compared to single-stranded vector genomes. The rAAV2 RSM can be used to normalize "in-house" reference standards for single-stranded vectors as well as self-complementary vectors by using a $0.5\times$ conversion factor.

The AAV2RSWG set a goal of generating an AAV2 RSM at a scale of 2×10^{15} vector genomes (vg) and 5000–10,000 vials at a concentration of 2×10^{11} vg/ml. Production and purification of the AAV2 RSM was carried out under non-GMP conditions using helpervirus-free transient transfection of 1.5×10^{11} cells and a three-column chromatographic purification (Potter et al., 2008). Approximately 150 ml of AAV2-GFP $= 5.69 \times 10^{14}$ vg ($= 3.79 \times 10^{12}$ vg/ml) was produced. The AAV2 RSM was diluted to the proper concentration (2×10^{11} vg/ml), filtered, and divided into two 1.3-l bulk portions. After the vial label was finalized, one of the 1.3-l portions of the bulk AAV2 RSM was filled by the ATCC into 2087 vials to produce ATCC VR-1616™. Each vial of the AAV2 RSM was filled with 0.5 ml at a target dose of 1×10^{11} vector genomes. The vials are frozen in the repository at ATCC and are available upon request. The other 1.3-l bulk is being held and will be filled at a later date if demand warrants.

In parallel, members of the AAV2RSWG submitted their protocols to the Quality Control subcommittee for the various assays that were used to characterize the AAV2 RSM. These protocols were reviewed and a lead protocol chosen for each assay. The assays include: (1) confirmation of the serotype and capsid titer by A20 ELISA (Progen); (2) evaluation of the purity, capsid subunit stoichiometry, and chemical integrity of the capsid by SDS-PAGE; (3) vector genome titer by qPCR; (4) infectious titer by $TCID_{50}$ with qPCR read-out and by transduction (GFP read-out); and (5) sequencing of the vector. In addition, a long-term stability study comprised of the infectivity and identity/purity assays is being performed according to a schedule. The lead protocols were beta-tested at the University of Pennsylvania's Gene Therapy Program and the protocols were posted on the ISBioTech website. In addition, the pTR-UF-11 plasmid was deposited and banked at ATCC and is available as ATCC# MBA-331™, and the cell substrate for infectious titering (HeRC32 cells; Chadeuf et al., 2000) was deposited and banked at the ATCC and is available upon request as ATCC# CRL-2972™. Vials of the adenovirus reference standard have been made available for use in titering the AAV2 RSM. Vials of the AAV2 RSM were distributed for characterization to 16 laboratories that volunteered to conduct single or multiple assays, followed by statistical analysis. The consensus titers and other characterization data from the multiple laboratories were compiled and published (Lock et al., 2010). To date, demand has been strong for the AAV2 RSM.

5. THE AAV8 REFERENCE STANDARD MATERIAL

The framework used in the development of the AAV2 RSM established the roadmap for developing RSMs based on other AAV serotypes. Most of the 81 clinical trials to date have involved AAV serotype 2 vectors (Gene Therapy Clinical Trials Worldwide, June 2011), but vector systems based on other AAV serotypes (Gao et al., 2002) are being developed rapidly and used in the clinic (Allay et al., 2011; Brantly et al., 2009; Gao et al., 2002; Mingozzi et al., 2009). AAV8 is becoming widely utilized for gene transfer given its high transduction efficiency, ability for systemic delivery (Toromanoff et al., 2008; Wang et al., 2005), and distinct tropism for cells and organs that can be targeted for transduction.

The AAV8RSWG has produced and purified an AAV8 RSM by transient transfection and cesium chloride density centrifugation, and 4088 vials have been filled with 0.125 ml at a concentration of 2×10^{12} vg/ml that were deposited at ATCC as ATCC# VR-1816™ (unpublished data). To harmonize the two RSMs, the AAV8 RSM shares the same vector genome derived from pTR-UF-11 that was used for the AAV2 RSM. The characterization of the AAV8 RSM will begin soon and includes the same parameters as evaluated for the AAV2 RSM: confirmation of the serotype and capsid titer by ELISA using an AAV8-specific conformational antibody (J. Kleinschmidt and Progen); evaluation of the purity, capsid subunit stoichiometry, and chemical integrity of the capsid by SDS-PAGE; vector genome titer by qPCR; infectious titer by $TCID_{50}$ with qPCR readout and by transduction (GFP readout); sequencing of the vector; and stability. Assay protocols, standards, controls, and cell substrates used for the AAV2 RSM will also be utilized for the characterization of the AAV8 RSM. Physical assays (vector genome titer [hybridization, PCR, or spectrophotometry] and SDS-PAGE) can be shared between the two reference standards. However, the ELISA and infectious titer assays will need to be modified to test the AAV8 RSM. As for the AAV2 RSM, the lead protocols are being beta-tested at the University of Pennsylvania's Gene Therapy Program prior to posting to the ISBioTech website for use by the volunteer testing laboratories who will characterize the AAV8 RSM.

6. METHODS USED TO CHARACTERIZE THE AAV RSMs

Administered doses are usually based on vector genome titer because the defective nature of AAV makes determining vector infectious units difficult. Titering methods based on vector genomes (using hybridization, real-time PCR, or spectrophotometry) are more reliable but give no information as to the infectivity of the vector. Determining infectious titer is still critical, as the ratio of infectious virions to vector genome-containing virions helps to determine the potency and strength of the rAAV vector preparation. Although several assays are employed routinely to evaluate rAAV vectors used for research, additional assay development is required to achieve the appropriate level of assay performance for each phase of product development and ultimately product licensure. Testing and characterization of each lot of the rAAV vector product is performed under cGMPs with safety testing performed under Good Laboratory Practices (GLP) prior to product release. In general, the characterization tests are product-specific (i.e., not generically applicable to different vector genomes or serotypes). It might be sufficient to have preliminary validated assays applicable for products entering Phase I clinical trials, but a full validation package is recommended for all assays used in the quality control analyses of the products entering Phase II/III clinical trials. The GMP QC assay development activity involves generating product-specific standards and controls, writing test records and reagent preparation logs, and training laboratory analysts to perform these procedures. Internal standards can be calibrated to external RSMs if they are available to the community. After the methods are established, then the assays are qualified to demonstrate that they perform appropriately based on their intended use. Once produced and purified, the clinical vector lot is characterized using the qualified assays to ensure it meets preset specifications for safety, identity, purity, potency, and stability, and the data is included in the regulatory filing documents. The AAV2 RSM and AAV8 RSM can be utilized to support the calibration and qualification of QC assays used for clinical product release. The ATCC Product Information Sheet included with each shipment summarizes the testing methods, characteristics, and values of each AAV RSM.

6.1. Assays for protein purity and identity

The AAV2 RSM bulk (before it was vialed) was analyzed by silver and Coomassie blue staining and the final vialed AAV2 RSM was analyzed by fluorescent dye (SYPRO ruby) following separation of the viral capsid proteins on reduced and nonreduced SDS polyacrylamide gels. Three capsid proteins (VP1, 2, and 3) were visible in the correct stoichiometry of approximately 1:1:10 and have the correct molecular weights (87, 72, and 62 kDa) and were free of non-AAV proteins (Lock et al., 2010). The rAAV8 RSM is yet to be tested.

6.2. Assays for rAAV vector genome titer

Real-time PCR-based assays are becoming the accepted standard to determine the vector genome titer (Clark et al., 1999; Drittanti et al., 2000; Lock et al., 2010; Veldwijk et al., 2002). Plasmid and unpackaged vector DNA external to the virion was digested for 1 h with nuclease, and the nuclease was inactivated by heat. Treated virus was added directly to the PCR reaction, alongside a dilution series of the vector plasmid DNA that was packaged (pTR-UF-11). The PCR primers and probe detect the viral vector DNA and plasmid standard curve. The vector DNA signal was compared to the signal generated from the plasmid DNA standard curve and extrapolated to determine a vector genome titer that was reported as vector genomes per ml (vg/ml). The same assay is used for the AAV2 RSM and AAV8 RSM. For the AAV2 RSM, although a titer of 2×10^{11} vg/ml was targeted when the vials were filled, when characterized the consensus titer was 3.28×10^{10} vg/ml (Lock et al., 2010). The rAAV8 RSM is yet to be tested.

6.3. Assay for infectious rAAV titer and the particle to infectivity ratio

To obtain an infectious titer, an end-point dilution based on a $TCID_{50}$ format was combined with real-time PCR. Rep/cap (Chadeuf et al., 2000) cells in a 96-well plate format were infected with adenovirus helper, and serial dilutions of rAAV were made in replicates. Following infection, the wells were evaluated by real-time PCR, and the Karber method was used to calculate the infectious titer (Lock et al., 2010). The assay measures the ability of the virus to infect *rep/cap* expressing cells (Clark et al., 1996; Zhen et al., 2004), unpackage, and replicate to give an accurate measurement of infectious virus regardless of the transgene or promoter used. The method, however, provides no information as to the functional status of the expression cassette delivered by the viral particle. The switch to *rep/cap* expressing cells or helperviruses also allows the elimination of wtAAV to supply *rep* and *cap* in the RCA (Snyder et al., 1996) and ensures that all cells are permissive for rAAV vector replication. Titer is reported in infectious units per ml (IU/ml). When the AAV2 RSM was characterized, the consensus titer was 4.37×10^9 IU/ml (Lock et al., 2010). The rAAV8 RSM is yet to be tested.

A comparison of the vector genome titer to the infectious titer produces the particle:infectious (P:I) ratio. For the AAV2 RSM the P:I ratio was 7.51 (vector genomes:infectious units; Lock et al., 2010); ratios below 10 are considered indicative of robust rAAV2 preparations. In some cases when the rAAV vector serotype has difficulty infecting cell lines *in vitro* to generate an infectious titer, the P:I ratio can be extremely high (Mohiuddin et al., 2005). However, as long as the ratio is consistent to show lot to lot product similarity and can be correlated to vector potency *in vivo*, *ex vivo*, or *in vitro*, then the vector lot can be considered suitable for use. For the AAV2 serotype vectors, HeLa cell-based rep/cap cell lines have been widely used (Chadeuf et al., 2000; Clark et al., 1996); however, other AAV serotypes and helperviruses infect HeLa, 293, and other cell lines inefficiently (Mohiuddin et al., 2005). Identifying cell lines that can be infected with different AAV serotypes and helperviruses efficiently is greatly needed, as this impacts not only the infectious titer values but also the particle to infectious ratio and transgene expression assays.

6.4. ELISA for determination of intact AAV virions

The AAV2 capsid ELISA assay was used to quantify total (infectious, noninfectious, and empty) viral particles. Grimm et al. (1999) reported that some preparations of rAAV may have a significant amount of empty particles. The A20 antibody recognizes assembled empty and full capsids, but not disrupted or partially assembled capsids (Wistuba et al., 1997). This assay can be used to determine what fraction of the rAAV2 particles are empty by subtracting the titer value of the full particles that contain vector genomes, but the use of the A20 antibody is limited to AAV2 and AAV3 serotypes (Grimm *et al.*, 2003). Recently, antibodies that recognize intact capsids of other AAV serotypes have been described (Grimm *et al.*, 2003). The A20 monoclonal is used in the ELISA assay for the AAV2 RSM and the ADK8 monoclonal antibody, specific for a conformational epitope on assembled AAV8 capsids, will be used for the AAV8 RSM capsid ELISA assay (Progen). These ELISAs will also serve in vector identity confirmation testing of the capsid serotype. When the AAV2 RSM was characterized, the titer consensus was 9.18×10^{11} particles/ml (Lock *et al.*, 2010). The rAAV8 RSM is yet to be tested.

6.5. Assay for transgene expression

The green fluorescent cell assay for transgene expression involves the analysis of cells from the infectious titer assay (described above), with cells expressing GFP being scored and a titer calculated based on the dilution factor. This titer determination is also frequently performed on naïve cells (cells without *rep* and *cap*) transduced with serial dilutions of the vector in the presence of a helpervirus. After 24 h of incubation at 37 °C, transduced cells were visualized and counted using a fluorescence microscope. Titer is reported in transducing units per ml (TU/ml). When the AAV2 RSM was characterized, the titer consensus was 5.09×10^{8} TU/ml (Lock *et al.*, 2010). The rAAV8 RSM is yet to be tested.

6.5.1. Safety testing

Safety testing includes a set of assays used to test preclinical and clinical vector product lots prior to use in animals or humans. Some of the same tests are performed to ensure that the RSMs are free of detectable contaminating agents or process residuals that could interfere with subsequent analytical testing. Safety tests can be developed for on-site testing, but there are several commercial vendors who offer GLP testing services. The primary safety tests for the AAV RSMs include:

(1) *Adventitious agents (in vitro and in vivo)*. These assays are designed to detect the presence of infectious viral agents of human or animal origin. A choice of cell lines (e.g., human, murine, bovine, porcine, etc.) is available for the *in vitro* assay, whereas rodents and embryonated hen's eggs are commonly used for the *in vivo* assay. In addition, a panel of PCR-based tests is used to detect the presence of specific viral agents. The cells used to produce the AAV2 RSM and AAV8 RSM are 293 cell lines that were banked under cGMP conditions for clinical vector manufacturing, and they tested negative for adventitious agents.
(2) *Mycoplasma*. The test for the presence of mycoplasma relies on the growth of mycoplasma during the expansion of indicator cells in

antibiotic-free conditions and detection of the organism using dye or PCR, as well as growth of mycoplasma in broth or on appropriate agar media. The AAV2 RSM production cell culture harvest and the purified formulated filtered bulk were tested. The final rAAV2 RSM product is negative for mycoplasma, although the harvest material was exposed to mycoplasma that was cleared and/or inactivated in the purification process, since the purified bulk tested negative for viable mycoplasma and mycoplasma DNA (Potter et al., 2008). Further, the GMP 293 cells used for the AAV2 RSM were negative for mycoplasma prior to transfer to the research laboratory where the AAV2 RSM was made. A summary of the mycoplasma result is included on the Product Information Sheet supplied with each shipment of the AAV2 RSM, and companies and institutions requesting the AAV2 RSM can make an informed decision about bringing it into their QC laboratories.
(3) *Endotoxin.* Endotoxin is detected using the Limulus Amebocyte Lysate Assay or by using a rabbit pyrogen test. The AAV2 RSM tested negative for endotoxin (Lock et al., 2010).
(4) *Sterility.* This assay determines the absence of bacterial or fungal organisms and must also include the performance of bacteriastasis and fungistasis analysis. The AAV2 RSM is sterile (Lock et al., 2010).

6.5.2. Stability studies

Short-term stability studies were designed to evaluate the stability of the rAAV2 vector during the filling of the vials at room temperature that can last several hours (Potter et al., 2008) and for determining the stability of the AAV2 RSM during the characterization phase where freezing and thawing of a vial needed to be avoided (Lock et al., 2010). Long-term stability studies were designed to generate data for the purified AAV2 RSM at the proper storage temperature, formulation, and fill volume, and in the storage container (Potter et al., 2008), where an average drop of 35% was observed that may be due to a combination of vector absorption to the container and vulnerability to freezing and thawing at dilute vector concentration. The ongoing long-term study is designed to demonstrate genetic and physicochemical stability and container integrity during the life of the AAV2 RSM. Stability samples are titered using the infectious titer assay and transgene expression assay to determine maintenance of the infectivity and genetic integrity of the RSM. Identity and protein purity testing is performed to ensure the chemical stability of the vector (absence of capsid protein degradation). Sterility testing is also performed to ensure that final product container integrity is maintained with storage over time. Testing of the AAV2 RSM is conducted annually, and the AAV8 RSM will follow a similar schedule.

7. Conclusions

Due to the wide variety of methods applied to vector characterization, and even the variability of data between laboratories using the same method, correlation of results from different sources is difficult. It is anticipated that the availability of RSMs will facilitate comparison of vector titers and vector performance. The rAAV RSMs are intended to be used to qualify and calibrate in-house reference materials and assays used in basic, preclinical and clinical research employing rAAV vectors, with the purpose of comparing data across laboratories. Each laboratory is encouraged to characterize and qualify its own in-house product-specific reference materials using the appropriate rAAV RSM. As the vials of rAAV RSMs are finite, this will conserve the AAV RSM since calibration of the in-house standard will need to be performed infrequently.

Following the example set by the ARMWG for developing the adenovirus reference material (Hutchins, 2002), the volunteer members of the AAV community have worked to develop two high-quality AAV RSMs. These will facilitate comparisons among nonclinical or clinical studies, aid in the manufacture of more consistent and higher quality vectors, and ultimately help formulate regulatory policy.

In the United States, the FDA CBER, Office of Cellular, Tissue, and Gene Therapies (OCTGT), and Division of Cellular and Gene Therapies (DCGT) recommend the use of reference materials. However, it is not the intent of the FDA to standardize assay methods across the field or to require that the values assigned to the RSM be duplicated during in-house assay validation. Further, there is no requirement in the United States to follow RSM testing procedures when characterizing preclinical or clinical vector product lots. Sponsors of rAAV vector INDs are encouraged to consult with the FDA/CBER or appropriate national agency for further guidance. Regulatory agencies, scientists, and medical professionals from academia and industry have recognized that the ultimate success of these AAV RSM efforts is to protect patients and contribute to the development and commercialization of AAV-based gene therapeutics.

ACKNOWLEDGMENTS

This work was supported by NIH grant U42RR11148 and Clinigene, an EC funded Network of Excellence. We acknowledge the generosity of members of the AAV RSM Working Groups, the laboratories who participated in the production of the AAV2 RSM and AAV8 RSM and characterization of the AAV2 RSM, along with ATCC, Nunc, Aldevron, Corning, Fisher Thermo Scientific, Indiana University Vector Production Facility, HyClone, Mediatech, Plasmid Factory, Progen, and ISBioTech. ROS may be entitled to royalties on technology discussed in this chapter. ROS owns equity in a gene therapy company that is commercializing AAV for gene therapy applications.

REFERENCES

Allay, J. A., et al. (2011). Good manufacturing practice production of self-complementary serotype 8 adeno-associated viral vector for a hemophilia B clinical trial. *Hum. Gene Ther.* **22,** 595–604.

Bainbridge, J. W., et al. (2008). Effect of gene therapy on visual function in Leber's congenital amaurosis. *N. Engl. J. Med.* **358,** 2231–2239.

Brantly, M. L., et al. (2009). Sustained transgene expression despite T lymphocyte responses in a clinical trial of rAAV1-AAT gene therapy. *Proc. Natl. Acad. Sci. USA* **106,** 16363–16368.

Burger, C., et al. (2004). Recombinant AAV viral vectors pseudotyped with viral capsids from serotypes 1, 2, and 5 display differential efficiency and cell tropism after delivery to different regions of the central nervous system. *Mol. Ther.* **10,** 302–317.

Carmen, I. H. (2001). A death in the laboratory: The politics of the Gelsinger aftermath. *Mol. Ther.* **3,** 425–428.

Chadeuf, G., et al. (2000). Efficient recombinant adeno-associated virus production by a stable rep-cap HeLa cell line correlates with adenovirus-induced amplification of the integrated rep-cap genome (In process citation). *J. Gene Med.* **2,** 260–268.

Cideciyan, A. V., et al. (2009). Human RPE65 gene therapy for Leber congenital amaurosis: Persistence of early visual improvements and safety at 1 year. *Hum. Gene Ther.* **20,** 999–1004.

Clark, K. R., Voulgaropoulou, F., and Johnson, P. R. (1996). A stable cell line carrying adenovirus-inducible rep and cap genes allows for infectivity titration of adeno-associated virus vectors. *Gene Ther.* **3,** 1124–1132.

Clark, K. R., Liu, X., McGrath, J. P., and Johnson, P. R. (1999). Highly purified recombinant adeno-associated virus vectors are biologically active and free of detectable helper and wild-type viruses. *Hum. Gene Ther.* **10,** 1031–1039.

Crystal, R. G., et al. (2004). Clinical protocol. Administration of a replication-deficient adeno-associated virus gene transfer vector expressing the human CLN2 cDNA to the brain of children with late infantile neuronal ceroid lipofuscinosis. *Hum. Gene Ther.* **15,** 1131–1154.

Donsante, A., et al. (2007). AAV vector integration sites in mouse hepatocellular carcinoma. *Science* **317,** 477.

Drittanti, L., Rivet, C., Manceau, P., Danos, O., and Vega, M. (2000). High throughput production, screening and analysis of adeno-associated viral vectors. *Gene Ther.* **7,** 924–929.

During, M. J., Kaplitt, M. G., Stern, M. B., and Eidelberg, D. (2001). Subthalamic GAD gene transfer in Parkinson disease patients who are candidates for deep brain stimulation. *Hum. Gene Ther.* **12,** 1589–1591.

Flotte, T. R., Burd, P., and Snyder, R. O. (2002). Utility of a recombinant adeno-associated viral vector reference standard. *Bioprocessing* **1,** 75–77.

Flotte, T. R., et al. (2003). Phase I trial of intranasal and endobronchial administration of a recombinant adeno-associated virus serotype 2 (rAAV2)-CFTR vector in adult cystic fibrosis patients: A two-part clinical study. *Hum. Gene Ther.* **14,** 1079–1088.

Flotte, T. R., et al. (2004). Phase I trial of intramuscular injection of a recombinant adeno-associated virus alpha 1-antitrypsin (rAAV2-CB-hAAT) gene vector to AAT-deficient adults. *Hum. Gene Ther.* **15,** 93–128.

Gao, G. P., Alvira, M. R., Wang, L., Calcedo, R., Johnston, J., and Wilson, J. M. (2002). Novel adeno-associated viruses from rhesus monkeys as vectors for human gene therapy. *Proc. Natl. Acad. Sci. USA* **99,** 11854–11859.

Gene Therapy Clinical Trials Worldwide. June 2011 Update. *J. Gene Med.*, http://www.wiley.co.uk/genmed/clinical/.

Grimm, D., Kern, A., Pawlita, M., Ferrari, F., Samulski, R., and Kleinschmidt, J. (1999). Titration of AAV-2 particles via a novel capsid ELISA: Packaging of genomes can limit production of recombinant AAV-2. *Gene Ther.* **6,** 1322–1330.

Grimm, D., Kay, M. A., and Kleinschmidt, J. A. (2003). Helper virus-free, optically controllable, and two-plasmid-based production of adeno-associated virus vectors of serotypes 1 to 6. *Mol. Ther.* **7,** 839–850.

High, K. A. (2007). Update on progress and hurdles in novel genetic therapies for hemophilia. *Hematol. Am. Soc. Hematol. Educ. Program* **2007,** 466–472.

Hutchins, B. (2002). Development of a reference material for characterizing adenovirus vectors. *Bioprocessing* **1,** 25–28.

Hutchins, B., et al. (2000). Working toward an adenoviral vector testing standard. *Mol. Ther.* **2,** 532–534.

Janson, C., et al. (2002). Clinical protocol. Gene therapy of Canavan disease: AAV-2 vector for neurosurgical delivery of aspartoacylase gene (ASPA) to the human brain. *Hum. Gene Ther.* **13,** 1391–1412.

Leger, A., et al. (2011). Adeno-associated viral vector-mediated transgene expression is independent of DNA methylation in primate liver and skeletal muscle. *PLoS One* **6,** e20881.

Lock, M., et al. (2010). Characterization of a recombinant adeno-associated virus type 2 Reference Standard Material. *Hum. Gene Ther.* **21,** 1273–1285.

Maguire, A. M., et al. (2008). Safety and efficacy of gene transfer for Leber's congenital amaurosis. *N. Engl. J. Med.* **358,** 2240–2248.

Manno, C. S., et al. (2003). AAV-mediated factor IX gene transfer to skeletal muscle in patients with severe hemophilia B. *Blood* **101,** 2963–2972.

Manno, C. S., et al. (2006). Successful transduction of liver in hemophilia by AAV-Factor IX and limitations imposed by the host immune response. *Nat. Med.* **12,** 342–347.

McCarty, D. M., Fu, H., Monahan, P. E., Toulson, C. E., Naik, P., and Samulski, R. J. (2003). Adeno-associated virus terminal repeat (TR) mutant generates self-complementary vectors to overcome the rate-limiting step to transduction *in vivo*. *Gene Ther.* **10,** 2112–2118.

Mendell, J. R., et al. (2010). Dystrophin immunity in Duchenne's muscular dystrophy. *N. Engl. J. Med.* **363,** 1429–1437.

Mingozzi, F., et al. (2009). AAV-1-mediated gene transfer to skeletal muscle in humans results in dose-dependent activation of capsid-specific T cells. *Blood* **114,** 2077–2086.

Mohiuddin, I., Loiler, S., Zolotukhin, I., Byrne, B. J., Flotte, T. R., and Snyder, R. O. (2005). Herpesvirus-based infectious titering of recombinant adeno-associated viral vectors. *Mol. Ther.* **11,** 320–326.

Moullier, P., and Snyder, R. O. (2008). International efforts for recombinant adeno-associated viral vector reference standards. *Mol. Ther.* **16,** 1185–1188.

Nakai, H., Montini, E., Fuess, S., Storm, T. A., Grompe, M., and Kay, M. A. (2003). AAV serotype 2 vectors preferentially integrate into active genes in mice. *Nat. Genet.* **34,** 297–302.

Penaud-Budloo, M., et al. (2008). Adeno-associated virus vector genomes persist as episomal chromatin in primate muscle. *J. Virol.* **82,** 7875–7885.

Ponder, K. P. (2011). Hemophilia gene therapy: A holy grail found. *Mol. Ther.* **19,** 427–428.

Potter, M., et al. (2008). Manufacture and stability study of the recombinant adeno-associated virus serotype 2 vector reference standard. *Bioprocessing* **7,** 8–14.

Snyder, R. O., and Flotte, T. R. (2002). Production of clinical-grade recombinant adeno-associated virus vectors. *Curr. Opin. Biotechnol.* **13,** 418–423.

Snyder, R. O., Xiao, X., and Samulski, R. J. (1996). Production of recombinant adeno-associated viral vectors. *In* "Current Protocols in Human Genetics," (N. Dracopoli, et al., eds.), pp. 12.11.11–12.11.24. John Wiley and Sons, New York.

Stedman, H., Wilson, J. M., Finke, R., Kleckner, A. L., and Mendell, J. (2000). Phase I clinical trial utilizing gene therapy for limb girdle muscular dystrophy: Alpha-, beta-, gamma-, or delta-sarcoglycan gene delivered with intramuscular instillations of adeno-associated vectors. *Hum. Gene Ther.* **11,** 777–790.

Sun, X., Lu, Y., Bish, L. T., Calcedo, R., Wilson, J. M., and Gao, G. (2010). Molecular analysis of vector genome structures after liver transduction by conventional and self-complementary adeno-associated viral serotype vectors in murine and nonhuman primate models. *Hum. Gene Ther.* **21,** 750–761.

Toromanoff, A., *et al.* (2008). Safety and efficacy of regional intravenous (RI) versus intramuscular (IM) delivery of rAAV1 and rAAV8 to nonhuman primate skeletal muscle. *Mol. Ther.* **16,** 1291–1299.

Veldwijk, M. R., *et al.* (2002). Development and optimization of a real-time quantitative PCR-based method for the titration of AAV-2 vector stocks. *Mol. Ther.* **6,** 272–278.

Wagner, J. A., *et al.* (1999). Safety and biological efficacy of an adeno-associated virus vector-cystic fibrosis transmembrane regulator (AAV-CFTR) in the cystic fibrosis maxillary sinus. *Laryngoscope* **109,** 266–274.

Wang, Z., *et al.* (2005). Adeno-associated virus serotype 8 efficiently delivers genes to muscle and heart. *Nat. Biotechnol.* **23,** 321–328.

Wistuba, A., Kern, A., Weger, S., Grimm, D., and Kleinschmidt, J. A. (1997). Subcellular compartmentalization of adeno-associated virus type 2 assembly. *J. Virol.* **71,** 1341–1352.

Zhen, Z., Espinoza, Y., Bleu, T., Sommer, J. M., and Wright, J. F. (2004). Infectious titer assay for adeno-associated virus vectors with sensitivity sufficient to detect single infectious events. *Hum. Gene Ther.* **15,** 709–715.

CHAPTER SIXTEEN

NIH Oversight of Human Gene Transfer Research Involving Retroviral, Lentiviral, and Adeno-associated Virus Vectors and the Role of the NIH Recombinant DNA Advisory Committee

Marina O'Reilly, Allan Shipp, Eugene Rosenthal, Robert Jambou, Tom Shih, Maureen Montgomery, Linda Gargiulo, Amy Patterson, *and* Jacqueline Corrigan-Curay

Contents

1. Introduction	314
2. A Brief History of the *NIH Guidelines* and Recombinant DNA Research Oversight	315
2.1. Development of the *NIH Guidelines*	315
2.2. The Recombinant DNA Advisory Committee (RAC) and human gene transfer	316
2.3. The RAC today	317
3. Role of the NIH and RAC in Review and Oversight of Human Gene Transfer Involving Retroviral, Lentiviral, and Adeno-associated Virus Vectors	322
3.1. Retroviral vector gene transfer	322
3.2. Lentiviral vector gene transfer	327
3.3. Adeno-Associated Virus (AAV) vector gene transfer	330
4. Conclusion	334
References	334

Abstract

In response to public and scientific concerns regarding human gene transfer research, the National Institutes of Health (NIH) developed a transparent oversight system that extends to human gene transfer protocols that are either

Office of Biotechnology Activities, National Institutes of Health, Bethesda, Maryland, USA

Methods in Enzymology, Volume 507
ISSN 0076-6879, DOI: 10.1016/B978-0-12-386509-0.00016-8

conducted with NIH funding or conducted at institutions that receive NIH funding for recombinant DNA research. The NIH Recombinant DNA Advisory Committee (RAC) has been the primary advisory body to NIH regarding the conduct of this research. Human gene transfer research proposals that are subject to the *NIH Guidelines for Research Involving Recombinant DNA Molecules (NIH Guidelines)* must be submitted to the NIH Office of Biotechnology Activities (OBA), and protocols that raise novel scientific, safety, medical, ethical, or social issues are publicly discussed at the RAC's quarterly public meetings. OBA also convenes gene transfer safety symposia and policy conferences to provide a public forum for scientific experts to discuss emerging issues in the field. This transparent system of review promotes the rapid exchange of important scientific information and dissemination of data. The goal is to optimize the conduct of individual research protocols and to advance gene transfer research generally. This process has fostered the development of retroviral, lentiviral, and adeno-associated viral vector mediated gene delivery.

1. INTRODUCTION

The National Institutes of Health (NIH) review and oversight process for human gene transfer research was established in response to public and scientific concerns about the novel scientific, medical, ethical, and social considerations raised by this area of clinical research. As the major funder of human gene transfer research along with the basic science that underpins it, the NIH has an important responsibility for the appropriate stewardship of this area of scientific activity and has exercised oversight of such research through the establishment of the Recombinant DNA Advisory Committee (RAC) and the *NIH Guidelines for Research Involving Recombinant DNA Molecules (NIH Guidelines)* (http://oba.od.nih.gov/rdna/nih_guidelines_oba.html).

The *NIH Guidelines* apply to all recombinant DNA research that is conducted at, or sponsored by, institutions which receive NIH funding for recombinant DNA research, and compliance with the *NIH Guidelines* is a term and condition of those grants. The *NIH Guidelines* were developed in consultation with the RAC, a committee that advises the NIH Director on issues pertaining to recombinant DNA research. The *NIH Guidelines* set forth a specific oversight system for human gene transfer research, with responsibilities for oversight of such research being shared by the NIH Office of Biotechnology Activities (OBA) and the local Institutional Biosafety Committees (IBCs). Under the *NIH Guidelines*, human gene transfer research is defined as the deliberate transfer of recombinant DNA, or DNA or RNA derived from recombinant DNA, into one or more human research participants. Human gene transfer research proposals that are subject to the *NIH Guidelines* must be submitted to OBA and be reviewed by the members of the RAC to determine whether they raise novel scientific, safety, medical, ethical, or social issues that warrant in-depth review and public discussion at one of the RAC's quarterly public meetings.

The NIH review process allows for an in-depth examination of the issues associated with this technology in a setting that permits public input, and a webcast of the meeting and the meeting materials remain available on OBA's website. This format of an open discussion has two important benefits. First, it makes this information available to scientists who can then incorporate new scientific findings and ethical considerations into the design of their trials. The efficiency of the research system is improved by allowing scientists to build on a common foundation of new knowledge emanating from this ongoing process of analysis and assessment. Second, it creates enhanced public awareness and allows for a public voice in the review of the safety and ethics of gene transfer research. This helps assure the public that scientists are attending to these important matters and sustains confidence in this area of research.

2. A Brief History of the *NIH Guidelines* and Recombinant DNA Research Oversight

2.1. Development of the *NIH Guidelines*

The history of the development of oversight for recombinant DNA research has been well documented (Frederickson, 2001) and will only be briefly reviewed. With the advent of recombinant DNA techniques in the early 1970s, a debate arose in the scientific community regarding the potential health and environmental risks of genetically manipulated organisms. The public also became involved in this debate. The scientific community responded to these concerns by voluntarily imposing a moratorium on certain experiments with recombinant DNA until the risks could be further characterized and procedures developed to minimize the risks. Scientists called for the formation of a national oversight body to ensure public discussion and ongoing oversight of this emerging technology. One of the outcomes of this process was the formation in 1974 of the RAC, an NIH advisory committee, which was tasked with developing the *NIH Guidelines*. The *NIH Guidelines* articulate standards for investigators to follow and to ensure the safe handling and containment of recombinant DNA and products derived therefrom. The *NIH Guidelines* outline the requirements for institutional oversight, including the establishment of local IBCs, and describe the responsibilities and procedures of the RAC. At the federal level, the *NIH Guidelines* are administered by OBA within the Office of the NIH Director. While compliance with the *NIH Guidelines* is mandatory for all investigators at institutions receiving any NIH funds for research involving recombinant DNA, the *NIH Guidelines* have become a universal standard for safe scientific practice in this area of research and are followed voluntarily by many private entities and other institutions not otherwise subject to their requirements.

2.2. The Recombinant DNA Advisory Committee (RAC) and human gene transfer

The *NIH Guidelines* were developed prior to the advent of human gene transfer research. In the 1980s, President Carter requested the President's Commission for the Study of Ethical Problems in Medicine and Biomedical and Behavioral Research to examine the use of recombinant techniques for "genetic engineering" in humans (President's Commission, 1982). The Commission issued a report recommending a broadening of the RAC's responsibilities to include issues raised by clinical research using these techniques (President's Commission, 1982). In 1983, in response to the President's Commission Report (Rainsbury, 2000), the RAC established a working group, the Human Gene Therapy Subcommittee of the RAC, which issued guidance to investigators in the form of a "Points to Consider in the Design and Submission of Human Somatic-Cell Gene Therapy Protocols" (Areen and King, 1990). These Points to Consider evolved into Appendix M of the *NIH Guidelines*. Appendix M outlines the roles and responsibilities of individual investigators, NIH-funded institutions, IBCs, the RAC, OBA, and the NIH Director in the conduct of human gene transfer research. It also provides guidance for optimal design of preclinical and clinical research and standards for informed consent. Investigators submitting new human gene transfer research protocols to NIH must provide written responses to a series of questions outlined in the current version of Appendix M, Sections M-II through M-V, of the *NIH Guidelines*.

In 1988, the first human gene transfer protocol was reviewed by the RAC and ultimately approved by the NIH Director in 1989. Initially, protocols were reviewed by the Human Gene Therapy Subcommittee and their recommendations were conveyed to the RAC for consideration. In 1992, this process was streamlined and protocols were directly reviewed by the full RAC which made recommendations to the NIH Director regarding approval (57 FR 14774). As the understanding of human gene transfer research increased, the role of NIH and the RAC with respect to human gene transfer evolved.

In the mid-1990s, the NIH Director convened two advisory committees of outside experts to conduct comprehensive reviews of the RAC, its role in oversight of human gene transfer, and the state of human gene transfer research in general. These committees concluded, and the NIH Director concurred, that gene therapy continued to pose scientific, safety, and ethical issues warranting public consideration. They also concluded that the value of the RAC could be augmented by having it focus on gene transfer protocols that were novel in significant ways or that posed important scientific, safety, or ethical issues and whose public discussion would help inform oversight bodies (Federal Food and Drug Administration (FDA), Institutional Review Boards (IRBs), IBCs), as well as the scientific community at large (60 FR 20726). The RAC was also tasked with assisting NIH in identifying "broad scientific, safety, social, and ethical issues relevant to gene therapy research as potential Gene Therapy Policy Conference Topics" (*NIH Guidelines* Section IV-C-2-e).

2.3. The RAC today

As a reflection of the 360° examination of science, safety, and ethics in human gene transfer research, which the RAC is charged to undertake, the committee today consists of up to 21 experts from the fields of virology, biosafety, clinical oncology/hematology, cardiology, infectious diseases, statistics, law, and bioethics (the current RAC roster can be found at http://oba.od.nih.gov/rdna_rac/rac_about.html). This broad range of expertise is necessary given the diversity of the trials that the RAC reviews (Fig. 16.1). Of note, the RAC always includes a public representative who is not trained in one of the scientific fields related to human gene transfer.

As mentioned earlier, a key focus of the Committee is the review of protocols presenting novel scientific, safety, or ethical issues in human gene transfer research. Each year approximately 60 new gene transfer protocols are registered with OBA and reviewed by RAC members (Fig. 16.2). Approximately 15–20% of these protocols are selected for in-depth review and public discussion at the RAC's quarterly meetings. The majority of clinical trials reviewed by the RAC are designated as Phase I trials and are often the first clinical application of the gene transfer agent (Fig. 16.3).

At its meetings, which are not only open to the public but also webcast and archived on NIH/OBA's website (http://oba.od.nih.gov/rdna_rac/rac_meetings.html), the RAC examines the scientific, clinical, and ethical dimensions of the human gene transfer protocol and makes recommendations to the investigator regarding trial design. These recommendations are reviewed by OBA on behalf of the NIH Director, and a letter outlining the recommendations is sent to the investigator, sponsor, IBC, and IRB, as well as to the FDA and the Office of Human Research Subjects Protection, both of which have nonvoting representation on the RAC.

OBA, in consultation with the RAC, conducts ongoing analysis of the safety of human gene transfer research. In 2001, in the aftermath of the death of a young research participant enrolled in a gene transfer trial (NIH report, 2002), the Gene Transfer Safety Assessment Board (GTSAB), a formal subcommittee of the RAC, was established to enhance the identification of significant safety issues, increase public knowledge, and strengthen

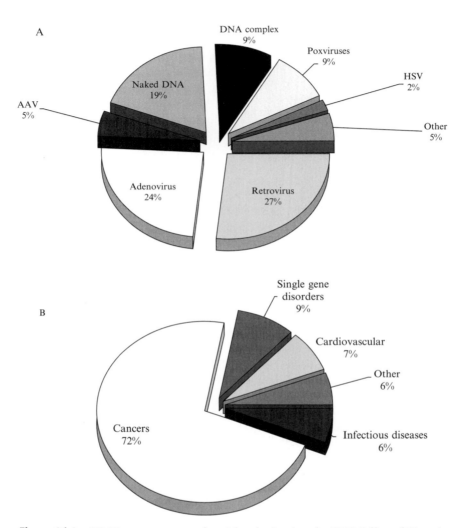

Figure 16.1 (A) Human gene transfer trials submitted to the NIH Office of Biotechnology Activities (OBA) by delivery system. (B) Human gene transfer trials by clinical indication.

the protection of research participants in human gene transfer research studies (66 FR 57970). The GTSAB consists of RAC members and meets quarterly with OBA staff in advance of each RAC meeting to review all serious adverse events (SAEs) that are judged to be possibly or probably related to the gene transfer. In addition, the GTSAB reviews summaries of all amendments to active gene transfer protocols and annual reports submitted to OBA each quarter. The GTSAB reports out to the RAC at each meeting, and in addition to informing the reviews of other protocols, this

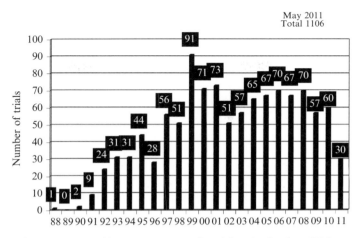

Figure 16.2 Number of human gene transfer trials submitted to OBA per year.

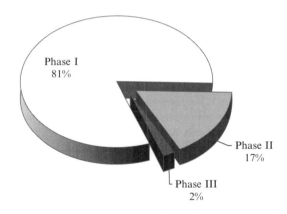

Figure 16.3 Human gene transfer trials by phase of trial.

activity often leads to presentations at RAC meetings by principal investigators regarding important developments in their protocol or leads to NIH safety symposia and other policy conferences that address issues common to a number of protocols. A public summary of the material reviewed by the GTSAB is made available by OBA after the meeting. Summaries of adverse events are posted as Data Management Reports along with the RAC meeting material. Public summaries of amendments and annual reports are included in the protocol information in the Genetic Modification Clinical Research Information System (GeMCRIS), a database that provides public information on over 1000 gene transfer protocols (http://www.gemcris.od.nih.gov/Contents/GC_HOME.asp).

As part of ongoing efforts to inform the scientific community about important issues pertinent to research participant safety, OBA periodically hosts gene transfer safety and policy symposia. These symposia are public forums for scientific experts to discuss emerging medical, scientific, ethical, and safety issues in clinical gene transfer research. By fostering discussion and information exchange, the symposia help to: (1) enhance understanding of the safety and toxicity of gene transfer, (2) identify critical gaps in current knowledge, (3) maximize research participant safety, (4) enhance informed consent processes, and (5) optimize the development of gene transfer clinical trials.

Finally, in addition to its responsibility in the review and oversight of clinical research, the RAC also advises the NIH on nonclinical research that is subject to the *NIH Guidelines*. This includes advising the NIH on certain experiments that raise important public health considerations, including certain research involving the introduction of drug resistance into human pathogens or research with highly pathogenic organisms. The RAC advises NIH on all amendments to the *NIH Guidelines*. Indeed, the evolution of the RAC's role in oversight of human gene transfer and the necessary amendments to the *NIH Guidelines* to effect these changes were all undertaken in consultation with the RAC.

An obvious question is how the role of the NIH, including the RAC, differs from that of FDA, which has regulatory authority over clinical human gene transfer research. While the two agencies share common concerns when reviewing any human gene transfer clinical trial, there are important differences. The RAC typically not only addresses the important safety issues in Phase I trials but also takes a more forward-looking assessment of the trial, challenging the investigator to think to the next stage in development and whether the trial, as designed, can answer the research questions necessary to continue translation of the particular agent. In addition, the RAC includes expertise from the biosafety community and addresses biosafety concerns relating to the administration of the human gene transfer product, including the safety of health care workers involved in the protocol, and the public. An in-depth review of the ethical issues raised by the trial design and the adequacies of the informed consent are also critical parts of the RAC's review. In fact, the review of many informed consent documents led to the development of an NIH Guidance on Informed Consent for Gene Transfer Trials that is available to all researchers; RAC input was critical to this document (see http://oba.od.nih.gov/oba/rac/ic/index.html). Finally, the RAC process is unique in its transparency. All proceedings are public and the webcast, minutes, and all presentations are available to investigators and the public through OBA's website. Investigators are able to benefit from the reviews of other gene transfer protocols and to understand emerging trends in clinical trial design. As discussed in more detail below, through the RAC, NIH has provided a forum for open debate on emerging issues in the field while educating the public on such research. These discussions help to inform the deliberations of the FDA, the Office of Human Research Protections, IRBs, and IBCs, enhancing their oversight activities.

There are also ongoing collaborations between FDA, NIH/OBA, and the RAC. As mentioned above, a member of FDA's Center for Biologics Evaluation and Research is one of the nonvoting federal representatives to the RAC. This representative receives copies of all reviews by the RAC and interacts with the Committee during meetings. In addition, the FDA is included in the meetings of the GTSAB. OBA seeks FDA input when planning safety symposia and policy conferences, and the FDA representative to the RAC keeps the RAC abreast of relevant new regulatory developments. Similarly, NIH/OBA provides comment on proposed changes to FDA regulations and has worked with FDA to harmonize the agencies' requirements with respect to the conduct of clinical trials. For example, in 2001, the *NIH Guidelines* were amended so that reporting of safety data under the *NIH Guidelines* is harmonized with the regulatory reporting requirements of FDA (66 FR 57970). In November 2006, the FDA issued the *Guidance for Industry: Gene Therapy Clinical Trials-Observing Subjects for Delayed Adverse Events* (http://www.fda.gov/BiologicsBloodVaccines/GuidanceComplianceRegulatoryInformation/Guidances/CellularandGeneTherapy/ucm072957.htm), and OBA adopted the recommendations for evaluating the adequacy of long-term follow-up under the *NIH Guidelines*.

In addition to the exchanges of information between the RAC and FDA, NIH and FDA collaborated on the development of GeMCRIS, an electronic data system that is used by OBA, the RAC, the public, and FDA. GeMCRIS has a publicly available interface which provides information on over 1000 human gene transfer protocols. In addition to providing the public a database in which they can search for gene transfer protocols based on title, clinical indication, investigator, phase of the trial and status, GeMCRIS contains detailed information on gene transfer vectors. Investigators and the public can search for protocols using a particular vector, the transgene expressed by the vector, or the disease indication.

The database contains safety data on all gene transfer clinical trials registered with OBA and reviewed by the RAC. Investigators can electronically submit SAEs to GeMCRIS and satisfy the reporting requirements under the *NIH Guidelines*. A copy of that report can also be submitted to FDA to satisfy their safety reporting requirements. Only OBA and FDA staff have access to all reports that are submitted to GeMCRIS and this relational database allows OBA and FDA to search adverse event reports across protocols based on the clinical application, vector, transgene, or the nature of the clinical event.

In summary, the NIH's and RAC's role has evolved with the field of human gene transfer research. While the RAC began as a body that reviewed all protocols and made recommendations regarding approval to the NIH Director, the current role of the RAC is more focused on identifying new scientific, clinical, and ethical issues and advancing the development of the field through an open, transparent review of new data. The specific role of NIH and the RAC in the review and oversight of gene transfer research involving retroviral, lentiviral, and adeno-associated virus (AAV) vectors is outlined below.

3. Role of the NIH and RAC in Review and Oversight of Human Gene Transfer Involving Retroviral, Lentiviral, and Adeno-associated Virus Vectors

3.1. Retroviral vector gene transfer

A retroviral vector was used in the first clinical gene transfer protocol approved by the NIH in 1989. They were the only vector system used for the subsequent three years (Fig. 16.4). Retroviral vectors were proposed for the first marking protocol (i.e., using the vector to track an intervention), the first protocol for cancer, for a monogenic disease (adenosine deaminase deficient-severe combined immunodeficiency, ADA-SCID) and for an infectious disease (human immunodeficiency virus-1, HIV) (Fig. 16.5). Retroviral vectors are now used for a wide range of clinical indications, accounting for 27% of the total protocols submitted to OBA. The majority of protocols involve *ex vivo* administration, that is, transduction of a cell by the vector and then administration of the modified cell to humans. Thirty-two percent of such protocols have used hematopoietic cells, often hematopoietic stem cells, in order to address genetic diseases characterized by defective development of one or more hematopoietic cell lines. However, retroviral vectors have also been used to transduce a variety of other target cell types and have been administered *in vivo*, that is, directly, in 6% of protocols (Fig 16.6).

A turning point in the use of retroviral vectors came in August 2002, when a French gene transfer team reported an SAE following gene transfer for X-linked severe combined immunodeficiency (X-SCID). X-SCID is one of several genetic diseases characterized by severe immunodeficiency due to a genetic mutation that prevents development of a functional immune system. In the absence of a bone marrow transplant, X-SCID leads to early mortality. In X-SCID, a mutation in the interleukin-2 receptor subunit gamma (*IL2RG*) gene, which encodes the cytokine receptor common gamma chain (γ), leads to a complete lack of T cells and natural killer cells. The French study involved administration of the research participant's $CD34^{+}$ hematopoietic stem cells that had been transduced

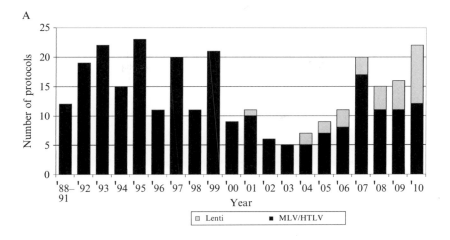

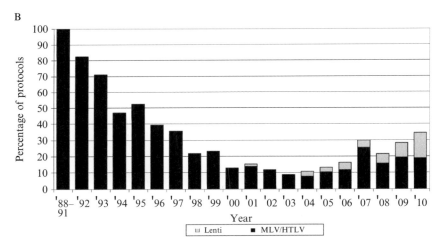

Figure 16.4 Trends in retroviral and lentiviral vector usage. (A) Number of protocols involving retroviral and lentiviral vectors submitted to OBA by year. (B) Protocols involving retroviral and lentiviral vectors as a percentage of total human gene transfer protocols submitted to OBA.

with a retroviral vector encoding the common gamma chain transmembrane protein subunit shared by receptors for interleukins 2, 4, 7, 9, 15, and 21. While nine research participants experienced significant restoration of their immune systems, one research participant developed leukemia approximately 3 years after gene transfer. In the expanded T cell clone, the retroviral vector had integrated into and activated LIM domain only-2 (*LMO-2*), a proto-oncogene (Hacein-Bey-Abina *et al.*, 2003). This case of leukemia, reported in August 2002, was the first report of cancer

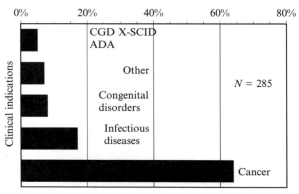

Figure 16.5 Retroviral vector protocols by clinical indication.

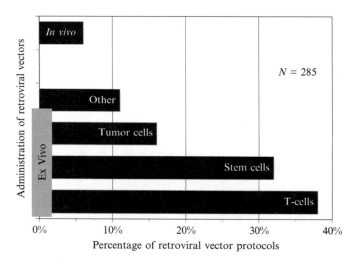

Figure 16.6 Administration of retroviral vectors.

development due to vector-mediated insertional mutagenesis in a human gene transfer trial.

In response to this report, in December 2002, the RAC convened a safety symposium to explore the adverse event and identify the knowledge gaps. The investigators presented the clinical case and the molecular analysis of the retroviral vector integration. Utilizing data from GeMCRIS, OBA provided an overview of all registered retroviral vector protocols and described GeMCRIS' utility in identifying similar events among these trials—particularly those in trials involving transduction of stem cells. While malignancies were identified in a small number of trials using retroviral vectors, in none of these cases could the malignancy be directly attributed to the integration of the vector, as in the X-SCID case. This finding underscored the novelty of the development of leukemia as a consequence of the gene transfer. The RAC discussion focused on developing points to consider for investigators preparing protocols involving retroviral vector gene transfer (addressing what additional preclinical studies might be used to help identify this risk), recommendations for subject monitoring, and changes to informed consent documents (http://oba.od.nih.gov/rdna_rac/rac_symposium_xscid.html).

Shortly after the December 2002 RAC meeting, a case of leukemia was reported in another research participant in the same trial (Hacein-Bey-Abina et al., 2003). OBA immediately sent a memorandum, in January 2003, to principal investigators and IBCs at institutions conducting human gene transfer trials that employ retroviral vectors to notify the field of the second SAE and to help promote the safe conduct of clinical trials and ensure that informed consent documents for these trials provided complete information regarding the risks for potential research participants (http://oba.od.nih.gov/oba/rac/SSMar05/pdf/XSCID_letter2003.pdf). At the time, the NIH urged investigators conducting retroviral vector-mediated gene transfer trials in hematopoietic stem cells to discontinue enrollment and vector administration until new data became available, and the possible etiology and risks of these events were considered by the appropriate federal advisory committees, including the RAC. The NIH coordinated with FDA and scientific associations to keep the investigator community informed. An unscheduled meeting of the RAC was convened in February 2003 to review the new data. The NIH Director accepted the RAC recommendations that retroviral gene transfer for X-SCID be limited to research participants who failed standard therapy (stem cell transplant) and that new trials be evaluated on a case-by-case basis with analysis of the risks and benefits of the trial, as well as the informed consent process (http://oba.od.nih.gov/rdna_rac/rac_symposium_xscid.html).

Subsequently, two more cases of leukemia were reported in the French trial and one in a similar X-SCID trial in the United Kingdom. All of the cases of leukemia appeared to share the common causative mechanism of insertional mutagenesis at or near oncogenes (Howe et al., 2008). OBA asked the RAC to review each of these events, which provided a forum for transparent and open discussions among investigators involved in research using retroviral vectors and the public. The most comprehensive discussion was held in March 2005. The two-day safety symposium included comprehensive updates on current US and international gene transfer trials for SCID diseases, including X-SCID, which employed similar vectors.

As new protocols for X-SCID were submitted to OBA, the RAC revisited its recommendations during the in-depth review and public discussion of these protocols. Recognizing that modifications in vectors may reduce the risk of insertional mutagenesis compared to the retroviral vectors used in the original X-SCID trials, the RAC recommended that while gene transfer should not yet be used in research participants who have an HLA-identical related donor for stem cell transplantation, in certain cases it could be used in lieu of haploidentical transplant given the relative risks and benefits of the two approaches.

The RAC's deliberations over time have provided an open forum for scientists, clinicians, regulators, and ethicists to share new data on these important events and have assured the public that the decision to continue using these vectors in both X-SCID and other gene transfer trials was carefully considered, science-based, and grounded in a careful ethical analysis of the risks and benefits to future research participants.

The clinical successes seen in the X-SCID trials have endured (Hacein-Bey-Abina et al., 2010) as have clinical benefits in other trials using these retroviral vectors, such as ADA-SCID (Auiti et al., 2009). Unfortunately, the risk of malignancy secondary to retroviral vector integration into the genome is not limited to the X-SCID trials. Two cases of myelodysplastic syndrome occurred in a German trial for chronic granulomatous disease (CGD), with one resulting in the death of the subject. Both of these events involved clonal myeloproliferation with vector insertion into proto-oncogenes (e.g., *MDS1-EVI1, PRDM16, SETBP1*; Stein et al., 2010). Most recently, a case of leukemia was reported in one research participant in a German trial for Wiskott-Aldrich Syndrome (Fischer et al., 2011) in which several participants had experienced reconstitution of the hematopoietic system (Boztug et al., 2010). In each of these trials, retroviral vectors were used to modify hematopoietic stem cells.

In light of the problem of insertional mutagenesis with these retroviral vectors, investigators have largely turned to lentiviral vectors or have modified γ retroviral vectors to decrease the risk of enhancer-mediated insertional mutagenesis. Critical questions remain regarding which preclinical models can best predict the safety of these new vectors, the optimum monitoring assays to use during trials, and how to best explain to research participants the potential long-term risks of these new vectors.

To assist in the review of new protocols, as well as promote the exchange of new data regarding these questions, in December 2010, OBA and the European Network for the Advancement of Clinical Gene Transfer and Therapy (CliniGene) cohosted a scientific symposium on "Retroviral and Lentiviral Vectors for Long-Term Gene Correction: Clinical Challenges in Vector and Trial Design." The discussion focused on the approaches being used for vector integration site detection, proposed safety modifications to retro and lentiviral vectors that may lower the risks of insertional mutagenesis, and the strengths and limitations of the available *in vitro* assays and animal models for predicting the human gene transfer experience. Recommendations regarding how best to monitor research participants for the development of a clonal dominance (which was a precursor to the development of the cases of leukemia in the previous trials) and criteria for determining stopping rules were also discussed. Ethical discussions focused on how to design trials with novel vectors or targets given the uncertainty in defining the risks of insertional mutagenesis. For example, what diseases should be selected and which population of research participants enrolled in the first trial with a new vector? What information should be provided in the informed consent process given the uncertainty regarding the risk of insertional mutagenesis due to the limits of preclinical models to compare vectors with proposed safety modifications to those used previously? The conference materials and a webcast are available on OBA's website (http://oba.od.nih.gov/rdna/rdna_symposia.html#CONF_003h), and a summary of the proceedings is being developed.

3.2. Lentiviral vector gene transfer

Consideration of the potential clinical advantages as well as the risks associated with the use of vectors derived from lentiviruses, such as HIV-1, occurred years in advance of the first submission of a protocol involving a lentiviral vector for human gene transfer. Compared to other retroviral vectors, lentiviral vectors offer the advantage of being able to transduce nondividing cells. However, there were some concerns with using a vector that is derived from a virus that causes serious disease in humans, including the potential for generation of replication competent virus. Such concerns led OBA to request that the RAC discuss these particular vectors prior to their proposed use in a clinical research trial. At the December 1997 RAC meeting, lentiviral vectors were included in a forum on new technologies, an initiative intended to address novel technologies in human gene transfer research. To inform the RAC's future reviews of lentiviral vector protocols, the meeting included an overview of the molecular genetics and biology of lentiviruses and the rationale for the development of vectors based on these viruses (http://oba.od.nih.gov/oba/rac/minutes/12151697.pdf).

In March 1998, an NIH Gene Therapy Policy Conference was held which devoted a more in-depth review of the issues associated with the future use of lentiviral vectors for human gene transfer. The conference included presentations covering the biology of lentiviral infections; the development of vector systems derived from HIV-1, equine infectious anemia virus (EIAV), and feline immunodeficiency virus (FIV); and the challenges of vector production and preclinical studies. The biosafety considerations pertinent to clinical applications of lentiviral vectors were additionally discussed, including the possibility of mobilization or recombination of HIV-1-based vectors in individuals who are concurrently HIV-1 positive or become HIV-1-infected postdelivery of HIV-1-based vectors. The risks of insertional mutagenesis and germ line transmission were also considered. Proceedings of the conference are available on the OBA website at http://oba.od.nih.gov/oba/rac/Lenti030998/gtpcsumm2.pdf.

To inform the field further and help the investigators and sponsors prepare the first submission of a protocol involving a lentiviral vector to OBA and FDA, in March 2001, the RAC discussed the design of an HIV-1-derived vector and clinical trial in advance of a formal submission of the protocol for RAC review (http://oba.od.nih.gov/oba/rac/minutes/march82001.pdf). The vector in question was designed to express an antisense RNA to the HIV-1 envelope; expression of the antisense RNA was controlled by the HIV long terminal repeat (LTR) so as to limit expression to HIV-1-infected cells expressing the viral regulatory proteins, Tat, and Rev. Only HIV-infected research participants were to be enrolled. After these discussions, the protocol was subsequently submitted to OBA and underwent in-depth review and public discussion by the RAC in September 2001 (http://oba.od.nih.gov/oba/rac/minutes/RAC%20Min%20090601.pdf). The investigators were invited in 2004 to return to update the committee on the results of the clinical trial by providing a summary of the preclinical, nonhuman animal data; participant characteristics; visit schedule; and monitoring of the research participants (http://oba.od.nih.gov/oba/rac/minutes/RAC_minutes_06-04.pdf).

While the first lentiviral vector protocols were initially proposed for administration to HIV-1-infected individuals, the vector was soon used in protocols for other disease indications, beginning with a protocol for mucopolysaccharidosis VII (MPS VII) in 2002. To date, HIV-1 or EIAV-derived vectors have been used in 30 protocols registered with OBA for such applications as melanoma, lymphoma, leukemia, glioblastoma, ovarian cancer, beta-thalassemia, Fanconi's anemia, sickle cell anemia, diabetes, and adrenoleukodystrophy. The reports of SAEs due to insertional mutagenesis caused by γ retroviral vectors accelerated the use of lentiviral vectors for protocols involving transduction of $CD34^+$ cells for diseases such as X-SCID, ADA-SCID, and Wiskott–Aldrich, based on the theory that the risk of insertional mutagenesis may be lower with lentiviral vectors due to the difference in integration site preferences (i.e., lentiviruses usually insert within genes as compared to γ retroviruses, which tend to insert near transcriptional start sites, thereby increasing the likelihood that the integration of the vector will lead to gene expression). Modifications were also made to lentiviral vectors to decrease this risk further, including, for example, the development of self-inactivating (SIN) vectors with deletions in the 3' LTR, resulting in transcriptional inactivation of the integrated vector (Gilboa et al., 1986).

However, in 2009, a French gene transfer trial for thalassemia that employed a lentiviral vector expressing the human *β-globin* gene in autologous hematopoietic stem cells reported that the integration of the vector led to the development of a partial clonal dominance. While this was not a clinical adverse event, this unexpected biological finding led to a halt in enrollment pending further analyses. Unlike the leukemia cases in the retroviral trials, this clonal population of cells had no apparent clinical consequences, and indeed the research participant no longer required regular blood transfusions after the gene transfer. Integration site analysis of the dominant clone revealed that the vector had inserted into the third intron of the high mobility group AT hook 2 (*HMGA2*) gene, causing aberrant splicing to a cryptic site within the vector insulator sequence (Cavazzano-Calvo et al., 2010). The truncated *HMGA2* transcript lacked binding sequences for *Let*-7 microRNA (miRNA). The binding of the miRNA is involved in downregulation of *HMGA2* mRNA. The result was overexpression of *HMGA2* in erythroid cells (Cavazzano-Calvo et al., 2010).

As this was the first report of a nonenhancer-mediated mechanism of insertional mutagenesis, OBA chose to inform the investigator community of this development despite uncertainty about the clinical significance. As the investigators noted in their subsequent publication about this event, it is unclear if the clonal population of cells will remain stable or even disappear, but there was the possibility that this change could be a "prelude to multistep leukaemogenesis" (Cavazzano-Calvo et al., 2010). In December 2009, the investigators on this trial presented the data at a public meeting of the RAC. The RAC discussion focused on risk:benefit considerations (e.g., insertional mutagenesis vs. the transfusion independence achieved by the research participant) and ability and limitations of preclinical studies to predict human experience (http://oba.od.nih.gov/oba/RAC/meetings/Dec2009/RAC_Minutes_12-09.pdf). This event was also an impetus for the organization of the December 2010 gene transfer policy conference on the use of retroviral and lentiviral vectors for long-term gene correction. The knowledge shared regarding vector design, preclinical studies, and clinical trial design will also inform the lentiviral vector field going forward.

In addition to the recommendations that the RAC review generates for individual clinical trials using lentiviral vectors, OBA worked with the RAC to develop the Biosafety Guidance for Research with Lentiviral Vectors to provide specific guidance for laboratory research with lentiviral vectors (http://oba.od.nih.gov/rdna_rac/rac_guidance_lentivirus.html). Because the *NIH Guidelines* do not explicitly address containment for research with lentiviral vectors, the document provides additional guidance for IBCs and investigators on how to conduct a risk assessment for lentiviral vector research to determine appropriate containment. The guidance includes information on when replication competent lentivirus testing is appropriate, assessment of containment for the administration of vectors to animals, and the housing and husbandry of animals that are or are not permissive for lentivirus replication, as well as examples of risk assessments for different types of research.

3.3. Adeno-Associated Virus (AAV) vector gene transfer

The first AAV vector protocol was reviewed by the RAC in 1994, and to date, 51 protocols using AAV vectors have been submitted to OBA and undergone RAC review. Unlike the gene transfer field as a whole, where the majority of protocols are for cancer, the majority of protocols using AAV vectors target nonmalignant conditions such as cystic fibrosis, hemophilia B, muscular dystrophy, arthritis, alpha-1 antitrypsin deficiency, cardiac failure, Pompe disease, and galactosialidosis. AAV vectors have been used frequently via localized delivery for neurodegenerative disorders (e.g., Parkinson's, Alzheimer's, and Canavan diseases and epilepsy) and to the eye for ocular disorders (e.g., Leber's Congenital Amaurosis (LCA) and macular degeneration) (Fig. 16.7).

NIH has provided a public forum for discussion of a number of issues that have arisen with the clinical use of AAV vectors. For example, in 2001, when preclinical data from one animal model suggested a possible association between AAV vector gene transfer and tumorigenesis, in particular liver tumors, OBA, in conjunction with the FDA, convened the fourth National Gene Transfer Safety Symposium, "Safety Considerations in the Use of AAV Vectors in Gene Transfer Clinical Trials." The preclinical research involved the use of AAV vectors to treat a mouse model of MPS VII, a lysosomal storage disease caused by a deficiency in the enzyme β-glucuronidase. One of the symposium goals was to evaluate the increased incidence of tumors seen in the one MPS VII mouse model in the context of the other long-term preclinical animal and human clinical data accumulated on AAV vector use. The potential for AAV vector gene transfer to contribute to tumor formation was compared to other possible contributing factors, including mouse tumor biology, age of the mice at vector administration, the particular disease being studied, and receptor interference. In order to provide guidance for future studies, discussion focused on the proper design of preclinical studies to address the tumorigenic potential of AAV vectors and the necessity for modifications to AAV vector gene transfer clinical trial design, including the informed consent process. The RAC stopped short of making any recommendations restricting the use of AAV vectors for certain conditions or in certain age groups, focusing the discussion instead on the relative risks and benefits that should be considered (http://oba.od.nih.gov/rdna/rdna_symposia.html#CONF_003c).

Another issue that arose early in the use of AAV vectors was the detection of this virus in the semen of research participants in a Phase I safety study for individuals with severe hemophilia B (Manno et al., 2006). The AAV vector expressed the gene for human Factor IX and was

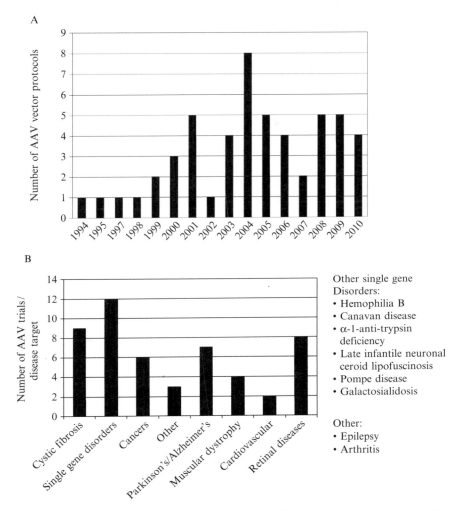

Figure 16.7 (A) Number of protocols involving AAV vectors submitted to OBA by year. (B) Protocols involving AAV vectors by clinical indication.

administered into the hepatic artery to maximize delivery to the liver. The possibility of horizontal or vertical transmission of gene transfer vectors was one factor that led to heightened scrutiny of gene transfer research; therefore, detection of the vector in semen was concerning. At the December 2001 (http://oba.od.nih.gov/oba/rac/minutes/Dec01minutes.pdf) and March 2002 RAC meetings (http://oba.od.nih.gov/oba/rac/minutes/March2002.pdf), RAC members and *ad hoc* experts discussed the finding of the AAV vector in semen of research participants. The data from the study were presented and analyzed, and the potential significance of these data for the risk of vertical transmission was discussed. The RAC provided guidance on potential assays and preclinical studies of germ line transmission, revisions to the protocol regarding sequential enrollment, and revisions to the informed consent document.

The hemophilia B trial recommenced enrollment and an adverse event involving an asymptomatic elevation in serum liver enzymes subsequently was reported in a research participant. This asymptomatic elevation in liver enzymes corresponded with a decrease in the transgene expression. The principal investigators hypothesized that a T cell response against the AAV capsid expressed in liver cells may have led to an immune response against the transduced liver cells (Manno et al., 2006). Similar findings were observed in a trial for lipoprotein lipase deficiency (Mingozzi et al., 2009). These findings had potential implications for gene transfer trials using AAV vectors, in particular systemic use, both for potential safety reasons as well as efficacy. The potential for cell-mediated immune responses to AAV vectors was examined in-depth at a safety symposium held by OBA in June 2007. The goal of the symposium was to increase awareness in the scientific community and the public of this emerging data and identify the surveillance and assays that could be used to better characterize this event. The presentations were designed to update the members of the RAC on the recent developments and to help inform the RAC's future reviews of clinical trials using AAV vectors as well as inform the field and the public (http://oba.od.nih.gov/oba/rac/meetings/jun07/AAV%20summary%20June%202007f.pdf).

Consideration of the potential T cell-mediated immune reaction to AAV vectors proved timely when, in July 2007, a participant in an arthritis trial that used an AAV vector died approximately three weeks after vector administration. The trial was a Phase I/II study of repeat intraarticular administration of an AAV vector encoding a tumor necrosis factor (TNF) receptor linked to the Fc fragment of an immunoglobulin molecule (IgG1) designed to block TNF-α activity. As per the sponsor of this trial, the DNA sequence of this transgene was identical to that used to produce etanercept (EMBREL®), an FDA-licensed and systemically administered TNF-α antagonist. Research participants were eligible for the trial if they had inflammatory arthritis and an inflamed joint despite current therapies, including the use of a systemic TNF-α antagonists such as etanercept.

The death received considerable media attention, including the question of whether it was due to the gene transfer (Weiss, 2007). Significant elevations in the subject's serum liver enzymes and reports of possible liver failure raised further questions regarding a possible role of the AAV vector in the death. To address these issues, in September 2007, OBA organized a special public meeting of the RAC to conduct an in-depth discussion of this event (http://oba.od.nih.gov/rdna_rac/rac_sae_protocol.html). Participants included the clinical team from the University of Chicago that had cared for the participant, outside experts, and the FDA. The RAC meeting provided a public overview of rheumatoid arthritis biology, the role of TNF-α antagonists, safety data from AAV vector-mediated gene transfer (in general and in the specific protocol), and the clinical case and autopsy data. The RAC assessed the possible role of gene transfer in the death and the need for revisions to this and other studies using AAV vectors, including the informed consent document and subject selection. The RAC concluded that the immediate cause of death was most likely a large retroperitoneal hemorrhage and disseminated histoplasmosis but also that additional data were needed to determine whether the gene transfer may have played any role in the clinical events leading up to her death.

OBA continued to work with members of the RAC GTSAB, the University of Chicago Hospital staff, the trial sponsor, FDA, and outside experts to gather and analyze new data. At the December 2007 RAC meeting, updated product testing data, additional quantitative PCR data regarding distribution of the vector, and information on the research participant's serum levels of the TNF-α antagonist were reviewed. Based on this additional data, the RAC concluded that none of the available laboratory evidence supported a conclusion that the intraarticular injection of the gene transfer vector contributed to the research participant's clinical course or death. The RAC confirmed its initial conclusion that the death of the research participant was primarily a result of an opportunistic infection, disseminated histoplasmosis with subsequent bleeding complications, and multiorgan failure. The apparent risk factor for such an infection was her systemic rheumatoid arthritis therapy, chiefly the TNF-α antagonist adalimumab. A possible immunologic response to the AAV vector could not be definitively ruled out due to a lack of data on AAV capsid-specific T cells, because blood samples to perform such testing were not available. Nonetheless, the RAC posited that even if such an immune response did occur, it was not causal in the death.

This event underscored the importance of collecting adequate blood and tissue samples for any research protocol, especially when the intervention targets complex physiologic systems such as the immune system or metabolic pathways. Redundancy in sample collection is important, since it may be difficult in advance to anticipate all of the tests that may be needed. This becomes particularly evident when it is necessary to determine causality of an adverse event. The RAC chair, Dr. Howard Federoff, noted that review of this unfortunate event could provide generalizable lessons to enhance the safety of other gene transfer trials and provide potential strategies to gather quickly the data needed for safety evaluations in case of an SAE (http://oba.od.nih.gov/rdna_rac/rac_sae_protocol.html).

In addition to the questions raised regarding the safety of AAV vectors, trials using AAV gene transfer research have also presented ethical issues.

The selection of pediatric diseases (e.g., Canavan disease, LCA) provided the impetus for several RAC general discussions regarding the ethics of gene transfer in pediatric populations (e.g., December 2005; http://oba.od.nih.gov/oba/rac/minutes/RAC_minutes_12-05.pdf). Other issues discussed include the incorporation of sham neurosurgical arms in the Parkinson's and Alzheimer's disease trials that used AAV vectors and involved intracranial delivery. The complex scientific, ethical, and social issues raised by the use of sham neurosurgical procedures, including trial design, subject recruitment, risk assessment, and informed consent were discussed at a safety symposium in June 2011 that was cosponsored by OBA and the NIH National Institute of Neurological Diseases and Stroke (http://oba.od.nih.gov/rdna_rac/rac_sham_con.html)

4. Conclusion

Through its work with the RAC, NIH continues to provide a public forum for discussion of individual gene transfer protocols and emerging issues of importance to the field. The RAC activities are designed to assist individual investigators in the design of their trials and to thereby optimize the safe and ethical conduct of gene transfer trials for research participants. The transparent exchange of information on individual protocols as well as the development of generalizable knowledge through special safety symposia and gene transfer policy conferences foster the dissemination of knowledge among scientists and promote the advancement of this field. The public is kept informed and is provided the opportunity to participate in discussions about a field that increasingly shows the potential for providing new therapeutics for unmet clinical needs. This transparency assures the public that scientific, clinical, and ethical issues raised by these interventions are being carefully considered and maintains confidence in this promising field of clinical research.

REFERENCES

Areen, J., and King, P. (1990). Legal regulation of human gene therapy. *Hum. Gene Ther.* **1,** 151–161.

Auiti, A., *et al.* (2009). Gene therapy for immunodeficiency due to adenosine deaminase deficiency. *N. Engl. J. Med.* **360,** 447–458.

Boztug, K., *et al.* (2010). Stem-cell gene therapy for Wiskott Aldrich syndrome. *N. Engl. J. Med.* **363,** 1918–1927.

Cavazzano-Calvo, M., *et al.* (2010). Transfusion independence and HMGA2 activation after gene therapy of human Beta-thalassemia. *Nature* **467,** 318–322.

Fischer, A., *et al.* (2011). Gene therapy for primary adaptive immune deficiencies. *J. Allergy Clin. Immunol.* **127,** 1356–1359.

Fredrickson, D. S. (2001). The Recombinant DNA Controversy: A Memoir, Science, Politics, and the Public Interest. ASM Press, Washington, DC, 1974–1981.

Gilboa, E., et al. (1986). Self-inactivating retroviral vectors designed for transfer of whole genes into mammalian cells. *Proc. Natl. Acad. Sci. USA* **83,** 3194–3198.

Hacein-Bey-Abina, S., et al. (2003). LMO-2 associated clonal T cell proliferation in two patients after gene therapy for SCID-X1. *Science* **302,** 415–419.

Hacein-Bey-Abina, S., et al. (2010). Efficacy of gene therapy for X-linked severe combined immunodeficieny. *N. Engl. J. Med.* **363,** 355–364.

Howe, S. J., et al. (2008). Insertional mutagenesis combined with acquired somatic mutations causes leukemogenesis following gene therapy in SCID-X1 patients. *J. Clin. Invest.* **118,** 3143–3150.

Manno, C., et al. (2006). Successful transduction of liver in hemophilia by AAV-factor IX and limitations imposed by the host immune response. *Nat. Med.* **12,** 342–347.

Mingozzi, F., et al. (2009). AAV-1 mediated gene transfer to skeletal muscle in humans results in dose-dependent activation of capsid-specific T cells. *Blood* **114,** 2077–2086.

NIH Report (2002). Assessment of adenoviral vector safety and toxicity: Report of the National Institutes of Health Recombinant DNA Advisory Committee. *Hum. Gene Ther.* **13,** 3–13.

President's Commission for the Study of Ethical Problems in Medicine and Biomedical and Behavioral Research (1982). Splicing Life: A Report on the Social and Ethical issues in Genetic Engineering with Human Beings. Washington, DC.

Rainsbury, J. M. (2000). Biotechnology on the RAC—FDA/NIH regulation of human gene therapy. *Food Drug Law J.* **55,** 575–600.

Stein, S., et al. (2010). Genomic instability and myelodysplasia with monosomy 7 consequent to EVI1 activation after gene therapy for chronic granulomatous disease. *Nat. Med.* **16,** 198–204.

Weiss, R. (2007). Death points to risks in research, one woman's experience in gene therapy highlights weaknesses in the patient safety net. *Washington Post.*

CHAPTER SEVENTEEN

Regulatory Structures for Gene Therapy Medicinal Products in the European Union

Bettina Klug,* Patrick Celis,[†] Melanie Carr,[†] *and* Jens Reinhardt*

Contents

1. Regulatory Requirements for ATMPs—Marketing Authorization	339
2. Annex I—Part IV to Directive 2001/83/EC	340
3. Definition of Gene Therapy Medicinal Products	341
4. Committee for Advanced Therapies	342
5. CAT–Stakeholder Interaction	345
6. Incentives—Small- and Medium-Sized Enterprise Office (SME Office) at EMA	346
7. National Support Structures	348
8. The Role of Patients' Organizations in the Process of Development of Gene Therapy Medicinal Products	349
9. Rare Diseases and Orphan Medicinal Products	350
10. The Role of European Research Networks	351
Acknowledgments	352
References	352

Abstract

Taking into account the complexity and technical specificity of advanced therapy medicinal products: (gene and cell therapy medicinal products and tissue engineered products), a dedicated European regulatory framework was needed. Regulation (EC) No. 1394/2007, the "ATMP Regulation" provides tailored regulatory principles for the evaluation and authorization of these innovative medicines. The majority of gene or cell therapy product development is carried out by academia, hospitals, and small- and medium-sized enterprises (SMEs). Thus, acknowledging the particular needs of these types of sponsors, the legislation also provides incentives for product development tailored to them.

The European Medicines Agency (EMA) and, in particular, its Committee for Advanced Therapies (CAT) provide a variety of opportunities for early

* Paul-Ehrlich-Institut, Langen, Germany
[†] European Medicines Agency, Canary Wharf, London, United Kingdom

interaction with developers of ATMPs to enable them to have early regulatory and scientific input.

An important tool to promote innovation and the development of new medicinal products by micro-, small-, and medium-sized enterprises is the EMA's SME initiative launched in December 2005 to offer financial and administrative assistance to smaller companies.

The European legislation also foresees the involvement of stakeholders, such as patient organizations, in the development of new medicines. Considering that gene therapy medicinal products are developed in many cases for treatment of rare diseases often of monogenic origin, the involvement of patient organizations, which focus on rare diseases and genetic and congenital disorders, is fruitful. Two such organizations are represented in the CAT.

Research networks play another important role in the development of gene therapy medicinal products. The European Commission is funding such networks through the EU Sixth Framework Program.

Scientific progress in cellular and molecular biotechnology has led to the development of "advanced therapies," offering treatment opportunities for a variety of inherited and acquired diseases, for example, in the field of hematology (Aiuti et al., 2002; Kang et al., 2010; Serana et al., 2010), neurological disease (Cartier et al., 2009; Kaplitt, 2007), metabolic disease, (Alexander, 2008), and muscular disease (Romero, 2004).

With the introduction of the concept of "Advanced Therapy Medicinal Products" (ATMPs), defined at the time when Gene Therapy Medicinal Products (GTMP) and somatic Cell Therapy Medicinal Products (CTMP) were introduced in the European legislation as early as 2003, it became clear that a dedicated regulatory framework for these medicinal products was necessary. Regulation (EC) No. 1394/2007 on Advanced Therapy Medicinal Products (hereafter referred to as "ATMP Regulation"), which entered into force on December 30, 2008, provided the necessary regulatory framework (Voltz-Girolt et al., 2011). The ATMP Regulation also introduced the product classes of tissue-engineered products and combined ATMPs.

At that time biotechnology products regulated at Community level were already subject to the centralized marketing authorization procedure, involving a single scientific evaluation of the quality, safety, and efficacy of the product carried out to highest possible standard by the Committee for Human Medicinal Products (CHMP) at the European Medicines Agency (EMA). The ATMP Regulation made the centralized procedure also compulsory for ATMP. The benefits for public health of a compulsory marketing authorization are obvious: a risk/benefit analysis by independent experts based on quality, nonclinical, and clinical data provides confidence to patients and health professionals in new medicinal products; the centralized marketing authorization procedure in addition permits pooling of expertise within the European Union. Both industry and patients also benefit from a facilitated access to the EU market via a single procedure. Manufacturers developing ATMPs to be marketed in the EU are provided with regulatory certainty for the development of their products and are ensured free movement of those products within the EU. Patients and physicians benefit from timely access to such innovative treatment options.

The ATMP Regulation provides tailored regulatory principles for evaluation and marketing authorization, for post-authorization follow-up and for traceability. It also established a new scientific committee, Committee for Advanced Therapies (CAT), at the EMA.

The EMA is a decentralized body of the European Union, located in London. The Agency can be considered as the "hub" of a European medicines network comprising over 40 national competent authorities in 30 EU and EEA–EFTA countries, the European Commission, the European Parliament, and a number of other decentralized EU agencies. The Agency works with a network of over 4500 "European experts" who serve as members of the Agency's scientific committees, working parties, or scientific assessment teams. These experts are made available to the Agency by the national competent authorities of the EU and EFTA states.

Taking into account the profile of the sponsors developing ATMPs (SMEs, hospitals), the ATMP Regulation provides incentives tailored to them. These include fee reductions for scientific advice and marketing authorization applications (MAAs) for ATMPs, ATMP classification and certification procedures.

1. REGULATORY REQUIREMENTS FOR ATMPs—MARKETING AUTHORIZATION

ATMPs have to fulfill the same scientific and regulatory requirements as all other medicinal products. The manufacture of ATMPs has to comply with the principles of good manufacturing practice (GMP), as set out in Commission Directive 2003/94/EC. Clinical trials have to be designed and performed according to the overarching principles and ethical requirements laid down in good clinical practice (GCP) as laid down in Commission Directive 2005/28/EC. Thus, requirements specified in the GMP and GCP guidelines have to be adapted to the technical specificities of ATMPs. An amendment of Annex 2: "Manufacture of biological medicinal products for human use" to the "EU Guideline to Good Manufacturing Practice (GMP) for medicinal products for human and veterinary use" is underway and will be published in EudraLex—Volume 4 GMP guidelines (http://ec.europa.eu/health/documents/eudralex/vol-4/index_en.htm). Likewise, detailed clinical guidelines for GCP have been adapted to the technical specificities of ATMPs and will be published in EudraLex—Volume 10

Clinical trial guidelines (http://ec.europa.eu/health/documents/eudralex/vol-10/index_en.htm).

Post-authorization requirements such as follow-up of adverse events, risk management, and traceability of the medicinal are likewise tailored to the specificities of the products (Jekerle et al., 2010).

In addition to this, as a special requirement, taking into account the novelty of these products, the ATMP Regulation introduced the possibility for post-authorization follow-up of long-term efficacy. In the field of gene therapy medicinal products, the necessity of long-term follow-up of safety and efficacy applies to all patients included in clinical studies. The technical requirements for such long-term surveillance are laid down in a specific guideline (EMEA/CHMP/GTWP/60436/2007). This approach serves the interest of the patients to benefit from timely access to innovative medicines and also benefit from a rigorous follow-up of safety and efficacy, as well as allowing, as a benefit to regulators and ultimately the scientific community, for the generation of extensive data on the given ATMP.

2. ANNEX I—PART IV TO DIRECTIVE 2001/83/EC

The ATMP Regulation not only provides the legal basis for the marketing authorization procedure for ATMPs but also incorporates these products in the legal framework of the pharmaceutical legislation, which is governed by Directive 2001/83/EC.

Annex I to this Directive lays down the technical requirements for all medicinal products specified in this Annex. Some medicinal products present specific features such that all the requirements for the marketing authorization application need to be adapted to these products.

Specific technical requirements with regard to the marketing authorization application are laid down for particular medicinal product groups, for example, plasma-derived medicinal products or vaccines. The ATMP Regulation provides the legal basis for the revision of part IV, Annex I which specifies the technical requirements for ATMPs. This approach of laying down tailored requirements does not only address the highly specific nature of the products but also allows a flexible approach to take into account the technological and scientific developments in this rapidly evolving field (Jekerle et al., 2010).

A further interesting feature is that with the revision of Annex I, the concept of a risk-based approach is introduced in the pharmaceutical legislation as an overarching principle for the development of ATMPs. At the beginning of and during the development program, a risk analysis is performed by the future applicant for a marketing authorization to determine the extent of quality, nonclinical, and clinical data to be included in the MAA. The risk-based approach is based on the identification of risk factors inherent to the nature of the ATMP in question and associated with its quality, safety, and efficacy. The risks associated with an ATMP are highly dependent on the biological characteristics of the product, for example, in case of gene therapy medicinal products, the biological characteristics of the vectors and the characteristics of protein expression as well as similar to all other products, its specific therapeutic use. The analysis of risk factors will allow the manufacturer to tailor the product development (including the nonclinical and clinical investigations) to the specificities of its product. The description of the risk analysis is part of the dossier for a marketing authorization application (EMA/CHMP/CPWP/708420/2009).

Annex I provides high-level technical requirements for the content of the MAA; these requirements are further substantiated by CHMP/CAT guidelines specific to gene therapy as well as to all aspects of the product development (http://ec.europa.eu/health/documents/eudralex/vol-3/index_en.htm).

3. DEFINITION OF GENE THERAPY MEDICINAL PRODUCTS

In order to define the borders between ATMPs and other medicinal products as well as between the different classes of ATMPs, specific definitions for gene therapy as well as for somatic cell therapy and tissue-engineered products had to be set up. These definitions were introduced in the ATMP Regulation and in the revised Annex 1, part IV to Directive 2001/83/EC.

Gene therapy medicinal products are defined as biological medicinal products which

(a) contain an active substance which contains or consists of a recombinant nucleic acid used in or administered to human beings with a view to regulating, repairing, replacing, adding, or deleting a genetic sequence;
(b) its therapeutic, prophylactic, or diagnostic effect relates directly to the recombinant nucleic acid sequence it contains, or to the product of genetic expression of this sequence.

Vaccines against infectious diseases are not included in the definition of gene therapy medicinal products, since an established regulatory pathway for vaccines against infectious diseases exists in the European Union. If required, the ATMP Regulation foresees the possibility that the CAT can be consulted by other scientific committees of the EMA, for example, by the CHMP, for the evaluation of any medicinal product which requires specific expertise falling within its area of competence.

Likewise, chemically synthesized nucleic acids (e.g., oligonucleotides) are also not falling under the definition of gene therapy.

4. COMMITTEE FOR ADVANCED THERAPIES

When developing the ATMP Regulation, it was realized that ATMPs pose new questions and challenges to both developers and regulators and that expertise to review ATMPs is not specifically represented in the main scientific committee at the EMA, the CHMP. Therefore, the ATMP Regulation identifies the necessary expertise, relevant to ATMPs, to be represented in the CAT: expertise on medical devices, tissue engineering, gene therapy, cell therapy, biotechnology, surgery, pharmacovigilance, risk management, and ethics. The CAT is the first scientific committee where the legislation requires that certain specific scientific expertise is represented. This is a clear acknowledgement of the specific nature of the ATMPs and the need for a multidisciplinary expert Committee to provide for an adequate, high-level evaluation of ATMPs. The outcome of a questionnaire completed by all CAT members and alternates showed that all fields of expertise required by the Regulation are indeed present (Celis, 2010; Fig. 17.1). It should be noted that most members indicated having expertise in more than one field of expertise (such as biotechnology and cell therapy).

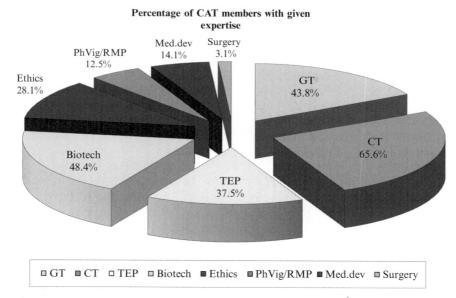

Figure 17.1 Percentage of the members of the CAT according to their product-related expertise (CAT members declared in most cases expertise in more than one area). Abbreviations used: GT, gene therapy; CT, cell therapy; TEP, tissue-engineered medicinal products; Biotech, Biotechnology; PhVig/RMP, Pharmacovigillance/Risk Management Plans; Med. dev, medical devices. (See Color Insert.)

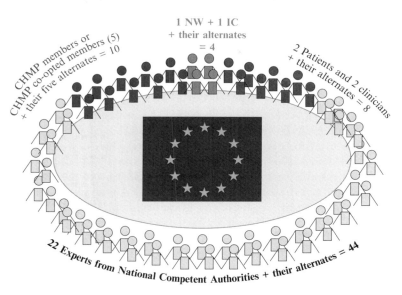

Figure 17.2 Composition of the CAT: Each figurine represents one of the CAT member or its alternates. The yellow figurines represent the members and alternates nominated by the national competent authorities. The red figurines are CHMP and CAT double members or co-opted members. The blue figures stand for the members of the clinicians' and the patients' organizations. Depicted in green are the representatives of Norway and Iceland and their alternates. (See Color Insert.)

The CAT is composed of experts nominated by the EU member states, the CHMP, and the European Commission (Fig. 17.2). Five members of the CHMP (together with five alternate members) are nominated by the CHMP to represent the CHMP in the CAT. Those members play an important communicators' role between both Committees. The Member States not represented by the five joint CHMP–CAT members each nominate a member and an alternate. Following a public call for expression of interest and after consultation of the European Parliament, the Commission appointed two members and two alternates in order to represent clinicians in the CAT, and two members and two alternates to represent the patients' organizations.

The CAT meets on a monthly basis (11 times a year). A summary of the CAT discussions is published in the CAT meeting reports, which can be found on the EMA Web site (CAT monthly reports published on the EMA website).

The main task of the CAT is to review MAAs for ATMPs. In addition, two new tasks have been assigned to the CAT, namely the ATMP classification procedure and the certification of ATMPs (more information on these two procedures is given below and can be found in the relevant guidelines published on the EMA Web site (EMA/CAT/99623/2008; EMA/CAT/418458/2008/corr).

The CAT is also routinely involved and consulted for scientific advice requests for ATMPs and can also be consulted by the CHMP on the review of non-ATMPs, especially when the CAT expertise is needed to complement that of the CHMP (e.g., on medicinal product–medical device combinations).

The MAA for ATMPs follows the principles of the centralized procedure for non-ATMPs (such as biotechnology-derived products and new chemical entities for certain indications): the evaluation will take a maximum of 210 days active review time (not including the time required by the applicant to respond to questions asked by the CAT), and the review will be performed by two independent review teams (a Rapporteur team and a Co-Rapporteur team, whereby the Rapporteur and Co-Rapporteur are CAT members). The CAT will prepare and adopt the milestone documents including the draft opinion (around day 200 of the procedure), which includes the product information (prescribers' and patients' leaflets), CAT assessment report, and Risk Management Plan. By day 210 of the procedure, the CHMP, on the basis of the draft opinion from the CAT, will adopt the final opinion for a given ATMP, which is thereafter transmitted to the European Commission, who will grant the EU marketing authorization. In order to minimize the risk that the CHMP would come to a different position than the CAT, a detailed procedure for evaluation of ATMPs has been developed which is built around full transparency between both committees (EMEA/630043/2008). It is still early days and the robustness of this procedure remains to be demonstrated.

The first new procedure, for which CAT has full responsibility, is a scientific recommendation on advanced therapy classification (EMA/CAT/99623/2008). This procedure, which has been included in the ATMP Regulation as an incentive for industry, gives the possibility for applicants developing products based on genes, cells, or tissues to request a recommendation from the CAT to determine whether their product would fulfill the definition of an ATMP. The outcome of the regulatory classification will allow the applicant to position their product in their future market, to develop it further in accordance with the applicable technical and regulatory requirements, and to benefit from specific incentives designed for ATMPs. The classification procedure is not a legal mandatory procedure before entering any other regulatory procedure at the EMA (such as scientific advice, orphan designation, ATMP certification, and marketing authorization). However, in case of doubt on the regulatory classification, EMA strongly encourages the applicant to submit a request for ATMP classification. The applicant can use this scientific recommendation from the CAT when applying for a legally binding product classification at the competent authority of a Member State. The Member State competent authority also provides a legally binding classification of a product as medicinal products or medical device.

The second new procedure under the CAT's responsibility, the ATMP certification procedure, is a scientific evaluation of early and late quality and nonclinical data, generated by small- and medium-sized enterprises (SMEs) developing ATMPs (EMA/CAT/418458/2008/corr). The evaluation of such data by the CAT will be followed by a certification provided to the company by the EMA. The CAT evaluation report will be attached to the certificate. This certification procedure could be very beneficial for the SME, as the review of early data will guide further pharmaceutical and nonclinical development. The certificate could also serve as a tool to enter into negotiations with other pharmaceutical companies or may allow SMEs to attract private funding to continue product development, especially for clinical trials.

5. CAT–Stakeholder Interaction

Whereas a lot of research and development in the field of ATMPs (including GTMPs) is being performed, only a limited number of these projects are translated into products for clinical development and marketing authorization. One of the contributing factors of this translational barrier is that developers of ATMPs tend to differ from those of established (chemical or biotechnological) medicines: the majority being academia, hospitals, and SMEs. These operators might not have easy access to the resources readily available to larger pharmaceutical companies such as regulatory experience/expertise, regulatory support structures, and funding. Such hurdles could limit the timely access to potential effective treatments to patients.

The EMA and the CAT, therefore, acknowledge the need for increased and early interactions and open dialog with the developers and manufacturers of ATMPs to cope with the regulatory and scientific challenges posed by these innovative medicines (Schneider et al., 2010). To that aim, manufacturers, academic and industrial developers as well as nonprofit organizations supporting ATMP development or acting as clinical trial sponsors are encouraged to interact with EMA/CAT via informal briefing meetings with EMA's Innovation Task Force (ITF), which facilitates early dialog and interaction with the EMA and CAT experts on procedural and scientific issues; interactions with the CAT in the framework of meetings with interested parties on topics of general interest; training sessions and workshops for industry, academia, and hospitals developing ATMPs, organized by CAT or its associated working parties (the Gene Therapy Working Party (GTWP) and the Cell-based Products Working Party (CPWP)); informal contacts with CAT members and CAT secretariat at scientific conferences and meetings; meetings with the SME Office of the EMA (see below); and requests for information addressed to the CAT secretariat via AdvancedTherapies@ema.europa.eu.

Early interactions can also take place in the form of meetings with the Secretariat of the Committee for Orphan Medicinal Products (COMP) in preparation of an application for Orphan Drug status, meetings with the Scientific Advice Working Party (SAWP) secretariat in preparation of a scientific advice request and presubmission meetings with EMA product team in preparation of an application for ATMP certification or marketing authorization. Also, a very useful means of establishing early contact with the CAT is to apply for a scientific recommendation on classification of ATMP, for which the developer/manufacturer can present early data on its product under development.

CAT values these early interactions with developers and manufacturers, as it will allow them to identify trends and raise awareness of new scientific or regulatory issues that require further considerations or development of guidance documents. In that respect, CAT has recently developed and published its workprogram for the 2010–2015 (EMA/CAT/235374/2010), which sets objectives and actions aiming to facilitate the access of ATMPs to patients.

6. Incentives—Small- and Medium-Sized Enterprise Office (SME Office) at EMA

Micro-, small-, and medium-sized enterprises operating in the pharmaceutical sector are often innovative companies active in the fields of gene or somatic cell therapy. Recognizing that a lack of experience with the EMA and its centralized marketing authorization procedure should not hinder an SME's ability to bring a new medicine to the market, the EMA launched an "SME Office" in December 2005 to offer financial and administrative assistance to these companies. The aim is to promote innovation and the development of new medicinal products by SMEs.

According to the current EU definition of an SME (Commission Recommendation 2003/361/EC), companies with less than 250 employees, and an annual turnover of not more than €50 million or an annual balance sheet total of not more than €43 million, are eligible for assistance from the EMA. Provided that the company fulfills the criteria and is established in the European Economic Area (this includes the EU Member States and Iceland, Liechtenstein, and Norway), the EMA will assign the company SME status. All companies with SME status can benefit from the same incentives whether they are classified as micro, small, or medium. Further information on how to apply for and maintain SME status is available on the EMA's Web site (http://www.ema.europa.eu Regulatory/Human Medicines/SME office).

The incentives, which focus on the main financial and administrative entry hurdles for SMEs in premarketing authorization procedures, include the following:

- Regulatory assistance from SME Office within the Agency;
- Fee reductions (90%) for scientific advice and inspections;
- Fee reduction for the application for marketing authorization for SMEs developing ATMPs or orphan medicinal products;
- Deferral of the fee payable for the application for marketing authorization or related inspection until after the grant of the marketing authorization;
- Conditional fee exemption where scientific advice is followed and the marketing application is unsuccessful;
- Certification of quality/nonclinical data for ATMPs intended for human use;
- Assistance with translations of the product information documents required for the grant of an EU marketing authorization.

Experience over the first 5 years that the SME office has been in operation (2006–2010) has shown SMEs to have a lower success rate through the centralized marketing approval procedure compared to non-SME companies. When analyzing the reasons for the negative outcomes, the Agency has noted that major objections ran particularly high in the areas of quality and clinical efficacy. In these key areas, it is therefore particularly important for SMEs to seek scientific advice from the Agency early in development.

Scientific advice will help a company ensure that the appropriate tests and studies are performed so that no major objections regarding the design of the tests are likely to be raised during evaluation of the marketing authorization application. The Agency offers prospective scientific advice to companies on all areas of development:

- Quality (chemical, pharmaceutical, and biological testing)
- Nonclinical safety (toxicological and pharmacological tests)
- Clinical aspects (clinical safety and efficacy)

The scientific advice procedure attracts a fee, which varies depending on the scope of the advice. Access for SMEs to the Agency's scientific advice, therefore, has been facilitated with a substantial 90% fee reduction. Further, a scientific evaluation of a marketing authorization application is more likely to be favorable where scientific advice has been sought from the agency (Regnstom et al., 2009). In the event of a negative outcome, a conditional exemption of the fee for the application for marketing authorization will be given to applicants who have requested such advice and who have actually taken it into account in the development of their medicinal product.

For SMEs developing ATMPs, the scientific advice process is complemented by the new certification procedure which was launched in 2009 (see above).

As development proceeds, the SME Office can facilitate contacts with the relevant scientific and regulatory staff within the Agency to address any questions that may arise, particularly in the run up to submitting a marketing authorization application.

To avoid placing additional financial constraints on a company in the run up to filing the application for marketing authorization, there is a possibility for SMEs to request deferral of payment of fees. Fee payment may be deferred up to 45 days after marketing authorization or within 45 days of withdrawal of the application or negative decision.

In addition, an application for marketing authorization for an advanced therapy medicinal product for human use from an SME applicant will have a 50% fee reduction (subject to proof that the product concerned has a particular public health interest in the Community). There is also a 50% fee reduction for post-authorization activities including the annual fee for the first year after marketing authorization. The fee reductions will be available for a limited period.

Additional financial incentives exist for SMEs developing medicinal products to treat rare diseases. SME sponsors of designated orphan medicinal products can benefit from a full (100%) waiver of the fee payable for the centralized marketing application and associated inspection fees. There is also a full fee waiver for post-authorization activities including the annual fee for the first year after marketing authorization.

Finally, because translating product information into all EU languages represents a considerable financial and administrative burden to SMEs entering the EU market, the EMA will provide translations of product information (summary of product characteristics, label and package leaflet, and Annexes A, II/IV) required to grant an EU marketing authorization. Translation into EU official languages will be provided free of charge by the Agency, it will be the responsibility of the applicant SME to provide Norwegian and Icelandic translations.

7. NATIONAL SUPPORT STRUCTURES

There are also some additional support measures available at national level. Innovation offices are established in several member states to give future applicants the opportunity for an early informative dialog with the National Competent Authorities.

In France, the Agence française de sécurité sanitaire des produits de santé (Afssaps) provides the opportunity of "innovation meetings" to open an early dialog on regulatory and scientific advice (http://www.afssaps.fr/Activites/Accompagnement-de-l-innovation/Afssaps-et-innovation/(offset)/0).

In Germany, the Paul-Ehrlich-Institute (PEI) has recently established an innovation office dedicated to ATMPs (innovation@pei.de). Stakeholders may seek advice in early stage of the development on quality, nonclinical as well as clinical development strategies. In a later stage, meetings are aimed at discussing an intended clinical trial authorization in Germany or in more than on EU Member State, the latter via the Voluntary Harmonization or Coordinated Assessment Procedure.

Meetings with one or more national competent authorities (http://www.hma.eu) may also be beneficial in preparing a European scientific advice through EMA.

8. THE ROLE OF PATIENTS' ORGANIZATIONS IN THE PROCESS OF DEVELOPMENT OF GENE THERAPY MEDICINAL PRODUCTS

The role of patients' organizations in the development of orphan medicinal products legislation may not be underestimated. Starting out to provide support to the families in which rare diseases occur, patients' organizations now also raise general public awareness to the plight of people suffering from disorders against which treatments are mostly not available and not likely to be developed. With benefit events (e.g., the Telethon television shows in France and in Italy) and support from celebrities, patients' organizations have raised enough money to be able to fund projects for the research and development of medicines for rare diseases. Moreover, they have eventually gained enough attention to foster the generation of the orphan drug legislation in the USA as well as in Europe.

In Europe, the role of the patients' organization is acknowledged by the fact that patients' organizations are represented in different structures of the EMA. Patients' organizations participate in the COMP, which recommends the orphan designation of a medicinal product (as described later in the text), with three members, which are designated by the European Commission. They also participate in the Paediatric Committee (PDCO), which is responsible for scientific assessment of pediatric investigation plans, with three members and three alternates. Moreover, patients' organizations participate with two members and two alternates in the CAT, as described above. Moreover, patients' organizations also participate with two members in the EMA Management Board.

The members of the patients' organizations have the same roles and responsibilities as the members that are nominated by the National Authorities of the Member States. They provide complimentary views in addition to the technical approaches of the regulatory bodies. Particularly, they represent the patients' interests and thereby add the viewpoint of those who will directly be affected by the regulatory decisions. Additionally to representing the patients' voice in the EMA, they also give information from the regulatory bodies back into the patients' community. Two important patient organizations in Europe are EURORDIS (the EURopean Organization for Rare DISeases; http://www.eurordis.org) and EGAN (the patients' network for medical research and health; http://www.egan.eu).

EURORDIS is a patient-driven nonprofit nongovernmental organization that was founded in 1997 as alliance of patient organizations. EURORDIS consists of 423 member organizations in 43 countries, comprising 34 European countries, of which 25 are Member States of the European Union. The activities of EURORDIS span from raising public awareness for rare diseases and empowerment of rare disease patient groups to the promotion of scientific and clinical research on rare diseases to the promotion of the development of rare disease treatments and orphan medicinal products

EGAN is an alliance of national genetic alliance sand European patient organizations. EGAN has a strong focus on genetic and congenital disorders. Therefore, different to EURORDIS, both common and rare disorders are in the center of their work.

From the beginning of 2009 to the end of 2011, both EURODIS and EGAN had one member and one alternate in the CAT.

9. Rare Diseases and Orphan Medicinal Products

Gene therapy is considered to be so far in most cases developed to treat diseases that fulfill the following two criteria: the disease is of monogenetic origin, that is, based on the mutation of a single gene, for which a replacement or repair may be envisaged, and the disease is so severe with no sufficient conventional treatment that an experimental therapy is the best possible, sometimes the only option.

Further, most of the diseases targeted by gene therapy are so-called rare diseases, which are defined as life-threatening or chronically debilitating conditions that affect no more than 5 in 10,000 people in the EU. They are in most cases not commercially interesting targets for conventional medicinal product development by the pharmaceutical industry.

Based on the positive experience that had been gained in the United States with the promotion of treatments for rare diseases (LIT), the European Commission established the Orphan Drug Regulation (Regulation (EC) No. 141/2000) to provide better treatments for the patients suffering from rare diseases. This Orphan Regulation proposes incentives for pharmaceutical enterprises that develop treatments for rare diseases. In order to benefit from these incentives, the pharmaceutical enterprises have to apply for a so-called orphan designation for their medicinal product.

The orphan designation is granted by the Commission based on a recommendation from the COMP, which is composed of experts from the Member States, as well as three representatives from patients' organizations and three further persons nominated by the EMA.

To qualify for orphan designation, a medicinal product must be intended for the diagnosis, prevention, or treatment of a life-threatening or chronically debilitating condition that is either affecting no more than 5 in 10,000 people in the EU at the time of submission of the designation application, or it is unlikely to generate enough revenue to cover the investment in its development without incentives after marketing.

In both cases, there must also be either no satisfactory method of diagnosis, prevention, or treatment of the condition concerned is authorized, or, if such a method does exist, the medicine must be of significant benefit to those affected by the condition. The orphan designation may be obtained at any stage of the development before the marketing authorization application.

Sponsors with an orphan designation are eligible to the following incentives: they are granted 10 years of market exclusivity in the EU after the marketing authorization is approved. The sponsors are also eligible to a special scientific advice procedure for the optimization of the developmental process and guidance on the dossier preparation as well as to aspects of the significant benefit that the product shows in comparison to other products already on the market. This so-called Protocol Assistance is free of charge. Moreover, an orphan drug designation also allows for the reduction of fees for all types of centralized regulatory activities such as marketing authorization, inspections, and variations. SMEs benefit from additional fee reductions (as described above). Moreover, the EU provides grants for the research and development of orphan medicines, for which sponsors with an orphan designation are eligible.

10. THE ROLE OF EUROPEAN RESEARCH NETWORKS

The development of gene therapies in most cases starts in clinical centers that are specialized in the treatment of rare diseases. For the involved clinicians and investigators, it is imperative to form networks in order to focus capacities, knowledge, and experience throughout Europe.

These networks are then able to form the critical mass needed to fulfill the requirements to perform clinical trials, for example, GMP-grade manufacture of the investigative new drug, nonclinical tests to support the proof of concept, and safety and efficacy.

One example for such a research network is CliniGene (European Network for the Advancement of Clinical Gene Transfer and Therapy). CliniGene has been funded and financed by EU Sixth Framework Program, designated as a European Network of Excellence.

CliniGene is working with academia, industry, regulatory bodies, clinics, and patients in order to integrate multidisciplinary research and development expertise with a focus to improve the safety and clinical efficacy of gene transfer/therapy medicinal products, that is, for clinical applications. Within the CliniGene network, seven technology platforms (adeno-associated viruses, retroviruses, lentiviruses, cell herapy, adenoviruses, nonviral methods (e.g., plasmids), and induced pluripotent stem cells) are developed. Research groups within the CliniGene network have successfully performed clinical trials with gene therapies, as can be seen on the CliniGene homepage (http://www.clinigene.eu/).

Another function of CliniGene is to gather and disseminate knowledge: the knowledge that is gained within their platforms is made publicly available on the CliniGene homepage as well as on publications and in scientific presentations, for example, given at the annual meetings of European society for Gene and Cell Therapy (ESGCT) (http://www.esgct.eu), in which CliniGene is involved.

ACKNOWLEDGMENTS

The authors thank the editors and Klaus Cichutek for the opportunity to present the European regulatory system for medicinal products. They also want to thank Klaus Cichutek for the critical review of the manuscript and his fruitful comments.

REFERENCES

Aiuti, A., et al. (2002). Correction of ADA-SCID by stem cell gene therapy combined with nonmyeloablative conditioning. *Science* **296**(5577), 2410–2413.

Alexander, I. E. (2008). Potential of AAV vectors in the treatment of metabolic disease. *Gene Ther.* **15**, 831–839.

Annex 2 Manufacture of biological medicinal products for human use. http://ec.europa.eu/health/files/eudralex/vol-4/pdfs-en/anx02en200408_en.pdf.

Cartier, N., et al. (2009). Hematopoietic stem cell gene therapy with a lentiviral vector in X-linked adrenoleukodystrophy. *Science* **326**, 818–823.

CAT monthly reports published on the EMA website. http://www.ema.europa.eu/ema/index.jsp?curl=pages/news_and_events/document_listing/document_listing_000196.jsp&murl=menus/about_us/about_us.jsp&mid=WC0b01ac05800292a8.

Celis, P. (2010). CAT—The new committee for advanced therapies at the European Medicines Agency. *Bundesgesundheitsblatt* **53,** 9–13.

Commission Directive 2003/94/EC of 8 October 2003 laying down the principles and guidelines of good manufacturing practice in respect of medicinal products for human use and investigational medicinal products for human use. *Official J. Eur. Union* **L 262,** 14 October 2003, p 24. http://ec.europa.eu/health/files/eudralex/vol-1/dir_2003_94/dir_2003_94_en.pdf.

Commission Recommendation 2003/361/EC of 6 May 2003 concerning the definition of micro, small and medium-sized enterprises. *Official J. Eur. Union* **L 124,** 20 May 2003, p. 36. http://eur-lex.europa.eu/LexUriServ/LexUriServ.do?uri=OJ:L:2003:124:0036:0041:en:PDF.

Commission Directive 2005/28/EC of 8 April 2005 laying down principles and detailed guidelines for good clinical practice as regards investigational medicinal products for human use, as well as the requirements for authorisation of the manufacturing or importation of such products. *Official J. Eur. Union* **L 91,** 9 April 2003, p. 13. http://ec.europa.eu/health/files/eudralex/vol-1/dir_2005_28/dir_2005_28_en.pdf.

Directive 2001/83/EC of the European Parliament and the Council of 13 November 2007 on the Community code relating to medicinal products for human use. *Official J. Eur. Union* **L 311,** 28 November 2001, p. 67. Consolidated version 05/10/2009. http://ec.europa.eu/health/files/eudralex/vol-1/reg_2008_1234/reg_2008_1234_en.pdf.

EMA/CAT/235374/2010: Committee for Advanced Therapies (CAT) Work Programme 2010–2015. http://www.ema.europa.eu/docs/en_GB/document_library/Work_programme/2010/11/WC500099029.pdf.

EMA/CAT/418458/2008/corr. Procedural advice on the certification of quality and non-clinical data for small and medium sized enterprises developing advanced therapy medicinal products. http://www.ema.europa.eu/docs/en_GB/document_library/Regulatory_and_procedural_guideline/2010/01/WC500070030.pdf.

EMA/CAT/99623/2008: European Medicines Agency (2009). Procedural advice on the provision of scientific recommendation on classification of advanced therapy medicinal products in accordance with article 17 of Regulation (EC) No. 1394/2007. http://www.ema.europa.eu/docs/en_GB/document_library/Regulatory_and_procedural_guideline/2010/02/WC500074745.pdf.

EMA/CHMP/CPWP/708420/2009: Concept paper on the development of a guideline on the risk-based approach according to Annex 1, Part IV of Dir. 2001/83/EC applied to advanced therapy medicinal products. http://www.ema.europa.eu/docs/en_GB/document_library/Scientific_guideline/2010/01/WC500069264.pdf.

EMEA/630043/2008: European Medicines Agency (2009). Procedural advice on the evaluation of advanced therapy medicinal products in accordance with article 8 of Regulation (EC) No. 1394/2007. http://www.ema.europa.eu/docs/en_GB/document_library/Regulatory_and_procedural_guideline/2010/02/WC500070340.pdf.

EMEA/CHMP/GTWP/60436/2007: Guideline on follow up of patients administered with gene therapy medicinal products. http://www.ema.europa.eu/docs/en_GB/document_library/Scientific_guideline/2009/11/WC500013424.pdf.

Innovation Task Force (ITF). http://www.ema.europa.eu/ema/index.jsp?curl=pages/regulation/general/general_content_000334.jsp&murl=menus/regulations/regulations.jsp&mid=WC0b01ac05800ba1d9.

Jekerle, V., et al. (2010). Legal basis of the Advanced Therapies Regulation. *Bundesgesundheitsblatt* **53,** 4–8.

Kang, E. M., et al. (2010). Retrovirus gene therapy for X-linked chronic granulomatous disease can achieve stable long-term correction of oxidase activity in peripheral blood neutrophils. *Blood* **115**(4), 783–791.

Kaplitt, M. G. (2007). Safety and tolerability of gene therapy with an adeno-associated virus (AAV) borne GAD gene for Parkinson's disease: An open label, phase I trial. *Lancet* **369,** 2097–2105.

Regnstom, J., et al. (2009). Factors associated with success of marketing authorization applications for pharmaceutical drugs submitted to the European Medicines Agency. *Eur. J. Clin. Pharmacol.* **66**(1), 39–48.

Regulation (EC) No. 1394/2007 of the European Parliament and of the Council of 13 November 2007 on advanced therapy medicinal products and amending Directive 2001/83/EC and Regulation (EC) No. 726/2004. *Official J. Eur. Union* **L 324,** 10 December 2007, p. 121.

Regulation (EC) No. 141/2000 of the European Parliament and of the Council of 16 December 1999 on orphan medicinal products. *Official J. Eur. Union* **L 18,** 22 January 2000, p. 1. http://ec.europa.eu/health/files/eudralex/vol-1/reg_2000_141/reg_2000_141_en.pdf.

Romero, N. B. (2004). Phase I study of dystrophin plasmid-based gene therapy in Duchenne/Becker muscular dystrophy. *Hum. Gene Ther.* **15,** 1065–1076.

Schneider, C. K., et al. (2010). Challenges with advanced therapy medicinal products and how to meet them. *Nat. Rev. Drug Discov.* **9**(3), 195–201.

Serana, F., et al. (2010). The different extent of B and T cell immune reconstitution after hematopoietic stem cell transplantation and enzyme replacement therapies in SCID patients with adenosine deaminase deficiency. *J. Immunol.* **185,** 7713–7722.

Voltz-Girolt, C., et al. (2011). The advanced therapy classification procedure; overview of experience gained so far. *Bundesgesundheitsblatt Gesundheitsforschung Gesundheitsschutz* **54,** 811–815.

Author Index

Note: Page numbers followed by "*f*" indicate figures, and "*t*" indicate tables.

A

Abbas, Z., 235
Abel, S. L., 189
Abel, U., 23, 111–112, 172–173
Acharya, S. A., 111
Acland, G. M., 242–243, 256, 261
Adam, M. A., 4–6, 36–37
Adams, S., 15–16, 19–20, 21, 22–23, 39, 60–61, 128–129, 172–173, 174
Ader, M., 256
Adler, R., 179–180
Adzhubel, A., 32–33
Afable, C., 90
Agbandje-McKenna, M., 230–231, 235–238, 240
Aghi, M., 219
Agricola, B., 1–4
Aguirre, G. D., 242–243, 256, 261
Ahn, C., 276
Ailles, L. E., 33–34
Aiuti, A., 15–16, 17–18, 21, 22–23, 36, 49, 60–61, 129, 142, 156–157, 172–173, 179–180, 190, 338
Akagi, K., 21
Aker, M., 15–16, 21, 22, 60–61, 129, 142, 172–173, 179–180
Albritton, L. M., 1–4
Aleman, T. S., 242–244, 256
Alexander, I. E., 338
Alexander, J. J., 256
Alexander, S. I., 19–20
Alfano, J., 11–12, 103–104
Al Ghonaium, A., 19–20, 21
Ali, I., 11–12, 102
Ali, R., 256
Allay, J. A., 302
Allen, H. D., 17
Allen, J. M., 38, 174, 233–234, 235
Allgayer, H., 78
Alling, D. W., 134–135
Allocca, M., 233–234, 256
Almarza, D., 157
Almarza, E., 36, 145–146, 147–148
Altman, D. G., 183
Alvira, M. R., 235, 245, 302
Alvord, W. G., 147
Amadio, S., 172–173
Amado, D., 261, 265

Ambrosi, A., 21, 172–175, 179–180, 181–183
Ameres, S. L., 278–279
Amigo, M. L., 45–46
Amora, R., 145–146, 148
Amrolia, P. J., 17
Anagnou, N. P., 109–110
Anand, V., 242–243, 257–258
Anderson-Cohen, M., 38, 135–136
Anderson, E. M., 277–279
Anderson, W. F., 1–6, 11–12, 60, 98, 102
Ando, D., 21–22
Andolfi, G., 15–16, 21, 22–23, 60–61, 129, 142, 172–173, 179–180, 190
Andrassi, M., 242
Andre, K., 90, 96–97
Andrews, R. G., 38
Andrieu-Soler, C., 256
Angell, D., 92, 103
Annett, G., 18, 129
Antoine, C., 17
Antonelli, A., 179–180
Antoniou, M. N., 39–40, 111
Antosiewicz-Bourget, J., 149
Ao, Z., 112–113
Appelt, J. U., 78
Apperley, J. F., 103–104
Aragon-Ching, J. B., 219
Areen, J., 316
Arens, A., 61, 78
Arias, A. A., 126–127
Armentano, D., 4–6
Arruda, V. R., 243–244
Arshad, S., 21–22
Arumugam, P. I., 39–40, 111
Asheuer, M., 190
Aslanidi, G., 230–231, 238–239
Asnafi, V., 39, 110–111, 172–173
Asokan, A., 235–237
Atchison, R. W., 230–231
Aubourg, P., 187–198
Aucoin, M. G., 233, 238–239
Audit, M., 40–41, 97, 98, 100
Auiti, A., 326
Aurias, A., 119–120
Auricchio, A., 242–244, 265, 267–268, 269
Azzi, A., 230–231

355

B

Bacchiega, D., 109–110
Bach, P., 36–37
Baiker, A., 60
Bailey, J., 173
Bainbridge, J. W., 256, 298
Bakar, Y., 245
Baker, D. J., 243–244
Baker, T. S., 230–231
Bakovic, S., 33–34
Balcik, B., 173
Baldi, L., 240
Bales, O., 39–40
Ball, C. R., 23, 60–61, 147, 180, 194
Banerjee, P. P., 23, 60–61, 147
Banfi, S., 242–244
Bantel-Schaal, U., 235
Bardenheuer, W., 36–37
Bargellesi, A., 109–110
Barington, T., 60–61, 128, 172–173, 190
Barraud, P., 39–40
Barrett, A. J., 127
Barsky, L. W., 18, 129
Barsov, E. V., 212
Bartholomae, C. C., 20–21, 23, 39, 60–61, 78–79, 111–112, 128–129, 172–175, 179–180, 181–183, 187–198
Bartlett, J. S., 231
Bartz, S. R., 90, 277–278
Bassin, R. H., 11–12, 102
Batten, M. L., 256
Battini, J. L., 40
Bauer, G., 32–33
Baum, C., 38, 39, 147–148, 172–173, 181
Bayford, J., 15–16, 19–20, 21, 22–23
Baylink, D. J., 6–9
Beach, D., 173–174
Beard, C., 149
Beard, G., 92, 103
Beard, P., 230–231
Beausejour, C. M., 21–22
Bechmann, I., 189
Behlen, L., 280–281
Behlke, M. A., 276–277
Beilin, C., 36, 145–146, 147–148
Beitzel, B. F., 179–180
Beldjord, K., 39, 60–61, 78–79, 110–111, 172–173
Bellinger, D. A., 243–244
Bell, M., 21
Bell, P., 265
Belmont, J. W., 1–4
Belohradsky, B. H., 15–16, 19–20, 21, 127
Benedicenti, F., 39, 172–175, 179–180, 181–183
Bengtson, K., 90
Benhamida, S., 190
Benjelloun, F., 61, 120, 180

Bennett, A., 230–231, 242–243
Bennett, J., 242–243, 255–274
Bennicelli, J. L., 242–244, 256, 261
Benninghoff, U., 15–16, 21, 22–23, 60–61, 129, 142, 172–173
Bercury, S. D., 245
Berenson, R. J., 110
Berges, C., 45–46
Berg, M., 172–173
Berlinger, D., 90, 96–97
Bermudez, E., 237
Bernardi, M., 181
Bernini, L. F., 109–110
Berns, A., 174
Berns, K. I., 230–231
Berry, C. C., 15–16, 19–20, 21, 61, 120, 179–180
Bertone, C., 242
Bestor, T. H., 39–40
Beuzard, Y., 109–124
Beyer, J., 45–46
Beyer, W. R., 90
Bhatia, M., 23–24
Bhatia, S., 230–231
Bianchi, P. E., 242
Biasco, L., 21, 60–61, 172–173, 179–180
Biasi, G., 201
Biederlack, S., 230–231
Bierhuizen, M. F., 37–38
Biffi, A., 172–173
Bilaniuk, L., 243–244
Binder, G., 90, 96–97, 100
Binley, K., 256
Biral, E., 181
Birmingham, A., 277–279
Bish, L. T., 298
Bishop, B. M., 233–234
Björgvinsdóttir, H., 134–135
Bjorkegren, E., 15–16, 19–20, 21, 22–23, 60–61
Black, M. A., 96
Blaese, R. M., 11–12, 18, 129, 139–140
Blake, B. L., 237–238
Blanche, S., 20, 127, 128, 189
Blankinship, M. J., 235
Blaser, S. I., 189
Blatt, P., 243–244
Blau, C. A., 61
Bleimer, U., 90
Bleu, T., 305
Blomer, U., 39, 111, 131, 147–148, 190
Bloom, E. T., 172–173
Bloom, M. L., 172–173
Blundell, M. P., 36, 39–40, 49, 145–146, 147–148
Blyth, K., 21
Boatright, J., 256
Bockstoce, D. C., 190–191
Boden, D., 276–277
Bodine, D. M., 1–4, 38, 135–136

Bodner, M., 1–4
Boehmer, J., 90
Boentert, M., 189
Böhm, M., 23, 60–61, 147
Bohne, J., 38, 181
Bohnlein, E., 11–12
Boitano, A. E., 23–24
Bollenbacher, J. M., 172–173
Bondanza, A., 181
Bonetti, C., 256
Boneva, R. S., 30
Bonham, L., 11–12, 103–104
Boni, J., 92, 103
Bonini, C., 181
Boon-Spijker, M., 244–245
Bordignon, C., 18, 172–173
Bordigoni, P., 127, 189
Borellini, F., 100, 102–103, 191
Boson, B., 38
Bottcher, B., 230–231
Boucher, D. W., 245
Bouchez, L. C., 23–24
Boudreau, R. L., 275–296
Bougnères, P., 187–198
Bouhassira, E. E., 111
Bouic, K., 6–9
Bouneaud, C., 60–61, 128
Bousso, P., 16–17, 18–19, 172–173, 190
Bovee, D., 61
Bowman, V. D., 230–231
Boye, S. E., 256
Boye, S. L., 242–244, 256
Boztug, K., 23, 60–61, 147, 156–157, 326
Bradburn, M. J., 183
Brady, R. O., 90
Brady, T., 111, 121
Bragadottir, R., 256
Brambilla, R., 172–173
Brambrink, T., 149
Brancati, F., 242
Brandt, S. J., 1–4
Brantly, M. L., 243–244, 298, 302
Braun, S. E., 45, 61, 180, 194
Bree, M. P., 245
Bremser, D., 242
Brendel, C., 36, 145–146, 147–148
Brenner, S., 33, 38, 128, 131, 135–137, 147–148
Brice-Ouzet, A., 49
Brigida, I., 15–16, 21, 22–23, 129, 142, 172–173
Brinkhous, K. M., 243–244
Brinkmann, A., 39, 129–130, 143, 147, 172–173
Brook, J. D., 231
Brooks, J., 18, 129
Brose, K., 4–6
Brouns, G., 60–61, 128, 172–173, 190
Brown, B. D., 36, 60
Brown, J. D., 36–37
Brown, L., 15–16, 19–20, 21, 22–23

Brown, M. R., 127, 139–140
Brown, P. O., 32–33
Brugman, M. H., 19–22, 39, 60–61, 78–79, 128–129, 172–173, 174, 179–180
Buchholz, C. J., 90
Buch, P. K., 256
Buck, C., 235
Buckley, R. H., 17
Bueren, J. A., 36, 145–146, 147–148
Buiting, K., 16–17, 18–19
Bukinsky, M. I., 32–33
Bukovsky, A. A., 103, 135–137, 191
Bukrinsky, M. I., 32–33
Bunting, K. D., 173
Burchard, J., 277–278
Burda, J. F., 233
Burd, P., 299–300
Burge, C. B., 276
Burger, C., 301
Burgess, S. M., 39, 179–180
Burns, J. C., 6–9
Burotto, F., 18, 129
Burrascano, M., 6–9
Burton, M., 243–244
Burwinkel, B., 60–61, 78–79, 129–130, 143, 147, 172–173
Bury, C., 36–37
Burzynski, B., 173
Bushman, F. D., 19–20, 32–33, 120, 156, 179–180
Bush, R. A., 256
Bu, W., 230–231
Byrne, B. J., 230–231, 235, 238–240, 242–244, 305
Byrne, E., 1–4

C

Caballero, M. D., 45–46
Caccavelli, L., 23, 39, 60–61, 78–79, 110–112, 172–173
Cai, X., 256
Calcedo, R., 235, 245, 298, 302
Callegari, J., 111
Callegaro, L., 15–16, 21, 22–23, 60–61, 129, 142, 172–173
Camaschella, C., 109–110
Cameron, E. R., 21
Campbell-Thompson, M., 238–239
Candotti, F., 16–18, 22
Cannon, P. M., 36, 90
Cant, A. J., 17
Caocci, G., 110
Caplen, N. J., 291
Cappellini, M. D., 109–110
Carbonaro, D., 18, 129
Carbonell, D., 111
Carlier, F., 60–61, 128

Carlucci, F., 15–16, 21, 22, 60–61, 129, 142, 172–173, 190
Carmen, I. H., 300
Carr, M., 337–354
Carter, A., 230–231
Carter, B., 243–244
Carter, C. S., 18, 127, 128, 129, 131–132
Cartier-Lacave, N., 60–61
Cartier, N., 23, 111–112, 156–157, 172–173, 187–198, 338
Carvalho, L. S., 256
Casanova, J. L., 15–27, 60–61, 172–173, 190
Case, S. S., 32–33
Caskey, C. T., 1–4
Casorati, G., 18, 190
Cassady, J. P., 149
Cassani, B., 15–16, 21, 22–23, 60–61, 129, 142, 172–173, 179–180, 181
Cass, B., 240
Castaneda, S. A., 110
Castanotto, D., 277–278
Casto, B. C., 230–231
Cattaneo, F., 15–16, 21, 22–23, 36, 49, 60–61, 129, 142, 172–173
Cattoglio, C., 21–22, 61, 156, 157, 161, 179–180
Cavallesco, R., 121
Cavazzana-Calvo, M., 15–27, 60–61, 120, 121, 128–129, 145–146, 147, 157, 172–173, 180, 187–198, 328–329
Cavrois, M. V., 32–33
Cayuela, J. M., 119–120
Cazzola, M., 172–173, 179–180
Cecchini, S., 238–239
Celis, P., 337–354
Certain, S., 15–27, 60–61, 172–173, 190
Cervenka, P., 235
Cesana, D., 171–185
Chadburn, A., 172–173
Chadderton, N., 256
Chadeuf, G., 302, 305
Chahal, P., 240
Chamberlain, J. S., 235, 256
Chandrasekaran, A., 145–146, 147
Chang, B., 256
Chang, H. Y., 277–278
Chang, L., 21, 129
Chang, N., 90, 96–97
Chang, T., 1–14
Chang, Y. N., 96–97
Chan, L., 96–97
Chanock, S. J., 126–127, 134, 148
Chan, R., 18, 129
Chao, H., 235
Chapel, H., 16–17
Chapman, M. S., 230–231
Charnas, L. R., 189

Charneau, P., 33–34, 190
Charrier, S., 36, 40–41, 49, 97, 98
Chatters, S. J., 20–21, 39, 60–61, 78–79, 128–129, 172–173, 174
Chaussain, J. L., 189
Chautard, H., 40–41, 97, 98
Chen, C. S., 237–239, 245, 286
Cheng, L., 145–146, 149
Chen, H., 179–180, 238–239
Chen, J., 100–101, 102
Chen, L., 233
Chen, R., 126–127, 148
Chen, S. T., 6–9, 90
Chen, Z., 90, 96–97
Cherepanov, P., 156
Cheshier, S., 39–40
Cheung, A. K., 231
Cheung, A. Y., 242
Chew, A. J., 243–244
Chiang, Y., 21, 129
Chico, J. D., 242
Chi, J. T., 277–278
Childs, R., 127
Chin, D., 32–33
Chinen, J., 128
Chin, L., 173–174
Chinnasamy, N., 90
Chin, P., 102–103, 191
Chiocca, E. A., 219
Chiodo, V., 235–237, 240
Chiorini, J. A., 230–231, 235
Chipchase, D., 92, 103
Choi, U., 36, 128, 130, 145–146, 147–148, 149
Chong, H., 103–104
Chou, B. K., 145–146, 149
Christensen, B. L., 126–127, 148
Christensen, H. O., 128, 172–173, 190
Christodoulopoulos, I., 90
Chui, D. H., 109–110
Chung, D. C., 242–243, 255–274
Chung, K. H., 276–277, 285
Ciceri, F., 110, 181
Cichutek, K., 90
Cideciyan, A. V., 242–244, 265, 298
Ciotti, F., 110
Ciuffi, A., 61, 120, 180
Civin, C. I., 135–137
Clappier, E., 39, 60–61, 78–79, 110–111, 119–120, 172–173
Clarke, J. T., 189
Clark, K. R., 304, 305
Clark, T. G., 183
Clerici, M., 21, 129
Cleveland, S. M., 21
Cleynen, I., 121
Closs, E. I., 1–4
Cloutier, D. E., 235

Cochrane, M., 245
Coffin, J. M., 156
Cohen, A., 109–110
Cohen, E. A., 112–113
Cohen, L. K., 243–244
Cohen, R., 90, 96–97
Colella, P., 233–234
Coleman, J. E., 256
Coley, M. E., 16–17
Colleoni, S., 21–22
Collins, G. S., 110
Collins, M. K., 38, 40, 96–97
Colomb, C., 109–124
Colombi, S., 181
Colosi, P., 232, 233–234
Conconi, F., 109–110
Conley, M. E., 16–17
Conley, S. M., 256
Conlon, T. J., 235, 242–244
Conrad, C. K., 243–244
Conrath, H., 232
Consiglio, A., 191
Conway, J. C., 235–237
Cooper, M., 235–237, 240
Cooper, R. M., 98, 190–191
Cooray, S., 15–16, 19–20, 21, 22–23, 29–57
Copeland, N. G., 21
Cordon-Cardo, C., 173–174
Cornetta, K., 1–4, 11–12, 78, 89–108, 113–114
Corral, M., 45–46
Corrigan-Curay, J., 313–335
Cortese, K., 256
Cosset, F. L., 38, 40
Costanzi, E., 4–6
Cournoyer, D., 1–4
Couto, L. B., 243–244
Couture, L. A., 102, 104
Cowan, M. J., 126, 189
Crameri, A., 237
Crawford, R. W., 235
Crinquette, A., 119–120
Crise, B., 39, 179–180
Criswell, H. E., 237–238
Critchley-Thorne, R. J., 201
Crittenden, M., 189
Crooks, G. M., 32–33, 129
Croop, J., 100–101
Cross, S. B., 96
Crystal, R. G., 298
Cuccuini, W., 119–120
Cullen, B. R., 280–281, 285
Culver, K. W., 18, 129
Cunningham, J. M., 1–4
Cunningham-Rundles, C., 16–17
Curtin, J. A., 19–20
Cutillo, L., 233–234

D

Daifuku, R., 243–244
Dai, X., 256
Dake, M., 243–244
Dalba, C., 201–202
Dal-Cortivo, L., 17, 23, 60–61, 111–112, 172–173
Dando, J. S., 11–12
Danos, O., 23, 36, 49, 130–131, 243–244, 304
Dantzer, J., 78
Darling, D., 96–97
Darlison, M., 109
Dave, U. P., 21
Davidoff, A. M., 243–245
Davidson, B. L., 275–296
Davies, E. G., 15–16, 17, 19–20, 21, 22–23, 128
Davis, B. R., 16–17, 96–97
Davis, D., 200
Davis, J., 128
Davis, R. W., 276–277
Dawood Markous, R. S., 110
de Alwis, M., 256
Deausen, E., 90, 96–97
de Boer, B. A., 189
Debree, M., 18
Debre, M., 17, 189
Decaluwe, H., 17
DeCarlo, E., 128, 131–132
De For, T., 189
de, F. P., 36–37
Deichmann, A., 19–20, 179–180
Deist, F. L., 172–173, 190
Dejneka, N. S., 242–243, 256
Delabesse, E., 39, 60–61, 78–79, 110–111, 172–173
Delattre, O., 119–120
Delavega, D. J., 1–4
Del Carro, U., 172–173
Del, C. C., 45–46
DeMatteo, R., 233
de Mendez, I., 133, 134
Dempsey, M. P., 32–33
Denaro, M., 111, 121
Denison, M. S., 23–24
de Noronha, C. M., 32–33
Denton, M. J., 242
Dent, P., 219
Deola, S., 15–16, 21, 22, 60–61, 129, 142, 172–173
De Paepe, A., 181–182
De Palma, M., 172–173
de Paula, E. V., 243–244
DePinho, R. A., 173–174
De, P. M., 39
Depolo, N. J., 1–4
Deprez, A., 230–231

De Ravin, S. S., 21–22, 126, 127, 128, 145–146, 147
de, R. D., 39
de Ridder, D., 20–21, 60–61, 78–79, 128–129, 172–173, 174
Derrow, C. W., 126
Desai, K., 102–103, 113–114
de Saint Basile, G., 15–27, 60–61, 172–173, 190
Desai, R. A., 245
Desch, J. K., 243–244
Deutsch, M. J., 60
De Villartay, J. P., 60–61, 128
Dewey, R. A., 23, 60–61, 147
Dezzutti, C. S., 30
Di Bartolomeo, P., 127
Dick, J. E., 23–24, 135–137
Dickson, J. G., 243–244
Diéz, I. A., 23, 60–61, 147
Dik, W. A., 119–120
Dimjati, W., 37–38
Dinauer, M. C., 23, 78, 133, 134–135
Dinculescu, A., 235–237, 240
Ding, C., 134–135, 238–239
DiPrimio, N., 235–237
Di Serio, C., 21, 172–175, 179–180, 181–183
Dishart, D., 276–277
Di Vicino, U., 233–234
Dock, G., 200
Doedens, M., 135–137
Doglioni, C., 172–175, 179–180, 181–183
Domenici, L., 256
Donahue, B. A., 243–244
Donello, J. E., 38
Dong, J. Y., 233–234
Donnelly, M. L., 36–37
Donnelly, O. G., 200–201
Donsante, A., 298
Doria, M., 233–234, 256
Douagi, I., 16–17, 18–19
Douar, A. M., 238–239
Doub, M., 90
Dougherty, J. P., 211–212
Douglas, M., 102, 104
Doulatov, S., 23–24
Douville, R. N., 201
Dowey, S. N., 145–146, 149
Downing, J. R., 21
Down, J., 121
Drenser, K. A., 256
Driever, W., 6–9
Drittanti, L., 304
Driver, D. A., 1–4
Dropulic, B., 90, 96–97, 100
Drouin, L. M., 237–238
Duan, D., 233–234, 260
Dubart-Kupperschmitt, A., 190
Dubielzig, R., 230–231
Dubois-Dalcq, M., 188

Ducroq, D., 242
Duffy, L., 98, 102, 104
Dufier, J. L., 242
Dugue, C., 40–41, 97, 98
Dullmann, J., 172–173
Dull, T., 49, 90, 93, 94, 103, 131, 135–137, 147–148, 191
Dunbar, C. E., 139–140
Dupre, L., 36, 49, 172–173
Dupressoir, T., 230–231
Dupuis-Girod, S., 60–61, 128
During, M. J., 298
Durocher, Y., 240
Durst, M., 230–231

E

Eager, R. M., 200–201
Eaves, C. J., 111
Eckenberg, R., 36, 49, 61
Ecker, J. R., 179–180, 276–277
Ecker, M., 78
Eckfeldt, C. E., 179–180
Eckstein, R., 45–46
Efficace, F., 110
Eglitis, M. A., 4–6, 189
Ehrbar, M., 230–231
Ehrhardt, A., 60
Eidelberg, D., 298
El-Beshlawy, A., 110
Elebute, M. O., 119
El Ghamarawy, M., 110
Ellard, F. M., 92
Elliger, C., 232
Elliger, S., 232
Ellis, J., 39–40, 111
Elwell, D., 243–244
Emerman, M., 32–33, 201
Emi, N., 90
Emshwiller, P., 134
Engel, B., 100
Engelhardt, J. F., 233–234, 235, 260
Engelman, A., 156
Engelstadter, M., 90
Enomoto, Y., 235
Eppert, K., 135–137
Epstein, N., 231
Escarpe, P., 102–103, 191
Espinoza, Y., 305
Etzioni, A., 16–17
Everhart, D., 256

F

Fabb, S. A., 243–244
Fabry, M. E., 111
Facchini, G., 179–180
Fairbanks, D., 60–61
Fairbanks, L. D., 18

Faller, D. V., 110
Fan, P. D., 233–234
Farson, D., 191
Farzaneh, F., 96–97
Fasano, S., 172–173
Fasth, A., 17
Fattore, S., 109–110
Fauchille, S., 40–41, 97, 98
Fazzi, E., 242
Fechner, H., 233
Fehse, B., 172–173
Feigenbaum, V., 188
Feld, J. J., 147
Feldman, S. A., 96
Felice, B., 156
Felsenfeld, G., 39–40
Ferguson, C., 179–180
Fernald, R. D., 181–182
Fernández-Klett, F., 189
Ferrari, F. K., 231, 233, 305–306
Ferrari, G., 18, 179–180, 190
Ferrua, F., 22–23
Ficara, F., 15–16, 21, 22, 60–61, 129, 142, 172–173, 190
Fichelson, S., 190
Fiebig-Comyn, A., 23–24
Fields, P. A., 243–244
Fife, K. H., 230–231
Figueredo, J., 245
Finckh, U., 242
Finke, R., 243–244, 298
Fiorelli, G., 109–110
Fire, A., 276–277
Fischer, A., 15–27, 61, 120, 128–129, 172–173, 180, 187–198, 326
Fischer, N., 179–180
Fischinger, P. J., 102
Fisch, P., 16–17, 18–19
Fisher, C., 110
Fisher, K. J., 233, 243–244
Fisher, T. C., 109–110
Flasshove, M., 36–37
Flavell, R. A., 201
Flaveny, C. A., 23–24
Fleisher, T. A., 21, 129, 134–135, 139–140
Flood, T., 127
Flotte, T. R., 240–244, 298, 299–300, 305
Flügel, A., 189
Foley, J. W., 245
Folks, T. M., 30
Follenzi, A., 33–34
Ford, M., 256
Forestell, S. P., 11–12
Fornerod, M., 38
Forster, A., 128–129
Forster, M., 173
Fortina, P., 109–110
Fossarello, M., 242

Fouquet, F., 190
Frane, J. L., 149
Frank, O., 173
Fraser, C. C., 19–20, 128–129, 172–173, 179–180
Freeman, S. M., 1–4
Freese, A., 243–244
Friborg, S., 243–244
Friedmann, T., 4–9, 60, 90, 200–201
Friedrich, W., 16–17, 18–19
Frizzell, R. A., 233–234
Frost, A. R., 39–40
Frost, L., 277–278
Frotscher, M., 189
Fruehauf, S., 78
Frugnoli, I., 181
Fucharoen, S., 109–110
Fuchs, E. J., 110
Fuess, S., 298
Fu, H., 233, 301
Fukumaki, Y., 109–110
Fulton, A., 242–244
Furushima, M., 242
Fusil, F., 111, 121
Fyffe, J., 96, 98

G

Gabriel, R., 61, 78
Gaburro, D., 109–110
Gaensler, K. M., 111
Gage, F. H., 39, 90, 111, 131, 147–148, 190
Gaken, J., 96–97
Gal, A., 242
Galantuomo, S., 242
Galea-Lauri, J., 96–97
Galimberti, S., 15–16, 21, 22–23, 60–61, 129, 142, 172–173
Galipeau, J., 96–97
Gallardo, H. F., 38
Gallay, P., 32–33, 111, 131, 147–148, 190
Galli, C., 21–22
Gallin, J. I., 126–127, 148
Galy, A., 36, 49
Gani, D., 36–37
Gansbacher, B., 172–173
Gan, Y., 90, 96–97
Gao, G. P., 233, 235, 245, 298, 302
Garcia-Hoyos, M., 233–234
Gardner, P., 243–244
Gargiulo, A., 256
Gargiulo, L., 313–335
Garrett, E., 103–104
Garrigue, A., 19–20, 23, 39, 60–61, 78–79, 110–111, 120, 172–173, 179–180
Gasmi, M., 90
Gaspar, B., 18

Gaspar, H. B., 15–16, 17–18, 19–20, 21–23, 39–40, 60–61, 128, 172–173, 190
Gaszner, M., 39–40
Gatti, R. A., 17
Gay, W., 38
Gaziev, J., 110
Geha, R. S., 16–17
Gendelman, H. E., 32–33
Gennery, A. R., 17
Genovese, P., 21–22
Gentile, A., 39
Gentleman, S., 242
Gentsch, M., 135–136, 147–148
Geny-Fiamma, C., 238–239
Georg-Fries, B., 230–231
Georgiadis, A., 256
Gerber, S., 242
Gerth, C., 242
Geuna, M., 33–34
Ghani, K., 96–97
Ghazi, I., 242
Giannakopoulos, A., 60–61
Giardina, P., 21–22
Giardine, B., 109–110
Gibbs, D., 233–234
Gibson, F. M., 119
Gifford, M. A., 134–135
Gilboa, E., 4–6, 328
Gillet-Legrand, B., 111, 121
Gilmour, K. C., 15–17, 18–20, 21, 22–23, 39, 60–61, 128–129, 172–173, 174, 190
Ginn, S. L., 19–20
Giordano, F. A., 78
Glader, B., 243–244
Glimm, H., 19–20, 23, 39, 59–87, 129–130, 143, 147, 172–173, 179–180, 194
Glynn, J., 90
Goff, S. L., 96
Goldfarb, D., 32–33
Goldfarb, O., 243–244
Goldman, A. I., 189
Goldman, J., 103–104
Goldman, R. D., 32–33
Gonzalez, M., 45–46
Goodman, S. A., 243–244
Good, P. D., 276–277, 291
Good, R. A., 17
Gordon, I., 139–140
Gordon-Smith, E. C., 119
Gottardi, E., 109–110
Govindasamy, L., 230–231, 235–237, 240
Gown, A. M., 243–244
Graham, F. L., 40–41
Grandin, L., 17
Grant, L., 21
Grant, R. L., 245
Grassman, E., 21–22, 39, 181
Gray, E., 237–238, 245

Gray, J. T., 145–146, 147, 244–245
Gray, S. J., 237–238
Greaves, D. R., 111
Greenberg, C. R., 16–17
Greenblatt, J. J., 129
Greene, M. R., 145–146, 147
Greene, W. C., 32–33
Gregorevic, P., 235
Gregory, P. D., 21–22
Gregory, R. I., 276
Grewal, S. S., 189
Grez, M., 23, 36, 145–146, 147–148
Grieger, J. C., 229–254
Grimm, D., 230–231, 232, 235, 277–278, 293–294, 305–306
Griscelli, C., 189
Grompe, M., 298
Gross, F., 15–27, 60–61, 172–173, 190
Grosveld, F., 111
Grosveld, G., 38
Grot, E., 235
Gruber, H. E., 199–228
Grund, N., 78
Guetard, D., 33–34
Guggino, W. B., 243–244
Gungor, T., 127
Guo, H., 276
Gurda, B., 230–231
Gurda-Whitaker, B., 230–231
Gu, S. M., 242
Guttinger, S., 276–277
Gu, Y., 126
Gyapay, G., 19–20, 179–180
Gyurus, P., 242

H

Haapala, D. K., 11–12, 102
Haas, C. A., 189
Haas, D. L., 32–33
Haase, R., 60
Haas, M., 4–6
Hacein-Bey-Abina, S., 15–27, 39, 60–61, 78–79, 110–112, 119–120, 128–129, 145–146, 147, 156–157, 172–173, 179–180, 187–198, 322–324, 326
Haggerty, S. A., 32–33
Hagstrom, J. N., 243–244
Haire, S. E., 256
Hajdari, P., 23
Halbert, C. L., 233–234, 235
Halene, S., 190–191
Halhal, M., 256
Hamdy, M., 110
Hammartrom, L., 16–17
Hammill, A. M., 200
Hammon, W. M., 230–231
Hanawa, H., 90, 111, 179–180, 243–244

Han, J., 276, 285
Han, L., 233–234
Hanna, J., 149
Hardison, R. C., 109–110
Hare, J., 230–231
Hargrove, P. W., 90, 111
Haria, S., 36, 145–146, 147–148
Harkey, M. A., 61
Harper, H., 235
Harper, S. Q., 235
Harrell, H., 239–240
Harris, C., 126
Harris, R., 189
Hart, A., 174
Hart, C. C., 276–277, 285
Hartley, J. W., 11–12, 98, 102
Hartley, O., 21
Haselgrove, J., 243–244
Hauck, B., 233, 242–243
Hauer, J., 15–16, 19–20, 21–22, 145–146, 147
Hauswirth, W. W., 231, 235–237, 240, 242–244, 256
Hawkins, T., 102, 104
Hawkins, T. B., 78
Hawley, R. G., 134
Hayakawa, T., 4–6, 36–37
Hay, B. N., 128
Hector, R. D., 21
Hegge, J., 235–237
Hehir, K., 109–124
Heidecke, V., 219
Heilmann, C., 17
Hein, A., 201
Heinzinger, N. K., 32–33
Hellemans, J., 181–182
Heller, G., 111
Heller, T., 147
Hematti, P., 156, 179–180
Heneine, W., 30
Hermonat, P. L., 233–234
Hernandez, R. J., 172–173
Hernandez-Trujillo, V., 181
Hershfield, M. S., 17–18, 22, 129
Herzog, R. W., 235–237, 240, 243–244
Hesemann, C. U., 61, 180
Heyward, S., 38
Hickstein, D. D., 126
Higashimoto, T., 39–40
High, K. A., 60, 242–244, 267–268, 269, 298
Hildinger, M., 38, 240
Hiraoka, K., 201–202
Hirsch, G., 36–37
Hirschhorn, R., 16–17
Hiscott, J., 201
Hiti, A. L., 109–110, 111
Hlavaty, J., 201–202, 219, 220
Hobbs, J. A., 235–237, 240
Hobson, M. J., 92, 101–102, 103

Hoffmann, G., 61, 180
Hoggan, M. D., 231
Holland, S. M., 126–127, 128, 131–132, 134–135, 148
Hollis, V. W. Jr., 102
Holmes, K. L., 38, 135–136
Holt, I. E., 179–180
Homoyounpour, N., 133, 134
Hong, B. K., 179–180
Hong, R., 17
Hoogerbrugge, P. M., 18
Hoots, K., 243–244
Hope, T. J., 32–33, 38, 49
Horneff, G., 16–17, 18–19
Horwitz, M. E., 38, 127, 135–136
Ho, T. T., 256
Hotz-Wagenblatt, A., 78
Housman, D. E., 231
Howe, S. J., 19–22, 29–57, 60–61, 78–79, 128–129, 147, 156–157, 172–173, 174, 179–180, 190, 325
Hsu, A. P., 128
Hsu, D., 102, 104
Huang, G. M., 111
Huang, Z., 233
Hubank, M., 20–21, 39, 60–61, 78–79, 128–129, 172–173, 174
Hue, C., 15–27, 60–61, 128, 172–173, 190
Hughes, L. E., 36–37
Hughes, S. H., 212
Huhn, D., 45–46
Hui, D., 242–243, 261, 265
Huie, M. L., 16–17
Hu, J., 19–20, 179–180
Hulsman, D., 173–174
Humeau, L., 90, 96–97
Humphries, M., 243–244
Humphries, R. K., 111
Hunter, L. A., 231
Hurh, P., 243–244
Hurlbut, G. D., 245
Huston, M. W., 21–22
Hutchins, B., 300, 308
Hutchison, S., 243–244
Hutvagner, G., 276
Hu, X., 237–238
Hwang, K. K., 239–240

I

Ichisaka, T., 149
Ikeda, Y., 38, 40
Imada, K., 172–173
Imai, Y., 190
Imanishi, Y., 256
Ingolia, N. T., 276
Iodice, C., 256

Isgro, A., 110
Ishii-Watabe, A., 4–6, 36–37
Iwaki, Y., 232

J

Jaalouk, D. E., 96–97
Jackson, A. L., 277–279
Jackson, S. H., 134–135
Jacob, D., 240
Jacobs, J. B., 242–243
Jacobs, M. A., 61
Jacobson, S. G., 242–244
Jaenisch, R., 149
Jakobsen, M. A., 128, 172–173, 190
Jakobsson, J., 39–40
Jambaque, I., 189
Jambou, R., 313–335
James, T. Jr., 256
Janson, C., 243–244, 298
Janssen, A., 256
Jasti, A., 102, 104
Jauch, A., 60–61, 78–79, 129–130, 143, 147, 172–173
Jayandharan, G., 235–237
Jeanson-Leh, L., 36, 49
Jekerle, V., 340
Jenkins, N. A., 21
Jenny, C., 40–41, 97, 98
Jia, B., 126
Jiang, B., 90, 96–97
Jiang, C. K., 16–17
Jiang, H., 243–244
Jiang, J., 237–238
Jin, M. J., 90
Jo, E. C., 240
Johansson, M., 38
Johnson, F., 243–244
Johnson, J. S., 231, 233, 237
Johnson, L. A., 96
Johnson, P. R., 304, 305
Johnson, T., 92, 101–103, 113–114
Johnston, J., 235, 245, 302
Jolly, C., 201
Jolly, D. J., 11–12, 90, 92, 103, 199–228
Jones, R. J., 110
Jones-Trower, A., 1–4
Jonsdottir, G. A., 149
Jordan, C. T., 32–33
Jurisica, I., 23–24

K

Kabat, D., 1–4
Kadota, K., 21–22
Kafri, T., 90
Kahl, C. A., 96
Kaimakis, P., 109–110
Kakakios, A., 19–20

Kalifa, G., 189
Kalota, A., 119–120
Kaludov, N., 230–231
Kamen, A. A., 96–97, 233, 238–239, 240
Kamino, K., 173
Kanda, T., 235
Kanegasaki, S., 133
Kang, E. M., 125–154, 338
Kang, W., 239–240
Kantoff, P. W., 4–6
Kaplan, J., 242
Kaplitt, M. G., 298, 338
Karaiskakis, A., 39–40
Karlsson, S., 172–173
Kasahara, N., 199–228
Kattamis, A. C., 109–110
Katz, M. L., 256
Kaul, R., 61
Kaushal, S., 242–244
Kavanaugh, M. P., 1–4
Kaye, R., 243–244
Kay, M. A., 233, 235, 243–244, 298, 305–306
Keerikatte, V., 61
Keiser, N. W., 235
Keller, G., 4–6
Kellner, J. D., 126–127
Kelly, E., 200
Kelly, M., 49, 90, 93, 94, 103, 131, 135–137, 191
Kelly, P. F., 38, 90, 135–136, 172–173
Kelsall, B. L., 172–173
Kemball-Cook, G., 244–245
Kempski, H., 20–21, 39, 60–61, 78–79, 128–129, 172–173, 174
Kende, M., 201
Kennan, A., 256
Kent, W. J., 194
Kepes, S., 111
Kern, A., 232, 305–306
Kewalramani, V., 32–33
Khvorova, A., 276, 278–279
Kiang, A. S., 256
Kiem, H. P., 11–12, 38, 103–104
Kiermer, V., 100, 102–103, 111–112, 172–173, 191
Kievit, E., 216
Kikuchi, E., 201–202
Kim, C., 126
Kim, F., 235
Kim, J. W., 1–4
Kim, K. A., 21–22
Kim, S. R., 233–234, 256
Kimura, T., 208–209
Kim, V. N., 90
Kim, Y. G., 240
King, A. A., 134
King, D. J., 20, 60–61, 128, 172–173, 179–180, 190
King, J. A., 230–231

King, P., 316
Kingsman, A. J., 90
Kingsman, S. M., 90, 92, 103
Kinner, A., 60–61, 78–79, 129–130, 143, 147, 172–173
Kinnon, C., 21–22, 36, 39–40, 145–146, 147–148
Kinsella, T. M., 6–9
Kintner, H., 239–240
Kirn, D., 201–202
Kjellstrom, S., 256
Kleckner, A. L., 243–244, 298
Kleinschmidt, J. A., 230–231, 232, 235, 305–306
Klug, B., 337–354
Klug, C. A., 39–40
Knipscheer, P., 174
Knoess, S., 173
Knop, D. R., 239–240
Knoss, S., 181
Kochenderfer, J. N., 96
Koehl, U., 39, 60–61, 129–130, 143, 147, 172–173
Kohlbrenner, E., 230–231, 238–239
Kohn, D. B., 4–6, 18, 32–33, 129, 130–131, 190–191
Kollia, P., 109–110
Kollias, G., 111
Komaromy, A., 242–243, 256, 261
Kondo, K., 235
Kong, J., 256
Konkle, B., 243–244
Kool, J., 174
Koop, S., 98, 102, 104
Kotin, R. M., 231, 235, 238–239
Kourilsky, P., 16–17, 18–19
Kouyama, K., 243–244
Kowal, K., 11–12
Kowolik, C. M., 90
Kozak, S. L., 1–4
Krall, W. J., 4–6, 130–131
Krämer, A., 60–61, 78–79, 129–130, 143, 147, 172–173
Kramer, B., 19–20
Kramer, M. J., 102
Krausslich, H. G., 38
Kreutzberg, G. W., 189
Krivit, W., 189
Krol, J., 276
Kronenberg, S., 230–231
Kühlcke, K., 39, 60–61, 129–130, 143, 147, 172–173
Kuhns, D. B., 130, 147
Kumar, A., 256
Kumaramanickavel, G., 242
Kumar-Singh, R., 256
Kume, A., 134, 238–239
Kung, S. H., 243–244
Kunkel, H., 39, 60–61, 129–130, 143, 147, 172–173

Kuribayashi, F., 133
Kurlandsky, L. E., 16–17
Kurre, P., 61
Kurtzman, G. J., 232
Kustikova, O. S., 172–173
Kutschera, I., 23, 60–61, 111–112, 172–173

L

Lagresle-Peyrou, C., 23, 61, 120, 180
Lamb, K., 238–239
Landais, P., 127
Landau, N. R., 90
Lane, M. D., 230–231
Langerak, A. W., 119–120
LaPorte, P., 6–9
Laroudie, N., 40–41, 97, 98
Larson, P. J., 243–244
Latham, M., 19–20
Latour, S., 15–16, 19–20, 21
Lau, C. J., 240
Laufs, S., 78
Laughlin, C. A., 231
Lau, K. H., 6–9
Laurenti, E., 23–24
LaVail, M. M., 256
Lavigne, M. C., 133, 134
Law, W., 1–4
Layh-Schmitt, G., 172–173
Leath, A., 89–108
Lebherz, C., 265
Leboulch, P., 20–21, 109–124, 128–129
Lechman, E. R., 135–137
Le Deist, F., 16–17, 18–19, 60–61, 128–129, 172–173
Lee, A. H., 145–146, 148
Lee, G., 21–22
Lee, H., 173–174
Lee, K., 240
Lee, M. A., 32–33
Lee, Y. L., 21–22, 276, 285
Leger, A., 298
Le, G. R., 38
Lehn, P., 4–6, 130–131
Lehrnbecher, T., 126–127, 148
Lei, B., 256
Lei-Butters, D. C., 235
Leichtle, S., 237
Leipzig, J., 61, 120, 180
Leitman, S. F., 127, 128, 131–132, 139–140
Lekstrom, K., 133, 134
Lemieux, C., 276–277
Lemke, N., 61, 180
Lenz, J., 173–174
Leonard, D. G., 243–244
Leonard, W. J., 18–19, 172–173
Leowski, C., 242
Leroy, S., 17

Leto, T. L., 133, 134
Leung, E., 145–146, 148
Leuschner, P. J., 278–279
Levaditi, C., 200
Levadoux-Martin, M., 23–24
Levasseur, D. N., 111
Levine, J. E., 126
Levy, J. R., 238–239
Levy, R., 61
Lewin, A. S., 235–237, 240, 256
Lewinski, M. K., 156
Lewis, B. C., 90
Lewis, B. P., 276
Lewis, D. B., 126–127
Lewis, M. A., 235
Lewis, P. F., 32–33, 201
Liang, F.-Q., 257–258
Li, B., 235–237, 240
Li, C., 232, 235–237
Li, E., 39–40
Li, F., 128, 131–132, 134–135
Li, J., 232, 237–238
Li, L. L., 134–135
Lim, A., 15–16, 19–21, 39, 60–61, 78–79, 110–111, 128–129, 172–173
Lim, B. P., 240
Li, M. Z., 276–277
Lin, H. F., 243–244
Lin, H. W., 32–33
Linsley, P. S., 278–279
Linton, G. F., 38, 127, 128, 130, 131–132, 133, 134, 135–136
Lipps, H. J., 60
Lipshitz, H. D., 39–40
Li, Q., 111, 235–237, 240, 256
Li, S., 111
Littera, R., 110
Littman, D. R., 90
Liu, A., 23
Liu, J., 239–240
Liu, X., 304
Liu, Y., 235
Li, W., 237
Li, X. J., 36–37, 126–127
Li, Y., 21–22, 39, 179–180
Li, Z., 172–173
Lobel, L. I., 202
Lockey, T., 145–146, 147
Lock, M., 302, 304, 305–307
Lockman, L., 189
Loedige, I., 276
Loes, D., 189
Logg, C. R., 199–228
Loiler, S., 305
Lo, M., 172–173
Lombardo, A., 21–22
London, I. M., 111
Long Priel, D. A., 130

Lorenz, B., 242
Lothrop, C. D. Jr., 243–244
Lo, T. W., 145–146, 148
Love, S. B., 183
Luban, J., 21
Lucarelli, G., 110
Luciw, P. A., 33
Lu, H., 233
Luk, A., 243–244
Lukason, M. J., 245
Luke, G., 36–37
Lund, A. H., 173–174
Lundberg, C., 39–40
Lund, E., 276–277
Luo, L., 256
Lusby, E. W., 230–231
Lu, T., 21–22, 145–146, 147
Lu, X., 90, 96–97, 100
Lu, Y., 235, 298
Lu, Z., 242
Luznik, L., 110
Luzzatto, L., 111
Lynn, A., 1–4
Lyubarsky, A., 242–244

M

MacIntyre, E., 39, 110–111, 172–173
MacKey, T., 96–97
Madden, V. J., 90
Maetzig, T., 173, 181
Maggioni, D., 18
Magnani, Z., 181
Maguire, A. M., 233–234, 242–244, 255–274, 298
Maguire, T. E., 134–135
Mah, C. S., 235–237, 240
Mahlaoui, N., 17, 111–112, 172–173
Malani, N., 21–22, 111
Malassis-Seris, M., 23
Malech, H. L., 33, 36, 38, 125–154
Malhaoui, N., 60–61
Malik, P., 39–40, 109–110, 111
Manceau, P., 304
Mancuso, K., 256
Mandelli, F., 110
Mandell, T., 239–240
Mandel, R. J., 49, 90, 93, 94, 103, 131
Manfredi, A., 256
Manley, S., 243–244
Manno, C. S., 243–244, 298, 330–332
Mansfield, P., 126
Mansour, M. R., 20–21, 39, 60–61, 78–79, 128–129, 172–173, 174
Mao, H., 256
Maples, P. B., 128, 131–132
Marangoni, F., 172–173
Marchal, C. C., 126–127

Marchesini, S., 172–173
Marciano, B. E., 126, 127, 147
Mardiney, M. III., 134–135
Marinello, E., 129, 142, 172–173
Markert, L., 17
Markoulaki, S., 149
Marktel, S., 110
Marrocco, E., 256
Marshall, J., 245
Marshall, K. A., 242–244
Marsh, J., 90, 96, 102–103, 113–114
Martinache, C., 15–16, 19–20, 21
Martin, F., 40, 96–97
Martin, J. N., 293–294
Martin-Rendon, E., 92
Martin, S., 49
Martins, I., 277–278, 293–294
Maruggi, G., 156–157, 181
Marzio, G., 212
Mas Monteys, A., 277–278, 286, 291, 293
Matano, T., 201
Mathias, L. A., 109–110
Matranga, C., 278–279
Matreyek, K. A., 156
Matsushita, T., 232
Matthes, M. T., 256
Matute, J. D., 126–127
Matveeva, O., 278–279
Mavilio, F., 155–170
May, C., 111, 172–173
Mayfield, T. L., 239–240
Ma, Z., 21–22, 145–146, 147
Mazarakis, N., 92
Mazurek, A., 276–277
Mazza, U., 109–110
Mazzolari, E., 18
McBride, J. L., 277–278, 293–294
McCall, A., 230–231
McCarty, D. M., 90, 233, 301
McCleland, M. L., 243–244
McClelland, A., 243–244
McCormack, J. E., 1–4
McCormack, M. P., 20–21, 128–129
McCormick, F., 200–201
McCowage, G. B., 19–20
McCown, T. J., 237–238
McCray, P. B., Jr., 278–279
McDermott, S. P., 135–137
McDonagh, K. T., 1–4
McGarrity, G. J., 1–4
McGrath, J. P., 304
McGuinness, R., 191
McIntosh, J., 244–245
McIntyre, E., 128–129, 172–173
McJunkin, K., 276–277
McKenna, R., 230–231
McLaughlin, S., 231
McPhee, S., 235–237, 243–244

Medico, E., 39
Mehrotra, A., 36–37
Meikle, S., 21
Meiselman, H. J., 109–110
Meissner, A., 149
Melnick, J. L., 245
Melo, J. V., 103–104
Mendell, J. R., 243–244, 298
Mendez, A. F., 238–239
Meng, X., 145–146, 148
Mento, S. J., 11–12
Merling, R., 96–97
Merten, O. W., 40–41, 97, 98, 238–239
Mertens, K., 244–245
Messner, A. H., 243–244
Meuse, L., 235, 243–244
Meuwissen, H. J., 17
Meyer, J., 172–173, 181
Meyers, E., 245
Mezey, E., 189
Miao, C. H., 243–244
Miccio, A., 179–180
Michalakis, S., 256
Mihelec, M., 256
Mikkers, H., 174
Miller, A. D., 1–6, 11–12, 18, 32, 36–37, 38, 40, 103–104, 129, 233–234, 235
Miller, A. R.-M., 103–104
Miller, D. G., 1–4, 235
Miller, E. B., 230–231
Miller, J. A., 127, 128, 131–132, 139–140
Miller, J. C., 145–146, 148
Miller, W., 109–110
Millington-Ward, S., 256
Mingozzi, F., 242–244, 261, 265, 298, 302, 332
Min, S. H., 235–237, 240, 256
Mirolo, M., 15–16, 21, 22–23, 60–61, 129, 142, 172–173, 179–180, 181
Mishra, A., 181
Miskin, J., 92, 103
Mitchell, A. M., 237–238, 240–242
Mitchell, M., 230–231
Mitchell, R. S., 23–24, 156, 179–180
Mitelman, F., 119–120
Mitrophanous, K. A., 40, 90, 92, 103
Mitts, K., 39–40
Miyanohara, A., 6–9
Miyoshi, H., 39, 90, 190, 256
Mizuguchi, H., 4–6, 36–37
Mizukami, H., 238–239
Moayeri, M., 145–146, 147
Mochizuki, H., 90
Mockaitis, K., 78
Modell, B., 109
Modlich, U., 156–157, 172–173, 181
Mody, D., 21–22, 145–146, 147

Mohiuddin, I., 305
Moiani, A., 155–170
Moir, R. D., 32–33
Molday, L. L., 256
Monahan, P. E., 233–234, 244–245, 301
Moninger, T. O., 230–231
Montagnier, L., 33–34
Montefusco, S., 256
Montgomery, M., 313–335
Montini, E., 39, 156–157, 171–185, 298
Montpellier, B., 119–120
Mooney, S., 78
Moore, K. A., 1–4
Moran, M. L., 243–244
Morecki, S., 15–16, 21, 22, 60–61, 129, 142, 172–173
Moreno-Carranza, B., 147–148
Morgan, G., 18
Morgan, R. A., 11–12, 90, 96, 98, 102
Morillon, E., 20–21, 23, 39, 60–61, 78–79, 110–111, 128–129, 172–173
Mori, S., 235
Moritz, T., 36–37
Mortellaro, A., 15–16, 21, 22, 60–61, 129, 142, 172–173, 190
Mortier, G., 181–182
Moshous, D., 17
Mosier, D. E., 190
Moss, R. B., 243–244
Motulsky, A. G., 200–201
Moullier, P., 297–312
Mount, J. D., 243–244
Mousses, S., 291
Mueller, C., 240–242
Mueller, P. R., 61
Mueller, S. N., 1–4
Muenchau, D. D., 1–4, 11–12, 98, 102
Muftin, G., 96–97
Muhlfriedel, R., 256
Mukherjee, S., 200
Müller, S., 17, 127
Mulligan, R. C., 4–6, 111, 130–131, 147–148, 190
Munnich, A., 242
Muraca, G. M., 110
Muraro, S., 181
Murthy, K. R., 242
Mussig, A., 61, 180
Mussolino, C., 256
Muzny, D. M., 1–4
Muzyczka, N., 230–231, 235–237, 238–239
Myers, L. A., 17

N

Naash, M. I., 256
Nadel, B., 119–120
Nagy, D., 243–244
Naik, P., 233, 301
Nairn, R., 40–41
Nakai, H., 298
Nakakura, T., 238–239
Nakamura, M., 133
Nalbantoglu, J., 240
Naldini, L., 32–34, 36, 39, 40–41, 49, 60, 90, 93, 94, 97, 98, 103, 111, 131, 147–148, 172–173, 181, 190, 191
Nam, H. J., 230–231, 235–237
Napoli, C., 276–277
Narfstrom, K., 256
Narita, M., 149
Nash, K., 238–239
Nash, Z., 256
Nathwani, A. C., 90, 243–245
Naud-Saudreau, C., 189
Naumann, N., 128
Naundorf, S., 39, 60–61, 129–130, 143, 147, 172–173
Naviaux, R. K., 4–6
Nechipurenko, Y., 278–279
Negre, D., 38
Negre, O., 109–124
Negrete, A., 238–239
Nemunaitis, J., 200–201
Nepomuceno, I. B., 243–244
Nerhbass, U., 33–34
Nesbeth, D., 96–97
Neven, B., 17
Ney, P. A., 1–4
Ng, C. Y., 244–245
Nguyen, K., 1–4
Nguyen, M., 49, 90, 93, 94, 103, 131
Nguyen, N., 11–12, 98, 102
Niamke, J., 239–240
Nichols, T. C., 243–244
Nicolau, S., 200
Nicoletti, A., 242
Nicolson, S. C., 237–238, 240–242
Niehues, T., 16–17, 18–19
Nie, J., 149
Nielsen, A. A., 21
Nienhuis, A. W., 1–4, 38, 90, 111, 243–244
Nieto, M. J., 45–46
Nietupski, J. B., 245
Nilsen, T. W., 157
Nobili, N., 18
Nolan, G. P., 6–9
Nolta, J. A., 129
Nomicos, E. Y., 128
Nomura, S., 102
Nonoyama, S., 16–17
Norbash, A. M., 243–244
Notarangelo, L. D., 16–18, 22, 190
Notta, F., 23–24
Nowaczyk, M. J., 189

Author Index

Nowrouzi, A., 21–22, 23, 60–61, 147
Nunoi, H., 133
Nusbaum, P., 15–27, 60–61, 172–173, 190

O

Obara, Y., 238–239
Ochs, H. D., 190
Ocwieja, K. E., 156
O'Dea, J., 1–4
O'Donnell, A., 110
Ogston, P., 230–231
Ohene-Frempong, K., 109–110
Ohnuki, M., 149
Oh, Y. H., 111
Okano, M., 39–40
Oliveira, N. M., 201
Olivieri, N. F., 110
Olson, J. S., 200
Olson, N. H., 230–231
Olsson, K., 38
Ongley, H. M., 230–231
Ono, F., 235
Opalka, B., 36–37
Opelz, G., 78
Opolon, P., 49
Orchard, P. J., 189
O'Reilly, M., 256, 313–335
Orfao, A., 45–46
Orkin, S. H., 134, 200–201
Ory, D., 38, 111, 131, 147–148, 190
Osborne, C. S., 20–21, 39–40, 128–129
Osborne, W. R., 4–6, 36–37
Ostertag, W., 90, 173
Ott, M. G., 39, 60–61, 78–79, 129–130, 143, 147, 156–157, 172–173
Otto, E., 1–4
Ouma, A. A., 145–146, 147
Ouyang, L., 90
Ozawa, K., 238–239
Ozelo, M. C., 243–244
Ozsahin, H., 127

P

Paar, M., 202
Pacak, C. A., 235
Padron, E., 230–231
Page, K. A., 90
Paige, C., 4–6
Palczewski, K., 242–243
Palfi, A., 256
Palu, G., 30–32
Pande, N. N., 245
Pang, J. J., 235–237, 240, 242–244, 256
Panina, P., 18
Pan, J., 145–146, 149
Pannell, D., 39–40
Pannetier, C., 16–17, 18–19

Pannicke, U., 16–17, 18–19
Papadaki, H. A., 119
Papapetrou, E. P., 21–22
Parker, A. E., 23–24
Parkhurst, M. R., 96
Park, J. Y., 240
Parks, W. P., 245
Parolin, C., 30–32
Parsley, K. L., 15–16, 19–20, 21, 22–23, 36, 60–61, 128, 145–146, 147–148, 172–173, 179–180, 190
Paruzynski, A., 59–87
Parziale, A., 109–110
Pasceri, P., 39–40
Passwell, J. H., 127
Patel, M. I., 218
Patel, S., 243–244
Patijn, G. A., 243–244
Patrinos, G. P., 109–110
Patterson, A., 313–335
Paul, C. P., 276–277, 291
Pawlik, K. M., 111
Pawlita, M., 305–306
Pawliuk, R., 20–21, 111, 128–129
Pawlyk, B. S., 256
Payen, E., 109–124
Pearce-Kelling, S. E., 242–243
Pearson, R. A., 256
Peccatori, J., 181
Pech, N., 134–135
Peebles, P. T., 11–12, 102
Pellin, D., 21
Penaud-Budloo, M., 298
Peng, H., 6–9
Perdew, G. H., 23–24
Perelman, N., 111
Perez-Simon, J. A., 45–46
Perignon, J. L., 18
Perrault, I., 242
Perrier, M., 233, 238–239
Perrine, S. P., 110
Persons, D. A., 90, 111, 172–173, 179–180
Perumbeti, A., 39–40
Pescarollo, A., 181
Petermann, K. B., 237
Peters, B., 78
Peters, C., 189
Petit, C., 33–34
Petrillo, M., 233–234, 256
Petroni, D., 109–110
Petrs-Silva, H., 235–237, 240
Petryniak, B., 126
Pfaffl, M. W., 181–182
Pfeifer, G. P., 61
Pflumio, F., 190
Pham, P. L., 240
Philbey, A., 21
Phillips, J. L., 235–237

Phillips, L. A., 102
Phuong, T. K., 1–4
Picard, C., 15–16, 17, 19–20, 21
Pickle, C. S., 145–146, 148
Pierce, E. A., 242–244, 267–268, 269
Pierce, G. F., 243–244
Pike, K. M., 147
Pike-Overzet, K., 20–21, 39, 60–61, 78–79, 128–129, 172–173, 174
Pilon, A. M., 38, 135–136
Pilz, I., 61, 180, 194
Pinkerton, T. C., 231
Pinkert, S., 233
Piras, E., 110
Pizzato, M., 30–32
Podsakoff, G. M., 232, 243–244
Poeppl, A., 23–24
Poliakov, E., 242
Poller, W., 233
Pollok, K., 98
Polo, J. M., 6–9
Poncz, M., 16–17
Ponder, K. P., 298
Pontremoli, S., 109–110
Ponzoni, M., 172–175, 179–180, 181–183
Porta, F., 17–18, 22, 127
Porter, M., 230–231
Pötgens, A. J., 133
Potter, J., 38
Potter, M., 306–307
Prass, K., 189
Preising, M., 242
Preiss, C., 129–130, 143, 147, 172–173
Premawardhena, A., 110
Preston, B. D., 211–212
Price, M. A., 32–33
Priel, D. A., 147
Priller, J., 189
Printz, M., 11–12
Prinz, C., 179–180
Prinz, M., 189
Proudfoot, N. J., 4–6
Provost, P., 276–277
Prum, B., 19–20, 179–180
Puck, J. M., 16–17, 20
Pu, D., 237–238
Pugh, E. N. Jr., 242–244
Pusch, O., 276–277
Puthenveetil, G., 111
Pyra, H., 92, 103

Q

Qiao, C., 237–238
Qin, Y., 276–277
Quattrini, A., 172–173
Quesada, O., 230–231
Qu, G., 242–243

Quirk, J. G., 233–234
Quiroz, D., 103
Qureshi, N., 111

R

Rabinowitz, J. E., 231, 232, 235–237
Rabreau, M., 230–231
Radcliffe, P. A., 92
Radeke, R., 191
Radford, I., 128–129, 172–173
Radrizzani, M., 40–41, 97, 98
Ragg, S., 61, 180
Ragland, A. M., 32–33
Ragni, M. V., 243–244
Raillard, S. A., 237
Rainov, N. G., 219
Rainsbury, J. M., 316
Raj, K., 230–231
Ralston, E. J., 145–146, 148
Ramesh, N., 4–6, 36–37
Rampalli, S., 23–24
Ramsay, N. K., 189
Ranzani, M., 172–174, 179–180, 181–183
Rasko, J. J., 243–244
Rathmann, M., 242
Ratner, L., 32–33
Ray, J., 242–243, 256, 261
Read, E. J., 127
Read, M. S., 243–244
Rebar, E. J., 145–146, 148
Recchia, A., 21, 60–61, 172–173, 179–180, 181
Redmond, D. E. Jr., 237
Redmond, T. M., 242–244, 256
Reeves, L., 96, 98, 100–101, 102
Regnstom, J., 338
Reichenbach, J., 23
Reik, W., 202
Rein, A., 11–12, 102
Reinhardt, J., 337–354
Reiser, J., 90
Reitsma, M. J., 32–33
Relander, T., 38
Rence, C., 230–231
Rengo, G., 232, 235–237
Rex, T. S., 233–234, 242–243
Reynolds, A., 276, 278–279
Reynolds, J., 11–12
Reynolds, T. C., 11–12, 32–33, 103–104, 243–244
Richter, J., 38
Rick, O., 45–46
Ridzon, D. A., 285
Riemer, C., 109–110
Rieux-Laucat, F., 15–16, 19–20, 21
Rigg, R. J., 11–12
Riggs, A. D., 61
Rissing, D., 102–103, 113–114

Risueno, R. M., 23–24
Rittner, K., 232
Rivella, S., 111, 172–173
Rivet, C., 304
Riviere, I., 4–6, 21–22, 130–131
Rizzo, S., 119
Robbie, S. J., 256
Robbins, J. M., 199–228
Robbins, P. B., 190–191
Roberts, J. L., 17
Robinson, N., 243–244
Rocchiccioli, F., 189, 190
Rocchi, M., 231
Roeder, I., 78
Roehl, H., 92, 103
Roesler, J., 135–137
Roe, T., 32–33, 201
Roetto, A., 109–110
Rohll, J. B., 92, 103
Rolland, M. O., 189
Rolling, F., 232
Roman, A. J., 242–244
Romeijn, L., 174
Romeo, R., 23–24
Roncarolo, M. G., 21, 36, 49, 110, 172–173, 181, 190
Ronen, K., 61, 111, 120, 180
Roos, D., 133
Rosenberg, S. A., 129
Rosenqvist, N., 39–40
Rosenstein, B., 243–244
Rosenthal, E., 313–335
Rosenthal, F. M., 60
Rosman, G. J., 32
Rossi, C., 18
Rossi, S., 242–244
Roth, S. L., 21–22, 111
Rottem, M., 16–17
Rozet, J. M., 242
Rudolf, E., 147–148
Rudolph, C., 173
Ruotti, V., 149
Rupin, J., 243–244
Ruscetti, S., 11–12, 102
Russell, D. W., 235
Russell, R., 173–174
Russell, S. J., 200
Russell, W. C., 40–41
Rutledge, E. A., 235
Rux, J. J., 235
Ryan, M. D., 36–37
Ryan, T. M., 111
Ryser, M. F., 135–136, 147–148

S

Sadelain, M., 38, 111, 172–173
Safer, B., 235
Sajjadi, N., 11–12

Sakai, Y., 230–231
Sakurai, K., 277–278
Salmon, P., 38
Salomoni, M., 181
Samara, M., 109–110
Samulski, R. J., 229–254, 301, 305–306
Samulski, T., 231, 233
Sanburn, N., 92, 100–101, 103
Sanders, D. A., 96
Sandrin, V., 38
Sanges, D., 256
Sanmiguel, J., 235, 245
San Miguel, J. F., 45–46
Santilli, G., 36, 145–146, 147–148
Santoni de Sio, F. R., 39
Santoni, F. A., 21
Sanvito, F., 172–175, 179–180, 181–183
Sarra, G.-M., 256
Sartori, D., 179–180, 181
Saslow, E., 243–244
Sastry, L., 90, 92, 98, 101–103, 113–114
Sata, T., 235
Sato, K., 235
Saunders, E. F., 189
Sbaiz, L., 109–110
Scadden, D., 1–4
Scaramuzza, S., 15–16, 21, 22–23, 36, 49, 60–61, 129, 142, 172–173, 179–180
Scarpa, M., 1–4
Schaffer, D. V., 96–97
Schambach, A., 21–22, 38, 39–40, 147–148, 173, 181
Schepers, A., 60
Schiedlmeier, B., 21–22, 39, 173
Schiff, R. I., 17
Schiff, S. E., 17
Schilz, A., 39, 60–61, 129–130, 143, 147, 172–173
Schlegelberger, B., 173
Schlehofer, J. R., 230–231
Schlesier, M., 16–17, 18–19
Schmidt, K., 230–231
Schmidt, M., 19–22, 23, 39, 59–87, 111–112, 128–130, 143, 147, 172–175, 179–180, 181–183, 187–198, 235
Schmidt, S., 173, 179–180
Schmitt, I., 90
Schneider, A., 36–37
Schneider, C. K., 345–346
Schneiderman, R. D., 38
Schnerch, A., 23–24
Scholes, J., 111
Scholz, S., 61, 78
Schonely, K., 90
Schroder, A. R., 156, 179–180
Schuesler, T., 173
Schultze-Strasser, S., 60–61, 78–79, 129–130, 143, 147, 172–173

Schupbach, J., 92, 103
Schwäble, J., 23, 60–61, 78–79, 129–130, 143, 147, 172–173
Schwartz, J. P., 90
Schwartz, S. B., 242–244
Schwarz, D. S., 276
Schwarzer, A., 23, 60–61, 147
Schwarz, K., 16–17, 18–19
Schwarzwaelder, K., 20–21, 39, 60–61, 78–79, 128–130, 143, 147, 172–173, 174, 179–180, 194
Schwein, A., 256
Schwella, N., 45–46
Scobie, L., 21
Scollay, R., 60
Scopes, J., 119
Scotti, M. M., 239–240
Scott, M. L., 242
Seamon, J. A., 201
Sebire, N. J., 21–22, 39
Seeber, S., 36–37
Seeliger, M. W., 256
Segal, B. H., 126
Segall, H. I., 103
Seger, R. A., 17, 23, 127
Seiler, M., 230–231
Sekhsaria, S., 128, 131–132, 134–135, 139–140
Selleri, S., 181
Sellers, S., 179–180
Selz, F., 15–27, 60–61, 172–173, 190
Semizarov, D., 277–278
Serana, F., 338
Sergi, L. S., 60, 172–175, 179–180, 181–183
Serke, S., 45–46
Serrano, M., 173–174
Servida, P., 18, 181
Sessa, M., 172–173
Setty, M., 21–22
Shah, N., 243–244
Shapiro, E. G., 189
Shapiro, H. M., 47
Sharma, Y., 179–180
Sharova, N., 32–33
Shaw, K. L., 18
Shearer, G., 21, 129
Shenk, T., 231, 233
Shen, Z., 256
Shera, D., 243–244
Sheridan, P. L., 1–4
Sherman, M. P., 32–33
Sherr, C. J., 174
Shigeoka, A., 18, 129
Shih, T., 313–335
Shin, K. J., 276–277
Shinn, P., 179–180
Shipp, A., 313–335
Shklyaev, S., 238–239
Shou, Y., 21

Siegert, W., 45–46
Signorini, S., 242
Siler, U., 39, 60–61, 129–130, 143, 147, 172–173
Silva, J. M., 276–277
Silver, J., 61
Silvin, C., 128
Siminovitch, K. A., 16–17
Simonelli, F., 242–244, 267–268, 269
Simons, D. L., 256
Sinclair, J., 20, 39–40, 60–61, 128, 172–173, 190
Siniscalco, M., 231
Sinnott, R., 235–237
Sipo, I., 233
Siritanaratku, N., 110
Sivera, P., 109–110
Skelton, D. C., 4–6, 130–131
Skeoch, C. H., 18
Slatter, M. A., 17
Slavin, S., 15–16, 21, 22, 60–61, 129, 142, 172–173, 179–180
Slepushkina, T., 90, 96–97
Slepushkin, V., 90, 96–97, 100
Slukvin, I. I., 149
Smiley, J., 40–41
Smith, A. J., 256
Smith, K. A., 190
Smith, P., 243–244
Smogorzewska, E. M., 18, 129
Smucker, B., 92, 101–102, 103
Smuga-Otto, K., 149
Smyth, C. M., 19–20
Snyder, R. O., 238–240, 243–244, 297–312
Sodani, P., 110
Solly, S. K., 201–202
Solomon, W., 111
Soltys, S., 232, 235–237
Somanathan, S., 235, 245
Somasundaram, T., 230–231
Sommer, J. M., 243–244, 305
Song, J., 32–33, 191
Sonntag, F., 230–231
Soohoo, C., 276–277, 291
Soong, B. W., 285
Sorensen, R., 60–61, 128–129
Sorrentino, B. P., 21, 145–146, 147
Souied, E., 242
Soule, B. P., 147
Soulier, J., 39, 60–61, 78–79, 110–111, 172–173
Souza, D. W., 245
Sparrow, J. R., 233–234
Speckmann, C., 16–17, 18–19
Speleman, F., 181–182
Spencer, C. T., 243–244
Spencer, L. T., 243–244
Spence, Y., 244–245
Spitz, L., 32–33
Spratt, S. K., 134–135, 243–244
Srikumari, C. R., 242

Srivastava, A., 230–231, 235–237, 240
Srivastava, D. K., 111
Staal, F. J., 172–173
Stafford, D. W., 243–244
Stambaugh, K., 1–4
Stanghellini, M. T., 181
Starkey, W., 103–104
Stedman, H., 243–244, 298
Steele, M., 126–127
Steigerwald, S. D., 61
Stein, S., 39, 60–61, 78–79, 129–130, 143, 147–148, 156–157, 172–173, 326
Stemmer, W. P., 237
Stephens, C., 256
Stephens, R., 21
Stern, M. B., 298
Stettler, M., 240
Stevenson, M., 32–33
Stewart, R., 149
Stitz, J., 90
St-Laurent, G., 240
Stockholm, D., 23
Stocking, C., 173
Storm, T. A., 233, 298
Streetz, K. L., 277–278, 293–294
Stripecke, R., 32–33
Stuhlmann, H., 202
Stull, N. D., 126–127
Suckau, L., 233
Sui, G., 276–277, 291
Sukonnik, T., 39–40
Sumaroka, A., 242–244
Sun, C. W., 149
Sun, D., 256, 257–258
Sun, J., 233–234, 242–245
Sun, X., 256, 298
Surace, E. M., 242–244, 256
Surrey, S., 109–110
Sutton, R. E., 32–33, 103
Sutton, S. E., 23–24
Sweeney, C. L., 145–146, 149
Swingler, S., 32–33
Szabo, E., 23–24

T

Taber, L. H., 245
Tabucchi, A., 15–16, 21, 22, 60–61, 129, 142, 172–173, 190
Tagliafico, E., 181
Tai, C. K., 201–202, 219, 220, 225
Tai, S. J., 243–244
Takada, Y., 256
Takahashi, K., 149
Takahashi, M., 39, 90, 256
Takefman, D., 100
Takeuchi, H., 201
Takeuchi, T., 235

Takeuchi, Y., 30–32, 38, 40
Tam, L. C., 256
Tanabe, K., 149
Tanabe, T., 256, 257–258
Tanaka, K., 90
Tan, C., 38
Tanimoto, N., 256
Tan, M., 256
Tan, Y., 189
Tao, W., 126
Tarantal, A. F., 235
Taupin, P., 17
Taylor, G., 243–244
Tein, I., 189
Testa, F., 242–244
Testaiuti, M., 243–244
Thattaliyath, B. D., 235
Theobald, N., 130
Thomas, C. E., 233
Thomas, D. L., 239–240
Thomas, G. F., 231
Thompson, A. R., 243–244
Thompson, D. A., 242
Thompson, L., 39–40
Thompson, M. A., 21
Thomson, J. A., 149
Thonglairoam, V., 109–110
Thornhill, S. I., 21–22, 39–40
Thrasher, A. J., 20, 21, 23, 29–57, 60–61, 128, 147–148, 179–180
Throm, R. E., 145–146, 147
Tian, S., 149
Tighe, R., 111
Tillson, D. M., 243–244
Tirunagari, L. M., 21–22
Tolstoshev, P., 21, 129
Tomari, Y., 278–279
Tomoda, K., 149
Torbett, B. E., 190
Toromanoff, A., 302
Toulson, C. E., 233, 301
Townes, T. M., 111, 149
Transfiguracion, J., 96–97
Traversari, C., 181
Travi, M., 109–110
Trifari, S., 172–173
Trifillis, P., 109–110
Tripathi, R., 21
Trono, D., 32–34, 38, 49, 90, 93, 94, 103, 111, 131
Trubetskoy, A., 173–174
Tsai, J. Y., 242
Tschernutter, M., 256
Tseng, L., 1–4
Tsou, W. L., 285
Tuan, D. Y., 111
Tubb, J., 243–244
Tu, D. C., 256

Tuddenham, E. G., 244–245
Turchetto, L., 181
Turner, G., 173–174

U

Uchida, E., 4–6, 36–37
Uchida, N., 32–33
Uckert, W., 90
Ulaganathan, M., 21–22, 39–40
Ulrick, J., 128
Umino, Y., 256
Urabe, M., 238–239
Urbinati, F., 21, 39–40, 60–61, 172–173, 179–180, 181
Urnov, F. D., 21–22
Uzel, G., 147

V

Vacca, A., 110
Vai, S., 190
Valacca, C., 21, 60–61, 172–173, 179–180
Valente, E. M., 242
Vallanti, G., 40–41, 97, 98
van Assendelft, G. B., 111
van Baal, S., 109–110
van Beusechem, V. W., 18
Vandenberghe, L. H., 235, 245, 265
Vandergriff, J., 90
Vandesompele, J., 181–182
Van de Ven, W. J., 121
Van Dyke, T., 233–234, 237, 244–245
Vanin, E. F., 1–4, 38, 90, 135–136, 243–244
van Lohuizen, M., 173–174
van Praag, H., 90
van Til, N. P., 21–22
Vargas, J. A., 243–244
Varmus, H. E., 90
Vazquez, L., 45–46
Vázquez, N., 126–127, 148
Veelken, H., 60
Vega, M., 304
Veldwijk, M. R., 304
Venneri, M. A., 60
Veres, G., 23, 60–61, 102–103, 111–112, 172–173, 191, 239–240
Verhoeven, E., 173–174
Verma, I. M., 4–6, 17–18, 39, 90, 111, 190, 243–244, 256
Vermeulen, A., 280–281
Veys, P., 17, 126
Viale, A., 21–22
Vidaud, M., 23, 60–61, 111–112, 172–173
Vigi, V., 109–110
Vigna, E., 172–173
Vile, R. G., 103–104
Villarreal, L., 232
Villeval, J. L., 128–129, 172–173

Vink, E., 174
Visigalli, I., 172–173
Visser, T. P., 21–22, 37–38
Vodicka, M. A., 90
Vodyanik, M. A., 149
Vogt, V. M., 30–32
Volpato, S., 109–110
Voltz-Girolt, C., 338
von Kalle, C., 20–21, 59–87, 128–129, 172–175, 179–180, 181, 182–183, 187–198
von Laer, D., 90
von Neuhoff, N., 173
von Ruden, T., 4–6
Vossen, J., 17
Voulgaropoulou, F., 305
Vowells, S. J., 128, 131–132, 134–135, 139–140

W

Waddington, S. N., 244–245
Wagemaker, G., 37–38, 172–173
Wagner, E. F., 4–6
Wagner, E. J., 276–277, 285
Wagner, J. A., 243–244, 298
Wahlers, A., 173
Wahn, V., 16–17, 18–19
Wainer, S., 235
Wajcman, H., 109–110
Walden, S., 243–244
Walker, J. R., 23–24
Wall, E. A., 276–277
Walrafen, P., 119–120
Walsh, C. E., 90, 235
Walters, R. W., 230–231
Walz, C., 230–231
Wang, C. H., 111
Wang, D. J., 243–244
Wang, D. Z., 237–238
Wang, G. P., 15–16, 19–20, 21, 39, 60–61, 78–79, 110–111, 120, 121, 156, 172–173, 180
Wang, H., 145–146, 149
Wang, J., 23–24, 233
Wang, L., 32–33, 190–191, 235, 239–240, 242–244, 245, 302
Wang, Q., 96
Wang, W. J., 61, 201–202, 219, 220, 225
Wang, X., 233
Wang, Z., 302
Wardell, T., 92, 103
Ward, F. E., 17
Warischalk, J. K., 237–238, 240–242
Warrington, K. H. Jr., 235–237, 238–239, 240
Wasi, P., 109–110
Waterhouse, C. C., 126–127
Watson, D., 19–20
Weatherall, D. J., 109–110
Weber, E. L., 36

Weerkamp, F., 172–173
Weger, S., 233, 305–306
Wehner, T., 189
Weigel-Van Aken, K. A., 235–237, 240
Weinberg, K. I., 109–110
Weissman, I. L., 39–40
Weiss, R. A., 40, 332–333
Weitzman, M. D., 17–18, 233
Wei, Z., 242–243
Wergedal, J. E., 6–9
Wernig, M., 149
Westerman, K. A., 111, 112–113, 121
Westerman, Y., 37–38
Westphal, M., 90
Wetzel, R., 243–244
Whiting-Theobald, N. L., 38, 128, 131–132, 135–137
Whitt, M. A., 38
Wicke, D. C., 173
Wiech, E., 16–17, 18–19
Wientjens, E., 173–174
Wilairat, P., 109–110
Wilcher, R., 231
Will, E., 173
Williams, D. A., 21–22, 39, 61, 100–101, 126, 172–173, 180
Williams, L. W., 17
Williams, M. L., 256
Wilson, C., 100
Wilson, J. M., 233, 235, 243–244, 245, 298, 302
Windsor, E. A., 242–244
Wine, J. J., 243–244
Winichagoon, P., 109–110
Winkler, A., 38
Winther, B., 243–244
Wissler, M., 61, 179–180
Wistuba, A., 305–306
Witte, D., 173
Witt, R., 191
Wodrich, H., 38
Wognum, A. W., 37–38
Wold, B., 61
Wolff, J. A., 4–6, 235–237
Wolf, J., 230–231
Wolf, S., 61, 78
Wolgamot, G., 11–12, 103–104
Wolken, N., 293–294
Wong, M., 19–20
Wood, A. J., 145–146, 148
Woodworth, L. A., 245
Wright, J. F., 242–244, 256, 261, 305
Wright, N. A., 126–127
Wrobel, I., 126–127
Wu, J., 235–237
Wu, L. C., 149
Wulffraat, N., 17, 20–21, 60–61, 128–129, 172–173

Wurm, F. M., 240
Wu, X., 39, 156, 179–180
Wu, Z., 233–234, 237, 244–245

X

Xiao, W., 232, 233
Xiao, X., 231, 232, 237–238, 305
Xiao, Y., 134
Xia, P., 39–40, 111
Xie, J., 233
Xie, Q., 230–231
Xin, K. Q., 238–239
Xu, D., 32–33
Xu, H., 96
Xu, J., 102–103, 256
Xu, L., 4–6
Xu, S. Q., 276–277
Xu, X., 256
Xu, Y., 90, 98, 103, 113–114
Xu, Z., 4–6, 36–37

Y

Yadav, S., 235–237
Yamada, K., 90
Yamanaka, S., 149
Yamauchi, A., 133
Yang, D. R., 16–17
Yang, E. Y., 243–244
Yang, F., 126
Yang, H., 233–234
Yang, J. C., 96, 233
Yang, L. C., 235, 237–239
Yang, Q., 157
Yang, Y., 38
Yan, K. P., 276
Yan, Z., 233, 235
Yao, J., 102, 104
Yao, S., 39–40
Yasumura, D., 256
Yates, F., 23
Yee, J. K., 1–14, 90, 111
Ye, G. J., 239–240
Yi, M., 21
Yin, C., 233–234, 244–245
Yin, F., 233–234, 244–245
Ying, G. S., 242–244
Yi, R., 276–277
Yoo, E., 103
Yow, M. D., 245
Yue, Y., 233–234
Yu, G., 32–33
Yu, J. H., 96–97, 149
Yu, L., 133
Yu, Q., 96–97
Yu, S. F., 4–6, 242
Yu, W., 126–127

Yu, X. J., 4–6, 32–33, 130–131
Yu, Y., 111
Yvon, E., 15–27, 60–61, 172–173, 190

Z

Zabner, J., 230–231
Zanta-Boussif, M. A., 40–41, 49, 97, 98
Zaoui, K., 61, 180, 194
Zarember, K. A., 126, 127, 147
Zayek, N., 102–103, 191
Zeitler, B., 145–146, 148
Zelenaia, O., 242–243
Zeller, W. J., 78
Zeng, L., 109–110, 111
Zeng, Y., 242–243, 256, 276–277, 280–281, 285
Zennou, V., 33–34
Zepp, F., 16–17, 18–19
Zhang, F., 15–16, 19–20, 21, 22–23, 39–40, 60–61, 179–180
Zhang, L., 145–146, 148
Zhang, Q., 242–243
Zhang, T., 233–234, 244–245
Zhang, W., 1–4
Zhang, Y., 235, 245
Zhao-Emonet, J. C., 190
Zhao, S., 181–182
Zhao, W., 235–237
Zhao, Y., 100
Zheng, Y., 126
Zhen, L., 134
Zhen, Z., 305
Zhi, Y., 245
Zhong, L., 235–237, 240
Zhou, J. F., 243–245
Zhou, S., 21–22, 145–146, 147
Zhou, X., 235
Zhu, S., 126–127, 148
Zhu, X. D., 231
Ziegler, R. J., 245
Ziegler, R. S., 189
Zielske, S. P., 45
Zincarelli, C., 232, 235–237
Zingale, A., 60
Zingsem, J., 45–46
Zitvogel, L., 201
Ziviello, C., 242
Zolotukhin, I., 235–237, 240, 305
Zolotukhin, S., 230–231, 235–237, 238–240
Zou, J., 145–146, 149, 256
Zufferey, R., 38, 49, 90, 93, 94, 102–103, 131, 191
Zujewski, J. A., 219
zur Hausen, H., 230–231, 235
Zwart, R., 38
Zwiebel, J. A., 1–4
Zychlinski, D., 181

Subject Index

Note: Page numbers followed by "*f*" indicate figures, and "*t*" indicate tables.

A

AAV. *See* Adeno-associated virus (AAV)
ADA. *See* Adenosine deaminase deficiency (ADA)
ADA-SCID. *See* Adenosine deaminase-deficient severe combined immune deficiency (ADA-SCID)
Adeno-associated virus (AAV)
  biology
    AAV2 genome organization, 230–231, 230*f*
    capsid proteins, 230–231
    latent infection, 231
    parvovirus family, 230–231
  clinical trials, rAAV
    hemophilia B, 243–245
    Leber's congenital amaurosis, 242–243
    rAAV utilization, 240–242, 241*f*
  and FIV, 293–294
  hemophilia B trial, 332
  NIH, 330
  pediatric diseases, 333–334
  protocols, 330, 331*f*
  rAAV manufacturing methods
    advantages, BEVS, 238–239
    BEVS and rHSV vectors, 238–239
    biology/vectorology, rAAV, 240
    gene therapy vector, multiple aspects, 238
    HEK293 cell line, 240
    novel approaches, 238
    production, 240
    rHSV infection systems use, 239–240
  RAC members, 330–332
  sample collection, 333
  TNF receptor, 332
  vectorology
    Ad helper plasmid, 232
    bottlenecks, 233
    capsids *in vivo*, 237–238
    disadvantages, 233–234
    HSV, 232
    important, 231–232
    kinetics, 233
    rAAV production, 232
    scAAV vectors, 233
    serotype *rep* sequence, 232
    strategy and serotypes, 235–237
    transduction efficiency, rAAV vectors, 233

Adenosine deaminase deficiency (ADA), 172–173
Adenosine deaminase-deficient severe combined immune deficiency (ADA-SCID), 129
Advanced therapy medicinal products (ATMPs)
  CAT, 344
  certification procedure, 345
  classification, 346
  development, 340–341
  and GCP, 339–340
  MAA, 344
  manufacture, 339–340
  post-authorization requirements, 340
  regulation, 339
  SMEs, 348
Artificial miRNA expression vectors
  considerations, 285–286
  materials, 291
  oligo design and cloning protocol
    anneal and polymers, 288
    backbone sequence, 286
    bacterial transformation, 290
    design and order, 288
    duplex instability, 288
    nucleotide manipulation, 286
    PCR purification, 288
    sense and antisense, 286
    Tb:mU6 plasmid, 289, 289*f*
    UNAfold, 288
ATMPs. *See* Advanced therapy medicinal products (ATMPs)

B

Baby hamster kidney (BHK) cells, 238, 239–240
Baculovirus expression vector system (BEVS)
  advantages, 238–239
  utilization, SF9, 238–239
BEVS. *See* Baculovirus expression vector system (BEVS)
BHK cells. *See* Baby hamster kidney (BHK) cells
Biotechnology Activities (OBA)
  GTSAB, 317–319
  NIH Guidelines, 315–316
  RAC, 317–319
Busulfan conditioning
  patient, 130
  single agent, 142–143

C

CAT. *See* Committee for advanced therapies (CAT)
Cell conditioning methods
  genomic structure, lentiviruses, 33, 34f
  lentiviral vectors, 33–34
  polypurine tract (PPT), 32
  "replication defective", 32
  retroviral and lentiviral vector production
    determination, infectious titer, 44–45
    harvesting virus, 43
    Phoenix producer cell lines, 37
    transient transfection, 42–43
    ultracentrifugation, 43–44
  retrovirus and lentivirus vector design
    gene silencing, 39–40
    marker genes, 37–38
    phenotoxicity, 34–36
    promoter, 36
    regulatory sequences, 38
    safety considerations, 39
    selectable markers, 37
    variations, 36–37
    virus envelope protein, 38
  RSV, 32
  transduction
    human HSCs, 46–48
    protein expression, 51–53
    real-time qPCR analysis, 48–51
  viral gene transcription, 30–32
  viral preintegration complex (PIC), 32–33
Central nervous system (CNS), 126
Certification testing
  amplification phase cell line
    highly infectable cell line, 101–102
    RCL amplification, 102
  indicator phase assays
    molecular assays, 103
    p24 capsid antigen, 102–103
    PERT, 103
    replication competent retroviruses, 102
  maximizing assay sensitivity
    cell to vector ratio, 104
    RCL testing, 104
  replication competent lentivirus detection assay, 100–101, 101f
  selection, positive control
    growth kinetics, 103–104
    RCL assay, 103
CGD. *See* Chronic granulomatous disease (CGD)
CG1711 hALD (MND-ALD) vector, 132f
Chronic granulomatous disease (CGD)
  gene therapy
    ADA-SCID, 129
    adults and older teenage patients, treatment, 128
    AR-CGD, 126–127
    autologous HSC transduced, 130
    chemotherapy, 128
    circulating neutrophils, 129–130
    clinical scale production, 137–139
    clinical trials, 181
    endeavors
      iPSC, 149
      lentivectors, 147–148
      X-CGD patients, 145–146
      zinc finger/TALENs, 148
    female carriers, X-CGD, 127
    gp91phox vector, 130
    GT deletion, 126–127
    HSC, 127
    myelodysplastic syndrome, 172–173
    myeloid marking, 129–130
    NADPH oxidase, 126
    patient care and monitoring, 126, 142–145
    p22 mutations, 126–127
    p40phox deficiency, 126–127
    producer cell line, vector production, 132–133
    target cell, 139–140
    transduction, clinical, 140–142
    transgene function and vector titer, preclinical test, 134–137
    vector design, 130–132
    vector insertional genotoxicity, 128–129
    X-SCID trials, 128
CliniGene, 352
Clonal repertoire, gene-corrected cells
  estimation, RT-PCR
    absolute quantification, 83–85
    agarose gel electrophoresis, 80–81
    dilution, standard and samples, 82–83
    purification, PCR product, 81–82
    standard preparation, 79–80
  gene therapy, 60
  high-throughput sequencing
    agarose gel electrophoresis, 76–77
    purification, PCR products, 73–74
    purification, third exponential PCR, 77–78
    Roche 454 platform, 78
    third exponential PCR, 75–76
  ligation-mediated PCR (LM-PCR), 61
  nrLAM-PCR
    agarose gel electrophoresis, 72–73
    amplification efficiency, 61
    concentration, microcon-50, 64
    exponential PCR, magnetic capture, 69–71
    first exponential PCR, 68–69
    ligation, single-stranded oligonucleotide, 66–68
    linear PCR, 62–64
    magnetic capture, 64–66
    second exponential PCR, 71–72
  retroviral integration sites, 60–61

Subject Index

CNS. *See* Central nervous system (CNS)
Committee for advanced therapies (CAT)
  ATMPs, 342
  certification procedure, 345
  CHMP and ATMPs, 344
  composition, 343, 343*f*
  EMA Web site, 343
  members percentage, 342, 342*f*
  procedure, 344–345
  stakeholder interaction
    COMP, 346
    EMA and CAT, 345–346
  task, 344

D

DMEM. *See* Dulbecco's modified Eagle's medium (DMEM)
Dulbecco's modified Eagle's medium (DMEM), 203, 217, 221

E

EMA. *See* European Medicines Agency (EMA)
Enzyme replacement therapy (ERT), 17
ERT. *See* Enzyme replacement therapy (ERT)
ESGCT. *See* European Society for Gene and Cell Therapy (ESGCT)
European Medicines Agency (EMA), 338–339
European Society for Gene and Cell Therapy (ESGCT), 352

F

FACS. *See* Fluorescence-activated cell sorting (FACS)
Fluorescence-activated cell sorting (FACS), 9–11, 160, 176, 178

G

GBM. *See* Glioblastoma multiforme (GBM)
GCP. *See* Good clinical practice (GCP)
G-CSF. *See* Granulocyte colony stimulating factor (G-CSF)
Gene therapy
  ADA deficiency
    antigen stimulation, 22–23
    transduced cells, 22
    transduction, HSC, 23
  CGD. *See* (Chronic granulomatous disease (CGD) gene therapy)
  SCID-X1
    ADA deficiency, 18–19
    DNA sequencing, 19–20
    gene mutation replacement, 21–22
    hematopoietic stem cells (HSCs), 20
    identification, retroviral integration site, 20–21
    PCR technology, 19–20
    proto-oncogenes, 21
    self-inactivating (SIN) vectors, 21–22
    thymopoiesis, 20
Gene therapy medicinal products (GTMP)
  CAT, 342–345
  definition, 341
Gene Transfer Safety Assessment Board (GTSAB)
  OBA, 321
  RAC members, 317–319
Genotoxicity assays
  adverse events, 172–173
  animal models, γRVs, 172–173
  gene expression and nearby vector integration sites, 180–183
  gene therapy vectors, 172–173
  transduction and transplantation, tumor-prone HSPCs
    advantage, 173–174
    bone marrow-derived lineage depleted cells, $Cdkn2a^{-/-}$ mice, 175–176
    Cdkn2a locus encodes, 174
    engraftment, 178–179
    ex vivo procedures, 176
    FACS analysis, 178
    in vitro cultures and *in vivo* marking, 176–178
    in vivo methods, 175
    LTR, 174–175
    oncogenic hit, 173
    PCR and genomic integration site analysis, 179–180
    tumor phenotype analysis, 178–179
    VCN measurement, 178–179
    vector integrations, 174–175
GFP. *See* Green fluorescence protein (GFP)
Glioblastoma multiforme (GBM)
  intracranial xenografts, 220
  primary brain tumors, treatment, 219
GLP. *See* Good Laboratory Practice (GLP)
GMP. *See* Good manufacturing process (GMP)
GMP production method
  evaluation, envelope pseudotypes, 100
  large-scale transfection procedures
    multitier cell factories (CF), 95
    preparation, 96
    required materials, 95
    viral supernatant, 96, 97*f*
  optimizing plasmid concentrations
    analysis, 95
    required materials, 93–94
    transfection reaction, 93
    transfections, 94
  packaging system, 93
  purification and concentration
    closed system, 97
    ion exchange, 98
    required materials, 97–98
    vector concentration and diafiltration, 99–100, 99*f*

Good clinical practice (GCP), 339–340
Good Laboratory Practice (GLP), 140
Good manufacturing process (GMP), 140
Graft-*versus*-host disease (GVHD)
   acute/chronic, 110
   allogeneic transplantation, 110–111
Granulocyte colony stimulating factor (G-CSF), 130–131
Green fluorescence protein (GFP)
   FACS analysis, 11–12
   gene encoding proteins, 11–12
   titer determination, 4–6
GTMP. *See* Gene therapy medicinal products (GTMP)
GTSAB. *See* Gene Transfer Safety Assessment Board (GTSAB)
GVHD. *see*Graft-*versus*-host disease (GVHD)

H

Hematopoietic stem and progenitor cells (HSPCs)
   leukemia, 172–173
   transduce, 172–173
   tumor-prone. *See* (Genotoxicity assays)
Hematopoietic stem cells (HSCs)
   allogeneic donor, 127
   autologous gene-corrected, 128, 129, 130
   $CD34^+$
      bulk culture, 139–140
      MFGS-gp91phox transduced human, 130–131
      patients' autologous, 128
      selection, 139
      transductions of human X-CGD, 133
      xenograft, 134–137
   gene transfer treatment, 129
Hematopoietic stem cell transplantation (HSCT)
   allogeneic, 132–133
   X-ALD, 130–132
Hemophilia B, 243–245
HSCs. *See* Hematopoietic stem cells (HSCs)
HSCT. *See* Hematopoietic stem cell transplantation (HSCT)
HSPCs. *See* Hematopoietic stem and progenitor cells (HSPCs)

I

Induced pluripotent stem cells (iPSC)
   CGD gene therapy, 149
   X-CGD patients, 145–146
Insertional mutagenesis
   cancer genes, identification, 174
   replication-competent γ-retroviruses, 181
   risks, 172–173
   severe adverse events, 172–173
Internal ribosome entry site (IRES)
   bicistronic vector, 4–6
   translation, downstream marker, 4–6

Ion exchange
   capsule equilibration, 98
   elute vector, 98
   load vector onto capsule, 98
   preparation, 98
iPSC. *See* Induced pluripotent stem cells (iPSC)
IRES. *See* Internal ribosome entry site (IRES)

L

LAM. *See* Linear amplification mediated (LAM)
Leber's congenital amaurosis (LCA) clinical trials, 242–243
Lentiviral gene therapy
   PEV, 111
   risk/benefit ratio, 110–111
   transduced autologous cells, 111–112
   vector-based transfer, 111
Lentiviral integration, human genome
   HIV PICs, 156
   MLV, 157
   preclinical study, 156–157
   retroviral integration, 156
   RNA sequence, 157
   seminal clinical study, 156–157
   SIN lentiviral vectors
      aberrant splicing events, 165–168
      cell transduction, cloning and mapping, 158–161
      chimeric transcripts, proviral and cellular gene sequence, 161–165
      description, 158
   transcriptional gene activation, 157
Lentivirus vectors, β-thalassemia
   cGMP production, 112–113
   cGMP transduction, $CD34^+$ Cells, 113
   clinical protocol
      bone marrow karyotype, 119–120
      exclusion criteria, 115–117
      globin chain production, 119
      inclusion criteria, 115
      integration site analysis, 120
      LTC-IC assays, 119
      patient follow-up, 117–120
      peripheral blood cells, 117
      primers, PCR, 117, 119*t*
      synopsis, 114
      transduced hematopoietic progenitors, 117–118
   clinical vector design (LentiGlobin^{TM}), 112
   gene therapy, 110–112
   GTP and GMO release testing, 113–114
   GVHD, 110
   monogenic disease, 109
   post-transplantation
      chromosomal IS analysis, 121
      erythroid lineage, 120–121
      lentiviral vector integration, 121
   RBC, 109–110

Subject Index

Ligation-mediated PCR (LM-PCR), 61
Linear amplification mediated (LAM)
  dangerous gene classes, 179–180
  PCR, 180
  vector integration sites, 179–180
Long terminal repeat (LTR) activity, 174–175

M

MAA. *See* Marketing authorization applications (MAA)
Marketing authorization applications (MAA), 339, 340
MLV. *See* Murine leukemia virus (MLV)
MMLV. *See* Murine Moloney leukemia virus (MMLV)
MTS assay
  procedures, 218–219
  required materials, 217–218
Multiplicity of infection (MOI), 90–91
Murine leukemia virus (MLV)
  based RRV, 215
  derived vectors, 156–157
  human gene therapy, 1–4
  γ-retroviruses, 1–4
  transactivation function, 4–6
  U3 sequence, 202, 206
  vector production, 1–4
Murine Moloney leukemia virus (MMLV), 130–131
Murine retrovirus vector
  gene therapy, 147
  MFGS
    gp91phox vector, 132*f*
    vector backbone, 130–131

N

National Institutes of Health (NIH) guidelines
  development, 315–316
  RAC, 316–317
Neutrophils
  circulating, 129–130, 144–145
  DHR
    assay, 135–136, 146*f*
    fluorescence, 134–135
  gene
    corrected iPSC, 149
    marked, 143
  gp91phox, 133
  human antigen CD13 positive, 136–137
  intracellular vesicles, 133
  oxidase-positive, 144–145, 145*f*
  X-CGD iPSC, 149
NHP. *See* Non-human primate (NHP)
Non-human primate (NHP)
  surgical procedures
    injection, 266–267
    required materials, 265–266
    transduction outcome determination, 267

Non-steroidal anti-inflammatory drug (NSAID), 221
Novel lentiviral vectors, clinical products
  certification testing
    amplification phase cell line, 101–102
    indicator phase assays, 102–103
    maximizing assay sensitivity, 104
    selection, positive control, 103–104
  development plan
    start, finish line, 90–92
    titer assessment, 92
  GMP production method
    evaluating envelope pseudotypes, 100
    large-scale transfection procedures, 95–96
    optimizing plasmid concentrations, 93–95
    purification and concentration, 96–100
  human cell types, 90
  molecular/immunologic detection assays, 100–101, 101*f*
  RCL, 100
  scale up vector production, 90
nrLAM-PCR
  agarose gel electrophoresis, 72–73
  amplification efficiency, 61
  concentration with Microcon-50
    procedure, 64
    required materials, 64
  first exponential PCR
    magnetic capture, 69–71
    procedure, 68–69
    required materials, 68
  ligation, single-stranded oligonucleotide
    procedure, 67–68
    required materials, 66
  linear amplification, 61
  linear PCR
    primers, 62–64, 63*t*
    procedure, 62–64
    required materials, 62
  magnetic capture
    procedure, 65–66
    required materials, 64–65
  MLV-based vector, 61
  second exponential PCR
    procedure, 71–72
    required materials, 71
NSAID. *See* Non-steroidal anti-inflammatory drug (NSAID)
Nucleic acid delivery, outer retina
  dog
    surgical procedures, 261–264
    transduction outcome determination, 264–265
  gene transfer, 256
  human
    injection, 269–270
    required materials, 268
    surgical procedure, 267–270

Nucleic acid delivery, outer retina (cont.)
　　transduction outcome determination, 270
　　mouse
　　　crystalline lens, 257–258, 257f
　　　retinal health/transduction outcome, 260–261
　　　surgical procedures, 258–260
　　NHP, 265–267
　　RPE cells, 256–257
　　toolkit, 256

O

OBA. See Office of biotechnology activities (OBA)
Office of biotechnology activities (OBA)
　　and FDA, 321
　　GTSAB, 317–319
　　RAC meeting, 317–319
　　X-SCID, 326
Oncolytic virotherapy, 200

P

PBMCs. See Peripheral blood mononuclear cells (PBMCs)
Peripheral blood mononuclear cells (PBMCs)
　　ALD protein expression, 145–149
　　patients, 130
PEV. See Position effect variegation (PEV)
Pharmaceutical Research and Manufacturers of America (PhRMA), 299
Phorbol myristate acetate (PMA) activation, 134
PhRMA. See Pharmaceutical Research and Manufacturers of America (PhRMA)
PICs. See Preintegration complexes (PICs)
Position effect variegation (PEV), 111
Preintegration complexes (PICs), 156
Prodrug activator gene therapy
　　RRVs encoding, 215–216
　　treatment, brain cancer, 201–202
Pseudotyped HIV-1 vectors development plan
　　start, finish line
　　　aliquoting strategy, 91–92
　　　clinical trial, 90
　　　in vitro virus assay, 91
　　　MOI, 90–91
　　titer assessment
　　　gene transfer, 92
　　　gold standard, 92
　　　vector potency (titer), 92

Q

q-PCR. See Quantitative PCR (q-PCR)
Quantitative PCR (q-PCR)
　　amplification, 209
　　analysis, 205, 209
　　assay, 207f
　　calculation, 176–177
　　gene expression assays, 181–182
　　genomic DNA, 177, 178–179
　　materials required, 208
　　perform reactions, 209–210
　　prepare standard curve, 208–209
　　reactions, 208
　　results, 210

R

rAAV vector. See Recombinant adeno-associated viral (rAAV) vector
RAC. See Recombinant DNA Advisory Committee (RAC)
RCL. See Replication competent lentivirus (RCL)
RCR. See Replication competent retrovirus (RCR)
Recombinant adeno-associated viral (rAAV) vector
　　assays
　　　infectious rAAV titer and infectivity ratio, 305
　　　protein purity and identity, 304
　　　rAAV vector genome titer, 304
　　　transgene expression, 306–307
　　clinical trials
　　　hemophilia B, 243–245
　　　Leber's congenital amaurosis, 242–243
　　　rAAV utilization, 240–242, 241f
　　description, 298
　　inject, 235
　　ITR, 233
　　manufacturing methods
　　　advantages, BEVS, 238–239
　　　BEVS and rHSV vectors, 238–239
　　　biology/vectorology, rAAV, 240
　　　gene therapy vector, multiple aspects, 238
　　　HEK293 cell line, 240
　　　novel approaches, 238
　　　production, 240
　　　rHSV infection systems use, 239–240
　　production, 232
　　reference standard material
　　　AAV2 RSM, 301–302
　　　AAV8 RSM, 302–303
　　reference standards
　　　AVV, 299–300
　　　pharmacopeia, 298–299
　　　PhRMA, 299
　　　RSMs, 299–300
　　transduction efficiency, 233
　　volunteer working groups
　　　description, 300
　　　effort, 300–301
　　　NIH, 300
Recombinant DNA Advisory Committee (RAC)
　　AAV vector, 330–334

Subject Index

biosafety community, 320–321
Director, 316–317
FDA, 321
GTSAB, 317–319
lentiviral vector
  HIV-1, 327–328
  retroviral, 327
  thalassemia, 328–329
NIH Office, 317, 318f
OBA, 317, 319f
Phase I trials, 317, 319f
protocol, 316
role, 320
safety data, 321
symposia, 320
tetroviral vector
  administration, 322, 324f
  GeMCRIS, 324–325
  leukemia, 325
  OBA, 326
  protocols, 322, 324f
  trends, 322, 323f
  X-SCID, 322–324
Reference standards materials (RSMs), 299
Regulatory structures, GTMP
  ATMPs, 339–340
  CAT, 342–345
  definition, 341
  Directive 2001/83/EC, 340–341
  diseases and orphan medicinal products
    COMP, 351
    gene therapy, 350
    sponsors, 351
  European research networks
    CliniGene, 352
    description, 351–352
    ESGCT, 352
  national support structures, 348–349
  patients' organization
    COMP, 349
    EGAN, 350
    EURORDIS, 350
    members, 350
    role, 349
  SME Office, 346–348
Replication competent lentivirus (RCL)
  assays, 130
  vector stock, 129–130
Replication competent retrovirus (RCR)
  detection
    amplication, 11–12
    hygromycin resistance, 11–12
  use, 200
  vector production, 4–6
Respiratory syncytial virus (RSV), 32
Retinal pigment epithelium (RPE) cells, 256–257
Retroviral and lentiviral vector production
  determination, infectious titer

flow cytometry, 44
qPCR, 45
harvesting virus, 43
packaging constructs, 40–41
phoenix producer cell lines
  gammaretroviral production, 41–42
  required materials, 41
transient transfection
  procedure, 43
  required materials, 42
ultracentrifugation, 43–44
Retroviral replicating vectors (RRV) cancer
  cell-killing function, 201–202
  CMV, 202
  development
    advantage, 212
    design and construction, prototype, 212
    genetic diversity, 211–212
    identification and cloning, mutations, 214
    molecular evolution and natural selection, 213
    prototype RRV, adoption, 215f
    virus production, 213
  gamma retroviruses, 200–201
  in vivo glioma model
    optimal therapeutic effect, 225
    procedures, 222–224
    required materials, 220–222
    time course, 224
  innate immunity, activation, 201
  MTS assay
    procedures, 218–219
    required materials, 217–218
    suicide gene, 215–216
  oncolytic virotherapy, 200
  oncolytic viruses, 201–202
  replicating viruses, 200–201
  titer and biodistribution study
    advantages, 205–206
    CMV promoter, 206
    illustrative results, 211
    preparation, template DNA, 206–208
    qPCR assay, 208–210
    "TaqMan", 205
  tumor microenvironment, 201
  vaccinia virus, 200
  virus production
    encodes, 205
    required materials, 203–204
    transfection procedure, 204–205
Retrovirus-based correction, SCID
  ADA deficiency patients, 17–18
  adaptive immunity, 16–17
  adenosine deaminase (ADA), 15–16
  Artemis deficiency, 16, 23
  cell types, 23–24
  ERT, 17
  ex vivo cell infection, 18

Retrovirus-based correction, SCID (cont.)
  gene therapy, ADA deficiency, 22–23
  gene therapy, SCID-X1, 18–22
  HSCT, 17
  immunodeficiency, 18
  Mendelian disorders, 16–17
  technological progress, 18
  vector design, 23
Retrovirus vector design
  construction, GOI
    CMV-LTR fusion, 4–6, 6f
    GFP, 4–6
    IRES, 4–6
    structure and types, 4–6, 5f
  MLV, 1–4
  RCR detection, 11–12
  structures, MLV genome and transcripts, 1–4, 3f
  titering, retroviral vectors
    antibiotics selection, 9–11
    FACS, 9–11
  transient production
    phoenix system, 8
    titer determination, 6–9
RNA interference (RNAi) vectors
  artificial miRNA expression vectors, 285–291
  description, 276
  gene silencing mechanisms, 276–277
  mRNAs and miRNA, 277–278
  screening
    eGFP reporter, 292, 292f
    endogenously expressed targets, 293
    gene silencing efficacy, 291
    information and recommendations, 293
  shRNAs expression cassettes, 280–285
  siRNA sequences
    22-nt target sites, 278–279
    target gene, 278
  viral
    AAV and FIV, 293–294
    RNAi expression cassettes, 293–294
RPE. See Retinal pigment epithelium (RPE)
RSMs. See Reference standards materials (RSMs)
RSV. See Respiratory syncytial virus (RSV)

S

scAAV vectors. See Self-complementary AAV (scAAV) vectors
Self-complementary AAV (scAAV) vectors
  FIX gene, 244–245
  produce, 233
  scAAV8 vector, 245
Severe, combined immunodeficiency (SCID)
  ADA deficiency, 18–19
  DNA sequencing, 19–20
  gene mutation replacement, 21–22
  hematopoietic stem cells (HSCs), 20

identification, retroviral integration site, 20–21
PCR technology, 19–20
proto-oncogenes, 21
self-inactivating (SIN) vectors, 21–22
thymopoiesis, 20
shRNA expression cassettes
  oligo design and cloning protocol
    DNA template, 281
    duplex instability, 281
    materials, 285
    PCR, 283
    sequence and order, 283
    UNAfold, 281
    U6 promoter, 283
  siRNA selection, 280–281
SIN lentiviral vectors
  aberrant splicing events
    materials, 166t
    protocol, 166t
  cell transduction, cloning and mapping
    disposables, 160t
    materials, 158t, 160t
    protocol, 160t
    structure, 159f
  chimeric transcripts, proviral and cellular gene sequence
    extraction and purification, poly(A)$^+$ RNA, 162–163
    5'RACE and RT-PCR, 163–165
  description, 158
Small-and medium-sized enterprises (SMEs)
  aim, 346
  development proceeds, 348
  EU marketing authorization, 348
  financial constraints, 348
  incentives, 347
  status, 346–347
Small inhibitory RNAs (siRNA), 275–296
SMEs. See Small-and medium-sized enterprises (SMEs)
SOP. See Standard operating procedure (SOP)
Spodoptera frugiperda (Sf9) cells, 238–239
Standard operating procedure (SOP), 140

T

TNF. See Tumor necrosis factor (TNF) receptor
Transduction
  human HSCs
    first round, 47–48
    harvesting, 48–51
    mitosis, 47
    prestimulation of cells, 47
    procedure, 47
    required materials, 46–47
    retronectin, 47–48
    second round, 48
    third round, 48

protein expression
   cell surface staining and preparation, 52–53
   flow cytometry analysis, 53
   fluorochrome-conjugated primary antibodies, 51
   real-time qPCR analysis
      packaging signal (ψ), 48–49
      required materials, 49–51
      standard curves, 49
Transgene expression assays
   safety test
      adventitious agents, 306
      endotoxin, 307
      mycoplasma, 306
      sterility, 307
   stability studies, 307
Tumor necrosis factor (TNF) receptor
   role, 332–333
   serum levels, 333

V

VCN. See Vector copy number (VCN)
Vector copy number (VCN)
   determination, 176–177
   per cell, calculation, 176–177
   tissues measurement, 178–179
   tumor, 178–179
Vector integration
   gene expression, 180–183
   genotoxic, 174–175
   retrieve, 180
   and virus, 179–180
Very-long-chain fatty acids (VLCFA), 126
Virus production
   encodes, 205
   required materials, 203–204
   RRV preparation, 213

transfection procedure, 204–205
VLCFA. See Very-long-chain fatty acids (VLCFA)

X

X-ALD. See X-linked adrenoleukodystrophy (X-ALD)
X-linked adrenoleukodystrophy (X-ALD)
   allogeneic HCT efficacy, mechanism, 132–133
   clinical protocol, 140–142
   definition, 126
   design and production, lentiviral vector, 134–137
   hematopoietic stem cell transplantation, 130–132
   lentiviral HSC gene therapy, 132–133
   neurological outcome, 149–150
   patient biological follow-up
      ALD protein expression, 145–149
      integration site (IS) characterization use, 149
      lentiviral vector copy number, 148
   release test, 139–140
   transduction, CD34$^+$ cells, 137–139
X-linked severe combined immunodeficiency (X-SCID)
   mutation, 322–324
   RAC's deliberations, 326
   retroviral gene transfer, 325
X-SCID. See X-linked severe combined immunodeficiency (X-SCID)

Z

Zinc finger nuclease
   application, 148
   clones, 149
   gene correction, 148

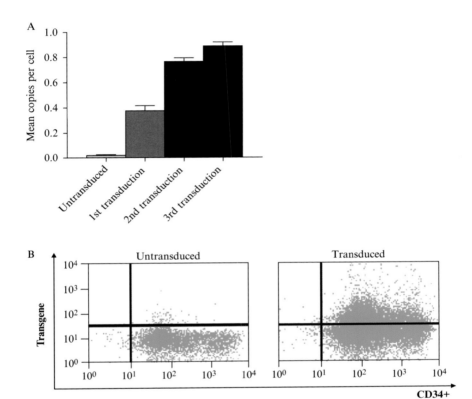

Samantha Cooray *et al.*, **Figure 3.4** Measurement of transduction efficiency. (A) Graph showing an example of the relative increases in the mean copy number per cell that can be obtained after one, two, and three successive rounds of transduction with a gammaretroviral vector. (B) Analysis of transgene expression by FACS in untransduced and transduced $CD34^+$ hematopoietic stem cells.

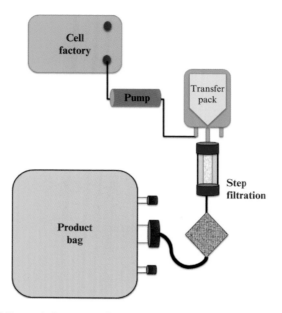

Anna Leath and Kenneth Cornetta, Figure 5.1 *Schematic of viral supernatant clarification.* Viral supernatant is pumped from the cell factory to a transfer pack. The supernatant is then filtered using multiple progressively smaller filters and collected in the product bag. This step filtration process assists in the prevention of filter clogging, which can lead to significant titer loss.

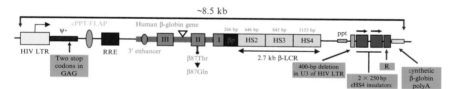

Emmanuel Payen et al., Figure 6.1 LentiGlobin™ vector and experimental design. (A) Diagram of the human β-globin ($β^{A-T87Q}$) lentiviral vector (LentiGlobin™, LG). The 3′β-globin enhancer, the 372-bp IVS2 deletion, the $β^{A-T87Q}$ mutation (ACA[Thr] to CAG[Gln]) and DNase I Hypersensitive Sites (HS) 2, HS3 and HS4 of the human β-globin locus control region (LCR) are indicated. Safety modifications including the two stop codons in the ψ+signal, the 400-bp deletion in the U3 of the right HIV LTR, the rabbit β-globin polyA signal, and the 2×250 bp cHS4 chromatin insulators are indicated. HIV LTR, human immunodeficiency type-1 virus long terminal repeat; ψ+, packaging signal; cPPT/flap, central polypurine tract; RRE, Rev-responsive element; βp, human β-globin promoter; ppt, polypurine tract.

Elizabeth M. Kang and Harry L. Malech, Figure 7.1 MFGS-gp91phox vector schematic. This vector is very similar to the large class of transfer vectors that have been designed from murine retroviruses. At the ends, the vector retains the long-terminal repeats that contain promoter activities and the sequence required for insertion into the target cell genome. The psi region has been altered by deletions and insertion of stop codons to prevent production of any gag or pol protein sequence but retains critical psi elements necessary for the vector RNA to be packaged in the amphotropic envelope producer line HEK-293-SPA into an infectious, but not replication-competent vector particle. The psi region has also retained splice donor and acceptor sequence that originally were used by the virus to enhance production of envelope protein. The envelope sequence has been replaced by cDNA encoding the gp91phox protein open reading frame. Not shown is that the same exact vector was also used to produce p47phox transgene in our 1995 clinical trial of gene therapy for p47phox-deficient AR-CGD. LTR, long-terminal repeat; ψ, packaging recognition signal; ORF, open reading frame; SD/SA, splice donor/splice acceptor sites.

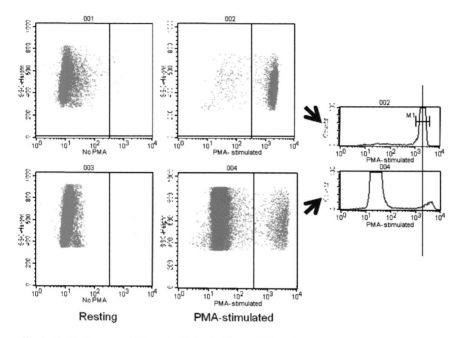

Elizabeth M. Kang and Harry L. Malech, Figure 7.3 Flow cytometry (DHR) assay of oxidase activity in peripheral blood neutrophils from Subject #1 of the 2006 trial of gene therapy for X-CGD at 38 months (more than 3 years) after gene transfer therapy. The upper panels show analysis of neutrophils from a normal control, while the lower panels show analysis of neutrophils from Subject #1. Not shown is that all of the analyses have already been gated (using forward and side scatter) to only include neutrophils from the peripheral blood. The large panels are dot plot format where the y-axis is measuring side-scatter and the x-axis is measuring DHR fluorescence which is an indicator of the intensity of oxidase activity. The right large dot plot panels show resting neutrophils that have not been activated by phorbol myristate acetate (PMA) and the neutrophils have low DHR fluorescence. Upon PMA stimulation, cells activate the oxidase and electrons flow to the DHR to increase fluorescence. Note that 99% of the activated neutrophils from the normal volunteer control are oxidase positive, and 0.7% of the activated neutrophils from X-CGD Subject #1 are oxidase positive. The dot plots of the activated neutrophil DHR analysis from control and Subject #1 are reformatted (arrows) to histogram plots at the right (y-axis is cell number; x-axis is DHR fluorescence), in order to line up the graphs to demonstrate that average oxidase activity in the 0.7% of neutrophils oxidase positive in Subject #1 are actually on average producing slightly more oxidants than the oxidase-positive neutrophils from the normal control. This indicates the high per cell efficiency of the MFGS vector in achieving functional correction of oxidase activity.

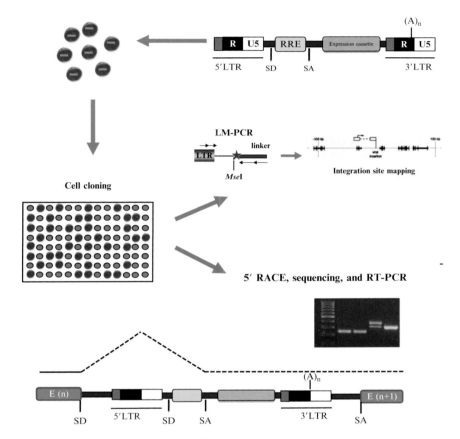

Arianna Moiani and Fulvio Mavilio, Figure 8.1 The experimental approach. A schematic structure of a SIN lentiviral vector is shown in the top right panel. RRE, rev responsive element; SD, splicing donor site; SA, splicing acceptor site; A_n, polyadenylation signal. 5′LTR and 3′LTR lack the U3 region containing the viral enhancer and promoter. The vector is used to transduce target cells, which are then cloned by limiting dilution. Vector integrations are mapped by linker-mediated (LM) PCR: genomic DNA is digested with a frequently cutting restriction enzyme (*Mse*I), ligated to a linker, amplified with vector- and linker-specific nested primers, and sequenced and aligned to the genome. Chimeric transcripts are amplified by RT-PCR or RACE and analyzed by gel electrophoresis, as showed in the middle panel. The bottom panel shows an example of an aberrant splicing event giving rise to a cellular–proviral chimeric transcript. The blue boxes represent the exons right upstream (E(n)) and downstream (E(n+1)) the provirus with their respective splice donor and acceptor sites (SA/SD). In this example, splicing occurs between the upstream exon SD site and the constitutive proviral SA site.

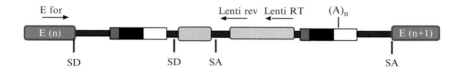

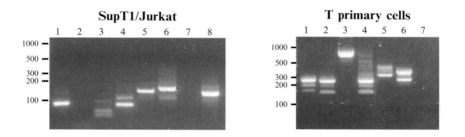

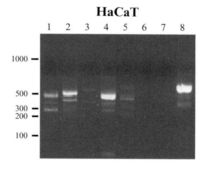

Arianna Moiani and Fulvio Mavilio, Figure 8.2 Experimental strategy used to detect aberrantly spliced transcripts by 5'RACE and RT-PCR. The upper panel shows a schematic view of a vector inserted in the intron between E(n) and E(n+1) and the position of the primers used in the 5'RACE and RT-PCR. Primers Lenti RT and Lenti rev anneal upstream the major splice acceptor site (SA) of the provirus, usually in the region preceding the expression cassette. The "E for" primer anneals in the exon upstream the integrated provirus. "Lenti RT" is used in the cDNA synthesis reaction. Aberrantly spliced products are detected by 5'RACE and RT-PCR using primer "Lenti rev" in combination with primer "E for." Example of 5'RACE and RT-PCR products detected on a 1% agarose gel electrophoresis and visualized by ethidium bromide staining in SupT1/JurkaT, HaCaT, and T primary lymphocytes is shown in the lower panels. A DNA molecular weight marker is shown at the left of each panel.

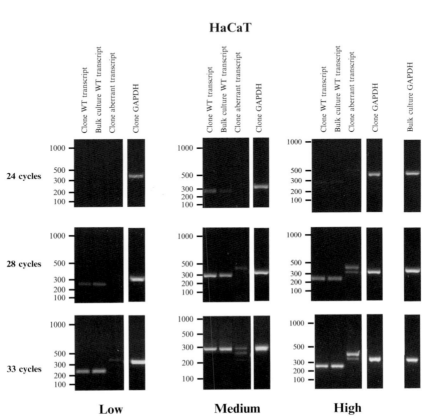

Arianna Moiani and Fulvio Mavilio, Figure 8.3 Semiquantitative analysis of aberrantly spliced transcripts. cDNA is prepared using random hexamers. Wild-type transcripts are amplified using combination of primer "E for" and "E rev" annealing in the exons immediately upstream and downstream the proviral integration as shown in the upper scheme. PCR products taken from different reactions (24, 28, and 33 cycles) are run on 1.5% agarose gel and visualized by ethidium bromide staining. Each gel panel shows, from left to right, the wild-type transcript obtained in a clone and a bulk culture, the aberrant transcript in the selected clone, and the GAPDH transcript in the selected clone and bulk culture, used for signal normalization. This analysis allows to define four arbitrary classes of relative transcript abundance, that is, High, when chimeric and wild-type transcripts are detected at the same amplification cycle (right panels); Medium, when chimeric transcripts are detected four PCR cycles later than wild-type transcripts (center panels); Low, when chimeric transcripts are detected eight PCR cycles later than wild-type transcript (left panels); and Rare, when aberrant transcripts are not detectable on RNA reverse transcribed by random hexamer primers, although they were identified by RNA reverse transcribed with a vector-specific primer (not shown).

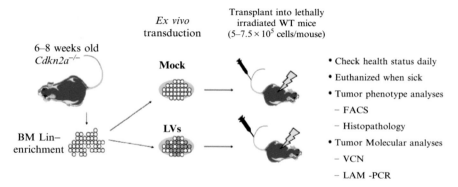

Eugenio Montini and Daniela Cesana, Figure 9.1 Experimental set up. Experimental strategy and schemes of the proviral form of the vectors used.

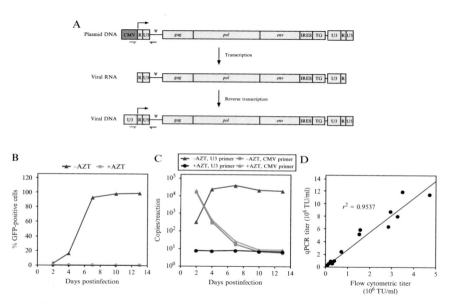

Christopher R. Logg et al., Figure 11.1 qPCR assay for determining provirus copy number and biological titer of RRV. (A) Strategy for detection of virus versus RRV plasmid contaminants. The stepwise changes in the RRV genome that occur through transcription of the plasmid and reverse transcription of the viral RNA are shown. Colored arrows indicate the binding location of primers that specifically amplify either RRV plasmid or virus. The use of a CMV-specific forward primer (red) will detect RRV plasmid in infected cells, either carried over from transfection or from another source, whereas a U3-specific primer (green) will detect only genuine reverse-transcribed RRV genomes. (B) RRV-mediated transmission of GFP expression following infection of cultured cells at low MOI with virus produced by transient transfection. In control cultures, AZT was included in the medium from the time of infection. GFP was detected by flow cytometry at 2, 4, 7, 10, and 13 days postinfection. (C) PCR quantitation of virus versus plasmid copies in genomic DNA from the same cultures as in panel B. The reactions employed a forward primer specific either for MLV U3 or for the CMV promoter. Note that the copy numbers per reaction below 50 are outside of the linear range of the assay. (D) Correlation between titer as determined by copy number determination by qPCR versus analysis of GFP expression by flow cytometry.

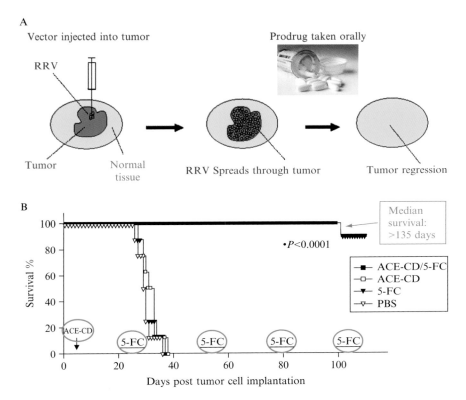

Christopher R. Logg et al., Figure 11.4 (A) The concept of Controlled Active Gene Transfer: Vector spreads through tumor but not normal tissue, hence the prodrug is activated into an anti-cancer drug only in the infected tumor cells. In the case of human clinical trials currently being conducted by Tocagen Inc., the RRV contains the cytosine deaminase (CD) gene from yeast. This enzyme converts the prodrug 5-FC, an FDA-approved anti-fungal compound that can be taken as an oral pill, into the anti-cancer drug 5-FU directly within the RRV-infected cancer cells. (B) RRV-mediated prodrug activator gene therapy can achieve long-term survival in a U-87 intracranial glioma model. PBS: control group injected with phosphate-buffered saline instead of vector and prodrug; 5-FC: control group injected with PBS instead of RRV, followed by systemic prodrug administration; ACE-CD: control group injected with RRV expressing CD gene, but without receiving prodrug afterward; ACE-CD/5-FC: treatment group receiving both RRV and prodrug. In the treatment group, only a single injection of vector was performed (ACE-CD, circled with downward arrow), followed by multiple cycles of prodrug (5-FC, circled with bars indicating duration of each cycle).

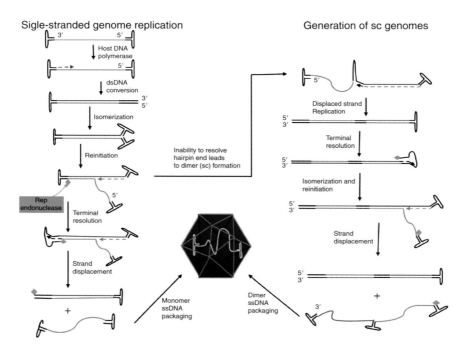

Joshua C. Grieger and R. Jude Samulski, Figure 12.2 AAV replication cycles. Replication cycles for the generation of (left) single-stranded genomes and (right) self-complementary genomes.

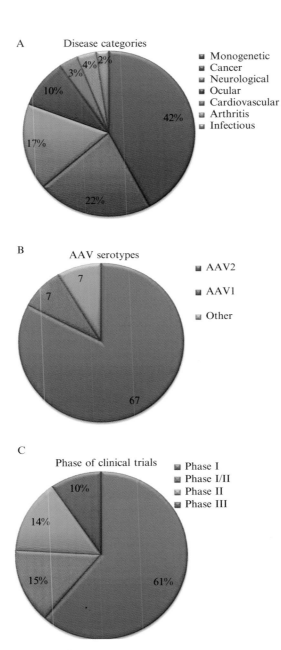

Joshua C. Grieger and R. Jude Samulski, Figure 12.3 Clinical trials utilizing rAAV. The percentage of clinical trials using rAAV (A) in different categories of disease, (B) in different phases, and (C) using capsids of different AAV serotypes. Other serotypes include AAV8, AAV2.5, AAV6, and AAVrh.10. Clinical trial data were compiled from the Gene Therapy Clinical Trials Worldwide Database (http://www.wiley.com/legacy/wileychi/genmed/clinical/).

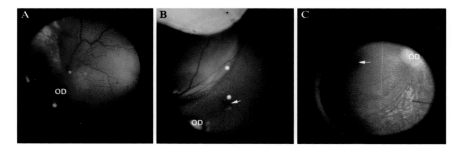

Jean Bennett et al., Figure 13.2 Appearance of the "bleb" immediately following subretinal injection in (A) dog, (B) non-human primate (NHP), and (C) human. OD, optic disc; arrow indicates the fovea in the NHP and the human. Panel (C) was taken from an intraoperative video recording.

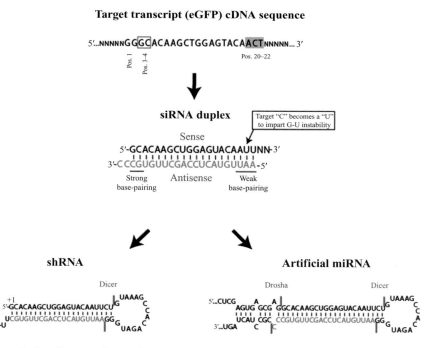

Ryan L. Boudreau and Beverly L. Davidson, Figure 14.1 siRNA sequence selection for hairpin-based vectors. Schematic depicting the design of an eGFP-targeted siRNA sequence which satisfies the rules outlined in Section 2.2. siRNA sequences may be embedded into shRNA or artificial miRNA scaffolds for expression.

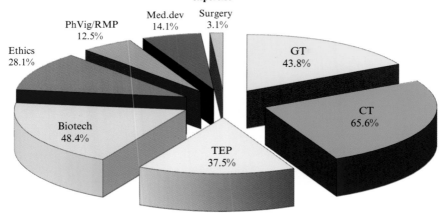

Bettina Klug et al., Figure 17.1 Percentage of the members of the CAT according to their product-related expertise (CAT members declared in most cases expertise in more than one area). Abbreviations used: GT, gene therapy; CT, cell therapy; TEP, tissue-engineered medicinal products; Biotech, Biotechnology; PhVig/RMP, Pharmacovigillance/Risk Management Plans; Med. dev, medical devices.

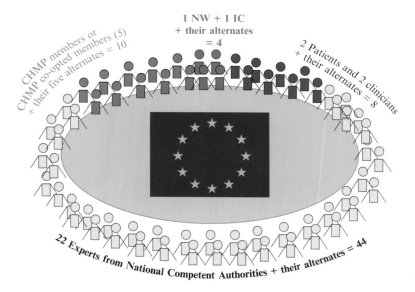

Bettina Klug *et al.*, Figure 17.2 Composition of the CAT: Each figurine represents one of the CAT member or its alternates. The yellow figurines represent the members and alternates nominated by the national competent authorities. The red figurines are CHMP and CAT double members or co-opted members. The blue figures stand for the members of the clinicians' and the patients' organizations. Depicted in green are the representatives of Norway and Iceland and their alternates.